Techniques and Concepts of High-Energy Physics III

NATO ASI Series

Advanced Science Institutes Series

A series presenting the results of activities sponsored by the NATO Science Committee, which aims at the dissemination of advanced scientific and technological knowledge, with a view to strengthening links between scientific communities.

The series is published by an international board of publishers in conjunction with the NATO Scientific Affairs Division

A Life Sciences **B Physics**	Plenum Publishing Corporation New York and London
C Mathematical and Physical Sciences	D. Reidel Publishing Company Dordrecht, Boston, and Lancaster
D Behavioral and Social Sciences **E Engineering and Materials Sciences**	Martinus Nijhoff Publishers The Hague, Boston, and Lancaster
F Computer and Systems Sciences **G Ecological Sciences**	Springer-Verlag Berlin, Heidelberg, New York, and Tokyo

Recent Volumes in this Series

Volume 126—Perspectives in Particles and Fields: *Cargèse 1983*
edited by Maurice Lévy, Jean-Louis Basdevant, David Speiser, Jacques Weyers, Maurice Jacob, and Raymond Gastmans

Volume 127—Phenomena Induced by Intermolecular Interactions
edited by G. Birnbaum

Volume 128—Techniques and Concepts of High-Energy Physics III
edited by Thomas Ferbel

Volume 129—Transport in Nonstoichiometric Compounds
edited by George Simkovich and Vladimir S. Stubican

Volume 130—Heavy Ion Collisions: *Cargèse* 1984
edited by P. Bonche, Maurice Lévy, Philippe Quentin, and Dominique Vautherin

Volume 131—Physics of Plasma-Wall Interactions in Controlled Fusion
edited by D. E. Post and R. Behrisch

Volume 132—Physics of New Laser Sources
edited by Neal B. Abraham, F. T. Arecchi, Aram Mooradian, and Alberto Sona

Series B: Physics

Techniques and Concepts of High-Energy Physics III

Edited by

Thomas Ferbel

University of Rochester
Rochester, New York

Plenum Press
New York and London
Published in cooperation with NATO Scientific Affairs Division

Proceedings of the Third NATO Advanced Study Institute on
Techniques and Concepts of High-Energy Physics III,
held August 2–13, 1984,
in St. Croix, Virgin Islands

Library of Congress Cataloging in Publication Data

NATO Advanced Study Institute on Techniques and Concepts of High-Energy Physics (3rd: 1984: Saint Croix, V.I.)
Techniques and concepts of high-energy physics III.

(NATO ASI series. Series B, Physics; v. 128)
"Proceedings of the Third NATO Advanced Study Institute on Techniques and Concepts of High-Energy Physics, held August 2–13, 1984, in St. Croix, Virgin Islands"—T.p. verso.
"Published in cooperation with NATO Scientific Affairs Division."
Bibliography: p.
Includes index.
1. Particles (Nuclear physics)—Congresses. I. Ferbel, Thomas. II. North Atlantic Treaty Organization. Scientific Affairs Division. III. Title. IV. Series.
QC793.N38 1984 539.7 85-25660
ISBN 0-306-42106-2

A Division of Plenum Publishing Corporation
233 Spring Street, New York, N.Y. 10013

Printed in the United States of America

Pstack SCIMON

PREFACE

The third Advanced Study Institute (ASI) on Techniques and Concepts of High Energy Physics was held at the Hotel on the Cay, in the scenic harbor of Christiansted, St. Croix, U.S. Virgin Islands. Christiansted was the site of the first ASI, and it was certainly a delight to return there again. As in the previous ASI's, the aim was to bring together a small group of promising young experimenters and several outstanding senior scholars in experimental and theoretical high energy physics in order to learn about the latest developments in the field and to strengthen contacts among scientists from different countries and different backgrounds. The institute was both a great scientific and a great social success; much of this was due to the beautiful setting and to the dedication of the Hotel management of Ray Boudreau and Hurchell Greenaway and their excellent staff.

The primary support for the meeting was once again provided by the Scientific Affairs Division of NATO. The ASI was cosponsored by the U.S. Department of Energy, by Fermilab, by the National Science Foundation, and by the University of Rochester. A special contribution from the Oliver S. and Jennie R. Donaldson Charitable Trust provided an important degree of flexibility, as well as support for worthy students from developing nations.

As in the case of the previous ASI's, the scientific program was designed for advanced graduate students and recent PhD recipients in experimental particle physics. The present volume of lectures should complement the material published in the first two ASI's, and again should prove to be of value to a wide audience of physicists. It is clear from the contents of this volume that the lecturers took great care to present their material in a coherent and inspiring way. Unfortunately, Alan Astbury could not provide a written version of his excellent lectures.

It is always a pleasure to acknowledge the encouragement and support that I have received from colleagues and friends in organizing this meeting. I am indebted to the members of my Advisory Committee for their infinite patience and superb advice. I am grateful to my distinguished lecturers for participating in the ASI. In particular, I thank Claudio Pellegrini and Mike Shaevitz for agreeing to speak on rather short notice. I also thank Giovanni Bonvincini and Phil Gutierrez for collaborating with Pellegrini on his lecture notes. Bob Wilson, one of the repeaters from the ASI at Lake George (1982), was very helpful in organizing the student presentations. I wish to thank Earle Fowler, Bernard Hildebrand and Bill Wallenmeyer for support from the Department of Energy, and David Berley for the assistance from the National Science Foundation. I thank Leon Lederman for providing me with access to the talents of Angela Gonzales and Jackie Coleman at Fermilab. At Rochester, I am indebted to Diane Hoffman, Judy Mack and Connie Murdoch for organizational assistance and typing. I also thank Hugh Van Horn, our departmental chairman, for his support. I wish to acknowledge the generosity of Chris Lirakis and of Mrs. Marjorie Atwood of the Donaldson Trust. Finally, I thank Drs. Craig Sinclair and Mario di Lullo of NATO for their continuing cooperation and confidence.

T. Ferbel
Rochester, N.Y.
June 1985

CONTENTS

TESTS AND PRESENT STATUS OF GAUGE THEORIES

R.D. Peccei

Max-Planck-Institut für Physik und Astrophysik

Munich, Federal Republic of Germany

PROLOGUE

In these lectures I want to discuss in a pedagogical way the predictions of the, so called, standard model for the strong, weak and electromagnetic interactions. For the benefit of the diligent student, I have included at the end of the lectures a set of exercises which I hope can be helpful for understanding better the material covered.

In the standard model the strong interactions are described by Quantum Chromodynamics (QCD) /1/ and the weak and electromagnetic interactions are given by the unified electroweak model of Glashow, Salam and Weinberg (GSW) /2/. The standard model is based on the gauge group:

$$G = SU(3) \times SU(2) \times U(1)$$

Here SU(3) is the exact gauge symmetry of QCD, in which the associated gauge bosons (gluons) are massless. In the electroweak case, the gauge group SU(2) x U(1) is spontaneously broken to $U(1)_{em}$:

$$SU(2) \times U(1) \longrightarrow U(1)_{em}$$

As a result of this phenomena, three out of the four gauge bosons of SU(2) x U(1) acquire mass ($W^{\pm}$ and Z^{o}) while one, associated with the photon (γ), remains massless.

Although the above features of the standard model are by now well known, it is useful, before embarking in a detailed examination of the standard model, to illustrate the crucial concepts with a simple example.

GAUGE THEORIES AT WORK: U(1) ABELIAN MODEL

Imagine two non interacting scalar particles of masses m_1 and m_2. Using the particle-field correspondence, we may describe this system by the Lagrangian

$$\mathcal{L} = -\frac{1}{2}\partial^\mu\phi_1(x)\partial_\mu\phi_1(x) - \frac{1}{2}m_1^2\phi_1^2(x) - \frac{1}{2}\partial^\mu\phi_2(x)\partial_\mu\phi_2(x) - \frac{1}{2}m_2^2\phi_2^2(x) \quad (1)$$

The fields $\phi_i(x)$, i = 1,2 are real fields since they each describe only one degree of freedom. From the Euler-Lagrange equations:

$$\partial_\mu \frac{\partial\mathcal{L}}{\partial\partial_\mu\phi_i(x)} - \frac{\partial\mathcal{L}}{\partial\phi_i(x)} = 0 \quad (2)$$

one obtains the equations of motion

$$(-\partial^2 + m_i^2)\,\phi_i(x) = 0 \quad (3)$$

for the fields ϕ_i. These equations are what one would expect for uncoupled excitations of masses m_i, i = 1,2.

If $m_1 = m_2$ the Lagrangian in (1) has an $O(2) \sim U(1)$ symmetry. That is, the Lagrangian remains invariant ($\mathcal{L} \to \mathcal{L}$) under the transformation:

$$\begin{pmatrix}\phi_1\\ \phi_2\end{pmatrix} \to \begin{pmatrix}\phi_1'\\ \phi_2'\end{pmatrix} = \begin{pmatrix}\cos\theta & \sin\theta\\ -\sin\theta & \cos\theta\end{pmatrix}\begin{pmatrix}\phi_1\\ \phi_2\end{pmatrix} \quad (4)$$

It is useful to write the real fields ϕ_1 and ϕ_2 in terms of a complex field

$$\phi = \frac{1}{\sqrt{2}}(\phi_1 + i\phi_2)$$
$$\phi^\dagger = \frac{1}{\sqrt{2}}(\phi_1 - i\phi_2) \quad (5)$$

In terms of this notation, the Lagrangian of Eq. (1) for $m_1 = m_2 = m$ becomes simply

$$\mathcal{L} = -\partial^\mu\phi^\dagger(x)\partial_\mu\phi(x) - m^2\phi^\dagger(x)\phi(x) \quad (6)$$

The U(1) invariance of this Lagrangian, under the transformation

$$\phi(x) \to \phi'(x) = e^{i\theta}\phi(x) \quad (7)$$

is obvious. For future use, I record here also the infinitesimal version of this transformation:

$$\phi'(x) = \phi(x) + \delta\phi(x)$$
$$\delta\phi(x) = i\,\delta\theta[1]\,\phi(x) \tag{8}$$

The transformation in Eq. (7), or Eq. (8), is a global transformation since the parameter θ, or $\delta\theta$, are independent of the space-time point x. Although the Lagrangian (6) is globally U(1) invariant, it is not invariant under local U(1) transformations, where $\theta = \theta(x)$ This is clear, since the derivatives in the kinetic energy terms contain an additional term when transformed:

$$\partial_\mu \phi(x) \longrightarrow \partial_\mu \phi'(x) = \partial_\mu \left(e^{i\theta(x)} \phi(x)\right)$$
$$= e^{i\theta(x)}\left(\partial_\mu \phi(x) + i(\partial_\mu \theta(x))\phi(x)\right) \tag{9}$$

It is possible to construct a locally U(1) invariant Lagrangian, from that of Eq. (6), by introducing into the theory compensating gauge fields. This, in fact, is a general result: a globally symmetric Lagrangian can always be made locally symmetric by introducing into the theory a gauge field for each of the local symmetries.

Consider the local transformation

$$\phi(x) \rightarrow \phi'(x) = e^{i\theta(x)} \phi(x)$$
$$A_\mu(x) \rightarrow A'_\mu(x) = A_\mu(x) + \frac{1}{g}\partial_\mu \theta(x) \tag{10}$$

Here $A_\mu(x)$ is a real gauge field - a Lorentz vector - and g is an arbitrary parameter which will eventually play the role of a coupling constant. Note that $A_\mu(x)$ transforms inhomogeneously under the U(1) transformation. It is easy to check that, precisely because of this inhomogeneous behaviour, the covariant derivative of the field $\phi(x)$:

$$D_\mu \phi(x) \equiv (\partial_\mu - ig A_\mu(x))\phi(x) \tag{11}$$

transforms homogeneously under local U(1) transformations:

$$D_\mu \phi(x) \rightarrow D'_\mu \phi'(x) = e^{i\theta(x)} D_\mu \phi(x) \tag{12}$$

In view of Eq. (12), it is clear that the Lagrangian

$$\mathcal{L} = -(D^\mu \phi(x))^\dagger (D_\mu \phi(x)) - m^2 \phi^\dagger(x)\phi(x) \tag{13a}$$

is locally U(1) invariant. Note that local invariance is achieved only by introducing interactions. Writing Eq.(13a) out in detail one has

$$\mathcal{L} = -\partial^\mu \phi^\dagger \partial_\mu \phi - m^2 \phi^\dagger \phi + \Big\{ g A^\mu [i(\partial_\mu \phi^\dagger)\phi - i\phi^\dagger(\partial_\mu \phi)] - g^2 A^\mu A_\mu \phi^\dagger \phi \Big\} \tag{13b}$$

The terms in the curly bracket above, which are the necessary additions to the Lagrangian of Eq. (6) to guarantee local U(1) invariance, represent interactions of the scalar fields ϕ and $\phi^\dagger$ with the gauge field A_μ. These interactions are depicted schematically in Fig. 1.

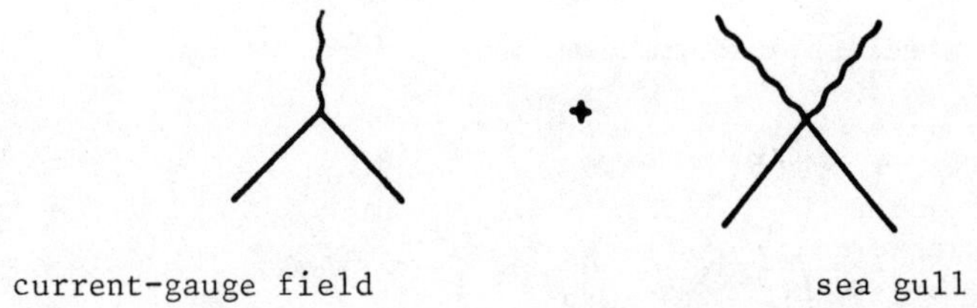

Fig. 1: Interactions which follow from the demand of local U(1) invariance

The "sea-gull" term above is necessary for scalar particles, because of their quadratic kinetic energy. For fermions only the "current-gauge field" term arises, from the demand of local invariance.

The Lagrangian (13) is not complete because the gauge field A_μ is only dynamical when kinetic terms for it are added. This is easily done, although the local U(1) invariance of Eq. (10) (gauge invariance) imposes certain restrictions. The field strength

$$F^{\mu\nu} = \partial^\mu A^\nu - \partial^\nu A^\mu \tag{14}$$

because of its curl-form is clearly invariant under local U(1) transformations:

$$F^{\mu\nu}(x) \to F'^{\mu\nu}(x) = F^{\mu\nu}(x) \tag{15}$$

Thus a sensible gauge field kinetic energy term is provided by

$$\mathcal{L}_{kin} = -\frac{1}{4} F^{\mu\nu} F_{\mu\nu} \tag{16}$$

However, no mass term for the gauge field A_μ is allowed by the local U(1) invariance.

To summarize, the Lagrangian

$$\mathcal{L} = -\frac{1}{4} F^{\mu\nu} F_{\mu\nu} - (D^\mu\phi)^\dagger (D_\mu \phi) - m^2 \phi^\dagger \phi \tag{17}$$

is invariant under the local U(1) transformations of Eq. (10). It describes a doublet of scalar fields which are degenerate in mass interacting (in a specific way) with a massless gauge field A_μ.

Quantum Chromodynamics (QCD) is an SU(3) generalization of the above procedure, with various important (but no in principle) differences. The fundamental matter fields in QCD, instead of the scalars above, are spin 1/2 colored quarks of various different types (flavors). A convenient notation for these fields is q_a^f, where f is a flavor index distinguishing the various type of quarks

$$q_a^f = \{u_a, d_a, c_a, s_a, t_a, b_a, \ldots\}$$

and a = 1,2,3 is an SU(3)-color-index. In what follows, unless it is needed for clarity, I shall suppress the flavor index. The quarks, q_a, transform as a $\underline{3}$ under the SU(3) group. Specifically, under infinitesimal SU(3) transformations one has

$$q_a \to q_a' = q_a + \delta q_a \tag{18a}$$

with

$$\delta q_a = i\, \delta\omega_i \left(\frac{\lambda_i}{2}\right)_{ab} q_b \tag{18b}$$

Here the $\delta\omega_i$, i = 1,2,..8 are the eight independent infinitesimal parameters characterizing SU(3) transformations (the analogs of $\delta\theta$ in Eq. (8)) and the matrices $\frac{1}{2}\lambda_i$ provide a 3 dimensional representation of the SU(3) generators G_i (the analog of 1, for the Abelian case of Eq. (8)). That is, the $\frac{1}{2}\lambda_i$ satisfy the SU(3) commutation relations:

$$\left[\frac{1}{2}\lambda_i, \frac{1}{2}\lambda_j\right] = i f_{ijk} \frac{1}{2}\lambda_k \tag{19}$$

where the f_{ijk} are the (totally antisymmetric) SU(3) structure constants.

To guarantee that the QCD Lagrangian constructed out of the quarks fields q_a is locally SU(3) invariant (invariant under transformations where $\delta\omega_i = \delta\omega_i(x)$) one must introduce a compensating gauge field (gluon field) for each of the eight parameters $\delta\omega_i$. These gauge fields $A_i^\mu(x)$ transform as the $\underline{8}$, or adjoint, representation of SU(3). Under infinitesimal local SU(3) transformations one has

$$A^\mu_j(x) \to A'^\mu_j(x) = A^\mu_j(x) + \delta A^\mu_j(x) \tag{20a}$$

where

$$\delta A^\mu_j(x) = i\,\delta\omega_i(x)(g_i)_{jk} A^\mu_k(x) + \frac{1}{g_3}\partial^\mu(\delta\omega_j(x)) \tag{20b}$$

The second term above is precisely the analog of the inhomogeneous term appearing in Eq. (10). The first term in Eq. (20b), which has no counterpart in the Abelian case, arises because the gluon fields transform non trivially under SU(3). Indeed the 8-dimensional matrices g_i are precisely those appropriate for the adjoint representation of SU(3):

$$(g_i)_{jk} = -i f_{ijk} \tag{21}$$

where the f_{ijk} are the SU(3) structure constants.

In analogy to what was done in the U(1) example, to construct the QCD Lagrangian one needs to know the quark fields covariant derivatives, $D_\mu q_a$, and a generalization of the gauge fields field strength, $F_i^{\mu\nu}$. The covariant derivative for the quark fields is easily found. It is easy to check that

$$D_\mu q_a(x) = \left(\partial_\mu \delta_{ab} - i g_3 \tfrac{1}{2}(\lambda_i)_{ab} A_{\mu i}(x)\right) q_b(x) \tag{22}$$

under local SU(3) transformations transforms in the same way as quarks field do (Eq. 18b):

$$\delta(D_\mu q_a(x)) = i\,\delta\omega_i(x)\left(\frac{\lambda_i}{2}\right)_{ab}(D_\mu q_b(x)) \tag{23}$$

The generalized field strengths $F_i^{\mu\nu}$ are also not difficult to find by the requirement that they transform homogeneously under local SU(3) transformations: That is, for infinitesimal transformations:

$$\begin{aligned}\delta F^{\mu\nu}_j(x) &= i\,\delta\omega_i(x)(g_i)_{jk} F^{\mu\nu}_k(x)\\ &= \delta\omega_i(x) f_{ijk} F^{\mu\nu}_k(x)\end{aligned} \tag{24}$$

A simple calculation gives the expression

$$F_i^{\mu\nu} = \partial^\mu A_i^\nu - \partial^\nu A_i^\mu + g_3 f_{ijk} A_j^\mu A_k^\nu \tag{25}$$

where the appearance of the last term is due precisely to the non Abelian nature of SU(3).

Putting all these ingredients together - and restoring the flavor index - gives the QCD Lagrangian

$$\mathcal{L}_{QCD} = -\bar{q}_a^f \gamma^\mu \frac{1}{i} D_\mu q_a^f - m_f \bar{q}_a^f q_a^f - \frac{1}{4} F_i^{\mu\nu} F_{i\mu\nu} \tag{26}$$

The principal structural difference of Eq. (26) from the U(1) example previously discussed resides in the interactions among the gauge fields, implicit in the $-1/4\, F^2$ term. Because of the quadratic term in Eq. (25), the gluon "kinetic energy" term contains both trilinear and quadrilinear interactions among the gluon fields, as depicted pictorially in Fig. 2.

Fig. 2: Additional interactions among gauge fields due to the non Abelian nature ($f_{ijk} \neq 0$) of SU(3)

In QCD the SU(3) local symmetry is realized in, what is called commonly, the Wigner-Weil way. That is, for each flavor of quark f one has a degenerate triplet of massive quark states, of mass m_f, interacting in the precise way specified by Eq. (26) with massless gluons, whose selfinteractions are also totally fixed. This, however, is not the only way in which symmetries can be realized in nature. Symmetries can also be realized in the, so called, Nambu-Goldstone way, in which the symmetry is a symmetry of the Lagrangian but not of the particle spectrum, because the vacuum state is not invariant under the symmetry. The GSW SU(2) x U(1) model is a theory in which the gauge symmetry is realized in the Nambu Goldstone way. The vacuum state is not SU(2)xU(1) invariant but only $U(1)_{em}$ invariant, so SU(2) x U(1) is spontaneously broken to $U(1)_{em}$. It is this phenomenon of spontaneous breakdown which provides the $W^{\pm}$ and Z^0 gauge bosons with a mass. To understand better what happens in the GSW model, it behoves us to study anew the simple U(1) model of before - with a modification which allows for the occurrence of spontaneous symmetry breakdown.

A slight generalization of the U(1) model of Eq. (17) is provided by the Lagrangian

$$\mathcal{L} = -\frac{1}{4} F^{\mu\nu} F_{\mu\nu} - (D^\mu \phi)^+ (D_\mu \phi) - V(\phi^+ \phi) \tag{27}$$

that is, the mass term for the scalar fields is replaced by a general function of $\phi^+\phi$, which can include terms of self interaction. Renormalizability restricts the "potential" $V(\phi^+\phi)$ to contain terms with at most four fields. The physics of the model in Eq. (27) is considerably different depending on whether the minimum of the potential V occurs for $\phi^+ \phi = 0$ or not. In the first case, the features of the model are essentially like those of the simple U(1) model of Eq. (17), except for some additional scalar self interactions. If $\phi^+\phi \neq 0$ at the minimum, however, the physics is completely different.

Classically, if the minimum of V occurs at $\phi^+ \phi = 0$ one has a unique minimum value, $\phi_{min} = 0$. On the other hand, if the minimum of V occurs at $\phi^+ \phi \neq 0$ one has an infinite set of equivalent parametrization for ϕ_{min}, namely

$$\phi_{min}(\alpha) = |\phi_{min}| e^{i\alpha} \tag{28}$$

Quantum mechanically, the first case corresponds to having a field ϕ whose vacuum expectation value $\langle \phi \rangle$ vanishes, while in the second case $\langle \phi \rangle \neq 0$. Such a non vanishing vacuum expectation value can only obtain if the vacuum itself is not invariant under U(1) transformations.

The physics of what is going on can be illustrated by picking V to have the specific form

$$V = \lambda \left(\phi^+ \phi - \frac{1}{2} v^2 \right)^2 \tag{29}$$

shown in Fig. 3. The potential V above breaks the U(1) symmetry spontaneously. Although the Lagrangian (27) is locally U(1) invariant, the vacuum state of the theory is not.

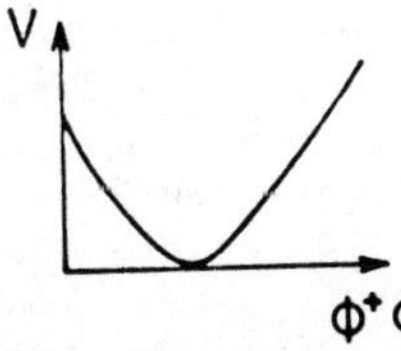

Fig. 3: The potential V of Eq. (29)

One can label the infinite set of vacua, corresponding to the classical configuration of Eq. (28), by the value of the angle α : $|\alpha\rangle$. Thus

$$\langle\alpha|\phi(x)|\alpha\rangle = \phi_{min}(\alpha) = e^{i\alpha}\frac{1}{\sqrt{2}}v \tag{30}$$

Clearly the U(1) transformation

$$\phi(x) \to \phi'(x) = U_\theta\,\phi(x)\,U_\theta^{-1} = e^{i\theta}\,\phi(x) \tag{31}$$

does not leave the vacuum invariant. In view of Eq. (30) one easily deduces that

$$U_\theta|\alpha\rangle = |\alpha-\theta\rangle \tag{32}$$

The spontaneous breakdown of the U(1) symmetry caused by the presence of the potential V, with its asymmetric minimum, has two effects:

(1) The U(1) degeneracy in the scalar sector of the theory is lifted.
(2) The U(1) gauge field acquires a mass (Higgs Mechanism).

The first phenomenon above occurs irrespective of whether the symmetry is a local or global symmetry of the Lagrangian. If the symmetry is global, as a result of the breakdown a zero mass excitation (or zero mass excitations, if more than one symmetry is spontaneously broken) appears in the theory (Goldstone boson). This Goldstone boson disappears if the Lagrangian symmetry is local. However, in this case, point 2) above obtains, in which the erstwhile massless gauge field gets a mass. Roughly speaking, one can think of the Goldstone excitation of the global symmetry as being the longitudinal component of the massive vector field.

These phenomena can be deduced by a somewhat more careful inspection of the Lagrangian (27). If the potential $V(\phi^+\phi)$ is as in Eq. (29), to do physics one must compute the excitations of the field around ϕ_{min} and not zero. Therefore, the Lagrangian in (27) ought to be reparametrized to take this fact into account. In effect, instead of dealing with a complex field ϕ, with non zero vacuum expectation value $\langle\phi\rangle \neq 0$, one should deal with the two real scalar fields, which characterize the excitations about ϕ_{min}. Although the physical phenomena described above obtain independently of how one parametrizes ϕ, a very convenient parametrization is provided by

$$\phi(x) = e^{i\xi(x)/v} \frac{1}{\sqrt{2}} (v + \rho(x)) \tag{33}$$

The real scalar fields $\xi(x)$ and $\rho(x)$, obviously, have zero vacuum expectation value.

A number of results are now easy to check:

(a) The potential $V(\phi^\dagger\phi)$ is obviously independent of ξ. If there were no gauge interactions, ξ would therefore be interpreted as the massless Goldstone boson.

(b) The "sea-gull" term $-g^2 A^\mu A_\mu \phi^\dagger\phi$, present in the scalar field "kinetic energy" term $-(D^\mu\phi)^\dagger(D_\mu\phi)$, generates a mass for the gauge field A_μ, when $\phi^\dagger\phi$ is replaced by its value at the minimum: $(1/2)v^2$,

$$\mathcal{L}_{mass} = -\frac{1}{2}(gv)^2 A^\mu A_\mu \tag{34}$$

This is the famous Higgs mechanism /3/.

(c) The covariant derivative $D_\mu\phi$, after a gauge transformation of the field A_μ, has a trivial ξ dependence. To wit, one has

$$\begin{aligned} D_\mu\phi &= (\partial_\mu - igA_\mu)\frac{1}{\sqrt{2}}(v+\rho)e^{i\xi/v} \\ &= e^{i\xi/v}(\partial_\mu - igB_\mu)\frac{1}{\sqrt{2}}(v+\rho) \end{aligned} \tag{35}$$

where

$$B_\mu = A_\mu - \frac{1}{gv}\partial_\mu\xi \tag{36}$$

From these observations one concludes that the Lagrangian of the theory is in fact independent of the field ξ. This Lagrangian describes the interaction of a massive vector field B_μ with a real scalar field ρ. No vestiges of the U(1) local symmetry remain, except for certain coupling relationships.(The derivation of the shifted Lagrangian is left as an exercise, Ex. 7.) It is worthwhile contrasting the particle spectrum in the two different possible realizations of the U(1) symmetry. In the Wigner-Weyl case one has a degenerate massive scalar doublet and a massless gauge field, in total four degrees of freedom. In the Nambu-Goldstone case these degrees of freedom are redistributed and one has a massive real scalar field along with a massive gauge field.

The Glashow-Salam-Weinberg SU(2) x U(1) model of electroweak interactions is a generalization of the above procedure. Although SU(2) x U(1) is a symmetry of the Lagrangian, it is not a symmetry of the spectrum of the theory. There is a breakdown of SU(2) x U(1) to $U(1)_{em}$, so that the only manifest multiplets of the theory are charge multiplets (the e^+ and e^- have the same mass, so do W^+ and W^-, etc.). The breakdown of SU(2) x U(1) $\rightarrow U(1)_{em}$ causes three of the gauge fields ($W^\pm$ and Z^o) to become massive, but the photon is massless, reflecting the unbroken $U(1)_{em}$ local symmetry.

The GSW Lagrangian contains three distinct pieces

$$\mathcal{L}_{GSW} = \mathcal{L}_{\substack{fermion\\gauge}} + \mathcal{L}_{\substack{Higgs\\gauge}} + \mathcal{L}_{\substack{Higgs\\fermion}} \tag{37}$$

The first term above contains all the weak and electromagnetic interactions of quarks and leptons and it is phenomenologically checked to high accuracy. The tests of the standard model, which I will describe in some detail in what follows, are really tests of this part of the GSW Lagrangian. The Higgs-gauge Lagrangian in Eq. (37) is put in essentially to precipitate the spontaneous breakdown of SU(2) x U(1) to $U(1)_{em}$. None of its detailed structure is really checked, except for the, so-called, $\Delta I_W = 1/2$ rule - which I will explain below. Finally, the last term in Eq. (37) is even more arbitrary. It is put in to allow for the appearance of quark and lepton masses, but it is really not checked at all otherwise. I discuss each of these terms in turn.

To construct $\mathcal{L}_{\substack{fermion\\gauge}}$ one needs to know the quantum number assignments of quarks and leptons under SU(2) x U(1). These assignments are dictated by phenomenology. For instance, the fact that the currents that participate in the weak interactions have a (V-A) form implies that only the left-handed quarks and leptons have non trivial SU(2) transformations. To be more precise the fermion structure of the GSW model is as follows:

(i) There is a repetive generation structure, with each generation having the same SU(2) x U(1) properties. At the moment we know of the existence of three families of quarks and leptons: the electron family (ν_e, e; u, d); the muon family $(\nu_\mu, \mu ; c, s)$; the tau family $(\nu_\tau, \tau ; t, b)$. There could be, however, more families yet to be discovered.

(ii) The quarks and leptons of each family have the SU(2) x U(1) assignments shown in Table I

Table I: SU(2) x U(1) quantum numbers of quarks and leptons. ψ_L, ψ_R are the helicity projections : $\psi_L = \frac{1}{2}(1-\gamma_5)\psi$; $\psi_R = \frac{1}{2}(1+\gamma_5)\psi$

states	$\begin{pmatrix} u \\ d \end{pmatrix}_L$	u_R	d_R	$\begin{pmatrix} \nu_e \\ e \end{pmatrix}_L$	e_R
SU(2)	2	1	1	2	1
U(1)	1/6	2/3	−1/3	−1/2	−1

No right-handed neutrinos are introduced into the model. If these states existed they would be totally SU(2) x U(1) singlets. The U(1) assignments of Table I follow from the formula for the electromagnetic charge in the model

$$Q = T_3 + Y \tag{38}$$

with Y being the U(1) quantum number

Armed with the SU(2) x U(1) transformation laws for the fermions of Table I, it is now straightforward to write down the appropriate covariant derivatives for these fields (cf Eq. (11) and (22) for an Abelian and a non Abelian symmetry, respectively). If W^μ_i are the three SU(2) gauge fields and Y^μ is the U(1) gauge field and if g and g' are their appropriate coupling constants one has for the covariant derivatives:

$$D_\mu \begin{pmatrix} u \\ d \end{pmatrix}_L = \left(\partial_\mu - ig'\frac{1}{6}Y_\mu - ig\frac{\tau_i}{2}W_{\mu i}\right)\begin{pmatrix} u \\ d \end{pmatrix}_L \tag{39a}$$

$$D_\mu u_R = \left(\partial_\mu - ig'\frac{2}{3}Y_\mu\right)u_R \tag{39b}$$

$$D_\mu d_R = \left(\partial_\mu + ig'\frac{1}{3}Y_\mu\right)d_R \tag{39c}$$

$$D_\mu \begin{pmatrix} \nu_e \\ e \end{pmatrix}_L = \left(\partial_\mu + ig'\frac{1}{2}Y_\mu - ig\frac{\tau_i}{2}W_{\mu i}\right)\begin{pmatrix} \nu_e \\ e \end{pmatrix}_L \tag{39d}$$

$$D_\mu e_R = \left(\partial_\mu + ig'Y_\mu\right)e_R \tag{39e}$$

The $\mathcal{L}_{\substack{\text{fermion}\\ \text{gauge}}}$ piece of $\mathcal{L}_{GSW}$ follows now immediately, exactly as in the U(1) example:

$$\mathcal{L}_{\substack{\text{fermion}\\\text{gauge}}} = -(\bar{u}\,\bar{d})_L\,\gamma^\mu \tfrac{1}{i} D_\mu \begin{pmatrix} u \\ d \end{pmatrix}_L - \bar{u}_R \gamma^\mu \tfrac{1}{i} D_\mu u_R$$

$$- \bar{d}_R \gamma^\mu \tfrac{1}{i} D_\mu d_R - (\bar{\nu}_e\,\bar{e})_L\,\gamma^\mu \tfrac{1}{i} D_\mu \begin{pmatrix} \nu_e \\ e \end{pmatrix}_L$$

$$- \bar{e}_R \gamma^\mu \tfrac{1}{i} D_\mu e_R - \tfrac{1}{4} Y^{\mu\nu} Y_{\mu\nu} - \tfrac{1}{4} W_i^{\mu\nu} W_{i\mu\nu} \tag{40}$$

In the above the field strengths $Y^{\mu\nu}$ and $W_i^{\mu\nu}$ are given by

$$Y^{\mu\nu} = \partial^\mu Y^\nu - \partial^\nu Y^\mu \tag{41a}$$

$$W_i^{\mu\nu} = \partial^\mu W_i^\nu - \partial^\nu W_i^\mu + g\,\epsilon_{ijk} W_j^\mu W_k^\nu \tag{41b}$$

I have written the above only for one family - for three families one just repeats the same structure three times. Also in writing Eqs. (39) and (40) I have suppressed the color indices of the quarks. These should be understood and summed over in Eq. (40).

There are a number of remarks that should be made at this point:

(1) No mass terms for leptons or quarks are allowed by the SU(2) symmetry and such terms are not included therefore in Eq. (40). A mass term involves a L-R transition

$$\mathcal{L}_{mass} = - m \bar{\psi} \psi = - m (\bar{\psi}_L \psi_R + \bar{\psi}_R \psi_L) \tag{42}$$

Since left-handed fields are doublets under SU(2) and right-handed fields are singlets, mass terms violate SU(2). Masses for the quarks and leptons will eventually be generated through a coupling to scalar fields, after SU(2) x U(1) is broken. Thus, for example, the masses m_f present in the QCD Lagrangian (26) have a non trivial weak interaction origin. I will return to this point again below.

(2) The Lagrangian of Eq. (40) contains explicit fermion-gauge field interactions which take the form

$$\mathcal{L}_{int} = g' J_Y^\mu Y_\mu + g J_i^\mu W_{i\mu} \tag{43}$$

where J_Y^μ is the "weak hypercharge" current and J_i^μ is the "weak isospin" current. A simple calculation - again written only for one family - yields the formulas

$$J_i^\mu = (\bar{u}\,\bar{d})_L \gamma^\mu \frac{\tau_i}{2} \begin{pmatrix} u \\ d \end{pmatrix}_L + (\bar{\nu}_e\,\bar{e})_L \gamma^\mu \frac{\tau_i}{2} \begin{pmatrix} \nu_e \\ e \end{pmatrix}_L \tag{44a}$$

$$J_Y^\mu = \frac{1}{6} (\bar{u}\,\bar{d})_L \gamma^\mu \begin{pmatrix} u \\ d \end{pmatrix}_L + \frac{2}{3} \bar{u}_R \gamma^\mu u_R - \frac{1}{3} \bar{d}_R \gamma^\mu d_R$$

$$- \frac{1}{2} (\bar{\nu}_e\,\bar{e})_L \gamma^\mu \begin{pmatrix} \nu_e \\ e \end{pmatrix}_L - \bar{e}_R \gamma^\mu e_R \tag{44b}$$

3) It is anticipated that, because of the SU(2) x U(1) $\rightarrow$ $U(1)_{em}$ breakdown and the fact that the charge operator is a linear combination of T_3 and Y (cf Eq. (38)), some linear combination of Y_μ and $W_{3\mu}$ will describe the massive Z^0_μ field, while the orthogonal combination will be the photon field, A_μ. Conventionally, one writes

$$\begin{aligned} W_3^\mu &= \cos\theta_W Z^\mu + \sin\theta_W A^\mu \\ Y^\mu &= -\sin\theta_W Z^\mu + \cos\theta_W A^\mu \end{aligned} \tag{45}$$

where θ_W is the, so called, Weinberg angle.

It is useful therefore to rewrite the interaction Lagrangian (43) in terms of the mass eigenstates A_μ and Z_μ, and to replace the hypercharge current J_Y^μ by

$$J_Y^\mu = J_3^\mu - J_{em}^\mu \tag{46}$$

A simple calculation then gives the formula

$$\begin{aligned} \mathcal{L}_{int} &= \frac{1}{2\sqrt{2}} g \left(W_+^\mu J_{-\mu} + W_-^\mu J_{+\mu} \right) \\ &+ \left\{ (g\cos\theta_W + g'\sin\theta_W) J_3^\mu - g'\sin\theta_W J_{em}^\mu \right\} Z_\mu \\ &+ \left\{ g'\cos\theta_W J_{em}^\mu + (g'\cos\theta_W - g\sin\theta_W) J_3^\mu \right\} A_\mu \end{aligned} \tag{47}$$

In the above we have made use of the definitions:

$$W_{\pm}^{\mu} = \frac{1}{\sqrt{2}}(W_1^{\mu} \mp i W_2^{\mu}) \tag{48a}$$

$$J_{\pm}^{\mu} = 2\,(J_1^{\mu} \mp i J_2^{\mu}) \tag{48b}$$

where the factor of 2 for the charged currents in Eq. (48b) is just a convention. Eq. (47) can be simplified by noting that if A^{μ} is indeed the photon field it must couple only to J^{μ}_{em} and this coupling must have the strength e, of the electromagnetic charge. This physical requirement gives the relation

$$g' \cos\theta_W = g \sin\theta_W = e \tag{49}$$

Eliminating g and g' in favour of e and the Weinberg angle yields for the interaction Lagrangian the, by now, standard form:

$$\mathcal{L}_{int} = e J^{\mu}_{em} A_{\mu} + \frac{e}{2\sqrt{2}\sin\theta_W}\left\{ J^{\mu}_{+} W_{-\mu} + J^{\mu}_{-} W_{+\mu} \right\} + \frac{e}{2\cos\theta_W \sin\theta_W} J^{\mu}_{NC} Z_{\mu} \tag{50}$$

where the neutral current, J^{μ}_{NC}, is given by

$$J^{\mu}_{NC} = 2\left[J^{\mu}_{3} - \sin^2\theta_W J^{\mu}_{em} \right] \tag{51}$$

Although Eq. (50) describes the electroweak interactions in the GSW model, for many purposes it suffices to work with a low energy approximation to this Lagrangian. Most weak interaction experiments deal with processes where the relevant momentum transfers are tiny compared to the values of the W and Z masses, (which - as will be seen soon - are of the order of 100 GeV). For $q^2 \ll M_W^2, M_Z^2$, it suffices to replace the W and Z propagators by

$$\langle T(W_+^{\mu}(q)\, W_-^{\nu}(0)) \rangle \simeq \frac{1}{i}\frac{\eta^{\mu\nu}}{M_W^2} \tag{52a}$$

$$\langle T(Z^{\mu}(q)\, Z^{\nu}(0)) \rangle \simeq \frac{1}{i}\frac{\eta^{\mu\nu}}{M_Z^2} \tag{52b}$$

where $\eta^{\mu\nu} = \begin{pmatrix} -1 & & & \\ & 1 & & \\ & & 1 & \\ & & & 1 \end{pmatrix}$ is the metric tensor. Using the approximation (52), the interaction Lagrangian (50) yields in second order perturbation theory an effective Lagrangian for the weak interactions of a current-current form:

$$\mathcal{L}^{Weak}_{eff} = \left(\frac{e}{2\sqrt{2}\sin\theta_W}\right)^2 \frac{1}{M_W^2} J_+^{\mu} J_{-\mu} + \frac{1}{2}\left(\frac{e}{2\cos\theta_W \sin\theta_W}\right)^2 \frac{1}{M_Z^2} J^{\mu}_{NC} J_{NC\mu} \tag{53}$$

For the charged current interactions this Lagrangian must agree with the Fermi theory

$$\mathcal{L}_{Fermi} = \frac{G_F}{\sqrt{2}} J_+^{\mu} J_{-\mu} \tag{54}$$

This then identifies the Fermi constant G_F as

$$\frac{G_F}{\sqrt{2}} = \frac{e^2}{8\sin^2\theta_W M_W^2} \tag{55}$$

It proves convenient to define the parameter ρ as

$$\rho = \frac{M_W^2}{M_Z^2 \cos^2\theta_W} \tag{56}$$

Then the effective interaction (53) takes the simple form

$$\mathcal{L}^{Weak}_{eff} = \frac{G_F}{\sqrt{2}} \left\{ J_+^{\mu} J_{-\mu} + \rho J^{\mu}_{NC} J_{NC\mu} \right\} \tag{57}$$

One sees from the above that the only free parameters to describe the (low energy) neutral current interactions in the GSW model are ρ and the Weinberg angle θ_W, which enters in the definition of $J_{NC\mu}$ of

Eq. (51). As I will discuss in much more detail later, experiments indicate that $\sin^2\theta_W \simeq 1/4$ and $\rho \simeq 1$. Using Eqs. (55) and (56) and the experimental value of the Fermi constant one can check that the W and Z boson masses are indeed of O (100 GeV). Thus the approximations that led to Eq. (57) are certainly justified for most "low energy" weak interaction experiments. The value $\rho \simeq 1$ is particularly noteworthy, since it gives some direct insight on how SU(2) x U(1) is broken down.

To examine this point it is necessary to consider how SU(2) x U(1) is broken down in the GSW model. This is the job of the Higgs-gauge piece of $\mathcal{L}_{GSW}$ in Eq. (37). Conventionally one introduces into the model some elementary scalar fields, whose self interaction causes (some of) them to have SU(2) x U(1) violating vacuum expectation values. Remarkably the result ρ = 1 ensues if the breakdown of SU(2) x U(1) $\rightarrow$ $U(1)_{em}$ is generated by scalar fields which are doublets under SU(2) x U(1). This result - which is also known as the $\Delta I_W = 1/2$ rule - will be demonstrated in detail by considering the simplest case of only one SU(2) doublet Higgs field.

Consider a complex SU(2) doublet scalar field

$$\Phi = \begin{pmatrix} \phi^0 \\ \phi^- \end{pmatrix} \tag{58}$$

With the charges as above indicated Φ has Y = - 1/2. Note that $\Phi^+ \neq \Phi$ so that Φ really describes four degrees of freedom. An SU(2) x U(1) locally invariant Lagrangian for Φ is immediate to write down:

$$\mathcal{L}_{\substack{Higgs\\gauge}} = -(D^\mu\Phi)^+(D_\mu\Phi) - V(\Phi^+\Phi) \tag{59}$$

where

$$D_\mu\Phi = (\partial_\mu + i\frac{g'}{2}Y_\mu - ig\frac{\tau_i}{2}W_{\mu i})\Phi \tag{60}$$

is the appropriate covariant derivative. If the potential V is chosen of the form

$$V = \lambda(\Phi^+\Phi - \tfrac{1}{2}v^2)^2 \tag{61}$$

SU(2) x U(1) will clearly break down. The choice for the vacuum expectation value of Φ :

$$\langle\Phi\rangle = \begin{pmatrix} \frac{1}{\sqrt{2}}v \\ 0 \end{pmatrix} \tag{62}$$

assures that the unbroken U(1) agrees precisely with $U(1)_{em}$,* for

$$Q\langle\Phi\rangle = \begin{pmatrix} 0 & 0 \\ 0 & -1 \end{pmatrix}\begin{pmatrix} \frac{1}{\sqrt{2}}v \\ 0 \end{pmatrix} = 0 \tag{63}$$

The masses for the gauge fields are obtained, as in the U(1) example, by replacing in the "seagull" terms the scalar field Φ by its vacuum expectation value $\langle\Phi\rangle$. This yields

$$\mathcal{L}_{mass} = -\left[\left(g\frac{\tau_i}{2}W_i^\mu - g'\frac{1}{2}Y^\mu\right)\langle\Phi\rangle\right]^\dagger\left[\left(g\frac{\tau_j}{2}W_{\mu j} - g'\frac{1}{2}Y_\mu\right)\langle\Phi\rangle\right]$$

$$= -\frac{1}{2}v^2(1,0)\left[\left(g\frac{\tau_i}{2}W_i^\mu - g'\frac{1}{2}Y^\mu\right)\left(g\frac{\tau_j}{2}W_{\mu j} - g'\frac{1}{2}Y_\mu\right)\right]\begin{pmatrix}1\\0\end{pmatrix} \tag{64}$$

$$= -\left(\frac{gv}{2}\right)^2 W_+^\mu W_{-\mu} - \frac{1}{2}\left(W_3^\mu\ Y^\mu\right)\begin{bmatrix} \frac{1}{4}g^2v^2 & -\frac{1}{4}gg'v^2 \\ -\frac{1}{4}gg'v^2 & \frac{1}{4}g'^2v^2 \end{bmatrix}\begin{pmatrix} W_{3\mu} \\ Y_\mu \end{pmatrix}$$

Clearly one sees that

$$M_W = \frac{1}{2}gv \tag{65a}$$

and that W_3^μ and Y^μ are not mass eigenstates. The 2x2 mass matrix in Eq. (64) is diagonalized by the transformation of Eq. (45) provided that the Weinberg angle θ_W obeys

$$\tan\theta_W = g'/g \tag{66}$$

This result agrees with that obtained earlier in Eq. (49), as it must. The eigenvalues of the 2x2 matrix are 0 and $\frac{1}{4}(g^2+g'^2)v^2$ corresponding to the mass squared of the photon and of the Z^0. Hence

$$M_Z = \frac{1}{2}\sqrt{g^2+g'^2}\,v \tag{65b}$$

Using Eqs. (65) and (66) it follows that the parameter ρ, defined in Eq. (56), is unity.

---

* This can always be chosen to happen for just one Higgs doublet by appropriate redefinitions. With more than one Higgs doublet one has to impose the preservation of $U(1)_{em}$ by the Higgs potential minimum.

In the case of the minimal Higgs model, with just one doublet Φ, after symmetry breakdown only one physical excitation remains. Three of the four degrees of freedom in Φ disappear physically, becoming the longitudinal component of the $W^{\pm}$ and Z^{0} bosons. The presence of a scalar particle - the Higgs boson - is an unavoidably consequence of the symmetry breakdown. One may parametrize Φ in the model, in an analogous way to that done for the U(1) model (c.f. Eq. (33)), so that $\mathcal{L}_{\text{Higgs-gauge}}$ just describes the interaction of this Higgs scalar with massive gauge fields. Consider

$$\Phi = e^{i\vec{\tau}\cdot\vec{\xi}/v} \begin{pmatrix} \frac{1}{\sqrt{2}}(v+H) \\ 0 \end{pmatrix} \tag{67}$$

Clearly the potential term $V(\Phi^{\dagger}\Phi)$ is independent of $\vec{\xi}$. The dependence of $D_{\mu}\Phi$ on $\vec{\xi}$ is also trivial, after a gauge transformation:

$$\begin{aligned} D_{\mu}\Phi &= \left(\partial_{\mu} - ig\frac{\tau_i}{2}W_{\mu i} + ig'\frac{1}{2}Y_{\mu}\right) e^{i\vec{\tau}\cdot\vec{\xi}/v} \begin{pmatrix} \frac{1}{\sqrt{2}}(v+H) \\ 0 \end{pmatrix} \\ &= e^{i\vec{\tau}\cdot\vec{\xi}/v}\Big[\partial_{\mu} - ig\Big\{e^{-i\vec{\tau}\cdot\vec{\xi}/v}\,\frac{\tau_i}{2}W_{\mu i}\,e^{i\vec{\tau}\cdot\vec{\xi}/v} \\ &\quad - \frac{1}{ig}e^{-i\vec{\tau}\cdot\vec{\xi}/v}\left(\partial_{\mu}e^{i\vec{\tau}\cdot\vec{\xi}/v}\right)\Big\} + ig'\frac{1}{2}Y_{\mu}\Big] \begin{pmatrix} \frac{1}{\sqrt{2}}(v+H) \\ 0 \end{pmatrix} \end{aligned} \tag{68}$$

Since the term in the curly brackets in Eq. (68) is just a gauge transformed field

$$e^{-i\vec{\tau}\cdot\vec{\xi}/v}\,\frac{\tau_i}{2}W_{\mu i}\,e^{i\vec{\tau}\cdot\vec{\xi}/v} - \frac{1}{ig}e^{-i\vec{\tau}\cdot\vec{\xi}/v}\left(\partial_{\mu}e^{i\vec{\tau}\cdot\vec{\xi}/v}\right) = \frac{\tau_i}{2}W'_{\mu i} \tag{69}$$

all $\vec{\xi}$ dependence in $D_{\mu}\Phi$ resides in the phase factor in front in Eq. (68). Thus only H enters in $\mathcal{L}_{\text{Higgs-gauge}}$ and one finds*

* In (70) we have, for simplicity, denoted the gauge transformed fields by the same symbol used before the gauge transformation

$$\mathcal{L}_{\substack{\text{Higgs}\\ \text{gauge}}} = -\frac{1}{2}\partial^\mu H\,\partial_\mu H - \lambda\left(vH + \frac{1}{2}H^2\right)^2$$

$$-\frac{1}{4}g^2(v+H)^2 W_+^\mu W_{-\mu} - \frac{1}{8}(g^2+g'^2)(v+H)^2 Z^\mu Z_\mu \tag{70}$$

I note two properties of the Higgs field **H**.
(1) The vacuum expectation value v has a magnitude fixed by the Fermi constant. Using Eq. (55) and (65a) one has

$$v = (\sqrt{2}\,G_F)^{-1/2} \simeq 250 \text{ GeV} \tag{71}$$

The Higgs mass, although proportional to v, is arbitrary since it depends on the unknown quartic Higgs coupling λ :

$$m_H^2 = 2\lambda v^2 \tag{72}$$

(2) The coupling of the Higgs meson to the W and Z, which can be read off Eq. (70), is proportional to the mass of the gauge bosons. Namely one finds

$$\mathcal{L}_{HZZ} = -\frac{1}{2}(g^2+g'^2)^{1/2} M_Z Z^\mu Z_\mu H \tag{73}$$

$$\mathcal{L}_{HWW} = -g M_W W_+^\mu W_{-\mu} H$$

I discuss now briefly the final ingredient in the GSW Lagrangian, the $\mathcal{L}_{\text{Fermion-Higgs}}$ term. Even though this sector is not checked at all, from its general structure an important property of the theory emerges. The existence of a doublet field Φ, transforming as (2,-1/2) with respect to SU(2) x U(1), and of its charge conjugate

$$\tilde{\Phi} = i\tau_2\Phi^* \sim (2, 1/2) \tag{74}$$

allows one to write <u>SU(2) x U(1) invariant interactions</u> which connect left and right quark and lepton fields. For instance, one can write the following SU(2) x U(1) invariant coupling involving u quarks

$$\mathcal{L}_{\text{Yukawa}} = -h\left\{(\bar{u}\,\bar{d})_L\begin{pmatrix}\phi^0\\ \phi^-\end{pmatrix}u_R + \bar{u}_R(\phi_0^*\;\phi^+)\begin{pmatrix}u\\ d\end{pmatrix}_L\right\} \tag{75}$$

Such a Yukawa interaction has the very nice property that it generates a mass for the u quark after SU(2) x U(1) breakdown. That is, the term in (75) when $\Phi \to \langle\Phi\rangle = \frac{1}{\sqrt{2}} v$ corresponds to a mass term

$$\mathcal{L}_{mass} = - \frac{hv}{\sqrt{2}} (\bar{u}_L u_R + \bar{u}_R u_L) \tag{76}$$

so

$$m_u = \frac{hv}{\sqrt{2}} \tag{77}$$

As in the Higgs case also here the mass for the fermions - in this case the u quark - is arbitrary, because the parameter h is. Using the parametrization (67) for the doublet Higgs field, and redefining the $\begin{pmatrix} u \\ d \end{pmatrix}_L$ field via an SU(2) rotation

$$\begin{pmatrix} u \\ d \end{pmatrix}_L \to e^{i \vec{\xi} \cdot \vec{\tau}/v} \begin{pmatrix} u \\ d \end{pmatrix}_L \tag{78}$$

gives

$$\mathcal{L}_{Yukawa} = - m_u \bar{u} u - \frac{m_u}{v} \bar{u} u H \tag{79}$$

One sees that the Higgs coupling is also here proportional to mass.

The analysis just discussed can be repeated for the d quark and the electron - using couplings involving $\tilde{\Phi}$ - and can be extended to families of quarks and leptons. In so doing the notion of mixing naturally emerges. Because this has important phenomenological consequences I want to explain clearly its origins. For the case of various families of quarks and leptons let us adopt the convenient notation:

$$\begin{aligned} Q_{iL} &= \left\{ \begin{pmatrix} u \\ d \end{pmatrix}_L , \begin{pmatrix} c \\ s \end{pmatrix}_L , \begin{pmatrix} t \\ b \end{pmatrix}_L , \dots \right\} ; \\ L_{iL} &= \left\{ \begin{pmatrix} \nu_e \\ e \end{pmatrix}_L , \begin{pmatrix} \nu_\mu \\ \mu \end{pmatrix}_L , \begin{pmatrix} \nu_\tau \\ \tau \end{pmatrix}_L , \dots \right\} ; \\ u_{iR} &= \{ u_R, c_R, t_R, \dots \} ; \quad d_{iR} = \{ d_R, s_R, b_R, \dots \} \\ \ell_{iR} &= \{ e_R, \mu_R, \tau_R, \dots \} \end{aligned} \tag{80}$$

Then the most general SU(2) x U(1) invariant Yukawa interaction reads

$$\mathcal{L}_{Yukawa} = \Big\{ -h^u_{ij} [\bar{Q}_{iL} \Phi u_{jR}] + h^d_{ij} [\bar{Q}_{iL} \tilde{\Phi} d_{jR}] + h^\ell_{ij} [\bar{L}_{iL} \tilde{\Phi} \ell_{jR}] + h.c. \Big\} \tag{81}$$

Note that in general there is no reason to suppose that there is vanishing intrageneration couplings. So $h_{ij} \neq 0$ also for $i \neq j$. When Φ and $\tilde{\Phi}$ are replaced by their vacuum expectation values, mass matrices

$$M^f_{ij} = h^f_{ij} \frac{v}{\sqrt{2}} \qquad \{ f = u, d, \ell \} \tag{82}$$

are generated which are arbitrary and non diagonal, since so are the h^f_{ij}.

Obviously, if the matrices M^f_{ij} are not diagonal one must make a basis change to deal with states of definite mass. This implies that the weak interaction eigenstates are in general not the mass eigenstates. Changing basis to mass eigenstates causes for the weak currents mixing between generations. This is the origin of the famous Cabibbo mixing angle! To be more precise, the basis change to mass eigenstates:

(i) Has no effect on leptonic currents, because $m_\nu = 0$.
(ii) Has no effect on neutral currents, because these are flavor diagonal. This is the well known GIM mechanism /4/.
(iii) Affects hadronic charged currents.

I will leave the proof of the first two assertions as a problem but will illustrate point (iii) explicitly. Consider specifically J^μ_-, which is given by

$$\begin{aligned} J^\mu_- &= 2 [\bar{u}_L \gamma^\mu d_L + \bar{c}_L \gamma^\mu s_L + \bar{t}_L \gamma^\mu b_L + \dots] \\ &= 2 (\bar{u}_L \, \bar{c}_L \, \bar{t}_L \dots) \gamma^\mu 1 \begin{pmatrix} d_L \\ s_L \\ b_L \\ \vdots \end{pmatrix} \equiv 2 (\bar{\psi}_u)_L \gamma^\mu 1 (\psi_d)_L \end{aligned} \tag{83}$$

A basis change to render M^f diagonal implies the replacements

$$(\psi_u)_L \to U^u_L (\psi_u)_L \quad ; \quad (\psi_d)_L \to U^d_L (\psi_d)_L \tag{84}$$

For J_-^μ this change gives

$$J_-^\mu = 2\,(\bar\psi_u)_L\,\gamma^\mu (U_L^u)^\dagger (U_L^d)(\psi_d)_L = 2(\bar\psi_u)_L\,\gamma^\mu\,\tilde C(\psi_d)_L \tag{85}$$

For n families the matrix $\tilde C$ is a unitary n x n matrix which has $\frac{1}{2}n(n-1)$ real angles and $\frac{n(n+1)}{2}$ phases. However, not all of these phases are physical since one can rotate (2 n-1) phases away, by redefinitions of the quarks fields $\psi \to e^{i\alpha}\psi$ which do not affect M^f_{diag}.*After this redefinition the matrix $\tilde C \to C$, which is known as the Cabibbo matrix. The Cabibbo matrix thus has $\frac{1}{2}n(n-1)$ real angles and $\frac{1}{2}(n-1)(n-2)$ physical phases. All of these angles and phases are undetermined in the model, since they originate from the unknown Yukawa couplings h^f_{ij}. Nevertheless, the model does predict that in general there should be a non trivial mixing matrix.

I end this section with three remarks on the mixing matrix C.
1) For the case of two generations there is only one angle and no phase:

$$C = \begin{pmatrix} \cos\theta_c & \sin\theta_c \\ -\sin\theta_c & \cos\theta_c \end{pmatrix} \tag{86}$$

where θ_c is the Cabibbo angle. The current J_-^μ reads

$$J_-^\mu = 2\{\bar u_L\gamma^\mu d_L\cos\theta_c + \bar u_L\gamma^\mu s_L\sin\theta_c - \bar c_L\gamma^\mu d_L\sin\theta_c + \bar c_L\gamma^\mu s_L\cos\theta_c\} \tag{87}$$

The first two terms above are the Cabibbo piece /5/ of the charged current, known already in the early 60's. The last two terms were added by Glashow, Iliopoulos and Maiani /4/ to guarantee that the "neutral current" J_3^μ, arising from the commutation algebra of J_-^μ and J_+^μ, be flavor diagonal. In the SU(2) x U(1) model the full (Cabibbo plus GIM) form of the current arises naturally, when changing to a basis in which the mass matrix of quarks is diagonal.

2) For three generations of quarks and leptons - as we apparently have - C is the Kobayashi Maskawa matrix /6/. This matrix has three real angles and one phase. The Kobayashi Maskawa phase allows for CP violation in the standard model. I will return to this point at the end of these lectures.

3) Even in the case of many families one can check that the coupling of the physical Higgs particle to the fermions is flavor diagonal and proportional to the fermion's mass:

---

* 2n-1 not 2n because one overall phase has no meaning.

$$\mathcal{L}^{y}_{Hff} = -\frac{m_f}{v}\bar{f} f H \tag{88}$$

DYNAMICS OF NAGT: ASYMPTOTIC FREEDOM, CONFINEMENT AND PARTON MODELS

Up to now I discussed only the symmetry properties of the standard model, without worrying too much about the dynamics. The dynamical issues in the electroweak theory, except for what really causes the SU(2) x U(1) breakdown, are not so difficult to tackle. Because the coupling constants g and g' are small, a perturbation treatment is warranted. For QCD, on the other hand, the dynamics is complex and needs to be discussed. In particular, the QCD Lagrangian is described in terms of quarks and gluons and yet, in the real world, we see only hadrons.

The important concept that allows one to make progress in QCD, even though its dynamics is complex, is the idea of the running coupling constant. To appreciate this concept it is particularly useful to refer back to a well known phenomenon in QED, that of vacuum polarization. Because of the possibility of virtual pair creation, the QED vacuum acts as a polarizable medium. This causes the effective charge of an electron to be screened at large distance. Because of this vacuum polarization effect the Coulomb potential is modified from its usual α/r form to:

$$V_{Coulomb}(r) = \frac{\alpha(r)}{r} \tag{89}$$

The function $\alpha(r)$, which plays the role of an effective coupling constant squared, because of the screening of the electron charge becomes smaller as r increases. This effect can be computed by considering the modification to the photon propagator due to the possibility of pair creation (see Fig. 4).

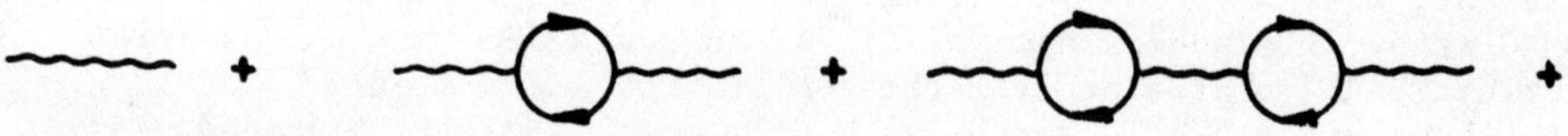

Fig. 4: Corrections to the photon propagator due to vacuum polarization

The result of this calculation, done first almost 50 years ago by Uehling /7/, reads for $q^2 \gg m^2$ and in momentum space:

$$\alpha(q^2) = \frac{\alpha}{1 - \frac{3\alpha}{\pi} \ln q^2/m^2} \tag{90}$$

The running coupling constant $\alpha(q^2)$ in QED increases as q^2 becomes large, reflecting the fact that the electron's charge is screened by the virtual e^+e^- pairs in the vacuum. Knowing $\alpha(q^2)$ as a function of q^2 one may compute the, so called, β-function, which is defined by:

$$\beta(q^2) = \frac{d\,\alpha(q^2)}{d \ln q^2} \simeq \frac{1}{3\pi} \left(\alpha(q^2)\right)^2 \tag{91}$$

The last result above, follows from Eq. (90) and is the lowest order approximation for β. This function is an intrinsic property of QED and it can be determined in a power series expansion in α.

$$\beta(\alpha) = b\,\alpha^2 + b'\alpha^4 + \dots \tag{92}$$

From the Uehling calculation one knows that $b = 1/3\pi$ and is therefore positive. The positivity of b, coupled with the definition (91), is enough to tell one immediately that there is screening in QED. That is, $\alpha(q^2)$ goes up as q^2 increases.

One can define an analogous β-function for QCD, which now depends on $\alpha_s = g_s^2/4\pi$, and calculate this function in perturbation theory. The result, up to terms of $O(\alpha_s^4)$, is

$$\beta(\alpha_s) = -\frac{(33-2n_f)}{12\pi}\,\alpha_s^2 \left\{ 1 + \frac{(153-19n_f)}{2\pi(33-2n_f)}\,\alpha_s^2 + \dots \right\} \tag{93}$$

In the above n_f stands for the number of flavors of quarks in QCD. The most remarkable feature of Eq. (93) is the negative sign in front of the α_s^2 term. This means that in QCD, in contrast to what happens in QED, there is antiscreening. This property was apparently known to 't Hooft, but the calculation of β to $O(\alpha_s^2)$ was carried out independently and first published by Politzer and Gross and Wilczek /8/. The terms of $O(\alpha_s^4)$ in Eq. (93) were calculated by Caswell and Jones /9/.

From the β-function of Eq. (93), provided that these are less than 17 flavors, one deduces that in QCD the effective coupling constant (running coupling constant) has the behaviour shown in Fig. 5.

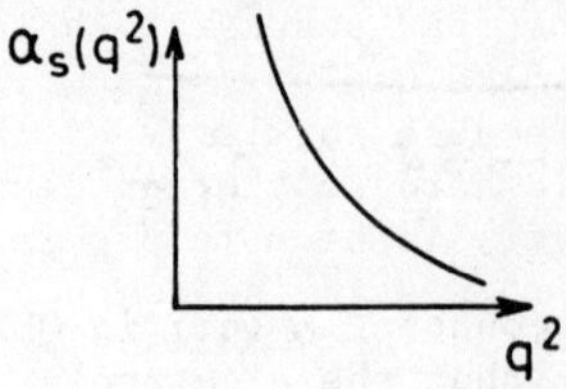

Fig. 5: Running coupling constant in QCD

The fact that $\alpha_s(q^2)$ decreases for large q^2 is due to the negative sign of the α_s^2 term in (93), whose cause can be traced to the gluons. These terms, which are absent in the Abelian QED case, contribute the factor of $-33/12\pi$. This number is due (in part) to the additional gluonic contributions to the gluon propagator, shown in Fig. 6.

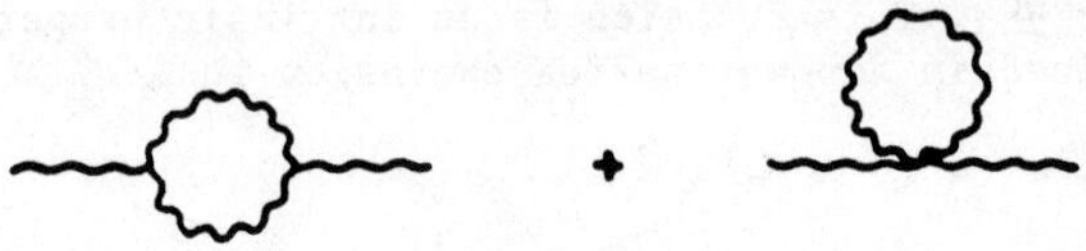

Fig. 6: Gluonic contributions to gluon propagator

The behavior of $\alpha_s(q^2)$ of Fig. 5 allows one to arrive at the following qualitative picture for the interactions of quarks and gluons. At small q^2 (large distances) the effective coupling constant $\alpha_s(q^2)$ becomes very large and colored objects bind strongly into color singlet states, the hadrons. This is the confinement property of QCD. For large q^2 (short distances), on the other hand, the effective coupling constant becomes very small and quarks and gluons act as quasi free objects. This is the famous asymptotic freedom of QCD. Confinement and asymptotic freedom explain two features of hadrons which were contradictory in the 1960's and early 1970's: Hadrons appear to be made up of quarks which are strongly bound together since they never seem to materialize in collisions (confinement) and yet when hadrons are probed at short distances they appear to be essentially composed of free quarks and gluons (Asymptotic freedom).

There are clearly two aspects of QCD dynamics which can be studied and probed. Either one examines the long distance behaviour of the theory, in which quark and gluons bind to make hadrons. This is the realm of hadronic spectroscopy and, theoretically, of lattice QCD or, one studies the short distance properties of the theory by examining "hard" collisions in which one is essentially dealing with scattering at the quark and gluon level. I shall, in the remainder of these lectures, only concentrate on this latter aspect of QCD, where one can proceed perturbatively.

For processes where hadrons scatter with large momentum transfer (hard scattering) one may make use of asymptotic freedom to compute the relevant cross sections. Because for large q^2 the effective coupling $\alpha_s(q^2)$ is small, hadronic hard scattering can be computed perturbatively at the quark and gluon level. To be more precise, three elements enter into these parton model hard scattering calculations:

1) Hadronic structure functions: These functions describe the distribution of quarks and gluons (partons) within the initial state hadrons.

2) Parton cross sections: These cross sections are calculable in QCD provided q^2 is large enough, so that a perturbation expansion in $\alpha_s(q^2)$ is tenable. It is this restriction that circumscribes parton model calculations of hadronic scattering processes to those involving large q^2. Only in this case can one compute - albeit approximately - the parton cross sections.

3) Hadronic fragmentation functions: These functions describe the distribution of hadrons expected from a given parton.

The calculation of a hadronic hard scattering process involves, in general, the convolution of the above three elements. Only the parton cross sections are calculable, so that only the overall structure but not the detailed form of the hadronic hard scattering reaction can be determined. Although the structure and fragmentation functions are, at present, uncalculable, because they involve the "soft processes" of transmutation of hadrons into quarks and vice versa, their universality makes the above procedure relevant. Because the distribution of quarks within a hadron, or the probability of creating a certain hadron from a quark, is process independent, once these distributions are known from some hadronic hard scattering process they allow one to predict what is to be expected in another hard scattering process.

These points are best illustrated by a practical example. Imagine trying to compute the deep inelastic production of pions in muon-proton scattering, for large momentum transfer, q^2. The process $\mu p \to \mu \pi X$ can be computed by convoluting three separate elements, as shown graphically in Fig. 6. The cross section involves the probability $f_q(\xi;q^2)$ (Box 1 in Fig. 6) of finding a quark with fractional momentum ξ within the proton, convoluted with the (parton) cross sections for the elastic scattering of this quark with the incoming muon (Box 2 in Fig. 6), convoluted with the probability $D_q^{\pi}(\xi';q^2)$ that the outgoing quark becomes a pion with fractional momentum ξ' (Box 3 in Fig. 6)*

---

* This is a slight simplification. I will discuss later the role of gluons in deep inelastic processes.

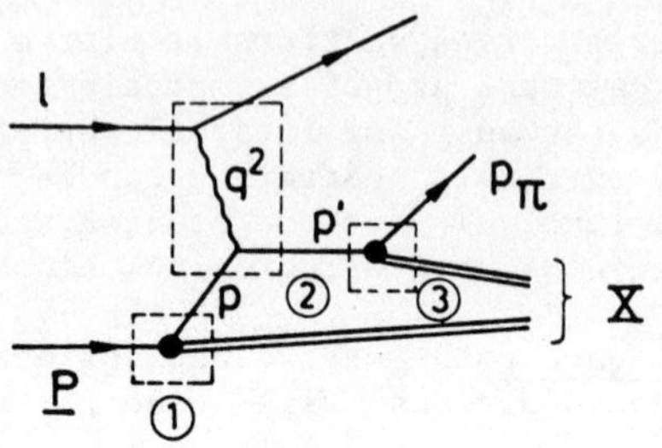

Fig. 6: Parton model breakup of deep inelastic pion production in μ - proton scattering

Let me be more specific. If P^μ and ℓ^μ are the initial momenta of the proton and muon, P_π^μ is the final momentum of the produced pion, and q^μ is the momentum transfer, it is usual to define the following kinematical variables:

$$x = -\frac{q^2}{2P\cdot q} \quad ; \quad y = \frac{P\cdot q}{P\cdot \ell} \quad ; \quad z = \frac{P\cdot P_\pi}{P\cdot q} \tag{94}$$

The quarks in Fig. 6 carry momenta p^μ and p'^μ and the fractions ξ and ξ' are defined by

$$p^\mu = \xi P^\mu \quad ; \quad P_\pi^\mu = \xi' p'^\mu \tag{95}$$

Thus at the parton level the kinematical variables x_p and z_p defined analogously to (94) are given by

$$x_p = \frac{-q^2}{2p\cdot q} = -\frac{q^2}{2P\cdot q\,\xi} = \frac{x}{\xi} \tag{96a}$$

$$z_p = \frac{P\cdot p'}{P\cdot q} = \frac{P\cdot (P_\pi/\xi')}{P\cdot q} = \frac{z}{\xi'} \tag{96b}$$

The inclusive cross section for the process $\mu p \rightarrow \mu \pi X$ computed from Fig. 6, assuming that the scattering proceeds as the sum of incoherent scatterings from each parton, is given explicitly by the formula

$$\frac{d\sigma^{\pi}}{dx\,dy\,dz} = \sum_q \int_x^1 \frac{d\xi}{\xi} \int_z^1 \frac{d\xi'}{\xi'}\, f_q(\xi;q^2) \left[\frac{d\sigma_{part}(q^2)}{dx_p\,dy\,dz_p}\right] D_q^{\pi}(\xi';q^2) \tag{97}$$

That is, the cross section is the sum of the convolutions of the probability $f_q(\xi;q^2)$ of having a quark of given fractional momentum within the proton scatter with the incoming muon, producing a final quark which has a probability $D_q^{\pi}(\xi';q^2)$ of producing the observed pion. The q^2 dependence of the structure and fragmentation functions, as well as that of the parton cross sections, is an effect of QCD, which I will discuss later in more detail. If one supposed that quarks were really free inside the proton at short distance $(\alpha_s(q^2) \rightarrow 0)$, as was assumed in the oldfashioned parton model of pre QCD days /10/, then this q^2 dependence would be omitted. The limits in Eq. (97) follow since x_p and z_p kinematically can reach only unity - neglecting mass effects - and the fractions ξ and ξ' can be at most one. The factors of ξ and ξ', are needed, furthermore, to transform x and z into x_p and z_p.

To lowest order in the electroweak interaction and in QCD (α_s^0) the parton cross section is given by the tree graph μ-quark scattering of Fig. 7. Except for very large q^2, where the Z^0 contribution is also important, the exchanged boson is just the photon and the cross section is proportional to $(q^2)^{-2}$.

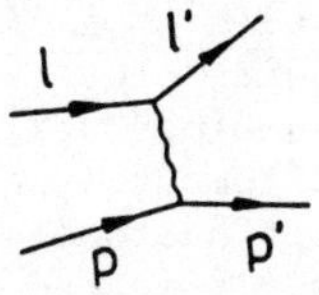

Fig. 7: Lowest order contribution to muon-quark scattering

Because one is dealing with a 2-2 scattering process it is easy to convince oneself, neglecting all masses relative to q^2 and P.q, that

$$\frac{d\sigma_{part}}{dx_p\,dy\,dz_p} \sim \delta(1-x_p)\,\delta(1-z_p) \tag{98}$$

Hence, in this approximation, the hadronic scattering process is just proportional to the product of the structure and fragmentation functions:

$$\frac{d\sigma^{\pi}}{dx\,dy\,dz} \sim \sum_{q} f_{q}(x)\, D_{q}^{\pi}(z) \tag{99}$$

It is useful to summarize the structure of the hadronic hard scattering cross sections by the symbolic formula:

$$d\sigma^{H} \sim f_{parton} \otimes d\sigma_{parton} \otimes D^{H}_{parton} \tag{100}$$

where f_{parton} and D^{H}_{parton} are process independent but uncalculable, while $d\sigma_{parton}$ is process dependent but calculable in a power series in α_s. Eq. (100) suggests already a general test of QCD, in the perturbative regime. Having found f_{parton} and D^{H}_{parton} from some process one may use them to compute another process. This procedure works rather well in practice. For instance, large $p_\perp$ jet production at the collider can be computed knowing the structure functions in deep inelastic scattering. There are, however, some subtleties and complications - like the K factor in Drell-Yan - which I will explain later on in these lectures.

In special circumstances, when one studies processes which are inclusive enough, one can altogether eliminate the fragmentation functions D^{H}_{parton}. The two best known examples are deep inelastic scattering processes (like $\mu p \to \mu X$) and e^+e^- annihilation into hadrons ($e^+e^- \to x$). Physically it is quite clear why no D-functions appear in these processes, since one is summing over all possible hadrons. In QCD this circumstance allows one to compute the process in a quark-gluon basis, because the probability of hadronization is unity. More formally, the fragmentation functions are eliminated by using the energy-momentum sum rule

$$\sum_{H} \int_0^1 d\xi'\, \xi'\, D_q^{H}(\xi'; q^2) = 1 \tag{101}$$

It is not hard to convince oneself that, for the processes at hand, precisely this weighted sum of the fragmentation functions enters.

Jet production, in which one is asking again for gross properties of a reaction, is also independent of the fragmentation functions, provided one is prepared to neglect effects due to the soft hadronization of quarks and gluons into hadrons *. To neglect the fragmentation func-

---

* More technically, as I will explain in a later section, one must also deal with processes which are infrared insensitive.

tions one needs to assume that the jet direction coincides with the parton direction. This then provides additional QCD tests, which are spectacularly successful qualitatively but less so quantitatively. I refer here to the presence of 3 jets at PETRA and at PEP and of large $p_\perp$ jets at the collider, whose really detailed comparisons with QCD, as I will explain, must still rely on some sensible hadronization model.

Before embarking in a more detailed analysis of QCD effects in hard scattering, it will prove useful to devote the next section to a discussion of deep inelastic scattering, calculated at zeroth order in QCD. This will allow me to interrelate a variety of processes and test thereby some of the predictions of the GSW model. In subsequent sections I will return to the QCD corrections to these results and see how these fare experimentally. Of course, this separate testing of aspects of the GSW model and of QCD is done here only for pedagogical purposes. In practice, one deals with both phenomena simultaneously.

DEEP INELASTIC PHENOMENA: 0th ORDER QCD TESTING THE GSW THEORY

I want to discuss in some detail the related deep inelastic processes

$$\ell^{\pm}\, p \rightarrow \ell^{\pm}\, X \tag{102a}$$

$$\begin{pmatrix} \nu_e \\ \bar{\nu}_e \end{pmatrix} p \rightarrow \begin{pmatrix} e^- \\ e^+ \end{pmatrix} X \tag{102b}$$

$$\begin{pmatrix} \nu_\ell \\ \bar{\nu}_\ell \end{pmatrix} p \rightarrow \begin{pmatrix} \nu_\ell \\ \bar{\nu}_\ell \end{pmatrix} X \tag{102c}$$

For the first process above both the electromagnetic and the weak neutral current contribute. The reaction (102b) is a purely weak charged current process, while (102c) occurs only because of the existence of weak neutral currents. Because all three of these processes are totally inclusive, the expression for the differential scattering cross section will not contain fragmentation functions. The differential cross section for these processes, therefore, takes the form (cf. Eq. (97))

$$\frac{d\sigma}{dx\,dy} = \sum_{\text{parton}} \int_x^1 \frac{d\xi}{\xi}\, f_{\text{parton}}(\xi; q^2)\, \frac{d\sigma_{\text{parton}}(q^2)}{dx_p\, dy} \tag{103}$$

As indicated above, I will compute Eq. (103) in this section only to lowest order in QCD. This means that all q^2-dependence in f_{parton} will be ignored and that the parton cross section corresponds to the simple quark lepton scattering process shown in Fig. 8.

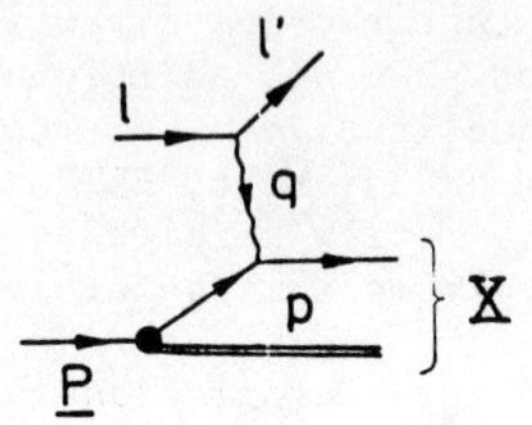

Fig. 8: Lowest order QCD contribution for deep inelastic scattering

To be a little more explicit, $d\sigma_{parton}$ will be calculated as the appropriate quark-lepton scattering processes, neglecting all fermion masses and spin averaging over the fermions. (Averaging over initial spins and summing over final spins). Simple kinematics gives for the 2-2 scattering process involved:

$$\frac{d\sigma_{parton}}{dx_p\, dy} = \frac{y}{16\pi q^2}\, \delta(1 - x_p)\, \langle |T|^2 \rangle \tag{104}$$

Here $\langle |T|^2 \rangle$ is the spin-averaged T-matrix element squared for lepton-quark scattering

$$\frac{1}{2} \sum_{s_q} \sum_{s_{q'}} \cdot \begin{pmatrix} 1/2 \\ 1 \end{pmatrix} \sum_{s_\ell} \sum_{s_{\ell'}} |T|^2 \equiv \langle |T|^2 \rangle \tag{105}$$

In the above the factor of 1/2 applies for initial state charged leptons, corresponding to the averaging over their initial polarizations. For neutrino initiated scattering, however, the corresponding factor is 1, since neutrinos, or antineutrinos, come in only one helicity.

For the reactions in Eqs. (102), the T matrix element can be calculated from the graph of Fig. 9. For moderate values of q^2, for reaction

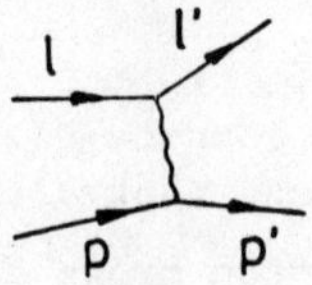

Fig. 9: Contribution to quark-lepton scattering

(102a) it suffices to consider only the photon exchange contribution. Similarly, for the charged current and neutral current reactions of Eqs. (102b) and (102c) one may replace the propagator by a contact interaction and therefore use the interaction Lagrangian of Eq. (57) to compute the T-matrix. One finds in this way the following parton T-matrices, for the corresponding processes of Eqs. (102):

$$T = -i \frac{e^2 e_q}{q^2} \left(J^{\mu}_{em} \cdot J_{\mu\, em} \right) \qquad (106a) \quad \text{EM}$$

$$T = i \frac{G_F}{\sqrt{2}} \left(J_+^{\mu} \cdot J_{-\mu} \right) \qquad (106b) \quad \text{CC}$$

$$T = i \frac{G_F}{\sqrt{2}} \rho \left(J^{\mu}_{NC} \cdot J_{\mu\, NC} \right) \qquad (106c) \quad \text{NC}$$

Here e_q is the quark charge in units of the positron charge. Hence it follows that

$$\langle |T|^2 \rangle = \frac{16\pi^2 \alpha^2 e_q^2}{(q^2)^2} L^{em}_{\mu\nu} Q^{\mu\nu}_{em} \qquad (107a) \quad \text{EM}$$

$$\langle |T|^2 \rangle = \frac{G_F^2}{2} L^{cc}_{\mu\nu} Q^{\mu\nu}_{cc} \qquad (107b) \quad \text{CC}$$

$$\langle |T|^2 \rangle = \frac{G_F^2}{2} \rho^2 L^{NC}_{\mu\nu} Q^{\mu\nu}_{NC} \qquad (107c) \quad \text{NC}$$

The tensor structures $L_{\mu\nu}$ and $Q_{\mu\nu}$ are straightforward to compute, for all processes, using the explicit form of the currents in the standard GSW model.

I will give an example of this computation for the charged current process $\nu_e d \to \ell u$. The T-matrix element is just

$$T = i \frac{G_F}{\sqrt{2}} \left[\bar{u}_\ell(\ell') \gamma^{\mu} (1-\gamma_5) u_{\nu_e}(\ell) \right] \cdot \left[\bar{u}_u(p') \gamma_\mu (1-\gamma_5) u_d(p) \right] \qquad (108)$$

The tensors $L^{cc}_{\mu\nu}$ and $Q^{cc}_{\mu\nu}$ come from the spin average of the factors in bracket above, corresponding to leptons and quarks, respectively. Thus

$$L^{cc}_{\mu\nu} = \sum_{s_\nu s_e} [\bar{u}_e(\ell')\gamma_\mu(1-\gamma_5)u_{\nu_e}(\ell)]\cdot[\bar{u}_e(\ell')\gamma_\nu(1-\gamma_5)u_{\nu_e}(\ell)]^*$$

$$= \mathrm{Tr}\ \gamma_\mu(1-\gamma_5)(-\gamma\cdot\ell)\gamma_\nu(1-\gamma_5)(-\gamma\cdot\ell')$$

$$= 2\ \mathrm{Tr}\ \gamma_\mu\ \gamma\cdot\ell\ \gamma_\nu\ \gamma\cdot\ell'(1-\gamma_5)$$

$$= 8[\ell_\mu\ell'_\nu + \ell'_\mu\ell_\nu - \ell\cdot\ell'\eta_{\mu\nu} - i\epsilon_{\mu\nu\alpha\beta}\ell^\alpha\ell'^\beta]$$

$$\equiv 8[\ell^S_{\mu\nu} - \ell^A_{\mu\nu}] \tag{109}$$

Note that the spin averaging in $L^{cc}_{\mu\nu}$ has no factor of 1/2 because one deals with initial neutrino (c.f. the discussion after Eq. 105). Also the functions $\ell^S_{\mu\nu}$ and $\ell^A_{\mu\nu}$ are obviously defined as the symmetric and antisymmetric terms in ℓ and ℓ' appearing in the next to last line in Eq. (109). Similarly one calculates

$$Q^{cc}_{\mu\nu} = \frac{1}{2}\sum_{s_d s_u} [\bar{u}_u(p')\gamma_\mu(1-\gamma_5)u_d(p)]\cdot[\bar{u}_u(p')\gamma_\nu(1-\gamma_5)u_d(p)]^*$$

$$= \frac{1}{2}\mathrm{Tr}\ \gamma_\mu(1-\gamma_5)(-\gamma\cdot p)\gamma_\nu(1-\gamma_5)(-\gamma\cdot p')$$

$$= \mathrm{Tr}\ \gamma_\mu\ \gamma\cdot p\ \gamma_\nu\ \gamma\cdot p'(1-\gamma_5)$$

$$= 4[p_\mu p'_\nu + p'_\mu p_\nu - p\cdot p'\eta_{\mu\nu} - i\epsilon_{\mu\nu\alpha\beta}p^\alpha p'^\beta]$$

$$\equiv 4[q^S_{\mu\nu} - q^A_{\mu\nu}] \tag{110}$$

where $q^S_{\mu\nu}$ and $q^A_{\mu\nu}$ are defined in an analogous way to $\ell^S_{\mu\nu}$ and $\ell^A_{\mu\nu}$.

Performing similar calculations for all the parton subprocesses, involving lepton scattering off both quarks and antiquarks, one secures the following list of formulas:

$$\left(L_{\mu\nu}\right)_{em}^{\ell\to\ell} = \left(L_{\mu\nu}\right)_{em}^{\bar{\ell}\to\bar{\ell}} = 2\left[\ell_{\mu\nu}^{S}\right] \tag{111a}$$

$$\left(L_{\mu\nu}\right)_{CC}^{\nu\to\ell} = \left(L_{\mu\nu}\right)_{NC}^{\nu\to\nu} = 8\left[\ell_{\mu\nu}^{S} - \ell_{\mu\nu}^{A}\right] \tag{111b}$$

$$\left(L_{\mu\nu}\right)_{CC}^{\bar{\nu}\to\bar{\ell}} = \left(L_{\mu\nu}\right)_{NC}^{\bar{\nu}\to\bar{\nu}} = 8\left[\ell_{\mu\nu}^{S} + \ell_{\mu\nu}^{A}\right] \tag{111c}$$

$$\left(Q_{\mu\nu}\right)_{em}^{q\to q} = \left(Q_{\mu\nu}\right)_{em}^{\bar{q}\to\bar{q}} = 2\left[q_{\mu\nu}^{S}\right] \tag{111d}$$

$$\left(Q_{\mu\nu}\right)_{CC}^{q\to q'} = 4\left[q_{\mu\nu}^{S} - q_{\mu\nu}^{A}\right] \tag{111e}$$

$$\left(Q_{\mu\nu}\right)_{CC}^{\bar{q}\to\bar{q}'} = 4\left[q_{\mu\nu}^{S} + q_{\mu\nu}^{A}\right] \tag{111f}$$

$$\left(Q_{\mu\nu}\right)_{NC}^{q\to q} = 4\left(Q_{qL}^{NC}\right)^2\left[q_{\mu\nu}^{S} - q_{\mu\nu}^{A}\right] + 4\left(Q_{qR}^{NC}\right)^2\left[q_{\mu\nu}^{S} + q_{\mu\nu}^{A}\right] \tag{111g}$$

$$\left(Q_{\mu\nu}\right)_{NC}^{\bar{q}\to\bar{q}} = 4\left(Q_{qL}^{NC}\right)^2\left[q_{\mu\nu}^{S} + q_{\mu\nu}^{A}\right] + 4\left(Q_{qR}^{NC}\right)^2\left[q_{\mu\nu}^{S} - q_{\mu\nu}^{A}\right] \tag{111h}$$

In the above the "charges" Q_{qL}^{NC} and Q_{qR}^{NC} which characterize the strength of the neutral current coupling to left- and right-handed quarks, are given in the GSW model by

$$Q_{qL}^{NC} = \left(\frac{\tau_3}{2}\right)_q - e_q \sin^2\theta_W \tag{112a}$$

$$Q_{qR}^{NC} = -e_q \sin^2\theta_W \tag{112b}$$

We note that in the above formulas going from particles to anti-particles corresponds to the changes of $S \leftrightarrow S$ and $A \leftrightarrow -A$. This follows since this change corresponds precisely to the interchange of the roles of $\ell \leftrightarrow \ell'$ and/or $p \leftrightarrow p'$. Note that these same changes also occur in flipping (V-A) into (V+A) interactions. That is, going to the antiparticle is akin to flipping helicity.

To compute the parton cross sections it remains to evaluate the contraction of $L_{\mu\nu}$ with $Q^{\mu\nu}$, for each of the cases in question. Simple kinematics yields the formulas

$$\ell^{S}_{\mu\nu}\, q^{\mu\nu}_{A} = \ell^{A}_{\mu\nu}\, q^{\mu\nu}_{S} = 0 \tag{113a}$$

$$\ell^{S}_{\mu\nu}\, q^{\mu\nu}_{S} = \frac{(q^2)^2}{2y^2}\left[1 + (1-y)^2\right] \tag{113b}$$

$$\ell^{A}_{\mu\nu}\, q^{\mu\nu}_{A} = \frac{(q^2)^2}{2y^2}\left[1 - (1-y)^2\right] \tag{113c}$$

It follows therefore that the y-dependence for the parton subprocesses is simple. To wit:

1) For electromagnetic processes, since only S-terms enter one has

$$\left(\frac{d\sigma}{dy}\right)^{em} \sim \left[1 + (1-y)^2\right] \tag{114}$$

2) for scattering of a left-handed quark with a left-handed lepton (or right-handed quark with a right-handed lepton) one has

$$\frac{d\sigma_{LL}}{dy} \sim \frac{d\sigma_{RR}}{dy} \sim \left[\ell^{S}_{\mu\nu} \mp \ell^{A}_{\mu\nu}\right]\cdot\left[q^{S\mu\nu} \mp q^{A\mu\nu}\right] \sim 1 \tag{115}$$

3) Mixed scattering, on the other hand, gives

$$\frac{d\sigma_{LR}}{dy} \sim \frac{d\sigma_{RL}}{dy} \sim \left[\ell^{S}_{\mu\nu} \mp \ell^{A}_{\mu\nu}\right]\cdot\left[q^{S\mu\nu} \pm q^{A\mu\nu}\right] \sim (1-y)^2 \tag{116}$$

4) Scattering leptons off antiquarks is equivalent to the scattering of leptons off quarks of the opposite helicity.

These results can be understood qualitatively in a simpler way /11/. In all the parton processes considered one is dealing with vector interactions which conserve helicity at each vertex. Consider then scattering of two left handed objects in the CM system, as shown in Fig. 10a. The initial state has J = o and so one expects that $(d\sigma/d\Omega)_{CM}$ be isotropic. This implies directly Eq. (115). If a right- and a left-handed state scatter, as in Fig. 10b, then J = 1 and backward scattering is forbidden, since it corresponds to $\Delta J = 2$. So $\left(\frac{d\sigma}{d\Omega}\right)_{CM} \sim (1+\cos\theta)^2$ which is equivalent to Eq. (116).

Fig. 10: Scattering of LL(a) and LR(b) fermions in the CM, with vector interactions.

Armed with Eqs. (103) and (104), and the explicit forms for $\langle |T|^2 \rangle$ which follow from Eqs. (107), (111) and (113), it is now straightforward to derive the parton model (0th order QCD) expressions for the reactions (102). In the laboratory system one finds:

$$\left(\frac{d\sigma}{dx\,dy}\right)^{EM} = \frac{8\pi\alpha^2 M E_e}{(q^2)^2}\left[\frac{1+(1-y)^2}{2}\right]\cdot\left\{\sum_q e_q^2\, x\left(f_q(x)+f_{\bar{q}}(x)\right)\right\} \qquad (117)$$

$$\left(\frac{d\sigma}{dx\,dy}\right)^{cc}_{\nu\ \text{induced}} = \frac{G_F^2 M E_\nu}{\pi}\cdot\left\{\sum_i 2x\, f_{d_i}(x)\,[1] + \sum_i 2x\, f_{\bar{u}_i}(x)\,[1-y]^2\right\} \qquad (118a)$$

$$\left(\frac{d\sigma}{dx\,dy}\right)^{cc}_{\bar{\nu}\ \text{induced}} = \frac{G_F^2 M E_\nu}{\pi}\cdot\left\{\sum_i 2x\, f_{\bar{d}_i}(x)\,[1] + \sum_i 2x\, f_{u_i}(x)\,[1-y]^2\right\} \qquad (118b)$$

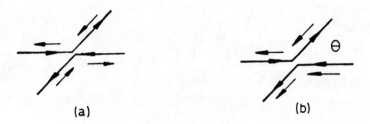

$$\left(\frac{d\sigma}{dx\,dy}\right)^{NC}_{\nu\ \text{induced}} = \frac{G_F^2 M E_\nu \rho^2}{\pi} \cdot \left\{ \sum_i 2x\, f_{q_i}(x) \left[(Q^{NC}_{q_i L})^2 + (Q^{NC}_{q_i R})(1-y)^2 \right] + \sum_i 2x\, f_{\bar{q}_i}(x) \left[(Q^{NC}_{q_i R})^2 + (Q^{NC}_{q_i L})(1-y)^2 \right] \right\} \tag{119a}$$

$$\left(\frac{d\sigma}{dx\,dy}\right)^{NC}_{\bar{\nu}\ \text{induced}} = \frac{G_F^2 M E_\nu \rho^2}{\pi} \cdot \left\{ \sum_i 2x\, f_{q_i}(x) \left[(Q^{NC}_{q_i R})^2 + (Q^{NC}_{q_i L})^2 (1-y)^2 \right] + \sum_i 2x\, f_{\bar{q}_i}(x) \left[(Q^{NC}_{q_i L})^2 + (Q^{NC}_{q_i R})^2 (1-y)^2 \right] \right\} \tag{119b}$$

These equations simplify in the, so called, valence quark approximation in which one neglects altogether the contributions of antiquarks in the nucleon. For lepton proton scattering, in this case, the curly brackets in Eqs. (117) - (118) become

$$\{\ \}^{EM}_{ep} \xrightarrow[\text{Valence}]{} \frac{4}{9}\, x\, u_v(x) + \frac{1}{9}\, x\, d_v(x) \tag{120}$$

$$\{\ \}^{CC}_{\nu p} \xrightarrow[\text{Valence}]{} 2x\, d_v(x) \tag{121a}$$

$$\{\ \}^{CC}_{\bar{\nu} p} \xrightarrow[\text{Valence}]{} 2x\, u_v(x)\, (1-y)^2 \tag{121b}$$

Here $u_v(x)$ and $d_v(x)$ are the distribution functions of up and down (valence) quarks in a proton. By charge symmetry they describe, respectively, also the distribution of down and up quarks in a neutron. Hence for lepton neutron scattering the curly brackets in Eqs. (117) and (118), in the valence approximation, become

$$\left\{ \right\}^{EM}_{eN} \xrightarrow[\text{Valence}]{} \frac{4}{9} \times d_v(x) + \frac{1}{9} \times u_v(x) \tag{122}$$

$$\left\{ \right\}^{CC}_{\nu N} \xrightarrow[\text{Valence}]{} 2x\, u_v(x) \tag{123a}$$

$$\left\{ \right\} \xrightarrow[\text{Valence}]{} 2x\, d_v(x)\,(1-y)^2 \tag{123b}$$

For an isoscalar target $N = \frac{1}{2}(n+p)$ the relevant formulas are simply:

$$\left\{ \right\}^{EM}_{eN} \xrightarrow[\text{Valence}]{} \frac{5}{18} \left[x\, u_v(x) + x\, d_v(x) \right] \tag{124}$$

$$\left\{ \right\}^{CC}_{\nu N} \xrightarrow[\text{Valence}]{} \left[x\, u_v(x) + x\, d_v(x) \right] \tag{125a}$$

$$\left\{ \right\}^{CC}_{\bar{\nu} N} \xrightarrow[\text{Valence}]{} \left[x\, u_v(x) + x\, d_v(x) \right] (1-y)^2 \tag{125b}$$

Data from the CHARM experiment at CERN /12/, shown in Fig. 11, for neutrino and antineutrino scattering on an isoscalar target shows that the characteristic y-dependence predicted in the valence approximation ($d\sigma_{\nu N}/dy \sim 1$; $d\sigma_{\bar{\nu} N}/dy \sim (1-y)^2$) is satisfied to within perhaps 10 % - 15 %. Furthermore, charged current and electromagnetic deep inelastic processes are interrelated in this approximation. From Eqs. (124) and (125a) one predicts that

$$\frac{\left\{ \right\}^{CC}_{\nu N}}{\left\{ \right\}^{EM}_{eN}} \equiv \frac{F_2^{\nu N}(x)}{F_2^{eN}(x)} = \frac{18}{5} \tag{126}$$

This behaviour is also well reproduced by the data, as Fig. 12 shows.

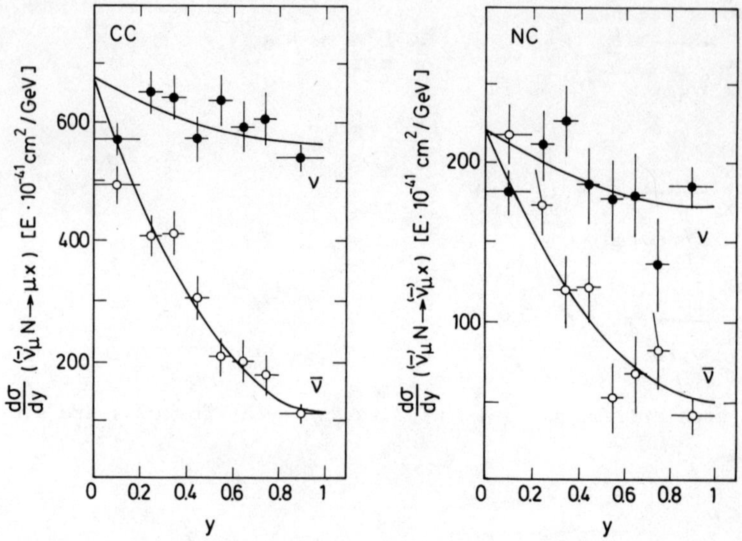

Fig. 11: ν_μ and $\bar{\nu}_\mu$ scattering on isoscalar targets. Data from Ref. 12

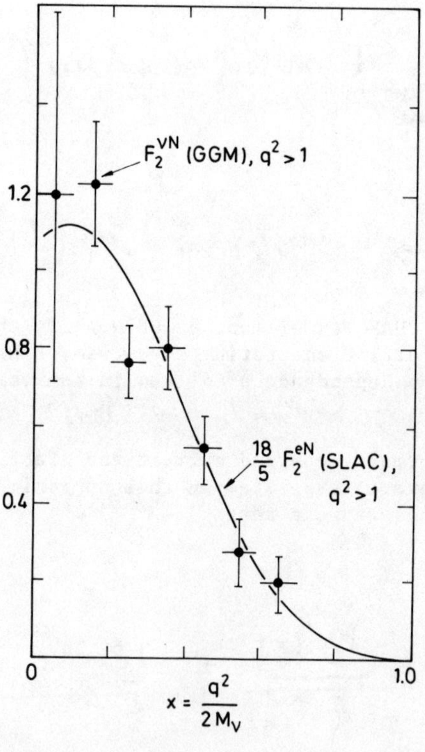

Fig. 12: Test of Eq. (126) using Gargamelle and SLAC data, from Ref. 13

In the valence quark approximation neutrino and antineutrino neutral current scattering off isoscalar targets also simplify. From Eqs. (119a) and (119b) one obtains

$$\{\ \}^{NC}_{\nu N} = \left[(Q^{NC}_{dL})^2 + (Q^{NC}_{uL})^2 + (1-y)^2 \left((Q^{NC}_{dR})^2 + (Q^{NC}_{uR})^2 \right) \right] x\left(d_v(x) + u_v(x)\right)$$

$$= \left[\left(\tfrac{1}{2} - \sin^2\theta_W + \tfrac{5}{9} \sin^4\theta_W \right) + (1-y)^2 \tfrac{5}{9} \sin^4\theta_W \right] \{\ \}^{CC}_{\nu N}$$

(127a)

$$\{\ \}^{NC}_{\bar{\nu} N} = \left[(Q^{NC}_{dR})^2 + (Q^{NC}_{uR})^2 + (1-y)^2 \left((Q^{NC}_{dL})^2 + (Q^{NC}_{uL})^2 \right) \right] x\left(d_v(x) + u_v(x)\right)$$

$$= \left[\tfrac{5}{9} \sin^4\theta_W \cdot \frac{1}{(1-y)^2} + \left(\tfrac{1}{2} - \sin^2\theta_W + \tfrac{5}{9} \sin^4\theta_W \right) \right] \cdot \{\ \}^{CC}_{\bar{\nu} N}$$

(127b)

These formulas allow one to compute the ratio of NC to CC processes, independently of the structure functions. Consider the ratio

$$R_\nu = \frac{\sigma_{NC}(\nu N)}{\sigma_{CC}(\nu N)} = \frac{\int_0^1 dx\,dy \left(\frac{d\sigma}{dx\,dy} \right)^{NC}_{\nu N}}{\int_0^1 dx\,dy \left(\frac{d\sigma}{dx\,dy} \right)^{CC}_{\nu N}} \tag{128}$$

Using Eq. (127a), and remembering that the NC cross section has an extra factor of ρ^2, yields:

$$R_\nu = \rho^2 \left[\tfrac{1}{2} - \sin^2\theta_W + \tfrac{20}{27} \sin^4\theta_W \right] \tag{129a}$$

A similar calculation for the antineutrino processes gives instead

$$R_{\bar{\nu}} = \rho^2 \left[\tfrac{1}{2} - \sin^2\theta_W + \tfrac{20}{9} \sin^4\theta_W \right] \tag{129b}$$

Recent experimental values from the CDHS /14/ and CHARM /15/ experiments at CERN give the values

$$R_\nu = 0.300 \pm 0.007 \qquad R_{\bar\nu} = 0.357 \pm 0.015 \qquad \text{(CDHS)} \tag{130a}$$

$$R_\nu = 0.320 \pm 0.010 \qquad R_{\bar\nu} = 0.377 \pm 0.025 \qquad \text{(CHARM)} \tag{130b}$$

Using our formulas and taking $\rho = 1$ and a value of $\sin^2\theta_W = 0.23$ implies

$$R_\nu = 0.31 \qquad R_{\bar\nu} = 0.39 \tag{131}$$

which are in nice agreement with Eqs. (130). A more careful analysis, which includes the contribution of the quark sea, carried out by Kim et al. /16/ yields

$$\sin^2\theta_W = 0.234 \pm 0.011 \tag{132}$$

for $\rho = 1$. Letting ρ vary also, the authors of Ref. 16 obtain

$$\sin^2\theta_W = 0.232 \pm 0.027 \qquad \rho = 0.999 \pm 0.025 \tag{133}$$

As I will show in the next section, a variety of other experiments involving neutral currents give values of $\sin^2\theta_W$ and ρ in agreement with the above. All these experiments taken together provide strong evidence for the validity of the GSW model - at least in its fermion-gauge sector.

It is conventional to describe deep inelastic experiments in terms of structure functions. For processes involving parity violation one can write on general grounds /17/, for $m_\ell = 0$ and $E_\ell >> M$, the differential cross section in terms of three independent structure functions:

$$\left(\frac{d\sigma}{dx\,dy}\right)^{CC/NC}_{\nu,\bar\nu} = \frac{G_F^2 M E_\ell}{\pi}\left\{\frac{1}{2}(2xF_1 \pm xF_3) + \frac{1}{2}(1-y)^2(2xF_1 \mp xF_3) + (1-y)(F_2 - 2xF_1)\right\} \tag{134}$$

For the electromagnetic processes the F_3 structure function does not contribute and one has

$$\left(\frac{d\sigma}{dxdy}\right)_{\ell}^{EM} = \frac{8\pi\alpha^2 M E_\ell}{(q^2)^2}\left[\frac{(1+(1-y)^2)}{2}2xF_1 + (1-y)(F_2 - 2xF_1)\right] \tag{135}$$

In general the structure functions F_i are functions of both x and q^2. However, in the parton model (0th order QCD) approximation employed to derive Eqs. (117) - (119), the structure functions depend only on x. Furthermore, as can be seen from these equations, there is a relation between F_1 and F_2, since no terms proportional to (1-y) enter. This relation, first derived by Callan and Gross /18/, is a result of having spin 1/2 partons and reads

$$F_2 = 2xF_1 \tag{136}$$

The structure functions F_2 and F_3 are easily extracted from Eqs. (117) - (119). One has

$$F_2^{EM}(x) = x\sum_q e_q^2\,(f_q(x) + f_{\bar q}(x)) \tag{137a}$$

$$\left(F_2^{NC}(x)\right)_\nu = \left(F_2^{NC}(x)\right)_{\bar\nu} = 2\rho^2 x\sum_q\left[(Q_{qL}^{NC})^2 + (Q_{qR}^{NC})^2\right](f_q(x) + f_{\bar q}(x)) \tag{137b}$$

$$\left(F_3^{NC}(x)\right)_\nu = \left(F_3^{NC}(x)\right)_{\bar\nu} = 2\rho^2\sum_q\left[(Q_{qL}^{NC})^2 - (Q_{qR}^{NC})^2\right](f_q(x) - f_{\bar q}(x)) \tag{137c}$$

$$\left(F_2^{CC}(x)\right)_\nu = 2x\sum_i(f_{d_i}(x) + f_{\bar u_i}(x)) \tag{137d}$$

$$\left(F_2^{CC}(x)\right)_{\bar\nu} = 2x\sum_i(f_{u_i}(x) + f_{\bar d_i}(x)) \tag{137e}$$

$$\left(F_3^{CC}(x)\right)_\nu = 2\sum_i(f_{d_i}(x) - f_{\bar u_i}(x)) \tag{137f}$$

$$\left(F_3^{CC}(x)\right)_{\bar\nu} = 2\sum_i(f_{u_i}(x) - f_{\bar d_i}(x)) \tag{137g}$$

Note that the structure function F_3 contains differences between quark and antiquark distributions, so that in principle it is sensitive to the valence content in the nucleon. This, non singlet, property is made apparent by considering, for charged current scattering, the case of 4 flavors, for deep inelastic scattering off an isoscalar target. One finds, using the fact that the distribution of charmed and strange quarks in proton and neutrons are the same and equal to the distribution of their antiparticles,

$$\left(F_3^{cc}(x)\right)_{\nu N} = \left[d(x) + u(x) - \bar{d}(x) - \bar{u}(x)\right] + 2\left[s(x) - c(x)\right] \tag{138a}$$

$$\left(F_3^{cc}(x)\right)_{\bar{\nu} N} = \left[d(x) + u(x) - \bar{d}(x) - \bar{u}(x)\right] + 2\left[c(x) - s(x)\right] \tag{138b}$$

The first bracket in Eqs. (138) measures just the distribution of valence d and u quarks. The small sea contribution in the last term can be eliminated by averaging the result of (138a) and (138b). Hence, indeed,

$$\bar{F}_3^{cc}(x) = \frac{1}{2}\left[\left(F_3^{cc}(x)\right)_{\nu N} + \left(F_3^{cc}(x)\right)_{\bar{\nu} N}\right] = d_v(x) + u_v(x) \tag{139}$$

measures only the valence quark distribution.

Because $\bar{F}_3^{cc}$ measures the distribution of valence quarks, clearly its integral gives the total number of valence quarks in a nucleon and so one expects

$$\int_0^1 dx\, \bar{F}_3^{cc}(x) = \int_0^1 dx\,\left(d_v(x) + u_v(x)\right) = 3 \tag{140}$$

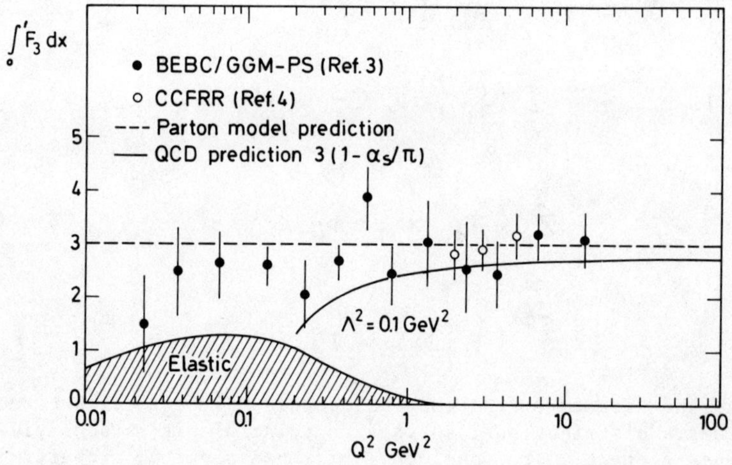

Fig. 13: Experimental status of the Gross-Llewellyn Smith sum rule. From Ref. 20

This result is the Gross-Llewellyn Smith sum rule /19/. The value of $\int_0^1 dx\, \bar{F}_3^{cc}(x)$, computed from older BEBC-Gargamelle data and from more recent CCFRR data, is plotted as a function of q^2 in Fig. 13 For large q^2 the sum rule appears to be reasonably well satisfied experimentally. The deviation for low q^2 could be a QCD effect. In fact, one can show /21/ that the Gross Llewellyn Smith sum rule is not exact in QCD but gets modified, to $O(\alpha_s)$, to

$$\int_0^1 dx\, \bar{F}_3^{cc}(x;q^2) = 3\left[1 - \frac{\alpha_s(q^2)}{\pi}\right] \tag{141}$$

The data is too uncertain - and the q^2 range probably too low - to test quantitatively this modification. However, qualitatively the corrections go in the right direction, suppressing the integral for smaller values of q^2.

For scattering off isoscalar targets the F_2 structure functions for neutrino and antineutrino scattering are the same. One finds, again for 4 flavors,

$$\left(F_2^{cc}(x)\right)_{\nu N} = \left(F_2^{cc}(x)\right)_{\bar{\nu} N} = x\left\{ \left(d(x) + u(x) + s(x) + c(x)\right) + \left(\bar{d}(x) + \bar{u}(x) + s(x) + c(x)\right)\right\} \tag{142}$$

Thus the difference in cross section between neutrino and antineutrino cc scattering off isoscalar targets gives immediately $\bar{F}_3^{cc}$. Using Eq. (134) one has

$$x\bar{F}_3^{cc}(x) = \frac{\left(\frac{d\sigma}{dxdy}\right)^{cc}_{\nu N} - \left(\frac{d\sigma}{dxdy}\right)^{cc}_{\bar{\nu} N}}{\frac{G_F^2 M E_\nu}{\pi}\left[1-(1-y)^2\right]} = x\left(u_v(x) + d_v(x)\right) \tag{143}$$

which is a direct measure of the valence quark distribution in nucleons. By considering a slightly different difference in cross sections one may also extract experimentally the distribution of the sea quarks. Using Eq. (134) and the results (138) and (142) one can write the difference

$$\frac{\left(\frac{d\sigma}{dxdy}\right)^{cc}_{\bar\nu N} - (1-y)^2\left(\frac{d\sigma}{dxdy}\right)^{cc}_{\nu N}}{\left(\frac{G_F^2 M E_\nu}{\pi}\right)} = \Big[x\left(\bar d(x) + \bar u(x) + 2s(x)\right)$$

$$+ (1-y)^2 x\left(2c(x) - 2s(x)\right) - (1-y)^4\left(\bar d(x) + \bar u(x) + 2c(x)\right)\Big] \tag{144}$$

For large y, $y \gtrsim 1/2$, this expression allows a good experimental determination of the sea quark distribution

$$\bar q(x) = x\left(\bar d(x) + \bar u(x) + 2s(x)\right) \tag{145}$$

A plot of this distribution from the CDHS experiment along with that of xF_3 and F_2, taken from Eisele's rapporteur talk in Paris /22/, is shown in Fig. 14

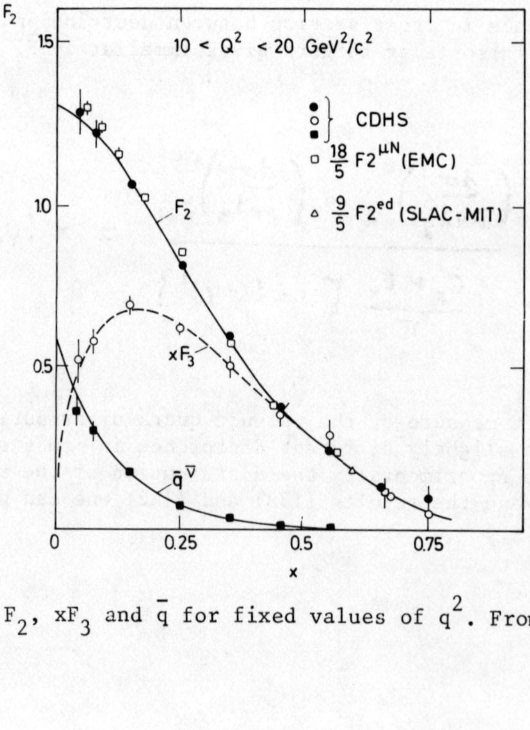

Fig. 14: F_2, xF_3 and $\bar q$ for fixed values of q^2. From Ref. 22

As can be seen the sea distribution is concentrated at small x.

A final comment is in order. The integral over $(F_2^{cc})_{\nu N}$ measures the momentum fraction of the charged constituents of the nucleon (quarks and antiquarks):

$$\int_0^1 dx \left(F_2^{cc}(x)\right)_{\nu N} = \int_0^1 dx \left(F_2^{cc}(x)\right)_{\bar{\nu} N} = \int_0^1 dx\, x \Big[\left(u(x) + d(x) + c(x) + s(x)\right) + \left(\bar{u}(x) + \bar{d}(x) + c(x) + s(x)\right) \Big] \tag{146}$$

This integral, neglecting the small contribution of the strange and charm sea, is just given by

$$\int_0^1 dx\; F_2^{cc}(x) = \frac{3\pi}{4G_F^2 M}\left[\frac{\sigma_{\nu N}^{cc} + \sigma_{\bar{\nu} N}^{cc}}{E_\nu}\right] \simeq 0.44 \pm 0.02 \tag{147}$$

The numerical result above follows from using for the average slope of the neutrino and antineutrino cross sections the value of Ref. 22

$$\frac{\sigma_{\nu N}^{cc} + \sigma_{\bar{\nu} N}^{cc}}{E_\nu} = (0.92 \pm 0.03) \times 10^{-38}\ \mathrm{cm}^2/\mathrm{GeV} \tag{148}$$

Eq. (147) implies that the gluons in the nucleon carry a substantial fraction of the constituent momentum ($\sim$ 50 %). The direct influence of the gluonic component of the nucleon is not felt in the lowest order QCD calculations I discussed in this section. However, to $O(\alpha_s)$ one "sees" the gluon structure function and one can try to extract it from the data. Before discussing the subject of QCD corrections to deep inelastic scattering, however, I examine some further tests of the GSW model involving neutral current processes.

SOME FURTHER TESTS OF THE GSW MODEL

In the previous section it was seen that deep inelastic neutral current scattering for neutrinos and antineutrinos gave the same value for $\sin^2\theta_W$. This is a first test of the validity of the GSW model in its fermion-gauge section. Deep inelastic charged lepton scattering, employing polarized leptons, provides further checks on the theory. In reality the process $\ell N \to \ell X$ has two contributions, as illustrated schematically in Fig. 15.

Fig. 15: Electromagnetic (1) and weak (2) contributions to deep inelastic lepton scattering

In the previous section we neglected altogether the effects of the weak neutral current interactions, because for the total rate its contribution for moderate q^2 are small. One can easily estimate that

$$\frac{|T_{weak}|^2}{|T_{EM}|^2} \sim \left(\frac{G_F q^2}{e^2}\right)^2 \sim 10^{-8}\left(\frac{q^2}{M^2}\right)^2 \tag{149}$$

However, if one studies the interference between the weak and electromagnetic contributions the effects can be reasonable:

$$\frac{|T_{weak} \cdot T_{EM}|}{|T_{EM}|^2} \sim \frac{G_F q^2}{e^2} \sim 10^{-4} \frac{q^2}{M^2} \tag{150}$$

Physically, the determination of such a weak-electromagnetic interference necessitates the measurement of some polarization asymmetry.

One can measure two types of asymmetries in polarized lepton deep inelastic scattering. Either one measures a cross section difference for scattering of leptons of different polarizations, but the same charge (A-asymmetry). Or one measures the difference in cross section between leptons and antileptons of a given polarization, or polarizations (B-asymmetry). Both effects are of order that estimated in Eq. (150). However, only the A-asymmetry is a purely parity violating effect. The A-asymmetry, defined by

$$A_{\ell^\pm;P_1,P_2} = \frac{d\sigma_{\ell^\pm;P_1} - d\sigma_{\ell^\pm;P_2}}{d\sigma_{\ell^\pm;P_1} + d\sigma_{\ell^\pm;P_2}} \tag{151}$$

has been measured at SLAC by scattering polarized electrons on a Deuteron target. The B-asymmetry, defined by

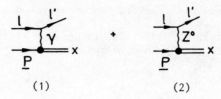

$$B_{\ell;P_1,P_2} = \frac{d\sigma_{e^+;P_1} - d\sigma_{e^-;P_2}}{d\sigma_{e^+;P_1} + d\sigma_{e^-;P_2}} \tag{152}$$

has been measured at CERN in deep inelastic scattering of polarized μ^+ and μ^- on a Carbon target. I will now detail what one expects for these asymmetries in the GSW model, again neglecting possible QCD corrections.

Consider first the A-asymmetry, which involves the difference between right and left polarized electron deep inelastic scattering:

$$A \equiv A_{e^-;R,L} = \frac{d\sigma_{e^-;R} - d\sigma_{e^-;L}}{d\sigma_{e^-;R} + d\sigma_{e^-;L}} \tag{153}$$

In a parton model to calculate A one must compute first the cross section for scattering e_R and e_L off quarks *. This is easily calculated from the T-matrix for e-q scattering, derived from the diagrams of Fig. 16.

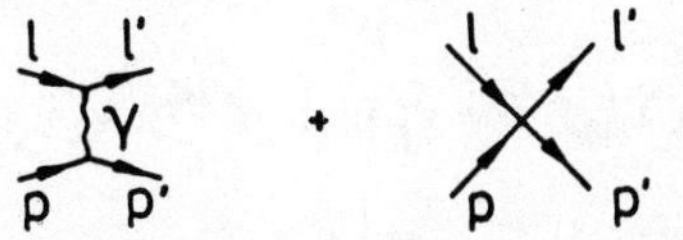

Fig. 16: e-q scattering, including weak NC effects

* In what follows I will compute in valence approximation and so will ignore antiquarks altogether.

$$T_{eq} = -\frac{4\pi\alpha e_q}{q^2}\,[\bar{u}(e')\gamma_\mu u(e)]\cdot[\bar{u}(q')\gamma^\mu u(q)]$$

$$+\frac{2G_F}{\sqrt{2}}\rho\,[\bar{u}(e')\gamma_\mu(Q^{NC}_{eR}(1+\gamma_5)+Q^{NC}_{eL}(1-\gamma_5))u(e)]$$

$$[\bar{u}(q')\gamma^\mu(Q^{NC}_{qR}(1+\gamma_5)+Q^{NC}_{qL}(1-\gamma_5))u(q)] \tag{154}$$

Here the charges Q^{NC}_{fR}, Q^{NC}_{fL} are those of the GSW model

$$Q^{NC}_{fR} = -e_f\sin^2\theta_W \quad ; \quad Q^{NC}_{fL} = \left(\frac{\tau_3}{2}\right)_f - e_f\sin^2\theta_W \tag{155}$$

and the factor of 2 in the second line arises because there are two e-q interactions in the product of $J^\mu_{NC}\cdot J_{\mu NC}$ of Fig. 16.

The calculation of the relevant cross sections is immediate once one decomposes also the electromagnetic contribution into right and left projections

$$\gamma^\mu = \frac{1}{2}\gamma^\mu(1+\gamma_5) + \frac{1}{2}\gamma^\mu(1-\gamma_5) \tag{156}$$

Then one has

$$T_{eq} = -\frac{\pi\alpha}{q^2}e_q\,[\bar{u}(e')(\gamma_\mu(1+\gamma_5)+\gamma_\mu(1-\gamma_5))u(e)]\cdot$$

$$\cdot[\bar{u}(q')(\gamma^\mu(1+\gamma_5)+\gamma^\mu(1-\gamma_5))u(q)]$$

$$+\sqrt{2}G_F\rho\,[\bar{u}(e')\gamma_\mu(Q^{NC}_{eR}(1+\gamma_5)+Q^{NC}_{eL}(1-\gamma_5))u(e)]$$

$$\cdot[\bar{u}(q')\gamma^\mu(Q^{NC}_{qR}(1+\gamma_5)+Q^{NC}_{qL}(1-\gamma_5))u(q)] \tag{157}$$

Recalling that (cf. Eqs. 115 and 116) $d\sigma_{LL} = d\sigma_{RR} \sim 1$, while $d\sigma_{LR} = d\sigma_{RL} \sim (1-y)^2$ it follows that

$$d\sigma_{e^-;R}(q_L) \sim \left| -\frac{\pi\alpha e_q}{q^2} + \sqrt{2}G_F\rho Q^{NC}_{eR} Q^{NC}_{qL} \right|^2 (1-y)^2 \tag{158a}$$

$$d\sigma_{e^-;R}(q_R) \sim \left| -\frac{\pi\alpha e_q}{q^2} + \sqrt{2}G_F\rho Q^{NC}_{eR} Q^{NC}_{qR} \right|^2 \tag{158b}$$

$$d\sigma_{e^-;L}(q_L) \sim \left| -\frac{\pi\alpha e_q}{q^2} + \sqrt{2}G_F\rho Q^{NC}_{eL} Q^{NC}_{qL} \right|^2 \tag{158c}$$

$$d\sigma_{e^-;L}(q_R) \sim \left| -\frac{\pi\alpha e_q}{q^2} + \sqrt{2}G_F\rho Q^{NC}_{eL} Q^{NC}_{qR} \right|^2 (1-y)^2 \tag{158d}$$

The desired asymmetry A is just the sum of Eqs. (158a) and (158b) minus the sum of Eqs. (158c) and (158d), weighted by the appropriate parton distributions, all normalized by the sum of all four terms in Eqs. (158), weighted by the relevant parton distributions.

After a little algebra one finds, in the valence quark approximation we are working in and dropping terms of $O(G_F^2)$

$$A = \frac{q^2 \left\{ \sum_q e_q\, q(x) \left[a_{1q} + a_{2q} \frac{[1-(1-y)^2]}{[1+(1-y)^2]} \right] \right\}}{\left\{ \sum_q e_q^2\, q(x) \right\}} \tag{159}$$

where

$$a_{1q} = -\frac{G_F\rho}{\sqrt{2}\pi\alpha} \left[Q^{NC}_{eR} - Q^{NC}_{eL} \right] \cdot \left[Q^{NC}_{qR} + Q^{NC}_{qL} \right] \tag{160a}$$

$$a_{2q} = -\frac{G_F\rho}{\sqrt{2}\pi\alpha} \left[Q^{NC}_{eR} + Q^{NC}_{eL} \right] \cdot \left[Q^{NC}_{qR} - Q^{NC}_{qL} \right] \tag{160b}$$

It is clear from the above that A is a purely parity violating asymmetry, since the coefficients a_{1q} and a_{2q} measure products of axial times vector couplings at the electron and quark vertices, respectively. For an isoscalar target like the Deuteron, employed in the SLAC experiments, one can easily check that the dependence on the structure functions in Eq. (159) cancels altogether. In this case the formula for the asymmetry reads, simply:

$$A = \left[a_1 + a_2 \frac{(1-(1-y)^2)}{(1+(1-y)^2)} \right] q^2 \tag{161}$$

where

$$a_1 = \frac{9}{5}\left[\frac{2}{3} a_{1u} - \frac{1}{3} a_{1d} \right] = -\frac{G_F \rho}{\sqrt{2}\pi\alpha} \left[\frac{9}{20} - \sin^2\theta_W \right] \tag{162a}$$

$$a_2 = \frac{9}{5}\left[\frac{2}{3} a_{2u} - \frac{1}{3} a_{2d} \right] = -\frac{G_F \rho}{\sqrt{2}\pi\alpha} \left[\frac{9}{20} - \frac{9}{5}\sin^2\theta_W \right] \tag{162b}$$

Note that a_2 vanishes for $\sin^2\theta_W = \frac{1}{4}$. Since as we have seen from our discussion of neutrino deep inelastic scattering $\sin^2\theta_W \simeq 0.23$, it is clear that little y-dependence is to be expected for the A-symmetry.

The results of the SLAC-YALE experiment /23/ on the A-symmetry are shown in Fig. 17

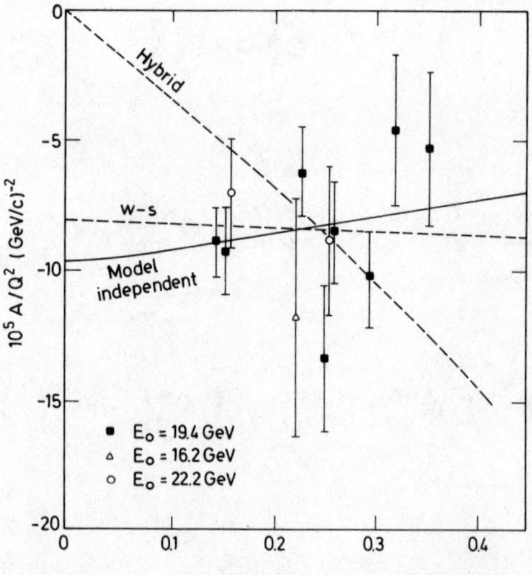

Fig. 17: Results of Ref. 23 on the A asymmetry in polarized eD deep inelastic scattering

The parameters a_1 and a_2 determined from the experiment are

$$a_1 = (-\,9.7 \pm 2.6) \times 10^{-5}\ \mathrm{GeV}^{-2}$$

$$a_2 = (4.9 \pm 8.1) \times 10^{-5}\ \mathrm{GeV}^{-2} \qquad (163)$$

A clear parity violation is seen and the y dependence of the data, within the large errors, is minimal as expected if $\sin^2\theta_W$ is near 1/4. Setting $\rho = 1$ this experiment yields a value for $\sin^2\theta_W$:

$$\sin^2\theta_W = 0.22 \pm 0.012 \pm 0.008 \qquad (164)$$

where the last error above is systematic. The agreement of this value with that obtained from deep inelastic neutrino scattering provides a further check on the GSW model. Note that this is a nontrivial check since different neutral current vertices are measured in the two different experiments.

Let me now discuss, with a little less detail, the B-asymmetry. The particular asymmetry measured in the CERN experiment on polarized μ-Carbon deep inelastic scattering /24/ is:

$$B \equiv B_{\mu;R,L} = \frac{d\sigma_{\mu^+;R} - d\sigma_{\mu^-;L}}{d\sigma_{\mu^+;R} + d\sigma_{\mu^-;L}} \qquad (165)$$

In a parton model and in the valence approximation therefore B is given by

$$B = \frac{\sum_q q(x)\,[d\sigma_{\mu^+;R}(q) - d\sigma_{\mu^-;L}(q)]}{\sum_q q(x)\,[d\sigma_{\mu^+;R}(q) + d\sigma_{\mu^-;L}(q)]} \qquad (166)$$

I computed already before the lepton-quark scattering cross sections (c.f. Eqs. (158))

$$d\sigma_{e^-;L}(q) = d\sigma_{e^-;L}(q_L) + d\sigma_{e^-;L}(q_R) \qquad (167a)$$

$$d\sigma_{e^-;R}(q) = d\sigma_{e^-;R}(q_R) + d\sigma_{e^-;R}(q_L) \qquad (167b)$$

and have remarked that the first terms in Eqs. (167) are proportional to 1 while the last terms in these equations are proportional to $(1-y)^2$. It is easy to convince oneself that the cross section $d\sigma_{\mu^+;R}(q)$ needed in Eq. (166) can be obtained from Eq. (167a) by the substitution

$$d\sigma_{e^+;R}(q) \leftrightarrow d\sigma_{e^-;L}(q) \qquad 1 \leftrightarrow (1-y)^2 \tag{168}$$

One now has all the elements for computing B and I just quote the result:

$$B = \frac{q^2 \left\{ \sum_q e_q\, q(x)\, b_q \right\} \left(1-(1-y)^2\right)}{\left\{ \sum_q e_q^2\, q(x) \right\} \left(1+(1-y)^2\right)} \tag{169}$$

with

$$b_q = -\frac{\sqrt{2}\, G_F\, \rho}{\pi\alpha}\, Q^{NC}_{eL} \left(Q^{NC}_{qR} - Q^{NC}_{qL}\right) \tag{170}$$

Here again only terms of $O(G_F)$ are retained.

A few comments are in order on this result. First, it is clear from Eq. (170) that the asymmetry is no longer a purely parity violating effect. In fact writing

$$Q^{NC}_{eL} = \tfrac{1}{2}\left[Q^{NC}_{eL} - Q^{NC}_{eR}\right] + \tfrac{1}{2}\left[Q^{NC}_{eL} + Q^{NC}_{eR}\right] \tag{171}$$

one sees that the asymmetry is almost purely a parity conserving effect since $Q^{NC}_{eL} + Q^{NC}_{eR} \simeq 0$ for $\sin^2\theta_W \simeq \frac{1}{4}$. This means that in comparing experimental results with the above formula, one should also subtract away some purely $O(\alpha^2)$ electromagnetic contributions to the asymmetry, before one tries to extract a value for $\sin^2\theta_W$. For the CERN experiment this has been done. Furthermore, since this experiment is done on Carbon, which is an isoscalar target, one can further simplify Eq. (168) and all structure function dependence drops out. A simple calculation gives

$$B = b\, q^2 \left[\frac{1-(1-y)^2}{1+(1-y)^2}\right] \tag{172}$$

with

$$b \simeq \frac{9}{5}\left(\frac{2}{3}b_u - \frac{1}{3}b_d\right) = \frac{-9G_F\rho}{10\sqrt{2}\,\pi\alpha}\left(1 - 2\sin^2\theta_W\right) \tag{173}$$

Numerically for $\sin^2\theta_W = 0.23$, $\rho = 1$ the coefficient b is

$$b = -1.51 \times 10^{-4}\ \mathrm{GeV}^{-2} \tag{174}$$

The measured B asymmetry in the CERN experiment /24/, after radiative corrections have been subtracted, for runs at 120 GeV and 200 GeV are shown in Fig. 18

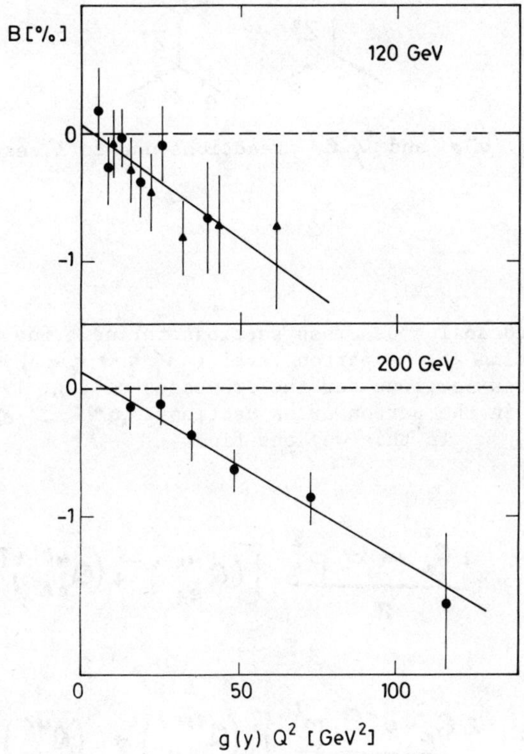

Fig. 18: q^2 dependence of the B asymmetry, with the kinematical factor $[1-(1-y)^2]/[1+(1-y)^2]$ removed, from ref. 24

The two slopes obtained

$$b(120\ \mathrm{GeV}) = (-1.76 \pm 0.75) \times 10^{-4}\ \mathrm{GeV}^{-2} \tag{175a}$$

$$b(200\mathrm{GeV}) = (-1.47 \pm 0.37) \times 10^{-4}\ \mathrm{GeV}^{-2} \tag{175b}$$

are in excellent agreement with the predictions of the GSW model, Eq.(174).

I would like to conclude this section with two further tests of the GSW model which involve purely leptonic process: $\nu_\mu e$ scattering and the measurement of forward-backward asymmetries in $e^+e^- \to \ell^+\ell^-$, with $\ell = \mu, \tau$. These processes need no parton model assumptions. However, the neutrino scattering experiments have somewhat limited statistics and the forward-backward asymmetry, although testing the GSW model, does not lead to a determination of $\sin^2\theta_W$. Nevertheless, these processes complement very nicely our discussion of deep inelastic scattering.

The reactions $\nu_\mu e \to \nu_\mu e$ and $\bar{\nu}_\mu e \to \bar{\nu}_\mu e$ are purely due to the weak neutral current, as shown in Fig. 19.

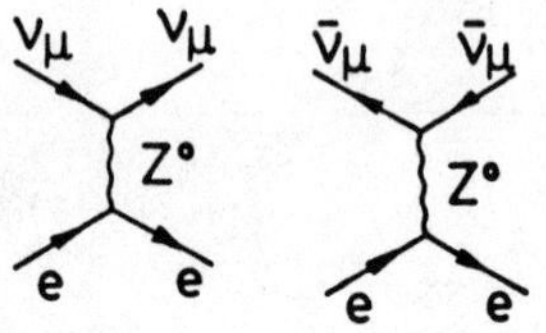

Fig. 19: $\nu_\mu e$ and $\bar{\nu}_\mu e$ reactions due to Z^o exchange.

We have calculated analogous cross sections for neutrino nucleon deep inelastic scattering at the parton level ($\nu_\mu q \to \nu_\mu q$; $\bar{\nu}_\mu q \to \bar{\nu}_\mu q$). To obtain the cross sections for the processes of Fig. 19 all we need to do is replace in the parton cross sections $Q^{NC}_{qL} \to Q^{NC}_{eL}$; $Q^{NC}_{qR} \to Q^{NC}_{eR}$. In this way one finds

$$\left(\frac{d\sigma}{dy}\right)^{\nu_\mu e} = \frac{2G_F^2 m E_\nu \rho^2}{\pi}\left[(Q^{NC}_{eL})^2 + (Q^{NC}_{eR})^2(1-y)^2\right] \tag{176a}$$

$$\left(\frac{d\sigma}{dy}\right)^{\bar{\nu}_\mu e} = \frac{2G_F^2 m E_\nu \rho^2}{\pi}\left[(Q^{NC}_{eR})^2 + (Q^{NC}_{eL})^2(1-y)^2\right] \tag{176b}$$

where, recall,

$$Q^{NC}_{eR} = \sin^2\theta_W \quad ; \quad Q^{NC}_{eL} = -\frac{1}{2} + \sin^2\theta_W \tag{177}$$

One should note that these cross sections are tiny, since here, in contrast to the scattering off nucleons, the electron mass m_e appears. Typically one expects cross sections of the order of 10^{-42} E_ν(GeV) cm^2. For high energy neutrino scattering ($E_\nu \simeq 100$ GeV) the cross sections are thus around 10^{-40} cm^2. For reactor experiments (where actually $\bar{\nu}_e$ are scattered) with E_ν of order of MeV the cross sections are vanishingly small ($\sim 10^{-45}$ cm^2) and signals can only be detected because of the very intense neutrino fluxes.

The CHARM collaboration at CERN has measured both* $\bar{\nu}_\mu e$ and $\nu_\mu e$ scattering /25/ at high energy with "high" statistics. Because both processes were observed their ratio can be determined, which gets rid of the unknown ρ parameter. Furthermore, in this way also various systematic errors can be eliminated. Integrating Eqs. (176) over y and taking their ratio, one predicts for the standard model

$$R = \frac{\sigma_{\nu_\mu e}}{\sigma_{\bar{\nu}_\mu e}} = \frac{3 - 12\sin^2\theta_W + 16\sin^4\theta_W}{1 - 4\sin^2\theta_W + 16\sin^4\theta_W} \tag{178}$$

which is a pure function of the Weinberg angle. The CHARM collaboration finds /25/

$$R = 1.37 \begin{smallmatrix} +0.65 \\ -0.44 \end{smallmatrix} \tag{179}$$

which implies a value for $\sin^2\theta_W$,

$$\sin^2\theta_W = 0.215 \pm 0.032 \pm 0.012, \tag{180}$$

which is in very good agreement with that determined in deep inelastic scattering. I should remark that, even though the errors in R are large, because the function in Eq. (178) is rapidly changing near $\sin^2\theta_W \simeq 1/4$ the errors for $\sin^2\theta_W$ are rather small. This is shown graphically in Fig. 20.

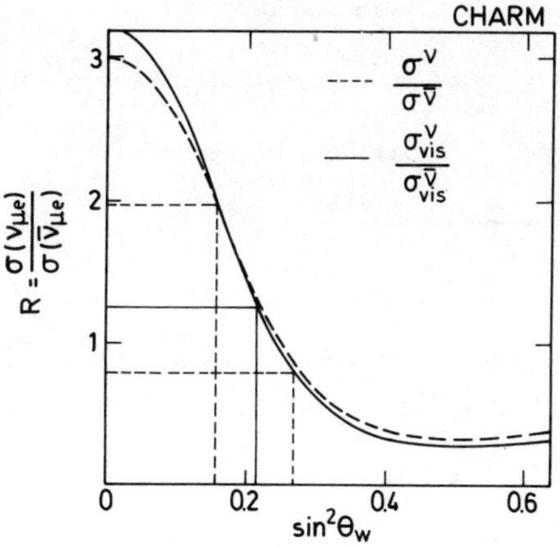

Fig. 20: Plot of R versus $\sin^2\theta_W$, showing the range measured by the CHARM experiment.

---

* "High" statistics in the field means of the order of 100 events!

The CHARM collaboration gives also values for the slope of the cross section rise with energy, for both ν_μ and $\bar{\nu}_\mu$ scattering /25/:

$$\sigma(\nu_\mu e)/E_\nu = (1.9 \pm 0.4 \pm 0.4) \times 10^{-42}\ \mathrm{cm}^2/\mathrm{GeV}$$

$$\sigma(\bar{\nu}_\mu e)/E_\nu = (1.5 \pm 0.3 \pm 0.4) \times 10^{-42}\ \mathrm{cm}^2/\mathrm{GeV} \qquad (181)$$

From these numbers one can extract a value for the ρ parameter, and they find:

$$\rho = 1.09 \pm 0.09 \pm 0.11 \qquad (182)$$

This value agrees within errors with the prediction of the standard model, with doublet Higgs breaking, $\rho = 1$. Very recently the CHARM results have been corroborated by a Brookhaven experiment /26/. They find

$$\sigma(\nu_\mu e)/E_\nu = (1.60 \pm 0.29 \pm 0.26) \times 10^{-42}\ \mathrm{cm}^2/\mathrm{GeV}$$

$$\sigma(\bar{\nu}_\mu e)/E_\nu = (1.16 \pm 0.20 \pm 0.16) \times 10^{-42}\ \mathrm{cm}^2/\mathrm{GeV} \qquad (183)$$

The last test of NC phenomena which I want to discuss concerns the process $e^+e^- \to \ell^+\ell^-$. As shown in Fig. 21, besides the usual electromagnetic contribution one expects also a Z^o contribution. For the values of $s = -q^2 = -(\ell + \bar{\ell})^2$ relevant to PETRA and PEP

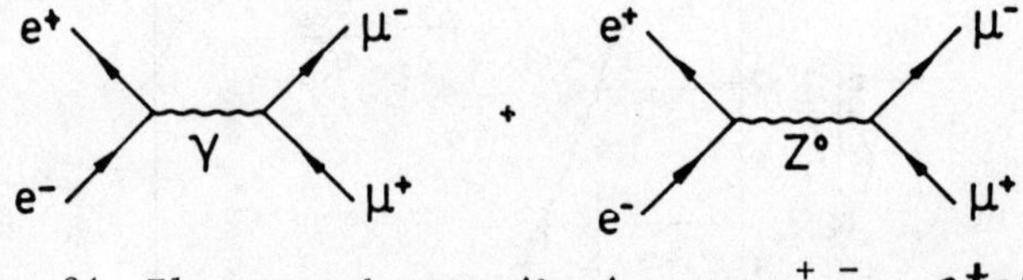

Fig. 21: Electroweak contributions to $e^+e^- \to \ell^+\ell^-$

it is still reasonable to approximate the second term in Fig. 21 by its current x current form. Hence, the relevant T-matrix for the process can be written as

$$T = i\left\{ \frac{e^2}{q^2}[\bar{U}(p)\gamma^\mu V(\bar{p})]\cdot[\bar{V}(\bar{\ell})\gamma_\mu U(\ell)] \right.$$

$$+ \frac{2G_F\rho}{\sqrt{2}}[\bar{U}(p)(\gamma^\mu g_{ev} + \gamma^\mu\gamma_5 g_{eA})V(\bar{p})]\cdot$$

$$\cdot[\bar{V}(\bar{\ell})(\gamma_\mu g_{ev} + \gamma_\mu\gamma_5 g_{eA})U(\ell)] \tag{184}$$

where the vector and axial leptonic coupling constants g_{ev} and g_{eA} are given by

$$g_{ev} = 2\sin^2\theta_W - \frac{1}{2} \quad ; \; g_{eA} = \frac{1}{2} \tag{185}$$

with $g_{ev} \ll g_{eA}$ for $\sin^2\theta_W = 0.23$. Since one is interested in the corrections only to $O(G_F)$ due to the weak neutral current one can write the spin averaged T-matrix as

$$\langle|T|^2\rangle \simeq \langle|T_{em}|^2\rangle + \langle|T_{em}\cdot T^*_{weak} + T^*_{em}\cdot T_{weak}|\rangle \tag{186}$$

It is not difficult to convince oneself that in the interference contributions only terms proportional to g^2_{ev} or g^2_{eA}, but not terms proportional to $g_{ev}\,g_{eA}$, appear. Furthermore the term proportional to g^2_{ev} gives only a correction to the total rate and is numerically small*. Hence I shall neglect it altogether. Focussing on the g^2_{eA} term only, it is easy to see that its contribution is

$$\langle|T|^2\rangle\Big|_{g^2_{eA}} = \frac{e^2}{q^2}\frac{G_F}{\sqrt{2}}\rho\, g^2_{eA}\,\mathrm{Tr}\,\gamma^\mu\gamma_5\,\gamma\cdot\ell\,\gamma^\nu\,\gamma\cdot\bar{\ell}\,.\,\mathrm{Tr}\,\gamma_\mu\gamma_5\,\gamma\cdot\bar{p}\,\gamma_\nu\,\gamma\cdot p$$

$$= -\frac{4\sqrt{2}e^2G_F\rho}{s}\left((\ell\cdot\bar{p})(\bar{\ell}\cdot p) - (\ell\cdot p)(\bar{\ell}\cdot\bar{p})\right)$$

$$= -\sqrt{2}\,e^2G_F\rho\cos\theta \tag{187}$$

---

* It is for this reason that these asymmetries at PETRA/PEP energies do not provide very useful information on $\sin^2\theta_W$.

where the last line applies in the e^+e^- CM-system and θ is the angle of the outgoing lepton with respect to the electron's direction.

From the above one sees that the effect of the weak neutral current is to produce a term linear in $\cos\theta$ in the angular distribution. Since the, lowest order, electromagnetic distribution is symmetric in $\cos\theta$, it is clear that the interference term can be selected by measuring the forward-backward asymmetry:

$$A_{F-B} = \frac{\int_0^1 d\cos\theta \left(\frac{d\sigma}{d\cos\theta}\right) - \int_{-1}^0 d\cos\theta \left(\frac{d\sigma}{d\cos\theta}\right)}{\sigma(e^+e^- \to \ell^+e^-)} \qquad (188)$$

Using that

$$\frac{d\sigma}{d\cos\theta} = \frac{1}{32\pi s} \langle |T|^2 \rangle \qquad (189)$$

and the standard result

$$\sigma(e^+e^- \to \ell^+\ell^-) = \frac{4\pi\alpha^2}{3s} \qquad (190)$$

it is now immediate to obtain the coefficient A_{F-B} predicted by the GSW model, for $\sqrt{s} \ll M_Z$

$$A_{F-B} = -\left[\frac{3G_F\rho}{16\pi\sqrt{2}\,\alpha}\right] s \qquad (191)$$

I end this section with a few remarks on Eq. (191):

(1) The coefficient of s in the above equation is of order 7 x 10^{-5} GeV^{-2}. Hence at the upper ranges of PEP and of PETRA ($s \sim 10^3$ GeV^2) the asymmetry should be a sizable effect.

2) Because A_{F-B} is not a parity violating effect - recall it was proportional to g_{eA}^2 - before comparing Eq. (191) with experiment one must subtract purely electromagnetic $O(\alpha^3)$ corrections to the asymmetry.

(3) At the highest PETRA energies the Z^o propagator begins to be felt. Including the propagator corresponds to the change

$$s \rightarrow \frac{s M_Z^2}{M_Z^2 - s} \qquad (192)$$

in Eq. (191), which increases the asymmetry.

(4) The asymmetry is clearly seen in the data of $e^+e^- \rightarrow \mu^+\mu^-$ shown in Fig. 22. This figure is taken from Naroska's talk /27/ at the 1983 Cornell conference. Essentially the same result /2/ obtains for $e^+e^- \rightarrow \tau^+\tau^-$, but the errors are naturally somewhat bigger.

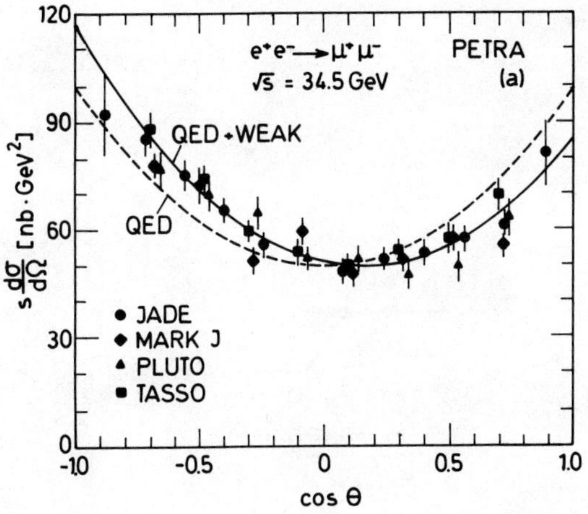

Fig. 22: A_{F-B} in $e^+e^- \rightarrow \mu^+\mu^-$ corrected for QED effects to $O(\alpha^3)$. From Ref. 27.

(5) There is good agreement of experiment with the GSW prediction. For the data at PETRA, taken at an average energy of $\sqrt{s}$ = 34.5 GeV the results are /27/:

$$\begin{aligned} (A_{F-B})_{\mu^+\mu^-} &= -10.8 \pm 1.1 \% \\ (A_{F-B})_{\tau^+\tau^-} &= -7.6 \pm 1.9 \% \end{aligned} \qquad (193a)$$

to be compared with the GSW prediction of - 9.4 %. Data taken at the highest PETRA energies is beginning to show the effect of the substitution (192) although, because of lower statistics, the errors are somewhat bigger. The average value of the $\mu^+\mu^-$ asymmetry for all four experiments at $\sqrt{s}$ = 41.6 GeV is /28/

$$(A_{F-B})_{\mu^+\mu^-} = -14.7 \pm 3.1 \% \qquad (193b)$$

while the expected asymmetry with (without) the substitution of Eq. (192) is 14.5 % (11.7 %).

W AND Z PRODUCTION: REFINED GSW TESTS AND BASIC QCD TEST

The most far reaching prediction of the GSW model, when it was proposed as a model of the weak interactions, was that these interactions are mediated by heavy vector bosons. Furthermore, given a measurement of $\sin^2\theta_W$ and of ρ, the masses of these bosons are predicted (cf Eqs. (55) and (56)). The observation at the CERN collider of both W /29/ and Z /30/ bosons, with masses in the predicted range, is thus a significant triumph of the model. I would like in this section to discuss these results first as a QCD test and then as a more refined test of the GSW model.

In lowest order QCD (parton model) one can compute the production of the W and Z bosons in pp annihilation from the diagram schematically shown in Fig. 23.

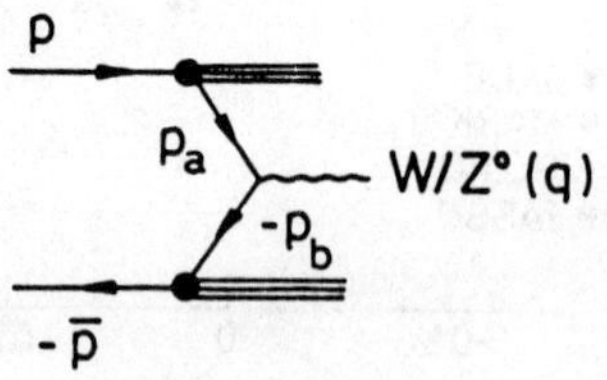

Fig. 23: Parton model diagram for W and Z boson production in $p\bar{p}$ annihilation

Essentially the same diagram applies, with trivial changes, for the Drell-Yan process $p\bar{p} \to \mu^+\mu^- X$. If one writes

$$p_a = \xi_a P \quad ; \quad p_b = \xi_b \bar{P} \tag{194}$$

kinematics fixes the product of the parton momentum fractions to be:

$$\xi_a \xi_b = \frac{M_V^2}{s} \tag{195}$$

The parton model production cross section for W or Z then is simply

$$\sigma(p\bar{p} \to W/Z\ X) = \sum_{a,b} \int d\xi_a\, d\xi_b\, f_{q_a}(\xi_a)\, f_{q_b}(\xi_b) \cdot \frac{1}{3}\, \sigma_{\text{Parton}}(q_a q_b \to W/Z) \tag{196}$$

The factor of $\frac{1}{3}$ above appears since for the quarks q_a, q_b to be able to annihilate into a W or a Z they must have the same color.

I shall compute Eq. (196) in valence approximation. For W^- production, for instance, this means that

$$f_{q_a}(\xi_a) = d_v(\xi_a)\ ; \quad f_{q_b}(\xi_b) = u_v(\xi_b) \tag{197}$$

That is the W^- is produced by annihilation of the valence d quark in the proton with the valence $\bar{u}$ quark in the antiproton (whose distribution is the same as that of the valence u quark in the proton). The elementary cross section $d + \bar{u} \to W^-$ is readily computed from Eq. (50), taking into account of Cabibbo mixing. The T-matrix is given by:

$$T = \frac{i e \cos\theta_c}{2\sqrt{2}\sin\theta_W}\, \bar{v}(p_b)\, \gamma^\mu (1-\gamma_5)\, u(p_a)\, \epsilon_\mu(q;\lambda) \tag{198}$$

where $\epsilon_\mu(q;\lambda)$ is the W-polarization tensor. Using that

$$\sum_\lambda \epsilon_\mu(q;\lambda)\,\epsilon_\nu(q;\lambda) = \eta_{\mu\nu} + \frac{q_\mu q_\nu}{M_W^2} \tag{199}$$

a simple calculation gives

$$\langle |T|^2 \rangle = \frac{1}{4}\sum_{s_a}\sum_{s_b} |T|^2 = \sqrt{2}\, G_F\, M_W^4 \cos^2\theta_c \tag{200}$$

The elementary cross section then follows from the standard formula

$$\sigma_{\text{Parton}}(d+\bar{u} \to W^-) = \frac{1}{(-4\, p_a \cdot p_b)} \int \frac{d^3 q}{(2\pi)^3\, 2q^0}\, (2\pi)^4\, \delta^4(q - p_a - p_b)\, \langle |T|^2 \rangle \tag{201}$$

which gives

$$\sigma_{parton}(d+\bar{u}\to W^-) = \sqrt{2}\,\pi\, G_F M_W^2 \cos^2\theta_c\, \delta(\xi_a \xi_b s - M_W^2)$$

(202)

Using Eq. (196), finally, this gives for the W^- production cross section

$$\sigma(p\bar{p}\to W^- X) = \frac{\sqrt{2}\,\pi}{3} G_F \cos^2\theta_c \left(\frac{M_W^2}{s}\right) \int_{M_W^2/s}^{1} \frac{d\xi}{\xi}\, d_v(\xi)\, u_v\left(\frac{M_W^2/s}{\xi}\right)$$

(203)

By charge symmetry exactly the same formula follows for W^+ production. For Z^o production a simple calculation gives

$$\sigma(p\bar{p}\to Z^o X) = \frac{\sqrt{2}\,\pi}{3} G_F \cos^2\theta_c \left(\frac{M_Z^2}{s}\right) \int_{M_Z^2/s}^{1} \frac{d\xi}{\xi}\, \mathcal{L}^o(\xi; M_Z^2/s)$$

(204)

where the luminosity function $\mathcal{L}^o$ is given by

$$\mathcal{L}^o(\xi; M_Z^2/s) = \left[\frac{1}{4} - \frac{2}{3}\sin^2\theta_W + \frac{8}{9}\sin^4\theta_W\right] u_v(\xi)\, d_v\left(\frac{M_Z^2/s}{\xi}\right)$$

$$+ \left[\frac{1}{4} - \frac{1}{3}\sin^2\theta_W + \frac{2}{9}\sin^4\theta_W\right] d_v(\xi)\, u_v\left(\frac{M_Z^2/s}{\xi}\right)$$

(205)

The above formulas predict the cross section for W and Z production at the collider, directly from a knowledge of the valence quark distribution functions obtained in deep inelastic scattering. In Fig. 24 I display some curves, obtained some time ago by Paige /31/, which illustrate the expected range of W and Z production at colliders. These cross sections include possible scale breaking effects typical of QCD, in which the structure functions "evolve" with q^2, but do not include effects of the, so called, K-factor. I will, in the next section, discuss these matters in more detail. Here I comment only that the curves

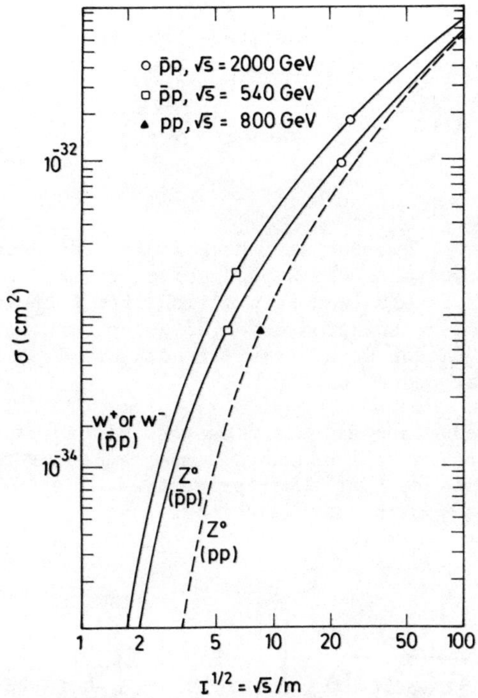

Fig. 24: Production cross sections for W and Z, from Ref. 31

in Fig. 25 probably underestimate the cross section by a factor of 1.5 - 2.

From Fig. 24 one sees that for the CERN collider ($\sqrt{s}$ = 540 GeV) one expects

$$\sigma(W^+) + \sigma(W^-) \approx 4 \text{ nb} \tag{206}$$

Experimentally the UA1 and UA2 collaborations have observed the W's decaying in the $e\nu$ mode and UA1 has also seen the $\mu\nu$ decay mode. For the cross section times branching ratio the collaborations give the values, /32/ /33/

$$\begin{aligned} (\sigma\cdot B)_{e\nu}\,(\text{UA2}) &= 0.53 \pm 0.10 \pm 0.10 \text{ nb} \\ (\sigma\cdot B)_{e\nu}\,(\text{UA1}) &= 0.53 \pm 0.008 \pm 0.09 \text{nb} \\ (\sigma\cdot B)_{\mu\nu}\,(\text{UA1}) &= 0.67 \pm 0.17 \pm 0.15 \text{ nb} \end{aligned} \tag{207}$$

The width for W's to decay into $\ell\nu$ is easily calculated

$$\Gamma(W \to \ell\nu) = \frac{G_F M_W^3}{6\sqrt{2}\pi} \tag{208}$$

This number can readily be seen to correspond to about 8 % of the total expected width. Thus, using Eq. (206), one would expect

$$(\sigma \cdot B)^{Th}_{e\nu} \simeq 0.32 \text{ nb} \tag{209}$$

This number is a little low, but as I have indicated above, the estimate (206) probably underestimates the W production cross section. At any rate, it is still remarkable that information from deep inelastic scattering can, via QCD, predict the production of heavy weak bosons in $p\bar{p}$ annihilation to within a factor of 2. This, if nothing else, attest to the validity of the perturbative QCD approach.

Although W/Z production gives a nice test of perturbative QCD, of course their discovery is of fundamental importance for the GSW model for it allows a precise test of their properties. Eq. (55) relates the Fermi constant to the W-mass and the Weinberg angle. It predicts for the W mass

$$M_W = \left[\frac{\pi\alpha}{\sqrt{2}G_F \sin^2\theta_W}\right]^{1/2} = \frac{37.28}{\sin\theta_W} \text{ GeV} \tag{210}$$

Using $\sin^2\theta_W = 0.23$ this gives for M_W:

$$M_W = 77.8 \text{ GeV} \tag{211a}$$

From the fact that $\rho \simeq 1$, the Z mass also follows

$$M_Z = \frac{M_W}{\cos\theta_W} = 88.7 \text{ GeV} \tag{211b}$$

These values for M_W and M_Z are somewhat below the average values obtained by the UA1 and UA2 collaborations /34/:

$$M_W = 82.1 \pm 1.7 \text{ GeV}$$
$$M_Z = 93.0 \pm 1.8 \text{ GeV} \tag{212}$$

This circumstance is actually expected since the values obtained in Eqs. (211) are computed without taking into account of radiative corrections. As I will show below, including radiative effects brings the theoretical and experimental values of the weak boson masses in complete quantitative agreement.

In lowest order the Weinberg angle was defined (cf Eq. (49)) in terms of the ratio of two coupling constants:

$$\sin \theta_W = \frac{e}{g} \tag{213}$$

However, radiative effects produce infinite corrections to the above formula, which must be removed by renormalization. After renormalization one is left with a $(\sin^2 \theta_W)_{Ren}$ which is perfectly finite and related to some physically measured parameters. In fact, one can define a variety of such $(\sin^2 \theta_W)_{Ren}$, all connected to one another by finite power series in α. Of course, also cross sections get modified to higher order and they contain various infinities that must be renormalized. The removal of these infinities must be done consistently and the radiative corrected formulas will differ depending on which way one has chosen to define certain parameters, including $(\sin^2 \theta_W)_{Ren}$.

The above is best illustrated by a concrete example. Consider again the ratio R of the $\nu_\mu e$ and $\bar{\nu}_\mu e$ cross sections. According to the lowest order calculation of the last section one had:

$$R = \frac{\sigma(\nu_\mu e)}{\sigma(\bar{\nu}_\mu e)} = \frac{3 - 12 \sin^2\theta_W + 16 \sin^4\theta_W}{1 - 4 \sin^2\theta_W + 16 \sin^4\theta_W} \tag{214}$$

which is a pure function of the (lowest order) Weinberg angle. Radiative corrections, an example of which is shown in Fig. 25, will modify our expectation for R. However, one expects that R will again be expressible in terms of a renormalized Weinberg angle by a formula analogous to Eq. (214), with coefficients that differ from those of Eq. (214) by O(α).

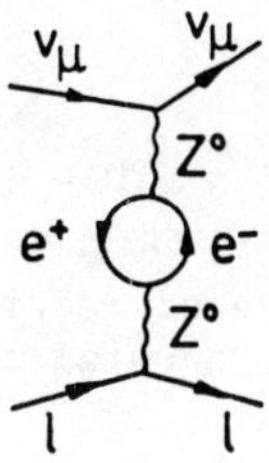

Fig. 25: A radiative contribution to $\nu_i \ell$ scattering

Including radiative corrections (to O(α)) one has

$$R = \frac{(3+a_i\alpha) - (12+b_i\alpha)(\sin^2\theta_W)^i_{Ren} + (16+c_i\alpha)(\sin^4\theta_W)^i_{Ren}}{(1+d_i\alpha) - (4+e_i\alpha)(\sin^2\theta_W)^i_{Ren} + (16+f_i\alpha)(\sin^4\theta_W)^i_{Ren}} \tag{215}$$

The various coefficients $a_i, b_i, \ldots, f_i$ above are calculable but their precise value depends on which definition for $(\sin^2\theta_W)^i_{Ren}$ one has chosen. In particular, one could define $(\sin^2\theta_W)^i_{Ren}$ by demanding that all the coefficients $a_i, \ldots, f_i$ vanish. However, such a definition is not particularly convenient if one wants to study what radiative shifts the theory predicts for the W and Z masses.

The most convenient choice for $(\sin^2\theta_W)_{Ren}$ is to define this parameter via the physical W and Z masses, by the relation

$$(\sin^2\theta_W)_{Ren} \equiv 1 - \frac{M_W^2}{M_Z^2} \tag{216}$$

Radiative corrections to neutrino neutral current deep inelastic scattering and polarized electron-deuteron scattering have been computed /35/ /36/ using this renormalization prescription and a value for $(\sin^2\theta_W)_{Ren}$ has been extracted. Before including radiative corrections the "best fit" value of Kim et al. /16/ for the combined NC neutrino data and the polarized e D data was (assuming $\rho = 1$)

$$\sin^2\theta_W = 0.229 \pm 0.009 \tag{217}$$

After including radiative corrections, these experiments imply for $(\sin^2\theta)_{Ren}$ of Eq. (216)

$$(\sin^2\theta_W)_{Ren} = 0.217 \pm 0.014 \tag{218}$$

Note that the effect of radiative corrections is not small, for this definition of $(\sin^2\theta_W)_{Ren}$, causing about a 6 % decrease in $\sin^2\theta_W$.

The formula for M_W which we previously used, Eq. (210), also gets modified by radiative corrections using the same definition of $(\sin^2\theta_W)$ one has /37/

$$M_W = \left[\frac{\pi \alpha}{\sqrt{2}\, G_\mu\, (\sin^2\theta_W)_{Ren}\, [1-\Delta r]} \right]^{1/2} \tag{219}$$

Here $G_\mu = (1.16637 \pm 0.00002) \times 10^{-5}\ \mathrm{GeV}^{-2}$ is the Fermi constant extracted from radiatively corrected μ-decay /38/ and Δr is a further radiative correction which can be computed /37/. Indeed, and this is what makes using the definition of $(\sin^2\theta_W)_{Ren}$ of (216) particularly useful, the correction Δr is essentially absorbed if one changes α in Eq. (219) to the running coupling constant evaluated at M_W^2; $\alpha(M_W^2)$. Marciano /37/ gives

$$(\Delta r)_{theory} = 0.0696 \pm 0.0020 \tag{220}$$

The change of α to $\alpha(M_W)$, on the other hand, already is equivalent to a 7.3 % change, so this is clearly the dominant effect.

Using Eqs. (220) and (218) in the formula (219) predicts for the W-mass, the radiatively corrected value

$$M_W = 83 \pm 2.5\ \mathrm{GeV} \tag{221a}$$

The Z^o radiatively corrected mass follows from

$$M_Z = \frac{M_W}{(\cos\theta_W)_{Ren}} = 93.8 \pm 2.5\ \mathrm{GeV} \tag{221b}$$

These values are in excellent agreement with the UA1 and UA2 averages of Eq. (212) and constitute an impressive success for the GSW model. One can, conversely, use the value of the W mass (or of the Z mass) and Eqs. (219) and (220) to extract $(\sin^2\theta_W)_{Ren}$. One obtains in this way

$$(\sin^2\theta_W)_{Ren} = 0.221 \pm 0.007 \tag{222}$$

This value of the Weinberg angle, extracted from properties of the W boson determined by the UA1 and UA2 collaborations is in excellent agreement with that of Eq. (218), which comes from the "low energy" νN and eD experiments.

PERTURBATIVE QCD-EVOLUTION EQUATIONS AND TESTS

Up to now I have discussed "hard scattering" processes only at the parton model level (0th order QCD). It is now time to discuss what corrections QCD gives to these parton model predictions. As I have mentioned earlier, QCD causes basically two different modifications:

(i) It makes the parton structure and fragmentation functions "run". That is, these functions now depend on q^2:

$$f_q(x) \rightarrow f_q(x;q^2)$$
$$D_q^H(z) \rightarrow D_q^H(z;q^2) \tag{223}$$

(ii) It changes $d\sigma_{parton}$ in a well defined way, which can be computed in a power series in $\alpha_s(q^2)$:

$$d\sigma_{parton} \rightarrow d\sigma_{parton}\left[1 + c\,\alpha_s(q^2) + \ldots\right] \tag{224}$$

with c a calculable coefficient.

The q^2-dependence of the structure and fragmentation functions is not calculable per se, since it arises from the "soft" processes by which hadrons get transmuted into partons and vice versa. However, one may compute the <u>evolution in q^2</u> of these functions. Given, for instance, $f_q(x,q_o^2)$ one can compute $f_q(x,q^2)$ for $q^2 > q_o^2$. The evolution of the structure and fragmentation functions can be computed by means of the, so called, Altarelli-Parisi equations /39/. This evolution is somewhat analogous to what happens with the running coupling constant. Given $\alpha_s(q_o^2)$ and the β-function one can compute $\alpha_s(q^2)$.

To derive the Altarelli-Parisi equations it proves useful to examine deep inelastic scattering processes again, to understand the physical reason for the running of the parton structure functions. I want to focus in particular on the quark + V^* scattering subprocess, shown in Fig. 26, where $V^* = \{\gamma, Z^0, W\}$ depending on what particular deep inelastic process one is considering

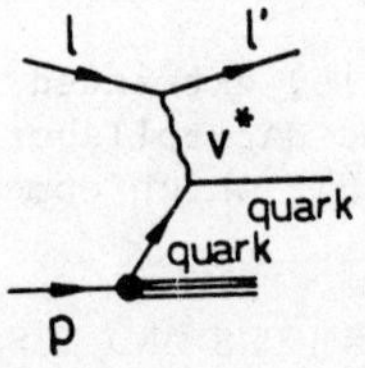

Fig. 26: Schematic depiction of a deep inelastic scattering process

The process $q + V^* \to q$ gets modified when one takes into account of QCD. There are both virtual modifications, which occur because quarks can emit and reabsorb gluons, and real modifications corresponding to the case when additional gluons are involved in the process. The relevant graphs for both contributions, to $O(\alpha_s)$, are shown in Fig. 27.

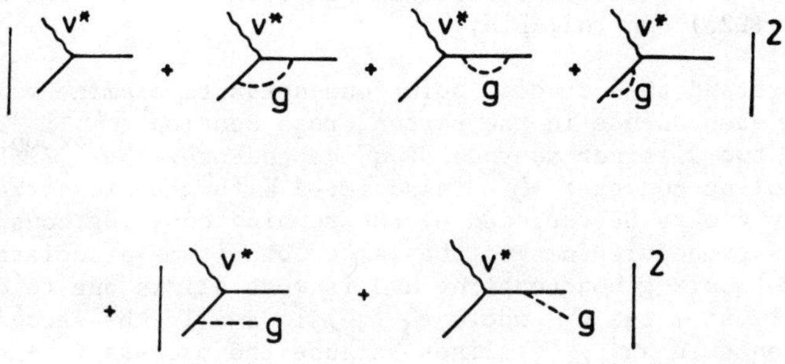

Fig. 27: Virtual and real QCD corrections to the process $q + V^* \to q$

Generically let me write the hadronic structure function corresponding to the unmodified process $q + V^* \to q$ as

$$F^o(x) = \int_x^1 \frac{d\xi}{\xi} f_q(\xi)\, \sigma^o_{parton}(x_p = \frac{x}{\xi})$$

$$= \int_x^1 \frac{d\xi}{\xi} f_q(\xi)\, \delta(\frac{x}{\xi} - 1) = f_q(x) \tag{225}$$

To lowest order, the parton cross section is just a δ-function, so that the hadronic structure function $F^o(x)$ is just the same as the quark distribution function, $f_q(x)$. If one takes into account of the QCD corrections of Fig. 27, the structure function $F^o(x)$ gets an additional q^2 dependent contribution. One can write the corrected structure function as

$$F(x;q^2) = \int_x^1 \frac{d\xi}{\xi} f_q(\xi) \left[\delta(\frac{x}{\xi} - 1) + \sigma^1_{parton}(\frac{x}{\xi}; q^2) \right] \tag{226}$$

As I will show below, Eq. (226) can be rewritten in terms of a running quark distribution function $f_q(x;q^2)$ plus a calculable correction of $O(\alpha_s(q^2))$:

$$F(x;q^2) = f_q(x;q^2) + O(\alpha_s(q^2)) \tag{227}$$

This separation, which may appear trivial at this stage, is important. Only by making the parton distribution functions run are the corrections in Eq. (227) controllable.

To understand this crucial point one needs to examine what are the sources of q^2-dependence in the parton cross section σ^1_{parton}. There are actually two distinct sources of q^2 dependence. The first is that the coupling constant α_s - associated with gluonic corrections - should really always be replaced by the running coupling constant $\alpha_s(q^2)$. This procedure sums up the large logarithms associated with the effective quark-gluon coupling and is what allows one to calculate perturbatively at large q^2, where $\alpha_s(q^2)$ is small. The second source of q^2 dependence in σ^1_{parton} arises because the process $V^* + q \to q + g$ contains an explicit logarithmic dependence on q^2. This logarithmic dependence is a typical results of the hard gluon bremsstrahlung spectrum, which is cut off by q^2: $\int^{q^2} dk_\perp^2/k_\perp^2 \to \ln q^2$ *. Schematically, therefore one expects for σ^1_{parton} the structure

$$\sigma^1_{parton}(x_p;q^2) = \frac{\alpha_s(q^2)}{2\pi}\left[P(x_p)\ln q^2 + R(x_p)\right] \tag{228}$$

where the functions $P(x_p)$ and $R(x_p)$ are calculable.

Eq. (228) shows that contrary to what one might have naively expected σ^1_{parton} is not small for large q^2! Even though $\alpha_s(q^2)$ vanishes, as $(\ln q^2)^{-1}$, in this limit, the hard gluon $\ln q^2$ factor compensates entirely this behaviour. To make sense of the parton model calculations I discussed in the previous sections, it is necessary that a reasonable perturbative contribution should emerge from σ^1_{parton}. Such a contribution indeed ensues once one factorizes some of the $\ln q^2$ factors in (228) into the parton structure function $f_q(\xi)$, thereby making this structure function run. Only by this factorization procedure does one obtain then a perturbative expansion in $\alpha_s(q^2)$.

To see what is involved, let me rewrite Eq. (226), using the expression (228) for σ^1_{parton}, in a suggestive way

---

* This result can also be understood in terms of mass singularities of the virtual quark exchanged in the process.

$$F(x;q^2) = \int_x^1 \frac{d\xi}{\xi} f_q(\xi) \left\{ \delta\left(\frac{x}{\xi}-1\right) + \frac{\alpha_s(q^2)}{2\pi} \left[P\left(\frac{x}{\xi}\right) \ln q^2 + R\left(\frac{x}{\xi}\right) \right] \right\}$$

$$\simeq \int_x^1 \frac{d\xi}{\xi} \left\{ f_q(\xi) + \frac{\alpha_s(q^2)}{2\pi} \int_\xi^1 \frac{d\xi'}{\xi'} P\left(\frac{\xi}{\xi'}\right) f_q(\xi') \ln q^2 \right\} .$$

$$\cdot \left\{ \delta\left(\frac{x}{\xi}-1\right) + \frac{\alpha_s(q^2)}{2\pi} R\left(\frac{x}{\xi}\right) \right\}$$

(229)

The factorized expression in the second line above is correct to $O(\alpha_s)$. Note that by this procedure I have split the contributions in σ^1_{parton} into a modification of the parton structure function $f_q(\xi)$ (first curly bracket) plus a true $O(\alpha_s(q^2))$ modification to σ^o_{parton} (second curly bracket). Let me define a running parton structure function $f_q(\xi;q^2)$, to this order in $\alpha_s(q^2)$, by

$$f_q(\xi;q^2) = f_q(\xi) + \frac{\alpha_s(q^2)}{2\pi} \ln q^2 \int_\xi^1 \frac{d\xi'}{\xi'} P\left(\frac{\xi}{\xi'}\right) f_q(\xi') + \ldots$$

(230)

Further let me define a perturbative parton cross section by

$$\sigma_{parton}(x_p;q^2) = \delta(x_p - 1) + \frac{\alpha_s(q^2)}{2\pi} R(x_p) + \ldots$$

(231)

Then the structure function $F(x;q^2)$ reads simply

$$F(x;q^2) = \int_x^1 \frac{d\xi}{\xi} f_q(\xi;q^2) \sigma_{parton}\left(\frac{x}{\xi}, q^2\right)$$

$$= f_q(\xi;q^2) + O(\alpha_s(q^2))$$

(232)

which is precisely the result I anticipated in Eq. (227). Crucial to obtaining this result is the factorization property which allows one to remove the unwanted $\ln q^2$ factors into a modified parton distribution function. One can show that this factorization is possible to all orders in α_s and furthermore that it is process independent /40/. That is, the transformation of parton distribution functions into running parton distribution functions is independent of the process in question. Of course the remaining perturbative parton cross section will in general depend on the process at hand.

The expression for $f_q(\xi;q^2)$ obtained in Eq. (230) is equivalent, to the order in α_s that I am working in, to the integro-differential equation.

$$\frac{df_q(\xi;q^2)}{d\ln q^2} = \frac{\alpha_s(q^2)}{2\pi}\int_\xi^1 \frac{d\xi'}{\xi'} f_q(\xi';q^2) P(\xi/\xi') \tag{233}$$

This is the Altarelli-Parisi equation /39/, which describes how the structure function $f_q(\xi;q^2)$ evolves with q^2. I have arrived at it by factorizing away the $\ln q^2$ pieces from σ^2_{parton} into a running structure function. There is a more physical way of obtaining this equation, which is the one originally followed by Altarelli and Parisi /39/. They argued that the distribution function of quarks, besides having an intrinsic component, has an additional contribution coming from the fact that partons of higher fractional momentum can always degrade their momentum by gluon (or quark) emission. That is, probing with higher q^2, one should always be able to "resolve" the quark density for instance into that of a quark plus a gluon. The "splitting function" $P(\xi/\xi')$ in Eq. (233) is precisely the probability function of finding within a quark of fractional momentum ξ' another quark of fractional momentum $\xi < \xi'$. As q^2 increases the physical effect which Eq. (233) portrays is a depopulation of the structure functions for large values of ξ and an increase of $f(\xi;q^2)$ for small values of ξ. This is shown schematically in Fig. 28

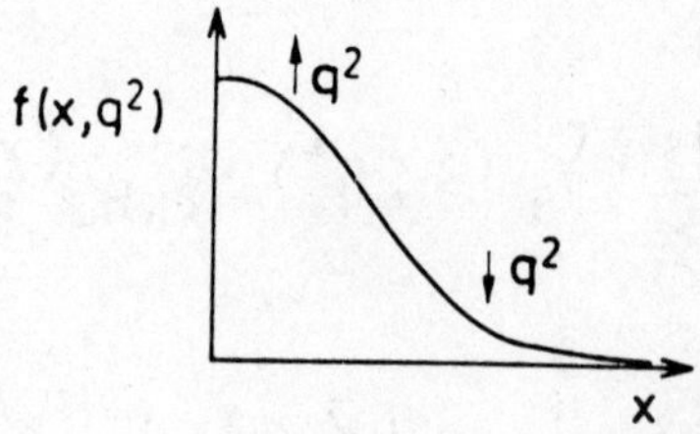

Fig. 28: Qualitative behaviour of $f(x;q^2)$ as a function of q^2

The physical effects of the running of the parton structure functions are most easily seen by focussing directly on the Altarelli-Parisi equations. Because the splitting function P is a function only of ξ/ξ', it is clear that the Altarelli-Parisi equation becomes simple if one considers not $f_q(\xi;q^2)$ directly but its moments:

$$M_n(q^2) = \int_0^1 d\xi\, \xi^{n-1} f_q(\xi;q^2) \tag{234}$$

Using Eq. (233) it follows immediately that the moments obey the differential equations

$$\frac{dM_n(q^2)}{d\ln q^2} = \frac{\alpha_s(q^2)}{2\pi} A_n M_n(q^2) \tag{235}$$

where

$$A_n = \int_0^1 d\xi\, \xi^{n-1} P(\xi) \tag{236}$$

These equations can be solved readily since the behaviour of the QCD coupling constant $\alpha_s(q^2)$ is known. Using the lowest order formula for the β-function, given in Eq. (93), one has

$$\frac{d\alpha_s(q^2)}{d\ln q^2} = \beta(\alpha_s) = -\frac{(33-2n_f)}{12\pi}\alpha_s^2(q^2) \tag{237}$$

which implies

$$\alpha_s(q^2) = \frac{[12\pi/33-2n_f]}{\ln q^2/\Lambda^2} \tag{238}$$

where Λ is a free parameter, which will be determined eventually from experiment.

Using Eq. (238), the q^2 behaviour of the moments $M_n(q^2)$ is easily computed from Eq. (235). One finds

$$M_n(q^2) \sim (\ln q^2)^{\frac{6A_n}{33-2n_f}}$$ (239)

Recalling Eq. (227), which related the behaviour of the hadronic structure functions to that of the parton structure functions, apart from corrections of $O(\alpha_s(q^2))$, it is clear that the main effect of QCD is to give logarithmic variations with q^2 to the moments of the experimentally measured structure functions. That is, QCD implies <u>scaling violations</u> in deep inelastic scattering. The moments of the structure functions will vary with $\ln q^2$ in a well defined way, $(\ln q^2)^{d_n}$, with the exponents d_n being calculable numbers.

Before going on to examine whether data in deep inelastic scattering really shows the scaling violations implied by QCD, I need to remedy a bit the above discussion, which contained some simplifications. To $O(\alpha_s)$ it is really not sufficient only to consider modifications of the process $q + V^* \rightarrow q$ in which gluons are emitted (e.g. $q + V^* \rightarrow q + g$). The graph of Fig. 2a, in which a gluon in the hadron pair produces a $q\bar{q}$ pair in the presence of the virtual vector boson V^* $(g + V^* \rightarrow q\bar{q})$ is of the same order in α_s and should be included in the, QCD corrected, computation of deep inelastic scattering.

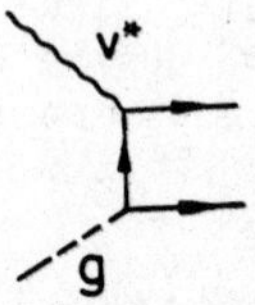

Fig. 29: Gluon pair production by V^*

The inclusion of gluons is rather easily done at the level of the Altarelli Parisi equation. The point is that the quark structure function as q^2 increases gets really two kind of additional contributions. One of these we already discussed. A quark of higher fractional momentum ξ' degrades, by emitting a gluon, into a quark of lower fractional momentum ξ. The splitting function $P(\xi/\xi') \equiv P_{qq}(\xi/\xi')$ basically is computed by calculating the probability of this transition – shown schematically by Fig. 30a. On the other hand, a quark of fractional momentum ξ can also ensue by pair production from a gluon. The splitting function for this process $P_{qg}(\xi/\xi')$ basically measures the probability for the process in Fig. 30b to happen. This additional contribution to the quark structure function obviously must be proportional to the distribution functions for gluons of fractional momentum ξ'.

Fig. 30: Processes giving rise to the splitting functions $P \equiv P_{qq}$ (a) and P_{qg} (b)

Thus the correct Altarelli-Parisi equation is not that of Eq. (233) but rather

$$\frac{d f_q(\xi;q^2)}{d \ln q^2} = \frac{\alpha_s(q^2)}{2\pi} \int_\xi^1 \frac{d\xi'}{\xi'} \Big[P_{qq}(\xi/\xi') f_q(\xi';q^2) + P_{qg}(\xi/\xi') g(\xi';q^2) \Big] \tag{240}$$

where $g(\xi';q^2)$ is the gluon distribution function. Because this equation involves also $g(\xi';q^2)$ it is not complete in itself. The gluon distribution function has an evolution equation which is similar to (240), except that it involves the splitting functions for a quark to become a gluon, P_{gq}, and for a gluon to become a gluon, P_{gg}. One has

$$\frac{d\, g(\xi;q^2)}{d \ln q^2} = \frac{\alpha_s(q^2)}{2\pi} \int_\xi^1 \frac{d\xi'}{\xi'} \Big[\sum_q P_{gq}(\xi/\xi') f_q(\xi';q^2) + P_{gg}(\xi/\xi') g(\xi';q^2) \Big] \tag{241}$$

The splitting functions P_{qq}, P_{qg}, P_{gq} and P_{gg} can be extracted by picking out the coefficient of the $\ln q^2$ terms in appropriate parton scattering processes. For future use, I record here the form for $P_{qq}(x_p)$ and give a very brief discussion of how this splitting function can be calculated. The result for $P_{qq}(x_p)$ one finds is that

$$P_{qq}(x_p) = \frac{4}{3}\left\{\frac{1+x_p^2}{(1-x_p)_+} + \frac{3}{2}\delta(1-x_p)\right\} \tag{242}$$

The form for $P_{qq}(x_p)$, for $x_p \neq 1$, is quite easily obtained by remarking that the $\log q^2$ term in the cross section for the process $q + V^* \to q + g$ arises essentially from an angle singularity in the differential cross section, in the limit in which all masses are set to zero. If θ is the angle, in the $q - V^*$ CM system, between the outgoing gluon and the incident quark direction, a simple calculation gives

$$\frac{d\sigma}{d\cos\theta}(q+V^* \to q+g) = \frac{1}{(1-\cos\theta)}\,\frac{\alpha_s}{2\pi}\left[\frac{4}{3}\,\frac{1+x_p^2}{1-x_p}\right] + \text{non singular terms} \tag{243}$$

thereby identifying $P_{qq}(x_p)$ for $x_p \neq 1$.

For $x_p = 1$ one has to proceed with some care /39/, /41/. The virtual gluon diagrams of Fig. 27 will give a contribution to the parton cross section which is proportional to $\delta(1 - x_p)$. The exact value of the coefficient of this δ-function depends on how one regulates the potentially divergent piece arising from soft gluon bremsstrahlung ($x_p \to 1$) /39/. The prescription adopted in Eq. (242) is to replace $(1 - x_p)^{-1}$ by $(1 - x_p)_+^{-1}$, which is really a principal value prescription. Basically it means that for any function $f(x_p)$ one has by definition

$$\int dx_p \frac{f(x_p)}{(1-x_p)_+} \equiv \int dx_p \frac{f(x_p) - f(1)}{1-x_p} \tag{244}$$

Given this definition, then it is not difficult to convince oneself that indeed the coefficient of $\delta(1 - x_p)$ in Eq. (244) must be 3/2. This follows since the first moment of $P_{qq}(x_p)$ is easily shown to vanish. The proof of this statement is left as an exercise,

$$\int_0^1 dx_p\, P_{qq}(x_p) = 0 \tag{245}$$

Although in general both quark and gluon structure functions enter in the Altarelli-Parisi equations (240) and (241), one may construct combinations of quark structure functions in which all dependence on the gluon structure functions vanishes. Basically one needs to construct non-singlet structure functions in which, obviously, the gluons contributions are missing. For instance, the valence quark distribution

$$V(x;q^2) = u_v(x;q^2) + d_v(x;q^2) \tag{246}$$

obeys the simple Altarelli-Parisi equation (233):

$$\frac{dV(x;q^2)}{d\ln q^2} = \frac{\alpha_s(q^2)}{2\pi}\int_x^1 \frac{d\xi}{\xi} P_{qq}\left(\frac{x}{\xi}\right) V(\xi;q^2) \tag{247}$$

Recalling Eq. (139) one sees that for the structure function $\overline{F}_3^{cc}(x;q^2) = \frac{1}{2}[(F_3^{cc}(x;q^2))_{\nu N} + (F_3^{cc}(x;q^2))_{\bar{\nu} N}]$ QCD predicts a simple behaviour for its moments. Since to leading order in $\alpha_s(q^2)$ (cf Eq. (227))

$$\overline{F}_3^{cc}(x;q^2) = V(x;q^2) + O(\alpha_s(q^2)) \tag{248}$$

it follows that

$$M_n^V(q^2) = \int_0^1 dx\, x^{n-1}\, \overline{F}_3^{cc}(x;q^2) \simeq \int_0^1 dx\, x^{n-1}\, V(x;q^2)$$
$$= C_n\, (\ln q^2/\Lambda^2)^{d_n} \tag{249}$$

Here the coefficient d_n is calculable in terms of the moments of P_{qq}, precisely as detailed in Eq. (239):

$$d_n = \frac{6}{[33-2n_f]}\int_0^1 d\xi\, \xi^{n-1} P_{qq}(\xi) \tag{250}$$

Using the explicit form of $P_{qq}(\xi)$ of Eq. (242) one finds readily that

$$d_n = \frac{8}{[33-2n_f]}\left[-\frac{1}{2} + \frac{1}{n(n+1)} - 2\sum_{j=2}^{n}\frac{1}{j}\right] \tag{251}$$

This result was obtained first, in quite a different fashion, by Georgi and Politzer /42/ and Gross and Wilczek /43/.

Singlet structure functions, like $F_2(x;q^2)$, have a somewhat more complicated behaviour with q^2 than non singlet structure functions. Because in these cases both the quark and gluon structure functions contribute, the moments of F_2 behave as the sum of two distinct $(\ln q^2)^{d_n^\pm}$ terms - with again the $d_n^\pm$ coefficients calculable. This makes the direct comparison of the QCD prediction with data more complicated and therefore I shall not further pursue this matter here. A useful discussion of how to proceed in these cases is provided by Buras /44/.

For the nonsinglet case a direct check of QCD, in principle, is provided by the, so-called, moment plot. Clearly a graph of $\ln M_n^{\nu}$ versus $\ln M_{n'}^{\nu}$, will have, according to Eq. (249), a slope of $d_n/d_{n'}$. Such a moment plot, taken from Altarelli's review /45/ is shown in Fig. 31 and, as can be seen, appears to be in excellent agreement with the QCD predictions

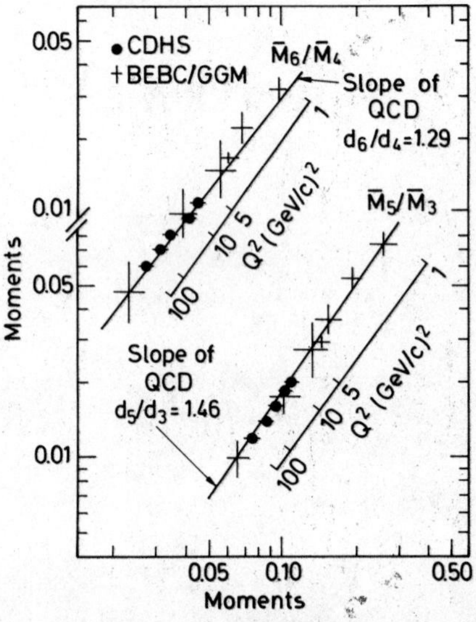

Fig. 31: Moment plot for the nonsinglet structure functions $\bar{F}_3^{cc}$, from Ref. 45

I should caution, however, that even though the agreement is impressive, there is really quite a bit of theory input that is included in this plot. In practice to compute the moments, since the measurements of F_3^{cc} do not cover all the range in x and q^2, one has to extrapolate the measured structure functions into the unmeasured regions. These extrapolations bias the ratios shown in Fig. 31.

A better approach to test QCD has been pursued lately. This consists of starting with a given set of structure functions at a fixed $q^2 = q_o^2$ and then using the Altarelli-Parisi equations directly to predict the behaviour of these structure functions for larger values of

q^2 ($q^2 > q_o^2$). The advantage of this approach over the moment plot is that, while the moments require a knowledge of $f_q(x;q^2)$ for all values of x, to compute the evolution of a structure function to $f_q(x;q^2)$ requires knowledge of $f_q(x_o,q_o^2)$ only for values of $x_o > x$.

In Fig. 32 I show a compilation of data on $xF_3^{\nu N}$, taken from the Cornell report of Dydak /46/. One sees clearly a nice qualitative agreement with the expectation of QCD. The structure function grows with q^2 for small values of x and decreases with q^2 for large values of x

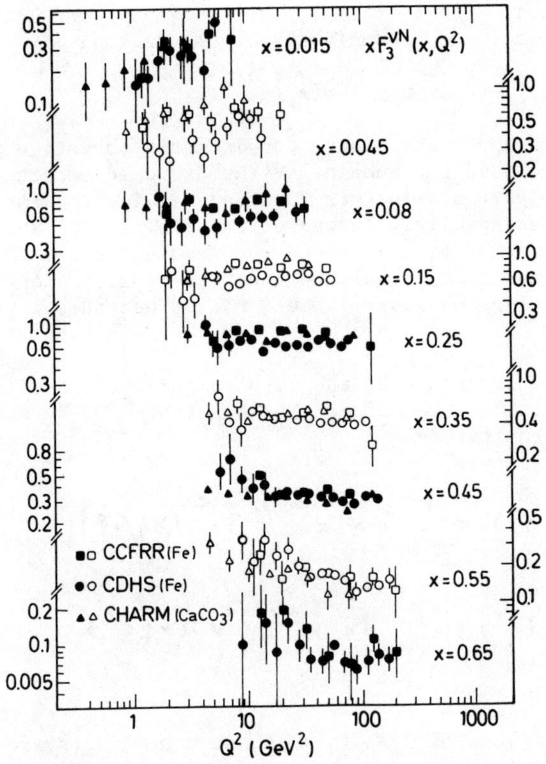

Fig. 32: Behaviour of $xF_3^{\nu N}$ as a function of q^2, from Ref. 46

This data can be used for a quantitative test of QCD by fitting it using the Altarelli Parisi equation. Such a fit determines the value for the only free parameter in the theory, the scale Λ of the QCD coupling constant (cf Eq. (238)). In practice, in performing these fits, also some higher order effects in $\alpha_s(q^2)$ are included and one determines a slightly different scale $\Lambda_{\overline{MS}}$, which is prescription dependent. I will discuss this, somewhat confusing, point below. However, let me first give the results. The best fit value for $\Lambda_{\overline{MS}}$ given by Dydak /46/ in his report is

$$\Lambda_{\overline{MS}} = (250 \pm 150) \text{ MeV} \tag{252}$$

This value is consistent with, although somewhat larger than, the previous best global fit value for $\Lambda_{\overline{MS}}$ of Buras /47/

$$\Lambda_{\overline{MS}} = (160 \,{}^{+\,100}_{-\,80}) \text{ MeV} \tag{253}$$

I end this section by commenting on the distinction between various QCD scale parameters. I will phrase my discussion in terms of moments of structure functions /44/ although an analogous discussion can be carried through in terms of Altarelli-Parisi equations, appropriately modified to include higher order QCD corrections /45/. Recall again our result (227) for the hadronic structure function

$$F(x;q^2) = f_q(x,q^2) + O(\alpha_s(q^2))$$

In the discussion up to now I have consistently forgotten the $O(\alpha_s(q^2))$ correction and equated the running of the hadronic structure function with that of the parton structure function. To include the $O(\alpha_s(q^2))$ terms necessitates specifying a way of renormalization and of calculating the evolution of $f_q(x;q^2)$ <u>beyond</u> the leading order. The evolution of $f_q(x;q^2)$ depends, in general, on the prescription adopted, although how $F(x;q^2)$ evolves obviously should not, since this latter quantity is measurable.

I detail the structure of the results for the non singlet $\bar{F}_3^{cc}$ structure function. Let me define the moment of $\bar{F}_3^{cc}(x;q^2)$ and of the valence parton density $V(x;q^2)$ by

$$M_n^V(q^2) = \int_0^1 dx\, x^{n-1}\, \bar{F}_3^{cc}(x;q^2) \tag{254}$$

$$\mathcal{M}_n(q^2) = \int_0^1 dx\, x^{n-1}\, V(x;q^2) \tag{255}$$

Neglecting corrections of $O(\alpha_s(q^2))$ these two moments are identical (c.f. Eq. (249)). Including these corrections, however, one has /44/:

$$M_n^V(q^2) = \mathcal{M}_n(q^2)\left[1 + \alpha_s(q^2) B_n\right] \tag{256}$$

where

$$\mathcal{M}_n(q^2) = \mathcal{M}_n(q_0^2)\left[\frac{\alpha_s(q^2)}{\alpha_s(q_0^2)}\right]^{d_n} \left\{1 + \left[\alpha_s(q^2) - \alpha_s(q_0^2)\right] Z_n\right\} \tag{257}$$

The coefficients B_n and Z_n depend on the renormalization scheme. However this scheme dependence cancels in Eq. (256), so that the physical moments are scheme independent.

Although the final formula is independent of the scheme, it does depend on the definition adopted for $\alpha_s(q^2)$ (i.e. on the Λ parameter used). Using the β-function computation to $O(\alpha_s^4)$, Eq. (93), one finds for $\alpha_s(q^2)$ the formula /44/:

$$\alpha_s(q^2) = \frac{12\pi}{[33-2n_f]} \frac{1}{\ln q^2/\Lambda^2} - \frac{72\pi(153-19n_f)}{[33-2n_f]^2} \frac{\ln(\ln q^2/\Lambda^2)}{\ln^2 q^2/\Lambda^2} \tag{258}$$

which contains corrections of $O\left(\frac{1}{\ln q^2}\right)$ to the lowest order formula, Eq. (238). Using this formula one may rewrite Eq. (256) as

$$M_n^V(q^2) = C_n \left[1 + \frac{R_n(q^2)}{(11-\frac{2}{3}n_f)\ln q^2/\Lambda^2} \right] \left(\ln q^2/\Lambda^2\right)^{d_n} \tag{259}$$

where

$$R_n(q^2) = B_n + Z_n - \frac{2(153-19n_f)}{(33-2n_f)} d_n \ln(\ln q^2/\Lambda^2) \tag{260}$$

Obviously, neglecting any $O(\alpha_s)$ corrections to the moments (B_n, $Z_n \to 0$) and dropping $(\ln q^2)^{-1}$ corrections to $\alpha_s(q^2)$, reduces Eq. (259) to the lowest order result, detailed in Eq. (249). Retaining the $R_n(q^2)$ corrections one must, however, specify which Λ one is dealing with. For if one changes Λ by

$$\Lambda \to e^K \bar{\Lambda} \tag{261}$$

in Eq. (259), one can reabsorb the change in a redefinition of R_n. To wit

$$R_n \to \bar{R}_n = R_n + \frac{2}{3}(33 - 2n_f) d_n K \tag{262}$$

To fit data using the corrected moment formula (259), one must specify which Λ one is dealing with. That is, what coefficients R_n one is really using. The $\Lambda_{\overline{MS}}$ scale which is now traditionally used is one for which certain constants, like ($\ln 4\pi - \gamma_E$), which normally arise in R_n, when one calculates using dimensional regularization, are removed. Obviously, there is nothing particularly sacred in this procedure. One should only take care that, if different physical quantities are analyzed, the same Λ scale is consistently used.

Higher order QCD corrections can be studied in other processes besides deep inelastic scattering. The factorization property discussed in the last section allows one to absorb uncontrollable $\ln q^2$ factors into running parton distribution functions, leaving over an, in principle, well defined perturbation expansion in $\alpha_s(q^2)$. I want to examine in this section the effects of QCD for Drell-Yan processes ($\mu^+\mu^-$ production or W/Z production in hadronic processes) since, as I have already mentioned, the QCD corrections for these processes appear to be significant. These effects are what lead to the K-factor ambiguity on the overall size of the production cross section.

The lowest order process $q + \bar{q}' \to V^*$ gets modified to $O(\alpha_s)$ by processes in which the vector boson V^* is accompanied in the final state either by a gluon ($q + \bar{q}' \to V^* + g$) or by a quark ($q + g \to V^* + q'$), as shown in Fig. 33. In addition there are virtual gluon corrections.

Fig. 33: $O(\alpha_s)$ real corrections to Drell-Yan processes

The effect of these additional contributions is two-fold: i) the original quark and antiquark distribution functions get replaced by running distribution functions. This is the leading QCD modification, totally analogous to what happened in deep inelastic scattering and is what is expected by the factorization property. (ii) Two different types of $O(\alpha_s(q^2))$ corrections ensue. One of these, subdominant, corrections basically arises from the second diagram of Fig. 33, and involves V^* production by quark-gluon fusion. This is a pure $\alpha_s(q^2)$ effect and is proportional to the gluon structure function. For the collider data this contribution is not particularly important and can be safely neglected. The second $O(\alpha_s(q^2))$ correction is directly a correction to the original parton process and, although formally of order α_s, turns out to be large. This correction is responsible for the K-factor ambiguity.

Let me specifically discuss W^- production, since I have already given the lowest order prediction for it in Eq. (203). Including QCD corrections, but neglecting the contribution of quark-gluon fusion, this formula is modified to

$$\sigma(p\bar{p}\to W^-X) = \frac{\sqrt{2}\pi}{3} G_F \cos^2\theta_c \frac{M_W^2}{s} \int d\xi\, d\xi'\, d_v(\xi; M_W^2)\, u_v(\xi'; M_W^2)$$

$$\cdot \left\{ \delta\left(\xi\xi' - \frac{M_W^2}{s}\right) + \theta\left(\xi\xi' - \frac{M_W^2}{s}\right) \frac{\alpha_s(M_W^2)}{2\pi} R\left(\frac{M_W^2}{\xi\xi' s}\right) \right\}$$

(263)

The parton distribution functions are running functions now, which are evaluated at M_W^2 which is the scale of the process. The explicit $O(\alpha_s)$ correction, denoted by R, has been calculated in the literature /45/ and is found to be very large. The origin of this unexpectedly large contribution can be traced to the fact that one is comparing space-like processes to time like processes. Let me try to explain this point.

The running parton distributions which enter in (263) are defined with respect to, the space-like, deep inelastic scattering. In factorizing away potentially dangerous $\ln q^2$ terms (or even $(\ln q^2)^2$ terms, which actually eventually cancel entirely between virtual and real contributions) q^2 is always positive. In the Drell-Yan process, on the other hand $q^2 < 0$. If one wants to reconstruct the running parton distribution of deep inelastic scattering, it is necessary to change the sign of q^2 and so some additional terms containing ln(-1), or even $(\ln(-1))^2$ are left over. The dominant term in R in Eq. (263) arises precisely in this way. The virtual correction for the vertex $V^* + q \to q$ and that for the vertex $q + \bar{q} \to V^*$, shown in fig. 34, both have a leading term proportional to $(\ln q^2)^2$, except that q^2 is positive in one case and negative in the other.

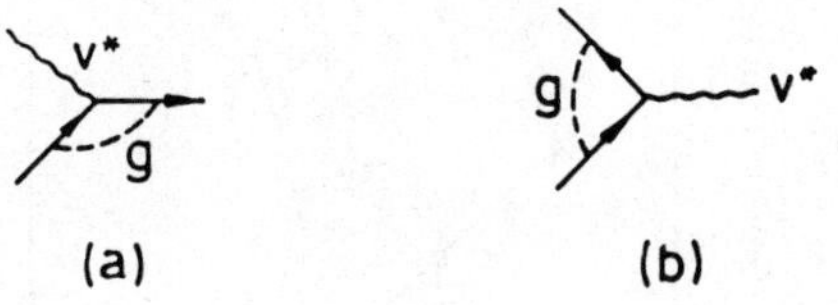

Fig. 34: Virtual gluonic corrections for deep inelastic (a) and Drell-Yan (b)

This leads to a contribution to R:

$$R_{continuation} = -\frac{\alpha_s}{2\pi}\left[\frac{4}{3}(\ln(-1))^2\right]\delta(\xi\xi' - \frac{M_W^2}{s})$$

$$= \frac{2\pi}{3}\alpha_s\,\delta(\xi\xi' - \frac{M_W^2}{s})$$

(264)

which is of the same order of magnitude as the leading term in Eq. (263). Note that this contribution just changes the value of the 0th order rate.

Because the $O(\alpha_s)$ corrections in R in Eq. (263) are so large, one cannot really trust the perturbative calculation. However, because these corrections come, mainly, from virtual effects one has at least a qualitative understanding of what QCD does. Basically, one expects that the usual parton model predictions obtain - including effects of the running of the structure functions - modified by some over all factor (the K-factor) which should be of the order of 1-2. For a fixed M_V^2 process - like W/Z production - this prediction is not terribly helpful, except that it allows one to understand why pure parton model predictions can be off by a factor of 2 or so. For $\mu^+\mu^-$ production, however, one has a stronger prediction. Namely that the parton model prediction for the shape of the differential cross section $d\sigma/dM_V^2$ should be manifest in the data. Fig. 35 shows data for the $\mu^+\mu^-$ mass distribution in the process $\bar{p}N \to \mu^+\mu^- X$, measured at Fermilab and CERN, compared to a Drell-Yan calculation using the CDHS parton structure functions, scaled by a K-factor of 2.3 /48/. It is clear that the agreement of the shape of the data with the parton model prediction is quite satisfactory.

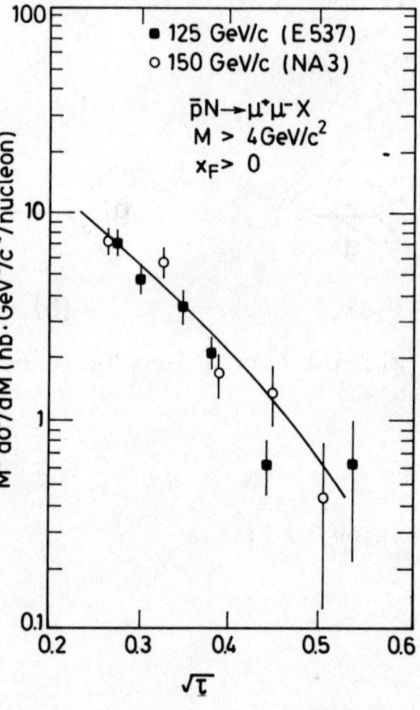

Fig. 35: Comparison of Drell-Yan data with the prediction of the parton model, scaled by a factor of 2.3. From ref. 48

JETS IN e^+e^- PHYSICS

The last item of QCD phenomenology that I want to discuss is hadronic jet production in e^+e^- annihilation*. Qualitatively one expects 2 jet production to dominate in e^+e^- collisions, with the jets following the direction of the $q\bar{q}$ pairs produced by the virtual photon. That is, the lowest order QCD diagram of Fig. 36 in the CM system should give rise to back to back hadronic jets, as illustrated in Fig. 37.

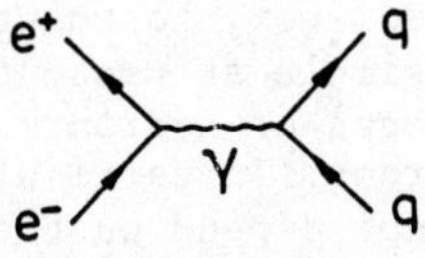

Fig. 36: Lowest order QCD diagram in e^+e^- annihilation

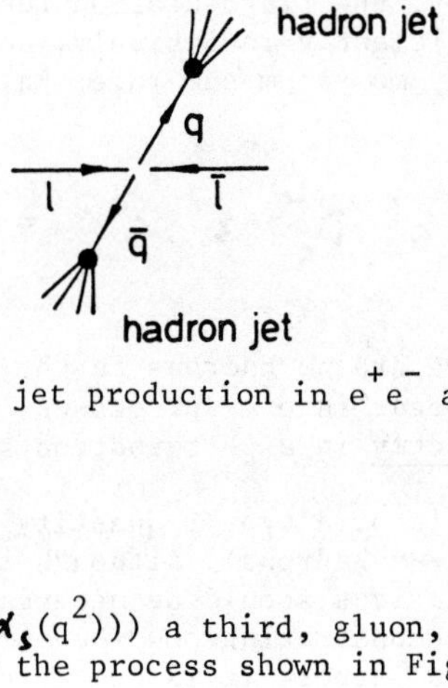

Fig. 37: Two jet production in e^+e^- annihilation

Occasionally (i.e. to $O(\alpha_s(q^2))$) a third, gluon, jet should also be produced corresponding to the process shown in Fig. 38.

---

* I will not discuss jet physics in hadronic collisions, since this item is very well covered in Astbury's lectures in these proceedings.

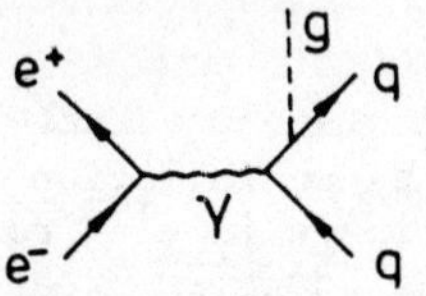

Fig. 38: $O(\alpha_s(q^2))$ QCD diagram in e^+e^- collisions, which gives rise tc a third hadronic jet

Both of these qualitative features are present in the SLAC and DESY data on $e^+e^- \rightarrow$ hadrons. However, to quantify these observations and compare to QCD is not so simple at present energies, because one cannot really neglect the process of hadronization by which quarks transmute themselves into hadrons. In particular, one would really like to study quantities which do not depend on the fragmentation functions $D_q^H(z)$, since these functions essentially parametrize our ignorance of the hadronization process. There are two general types of observables in e^+e^- processes where one can proceed, in principle, without a knowledge of the fragmentation functions. (I say, in principle, because in practice in most cases some mild information on hadronization remains necessary). Type I quantities consist of appropriately weighted cross sections, for which one can show that the D_q^H functions disappear. Type II quantities are appropriate averages which, because they are free of infrared singularities, should also be independent of the D_q^H functions.

For a type I quantity, the fragmentation functions do not enter because one is looking sufficiently inclusively, and with an appropriate weight, so that the energy momentum sum rule, Eq. (101), for the D_q^H functions can be used:

$$\sum_H \int d\xi' \, \xi' \, D_q^H(\xi'; q^2) = 1$$

Furthermore, because there are no hadrons in the initial state, obviously no structure functions appear in e^+e^- processes. Hence, type I quantities should be calculable <u>directly</u> in a perturbation series in $\alpha_s(q^2)$.

The best known example of a type I quantity is the total hadronic cross section, $\sigma_{tot}(e^+e^- \rightarrow$ hadrons). Although it is intuitively obvious that no fragmentation functions should be necessary in calculating σ_{tot}, it may prove worthwhile demonstrating how the sum rule (101) eliminates the D_q^H functions. For this purpose it is convenient to write σ_{tot} formally as a sum of inclusive cross sections

$$\sigma_{tot} = \sum_H \int dz \, z \, \frac{d\sigma^H}{dz} \tag{265}$$

Here z is the energy fraction carried by the hadron H in the final state. Note that the z weighing above is necessary; if not one would obtain, instead of σ_{tot}, the total cross section multiplied by the average hadron multiplicity. The inclusive cross section can be written in terms of the usual parton convolution

$$\frac{d\sigma^H}{dz} = \sum_{parton} \int_z^1 \frac{d\xi'}{\xi'} \left(\frac{d\sigma_{parton}}{dz_p} \right) D^H_{parton}(\xi', q^2)$$

$$= \sum_{parton} \int d\xi' \, dz_p \, \delta(z - \xi' z_p) \left(\frac{d\sigma_{parton}}{dz_p} \right) D^H_{parton}(\xi', q^2)$$

(266)

Inserting (266) into (265) and using Eq. (101) immediately gives

$$\sigma_{tot} = \sum_{parton} \int dz_p \, z_p \, \frac{d\sigma_{parton}}{dz_p} \tag{267}$$

which shows that the total hadronic cross section can be calculated at the parton level, without any need of fragmentation functions.

It is usual to define instead of σ_{tot} the quantity R which is the ratio of σ_{tot} to the μ-pair cross section

$$R = \frac{\sigma_{tot}(e^+e^- \to \text{hadrons})}{\sigma(e^+e^- \to \mu^+\mu^-)} \tag{268}$$

To lowest order on QCD, since the graph in Fig. 36 differs from that for μ-pair production only because quarks have different charges and a given color multiplicity, it is obvious that

$$R = 3 \sum_q e_q^2 \tag{269}$$

This equation receives QCD corrections and the result of a rather straightforward calculation gives /49/

$$R = 3 \sum_q e_q^2 \left\{ 1 + \frac{\alpha_s(q^2)}{\pi} + \dots \right\} \tag{270}$$

Clearly R with increasing energy approaches the parton model value from above. However, the correction is small and it is quite difficult to measure this effect given the errors in the data, shown in Fig. 39. Obviously, however, the parton model prediction, which says that R is essentially a constant, appears to be valid up to the highest measured energies. One should not forget that this is a non trivial consequence of QCD also!

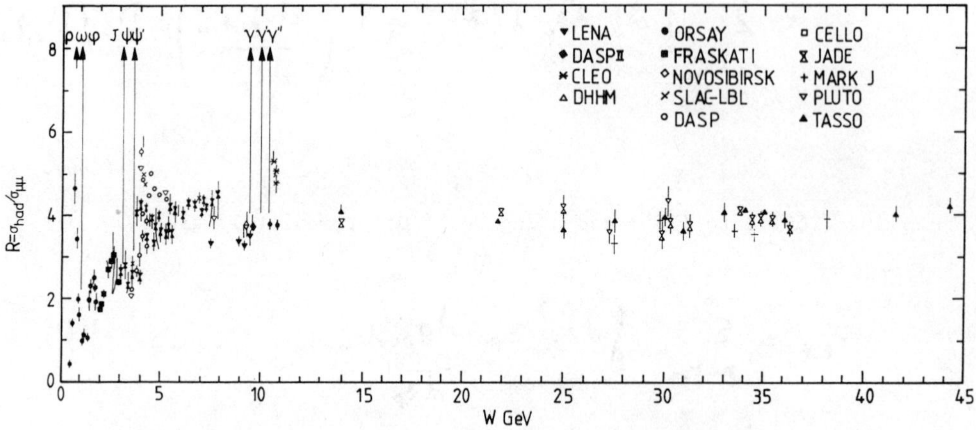

Fig. 39: Behaviour of R versus energy $W = \sqrt{-q^2}$, from Ref. 50

A less inclusive quantity than σ_{tot}, which is again of Type I, is the angular energy flow /51/ of hadrons in e^+e^- annihilation. It is not hard to convince one self that

$$\frac{d\Sigma}{d\Omega} = \sum_H \int dz\, z \frac{d\sigma^H}{d\Omega\, dz} = \sum_{parton} \int dz_p\, z_p \frac{d\sigma_{parton}}{d\Omega_p\, dz_p} = \frac{d\Sigma}{d\Omega_{parton}} \tag{271}$$

so that also $d\Sigma/d\Omega$ does not need any fragmentation function information and is calculable in a power series in $\alpha_s(q^2)$. Unfortunately, this quantity, as well as more complicated energy-energy correlation functions /52/, contains an implicit assumption which makes it necessary to use some hadronization information before it can be compared to data. Namely in (271) one assumed that the hadronic energy flow follows precisely that of the partons : $d\Omega = d\Omega_{parton}$. This is obviously an approximation. When the partons hadronize, in general the resulting hadrons will be produced with some fixed average spread in $p_\perp$, $\langle p_\perp \rangle$, with respect to the parton. This introduces uncertainties in the hadron direction of $O(\langle p_\perp\rangle^2/q^2)$ which tend to smear the resulting energy flow. With this caution in mind, I show in Fig. 40 some recent data on energy-energy correlations in e^+e^- collisions compared to what is predicted by QCD (dashed curve) and including some hadronization corrections (solid curve).

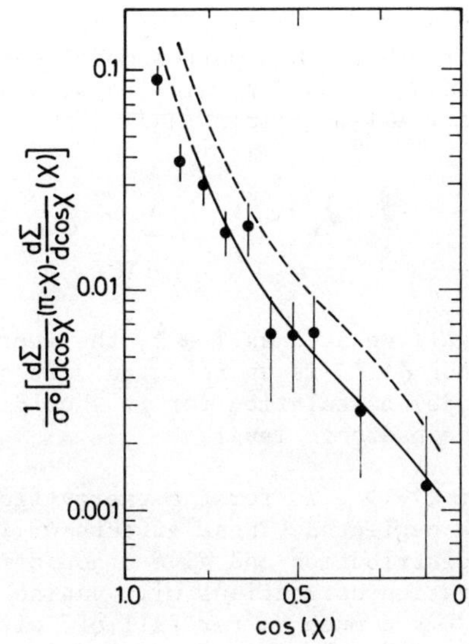

Fig. 40: Data on energy-energy correlations in e^+e^-, from Ref. 53

Type II quantities are averages of distributions in e^+e^- collisions which are well defined at the parton level. More precisely, they are quantities without any infrared singularities produced by collinear or soft gluon radiation. Because these quantities are calculable directly at the parton level they should give information on hadronic properties, which do not require introducing fragmentation functions. Roughly speaking, the fragmentation functions are put in, in a QCD calculation, to absorb the uncalculable soft processes by which partons turn into hadrons. The presence of infrared singularities at the parton level, which are then incorporated into the fragmentation functions, is precisely what makes these functions "run". If no infrared singularities are present, there should be no need to introduce fragmentation functions to absorb them!

Obviously all Type I quantities are also Type II, since they are calculable directly at the parton level. However, some Type II quantities sometimes prove to be more convenient to analyze. A good example, and the only one which I shall discuss, is Thrust. The Thrust variable is defined by

$$T = \frac{\sum_i |\vec{p}_i \cdot \hat{e}|}{\sum_i |\vec{p}_i|} \tag{272}$$

where Pi are the momenta of the outgoing hadrons and the vector $\hat{e}$ is varied until the maximum value for T is found. This is then the Thrust axis. The lowest order process $e^+e^- \rightarrow q\bar{q}$ obviously gives a Thrust distribution which is a δ-function at T = 1. Gluon radiation will give events which are not back to back and so, at the parton level, one has a spreading of this δ-function. The actual thrust distribution is, however, infrared singular at $T = 1$, reflecting the fact that gluons like to be soft and collinear. Thus, this parton model calculation cannot be translated directly into a hadronic prediction without introducing fragmentation functions. As $T \to 1$ one finds /54/

$$\frac{1}{\sigma_{tot}} \frac{d\sigma}{dT} \simeq \frac{8}{3} \frac{\alpha_s(q^2)}{\pi} \frac{1}{1-T} \ln \frac{1}{1-T} \qquad (273)$$

Although (273) is not well behaved as $T \to 1$, the average $\langle 1-T \rangle$ is perfectly well defined. Thus $\langle 1-T \rangle$ is an infrared safe quantity and in principle the parton model calculation for it should reflect directly what is happening at the hadronic level.

Of course also for $\langle 1-T \rangle$, as for the energy flow, not all effects of hadronization can be neglected. These effects will tend to smear the 0th order QCD $\delta(1-T)$ distribution and give a non zero value for $\langle 1-T \rangle$. However, these hadronization corrections will vanish as $\langle p_\perp \rangle^2/q^2$. What QCD predicts for $\langle 1-T \rangle$ is a much softer fall off with q^2, namely $\langle 1-T \rangle \sim \alpha_s(q^2) \sim (\ln q^2)^{-1}$. In Fig. 41 I show some recent TASSO data /55/ on the behaviour of $\langle 1-T \rangle$ versus $W = \sqrt{-q^2}$. The fit shown is the QCD prediction, in which hadronization is incorporated, using the independent jet model of Hoyer et al. /57/. The effect of hadronization can be clearly seen by focusing on the dotted line in the figure which represents just $e^+e^- \to q\bar{q}$. However, it is obvious that the expected QCD effect is clearly distinguishable, especially at higher W values. This is a nice way to demonstrate the existence of gluon radiation.

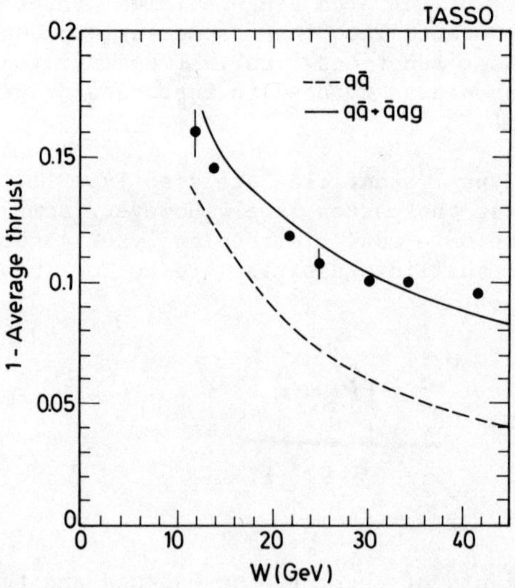

Fig. 41: Behaviour of $\langle 1-T \rangle$ versus $W = \sqrt{-q^2}$, from Ref. 55

WHERE ARE WE NOW: COMMENTS AND SPECULATIONS

I would like to end these lectures by discussing some of the open issues in the standard model. I hope that the discussion up to now has made it clear that the standard model does provide an excellent phenomenological description for the interactions of leptons and for hadronic, hard scattering processes. Nevertheless, there are some open questions that need to be resolved, and whose answers may point to a deeper synthesis. I begin by discussing QCD, since here there are fewer questions of principle left open.

QCD, in some sense, is better established than the GSW model. The success of the parton model are really QCD successes. Quantum chromodynamics passes all qualitative tests, with flying colors *: there is approximate scaling in R and in deep inelastic scattering; 2 jet and 3 jet phenomena are seen; color counting rules work; etc. In addition, QCD passes at least some semiquantitative tests: the structure functions evolve in the predicted way with q^2, growing for small x and decreasing at large x; there is evidence for gluon radiation in e^+e^- collisions, of the magnitude predicted; the data for R lies above the parton model predictions, as expected; there is a K-factor in Drell-Yan processes and the shape of the mass distribution of the lepton pairs agrees with the parton model; in addition, lattice calculations of the hadronic spectra are beginning to give encouraging results /57/.

It is clear that really quantitative tests of QCD will have to wait until a better understanding of how to calculate long distance properties is achieved. Although the lattice approach to QCD is beginning to tackle with success some of the static aspects of hadrons (masses, magnetic moments, charge radii), it is very difficult to see how it could provide, in the near future, some more complicated dynamical information, like the q^2 behaviour of structure functions. The hope is, however, that lattice calculations may give some insight on how to better tackle analytically some of the long distance aspects of QCD.

To my knowledge, there are really two open problems in QCD. One of these is why is there no strong CP violation /58/? This problem is somewhat technical and it would take me too far afield to discuss it properly here. Suffice it to say that the vacuum structure of QCD suggests than an additional term $\theta F\tilde{F}$, where $\tilde{F}$ is the dual of the color field strength, should be present in the QCD Lagrangian. This term induces a large electric dipole moment for the neutron, not seen experimentally, and thus θ must be very small ($\theta \lesssim 10^{-8} - 10^{-9}$). This is a mystery, unless one can find some plausible dynamical reason for θ to be so small. Very probably, at some deep level, the protective chiral symmetry suggested by H. Quinn and I long ago /59/ is the correct solution to this problem, although technically one runs into problems with axions.

The second and main problem of QCD is one of calculation. There are so many things one would like to calculate, but as of yet our theoretical tools are too primitive to achieve much success. I list below

---
* Sorry, I could not resist this pun.

some of the things that it would be nice to be able to compute, from first principles, from QCD. This list is ordered in a (probable) order of difficulty:

(i) Spectrum of hadrons: mesons, glueballs, baryons and exotic states ($(q\bar{q})^2$, etc.)

(ii) Structure functions and fragmentation functions

(iii) Soft processes, like elastic π p scattering and multiparticle production

(iv) Nuclear physics

Clearly the theoretical task ahead is immense and there will be little additional experimental input which can serve as illumination.

The open problems of the GSW model are of a different nature. As should have been clear from these lectures, the GSW model agrees beautifully with experiment. However, what is really tested is essentially only the gauge-fermion sector of the model. Radiative corrections do, in principle, depend on the Higgs sector; but for reasonable Higgs masses, the dependence in practice is well within the errors in $\sin^2\theta_W$. The principal information that we have on $\mathcal{L}_{\text{Higgs-gauge}}$ is that $\rho \simeq 1$, which means that the symmetry breaking is done by an SU(2) doublet object. Of $\mathcal{L}_{\text{Higgs-fermion}}$ we know even less, except that with three families the Cabibbo matrix can have a complex phase, so that there is a natural origin for CP violation in the model. Because the pieces of the GSW Lagrangian connected with the Higgs sector are largely unknown, they constitute a natural open problem. Actually, as I will explain below there is some theoretical prejudice <u>against</u> having $\mathcal{L}_{\text{Higgs-gauge}}$ and $\mathcal{L}_{\text{Higgs-fermion}}$ as given in the standard model. So not only are these pieces an experimental open problem, they are a theoretical one too!

I begin by discussing some potential theoretical difficulties of the pure Higgs sector. Recall that the breakdown of the SU(2) x U(1) symmetry was caused by introducing the Higgs potential (61):

$$V = \lambda\left(\Phi^\dagger\Phi - \tfrac{1}{2}v^2\right)^2$$

Although the parameter λ is free, the scale v in the potential V is fixed by the scale of the Fermi constant. Indeed, from Eq. (72), one has

$$v = (\sqrt{2}\,G_F)^{-1/2} \simeq 250 \text{ GeV}$$

Wilson /60/, was the first to point out that fixing v this way was unnatural, if one imagined that there is some cut off in the theory. Since radiative corrections for scalar fields are quadratically divergent, the parameter v can only be maintained at 250 GeV by carefully tuning the original parameters of the theory. Otherwise the natural value it would take would be that of the cut off.

I illustrate this by considering what happens to the mass of the physical Higgs field. Recall that to lowest order, Eq. (72), we found for the Higgs field mass

$$m_H^2 = 2\lambda v^2$$

This mass, however, gets shifted by radiative corrections, some examples of which are shown in Fig. 42

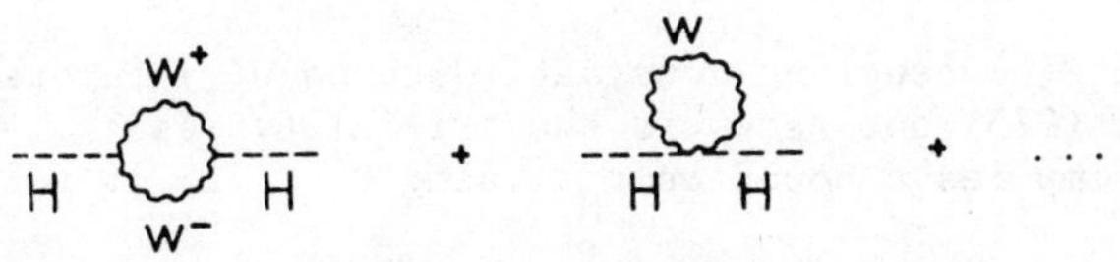

Fig. 42: Contributions to the radiative mass shift of the Higgs field

Since the graphs in Fig. 42 are quadratically divergent, evaluating them with a cut off, Λ_c, gives for the radiative corrected Higgs mass the formula

$$m_H^2 = 2\lambda v^2 + \alpha \Lambda_c^2 \tag{274}$$

Clearly the Higgs mass gets driven to Λ_c, if $\Lambda_c >> v$. To keep $m_H^2 << \Lambda_c$, the original parameter v has to also be very big and a careful cancellation (fine tuning) has to occur.

If one lets the parameter $\Lambda_c \to \infty$, this discussion ceases to make sense. Since the GSW model is renormalizable, all that really happens is that the Higgs mass has to be put in as an undetermined parameter, to absorb the relevant infinities. The same thing happens to the Fermi scale v. It also has to be put in as an input in the theory. Therefore, strictly speaking there is no problem in the GSW model in isolation. However, gravity really introduces a scale, because for momenta above $M_{Planck} \simeq 10^{19}$ GeV it cannot be ignored. Therefore the naturalness problem is a real problem and one should look for possible ways to avoid it. I will return to this point shortly, but first I want to mention a second possible difficulty connected with the Higgs potential-triviality.

Triviality is a property proven by mathematical physicists /61/ for pure $\lambda\phi^4$ field theory, with ϕ a real field. What was shown is that the only consistant version of this theory is the one for which $\lambda_{ren} = 0$, i.e. a theory with no interactions. Such a result, of course, is a disaster if it was applicable in general, for it is precisely the self interactions of the Higgs fields which are needed for the spontaneous breakdown of SU(2) x U(1)! Of course, it is a long way from the simple $\lambda\phi4$ theory to the GSW model, where there are besides scalar fields also gauge fields and fermions. In fact, it has been argued /62/ that perhaps the presence of the gauge fields - at least the U(1) gauge field - may stabilize the Higgs sector and avoid the triviality argument. Roughly speaking, one can think of the triviality result as being due to the fact that $\lambda = 0$ is the only stable point of the theory. If λ is positive*, it will be always driven with increasing q^2 to infinity since the β-function grows with λ. What was argued in Ref. 62, is that if there is a U(1) gauge field in the theory, this stabilizes the behaviour of λ provided that λ is small enough. Stability - at least in a perturbative sense - was found provided

$$\lambda \lesssim g'^2 \tag{275}$$

where g' is the U(1) coupling constant. With no U(1) interaction, it is clear that from (275) one recovers the triviality result, $\lambda \to 0$. Furthermore Eq. (275) implies a bound on m_H. Using Eq. (72) it follows that

$$m_H^2 \lesssim \frac{8}{\tan^2\theta_W} M_W^2 \tag{276}$$

which implies $m_H \lesssim 130$ GeV.

The triviality problem may be solved, therefore, if the coupling is small enough. A "perturbative" solution can also be found for the naturalness problem, but it involves a rather large speculative step. Namely, that nature is approximately supersymmetric. If all the particles in the standard model had supersymmetric partners then the mass shift in Eq. (274) would no longer be quadratically divergent. The W exchange graph, for example, would be accompanied by a Wino exchange graph (which is the spin 1/2 partner of the W), as shown in Fig. 43. These two contributions would cancel each other off, since the fermionic graph has a (-1) factor from Fermi statistics. This cancellation would be incomplete if supersymmetry were broken by having the W and Wino have different masses. Nevertheless, the radiative corrections to m_H^2 would depend only on $\ln\Lambda_c$ and not any more on Λ_c^2 and they would be therefore naturally under control, if α is small.

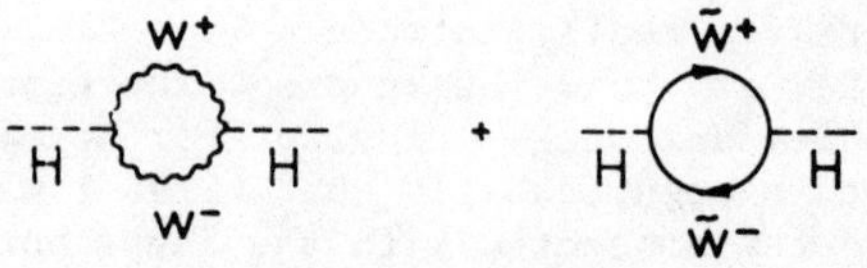

Fig. 43: Radiative corrections to the Higgs mass in a supersymmetric theory

* λ must be positive for the positivity of the potential V

If supersymmetry is the solution to the naturalness problem of the Higgs potential, there is an extraordinarily rich phenomenology of superpartners wanting to be discovered. These superpartners cannot be pushed too far away in mass from the Fermi scale v, since it is their existence which is supposed to render this parameter natural. However, there may be a different solution to naturalness than the "perturbative" supersymmetric one. Clearly there is no problem with naturalness in Eq. (274) if the cut off $\Lambda_c \sim v$, because the radiative corrections then are small. However, having a physical cut off of the order of v means really that the Higgs particle cannot be elementary! It must be a bound state of some underlying strongly interacting theory. This scenario is the Technicolor scenario /63/.

Technicolor is a non perturbative solution to the naturalness problem. One basically replaces the whole Higgs sector by an underlying Technicolor theory. What causes the breakdown of SU(2) x U(1) is the existence of SU(2) x U(1) breaking condensates in this theory, $\langle \bar{T}T \rangle \neq 0$. The scale of these condensates is related to the dynamical scale of the Technicolor theory Λ_{TC} - the analogue of Λ for QCD. Hence, the Fermi scale is also related to Λ_{TC} and its value is a purely dynamical issue. Obviously, if a Technicolor theory really exists and it replaces the Higgs potential, also the triviality problem is irrelevant.

There are, therefore, two possible options for removing from the Higgs sector of the GSW model the stigmas of unnaturalness and triviality. A "perturbative" option, in which one has a light Higgs boson and a doubling of all degrees of freedom, through supersymmetry*. Or a "non perturbative" solution, in which there is no Higgs potential at all but a dynamical breakdown of SU(2) x U(1) occurs due to condensate formation in some new underlying theory. In both cases there is considerable additional physics beyond the standard GSW model. If either of these options obtains, spectacular phenomena await discovery in the next generation of experiments, which will be probing the energies of the Fermi scale.

This is clearly not the place to discuss the phenomenology expected from these speculative extensions of the standard model. Nevertheless, let me give at least one example of the way in which the physics of the GSW model will change if one of these scenarios is true. If the Higgs sector is replaced by a strongly interacting theory, one predicts that at sufficient high energies the W-bosons will scatter strongly. This remarkable result can be understood as follows. The formation of technicolor condensates $\langle \bar{T}T \rangle \neq 0$, being a manifestation of spontaneous symmetry breakdown, causes the appearance of Goldstone excitations in the spectrum of the underlying technicolor theory. (An analogous phenomena actually happens in QCD. There one knows that quark condensates form $(\langle \bar{u}u \rangle \neq 0)$ In the limit of zero quark masses these condensates break an exact global symmetry and Goldstone pions emerge. Restoring the quark masses gives the pions a small mass). The W-bosons get a mass precisely by absorbing these excitations, since the $\langle \bar{T}T \rangle$ condensates are assumed to break SU(2) x U(1). Thus, roughly speaking, the longitudinal component of the W fields is made up by the Technicolor Goldstone excitations - Technipions. At scales of the order of the dynamical scale of Technicolor

---

* Supersymmetric extensions of the GSW model, in fact, require two Higgs multiplets. In general one of the five physical Higgs of these extended theories is light.

Λ_{TC}, these Technipions will scatter strongly, just as pions scatter strongly. More precisely, one can argue for pions that the strong scattering should occur at energies of the order $\sqrt{s} \simeq \sqrt{16\pi}\, f_\pi$ where $f_\pi \simeq 95$ MeV is the pion decay constant, which characterizes the breakdown which made pions, (approximate) Goldstone bosons in QCD. For the Technipions, the appropriate breakdown scale is just the Fermi scale $v \simeq 250$ GeV. Hence, one predicts that W should scatter strongly at energies of the order $\sqrt{s} \simeq \sqrt{16\pi}\, v \simeq 1.5$ TeV.

Up to now, I discussed potential problems associated with the Higgs potential. The Fermion-Higgs sector is also unsatisfactory theoretically. The introduction of Yukawa couplings between fermions and the Higgs doublet causes the appearance of both fermion masses and Cabibbo mixing angles. However, because these couplings are arbitrary, the masses and mixing angles are uncalculable. This is the price one must pay if one retains the quarks and leptons as elementary and introduces no other interactions. It is difficult to conceive that all the fermion masses and mixing angles are really free parameters. The known quarks and leptons now span a range in mass of almost five orders of magnitude and have certain characteristic patterns which cry out to be explained. For instance the mixing angles among quarks which are very well separated in mass is extremely small. Why is this so?

If there is no elementary Higgs boson as in the Technicolor option then one must also find a way to generate the masses of quarks and leptons. The most "reasonable" supposition in this case is that really - despite the evidence for elementarity proved by the incredible successes of QED - quarks and leptons are composite objects. Composite models of quarks and leptons /64/ offer the hope of providing an understanding of the quark and lepton mass patterns, since in principle these parameters are now calculable. However, these models are still in a very primitive state and face extremely difficult dynamical challenges. My own view, nevertheless, is that this is the correct way to go beyond the standard model. I speculate that the underlying theory that provides for the spontaneous breakdown of SU(2) x U(1) is also the same theory whose "light" bound states are the quarks and leptons.

Having discussed some of the open theoretical problems of the standard model, let me close these lectures by making some comments also on some recent experimental challenges to the standard model. I want to touch upon three subjects, very briefly.

(i) The EMC effect
(ii) The long B lifetime and CP violation
(iii) Exotic events at the collider

I do not believe that any of this new data necessarily puts the standard model in trouble. However, the CP violation data and some of the collider data could well be the first indication that some non standard physics is emerging.

The EMC effect /65/, named after the collaboration that found it, is the surprising observation that the x-dependence of structure functions seems to depend on the nucleon number A of the target. This is illustrated in Fig. 44.
Obviously if deep inelastic scattering in nuclear targets is off quarks, then there should be no A dependence. Fermi motion effects alter things

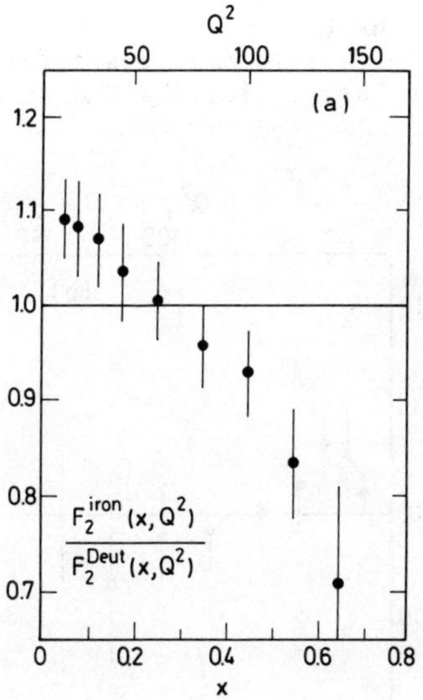

Fig. 44: Illustration of the EMC effect, from Ref. /65/

slightly, but go the other way. So the behaviour shown in Fig. 44 is really a challenge to QCD.

Because one is dealing with nuclei, a variety of "nuclear" explanations for the EMC effects have been proposed /66/. I am really not terribly competent in judging the reasonableness of these suggestions. However, I believe that a QCD inspired qualitative explanation by Jaffe /67/, and a more quantitative analysis by Close, Jaffe, Roberts and Ross /68/, are probably closer to the truth. The idea is very simple to grasp. The plot in Fig. 44 has the qualitative behaviour of what one expects in QCD if one plots the ratio of two structure functions taken at <u>different</u> q^2 values. For higher q^2 there is an increase at small x and a decrease at large x. The data of Fig. 44 looks precisely as if the x distribution in large A nuclei has been degraded to lower x. Such a circumstance could happen if quarks in nuclei are partially deconfined. That is, one has the beginning of a quark-gluon plasma in the presence of many nucleons, which means that the effective QCD scale Λ is slightly decreased in nuclei.

Specifically, Close et al. /68/ in their analysis showed that if the effective nucleon size $R \sim \frac{1}{\Lambda}$ in a nucleus becomes $R' \sim \frac{1}{\Lambda'}$, then one would expect for the nuclear structure function the relations:

$$F_2'(x;q^2) = F_2(x;\xi q^2) \tag{277}$$

where

$$\xi = \left[\left(\frac{R'^2}{R^2}\right)\right]^{\alpha_s(q_o^2)/\alpha_s(q^2)} \tag{278}$$

For reasonable range of parameters ($q_o^2 = 1$ GeV^2, $q^2 = 50$ GeV^2, $R'/R = 1.15$) one finds that $\xi \simeq 2$. As can be seen from Fig. 45, in the ratio of $F_2^{Iron}(x,q^2)$ to $F_2^{Deut.}(x,q^2/2)$, essentially the EMC effect has disappeared. Thus, even though I find the EMC effect interesting, I

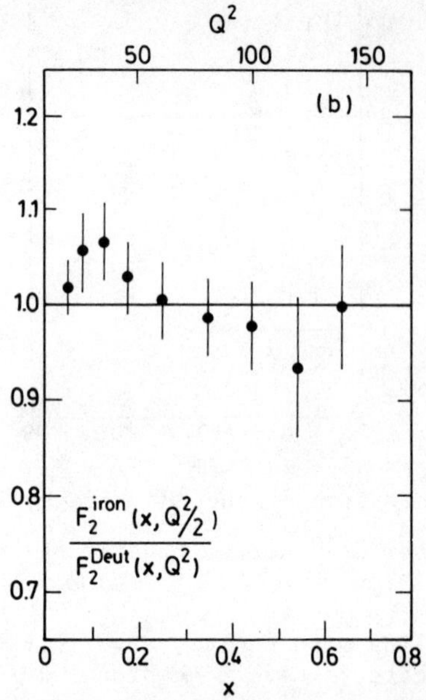

Fig. 45: Compensation of the EMC effect by degradation. From Ref. 68

believe it does not constitute a dangerous problem for QCD. The effective rescaling of the nuclear structure functions, given in Eq. (277), probably is a true reflection of what is really happening.

The standard model is perhaps slightly more challenged by recent measurements bearing on CP violation. The dominant source of CP violation in the kaon system is due to the mass mixing parameter ϵ in the K-$\bar{K}$ mass matrix. In the standard model, the major contribution to ϵ /69/ arises from the imaginary part of the box graph diagram of Fig. 46.

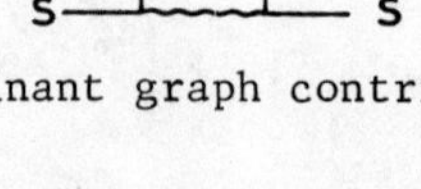

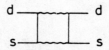

Fig. 46: Dominant graph contributing to ϵ

Since for two generations of quarks and leptons the Cabibbo mixing matrix is real, there is no CP violation unless all three generations of quarks participate in the diagram of Fig. 46. It follows therefore that, in the standard model, the parameter ϵ must be proportional to the product of the sines of all the real angles in C, times the sine of the CP violating phase δ :

$$\epsilon \sim \sin\theta_1 \sin\theta_2 \sin\theta_3 \sin\delta \tag{279}$$

where $\theta_1 = \theta_c$ is the Cabibbo angle.

Recent measurements on the lifetime of B mesons have shown /70/ that these states are surprisingly long lived. This can be understood only if the weak transitions $b \rightarrow c$ and $b \rightarrow u$, which depend on the values of θ_2 and θ_3, are very suppressed. In turn, small values of θ_2 and θ_3 force ϵ to be very small. In fact for θ_2 and θ_3 sufficiently small the parameter ϵ, calculated in the standard model, may in fact be smaller than the observed $\epsilon_{exp} \simeq 2.3 \times 10^{-3}$. This circumstance would necessitate introducing physics beyond the GSW model to explain CP violation.

The present situation is a little uncertain. To calculate ϵ one needs besides the product of angles of Eq. (279) also values for the top and charm quark masses. In addition one needs an estimate for the matrix element of $\bar{d}\gamma^{\mu}(1-\gamma_5)s\, \bar{d}\gamma_{\mu}(1-\gamma_5)s$ between K and $\bar{K}$. If $m_t \simeq 40$ GeV, as has been suggested recently by the UA1 collaboration /71/, probably the calculated value of ϵ can be made to agree with ϵ_{exp}, if δ is near 90° and the value of the $(\bar{d}s)^2$ matrix element is near that given by the vacuum insertion approximation /72/. However, the situation needs further clarification and could become worse if the B lifetime is shown to be even longer, or m_t is smaller than 40 GeV.

Besides ϵ, there is a further parameter, ϵ', which measures CP violation in the kaon system. ϵ' essentially measures how much CP violation is there in the $\Delta S = 1$ part of the weak Lagrangian. The ratio ϵ'/ϵ is calculable in the standard model and is considerably less dependent on the Cabibbo mixing angles, since these disappear in the ratio. New data on ϵ'/ϵ has become available this year, from two different experimente /72/ /73/. This ratio is still consistent with zero:

Ref. 72: $$\frac{\epsilon'}{\epsilon} = (-4.6 \pm 5.3 \pm 2.4) \times 10^{-3} \tag{280}$$

Ref. 73: $$\frac{\epsilon'}{\epsilon} = (4.5 \pm 8) \times 10^{-3} \tag{281}$$

The magnitude of this effect predicted by the standard model has a large uncertainty, in part due to the estimation of another hadronic matrix element. However, the predictions /74/ which are in the range $\epsilon'/\epsilon \sim (5-10) \times 10^{-3}$ are beginning to be seriously challenged by experiment. It will be interesting to see how this situation develops in the future.

The last, and perhaps most interesting, experimental anomalies are connected with data obtained at the CERN collider. A variety of unusual phenomena have been reported by the UA1 and UA2 collaborations in the last year, ranging from anomalous radiative decays /75/ to the productions of jets accompanied by large missing energy /76/, or by a lepton and large missing energy /77/. Because one is dealing with a very small sample of events, it is quite possible that all these indications of "new phenomena" may eventually just prove to be statistical fluctuations.

At any rate, this collider exotica has generated an enormous amount of theoretical activity. I shall not try here to review all the possible suggestions of new physics beyond the standard model, which have been adduced to explain this data /78/. Rather, I will just give one example to give a flavor of the possible physics which may be emerging from the collider.

If low energy supersymmetry is a reality, both quarks and gluons should have superpartners (squarks and gluinos). If these states are not too massive they should be abundantly produced at the collider. For instance, gluino pair production $q\bar{q} \rightarrow \tilde{g}\tilde{g}$ would follow from graphs which are analogous to the ones for gluon production. An example of this is illustrated in Fig. 47:

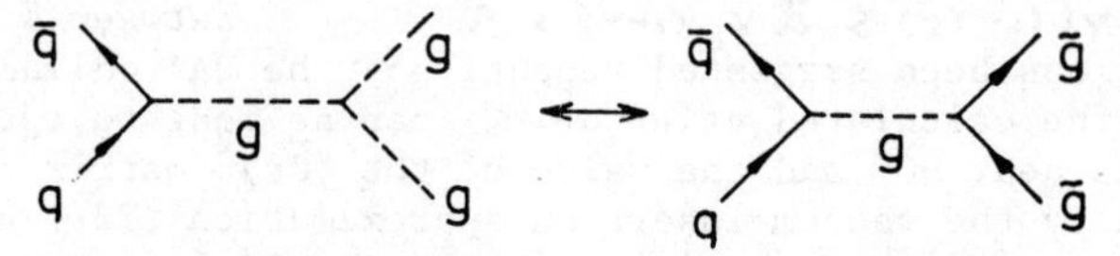

Fig. 47: Correspondence between gluon and gluino production graphs

Both squarks and gluinos can be the source of the jet plus missing energy events. Basically these particles in general must decay to the lightest supersymmetric particle, which in most models is the photino - the supersymmetric partner of the photon. The presumption is that the missing energy "seen" at the collider is due to photino production coming from squark or gluino decay. This scenario has been analyzed in detail by a number of authors /79/. The consensus is that if gluinos or squarks exist in the 30-40 GeV mass range then one would expect about the number of monojet events reported by the UA1 collaboration. Lighter gluinos or squarks would produce too much exotica. Again, more data forthcoming from the collider, should shed light on these speculations.

Let me conclude my lectures on the status of gauge theories of the strong,weak and electromagnetic interactions by emphasizing three main points:

(1) There is impressive evidence supporting certain pieces of the standard model ($\mathcal{L}_{\text{fermion-gauge}}$ and the short distance aspects of QCD).

(2) There are grey or unknown areas in the standard model ($\mathcal{L}_{\text{Higgs}}$ and the long distance aspects of QCD).

(3) Some of the tantalizing hints emerging from experiments now are bound to be reinforced with the advent of the new generation of colliders operating at or near the Fermi scale. Hopefully, new insights into gauge theories will emerge experimentally in the not too distant future, which should

help us clarify the deeper open questions of the standard model.

ACKNOWLEDGEMENTS

I am extremely grateful to Tom Ferbel for giving me the opportunity to lecture in such pleasant surroundings and to all the students of the school for their unbounded enthusiasm. I should also thank Sherwin Love for his assistance with some of the material discussed in these notes.

Exercises

1. Verify the transformation law, Eq. (12), for the covariant derivative $D_\mu \phi(x)$

2. Check that under local U(1) transformations

$$\mathcal{L}_{mass} = -\frac{1}{2} m_A^2 A^\mu(x) A_\mu(x) \not\rightarrow \mathcal{L}_{mass}$$

Is $\mathcal{L}_{mass}$ invariant under global transformations?

3. Check that the adjoint matrices for SU(2) $(g_i)_{jk} = -i\epsilon_{ijk}$ obey the SU(2) algebra $[g_i, g_j] = i\epsilon_{ijk} g_k$

4. Verify Eq. (23) for $\delta(D_\mu q_a(x))$

5. Prove that the field strengths $F_i^{\mu\nu}$ of Eq. (25) transform under infinitesimal transformations as

$$\delta F_j^{\mu\nu}(x) = \delta\omega_i(x) f_{ijk} F_k^{\mu\nu}(x)$$

6. Prove Eq. (32)

7. Show that for the U(1) model the shifted Lagrangian written in terms of the ρ and B_μ fields reads

$$\mathcal{L} = -\frac{1}{4} F^{\mu\nu} F_{\mu\nu} - \frac{1}{2} m_A^2 B^\mu B_\mu - \frac{1}{2} \partial^\mu \rho \partial_\mu \rho - \frac{1}{2} m_\rho^2 \rho^2$$

$$- \lambda(v\rho^3 + \frac{1}{4}\rho^4) - \frac{1}{2} g^2 B^\mu B_\mu (\rho^2 + 2\rho v)$$

where $m_A^2 = (gv)^2$ and $m_\rho^2 = 2\lambda v^2$

8. An exotic triplet of leptons $\ell_L \sim 3$, $\ell_R \sim 1$ are discovered. What are the appropriate covariant derivatives for these states, under SU(2) x U(1), if their charges are +1, 0 and -1

9. Verify Eq. (40)

10. Verify Eq. (47)

11. Show that Eq. (53) arises in second order perturbation theory. Pay

particular attention to factors of 2 and 1/2 and i

12. Derive Eq. (64)

13. Convince yourself that a factor of $\frac{1}{2} M^2$ is necessary for the mass term of a real field, but a factor of M^2 is correct for the mass term of a complex field

14. Check that for $\vec{\xi}/v$ infinitesimal, Eq. (69) reduces to the infinitesimal transformation law for the W_i^μ fields under SU(2) local transformations

15. Derive Eq. (70)

16. Show that the charge conjugate field

$$\tilde{\Phi} = i\tau_2 \Phi^* = \begin{pmatrix} \phi^+ \\ -\phi^{0*} \end{pmatrix}$$

transforms as an SU(2) doublet

17. Show that the basis change analogous to (84)

$$(\psi_f)_L \to U^f_L (\psi_f)_L \quad ; \quad (\psi_f)_R \to U^f_R (\psi_f)_R$$

has no effect on leptonic charged currents, if neutrinos are massless. Show further that such a change cannot affect weak neutral currents

18. Consider the Cabibbo matrix C for the case of two families of quarks and leptons, for the special case in which the up and down quark matrices have the form

$$M^u = \begin{pmatrix} 0 & \alpha \\ \alpha^* & \beta \end{pmatrix} \quad ; \quad M^d = \begin{pmatrix} 0 & \gamma \\ \gamma^* & \delta \end{pmatrix}$$

Can you say something special about the Cabibbo angle's relation to the quark masses in general. What about if $\alpha, \beta, \gamma, \delta$ are real

19. Show that even for the case of many families the coupling of the physical Higgs boson to fermion is diagonal. That is, it is given by Eq. (83)

20. Using the QED result for the β-function, show that in QCD the fermionic contribution to the β-function, for each fermion flavor, is

$$\beta_f^{QCD} = \frac{1}{6\pi} \alpha_s^2$$

21. Prove Eq. (98)

22. The deep inelastic cross section $\frac{d\sigma}{dx\,dy}$ obeys the, so called, energy momentum sum rule

$$\frac{d\sigma}{dx\,dy} = \sum_{H} \int_0^1 dz\, z \left(\frac{d\sigma^H}{dx\,dy} \right)$$

Use this result and the fact that no fragmentation functions enter in $d\sigma/dy\,dx$ to show that the fragmentation functions $D^H_{\text{parton}}(z';q^2)$ obey Eq. (101)

23. Neglecting all masses, but starting from first principles derive Eq. (104)

24. By using the relations

$$\sum_{\text{Spin}} u_\alpha(p)\,\bar{u}_\beta(p) = \sum_{\text{Spin}} v_\alpha(p)\,\bar{v}_\beta(p) = -(\gamma\cdot p)_{\alpha\beta}$$

$$\mathrm{Tr}\,\gamma_\lambda\gamma_\alpha\gamma_\nu\gamma_\beta = 4\,(\eta_{\lambda\alpha}\eta_{\nu\beta} - \eta_{\lambda\nu}\eta_{\alpha\beta} + \eta_{\lambda\beta}\eta_{\nu\alpha})$$

$$\mathrm{Tr}\,\gamma_\lambda\gamma_\alpha\gamma_\nu\gamma_\beta\gamma_5 = 4i\,\epsilon_{\lambda\nu\alpha\beta}$$

derive Eqs. (111)

25. Prove the kinematical relations (113). Note $\epsilon^{0123} = 1$ while $\epsilon_{0123} = -1$

26. Show that the Paschos-Wolfenstein ratio (Phys. Rev. D7 (1973) 91)

$$R_{PW} = \frac{\sigma_{NC}(\nu N) - \sigma_{NC}(\bar{\nu} N)}{\sigma_{NC}(\nu N) + \sigma_{NC}(\bar{\nu} N)}$$

in the GSW model is given by

$$R_{PW} = \rho^2 \left(\tfrac{1}{2} - \sin^2\theta_W\right)$$

27. Prove Eq. (138)

28. Show that the difference

$$\frac{1}{2x}\left[\left(F_2^{cc}(x)\right)_{\bar{\nu}p} - \left(F_2^{cc}(x)\right)_{\nu p} \right] = \left(u(x) - \bar{u}(x)\right) - \left(d(x) - \bar{d}(x)\right)$$

Then show that because protons (neutrons) have charge 1(0), it follows that

$$\int_0^1 \frac{dx}{2x} \left[(F_2^{cc}(x))_{\bar{\nu} p} - (F_2^{cc}(x))_{\nu p} \right] = 1$$

This is the Adler sum rule (Phys. Rev. 143 (1966) 1144)

29. Derive Eq. (144)

30. Derive Eqs. (160)

31. Verify that $d\sigma_{e^+;R}(q)$ can be gotten from $d\sigma_{e^-;L}(q)$ by the substitution of Eq. (168)

32. For $\nu_e e \to \nu_e e$ scattering also charged currents contribute. By making use of the Fierz identity

$$(\bar{\nu}_e \gamma^\tau (1-\gamma_5) e)(\bar{e} \gamma_\tau (1-\gamma_5) \nu_e) = (\bar{\nu}_e \gamma^\tau (1-\gamma_5) \nu_e)(\bar{e} \gamma_\tau (1-\gamma_5) e)$$

show that

$$\left(\frac{d\sigma}{dy}\right)^{\nu_e e} = \frac{2 G_F^2 m E_\nu \rho^2}{\pi} \left[(Q_{eL}^{NC} + \tfrac{1}{\rho})^2 + (Q_{eR}^{NC})^2 (1-y)^2 \right]$$

33. (Not for Greek students) Cross sections of 10^{-33} cm^2, 10^{-36} cm^2 and 10^{-39} cm^2 are in nanobarns, picobarns and fentobarns, respectively. What do 10^{-42} cm^2 and 10^{-45} cm^2 correspond to?

34. Show that because $\mathrm{Tr}\, \gamma_\mu \gamma \cdot p\, \gamma_\nu\, \gamma \cdot \bar{p}$ is symmetric in μ and ν, while $\mathrm{Tr}\, \gamma_\mu \gamma \cdot p\, \gamma_\nu\, \gamma \cdot \bar{p}\, \gamma_5$ is antisymmetric in μ and ν, then the interference term in $e^+e^- \to \ell^+\ell^-$ is proportional to g_{eV}^2 and g_{eA}^2 but not $g_{eV}\, g_{eA}$

35. Derive Eq. (187)

36. Derive Eq. (200)

37. Derive the expression for the Z^0 luminosity function of Eq. (205)

38. Derive Eq. (208) and show that the branching ratio $B(W \to e\nu) \simeq 8\ \%$ follows directly by counting possible decay channels of the W

39. Derive Eq. (235) for the moments of the running parton distribution function

40. Using Eq. (238) show that the solution of Eq. (235) is Eq. (239)

41. Interpret the splitting functions P_{gq} and P_{gg} diagramatically

42. Show that the valence quark sum rule

$$\int_0^1 d\xi \left[f_q(\xi;q^2) - f_{\bar{q}}(\xi;q^2) \right] = \nu_q$$

where ν_q is the number of valence quarks of type q, implies that

$$\int_0^1 d\xi \, P_{qq}(\xi) = 0$$

43. Derive Eq. (251) by using the explicit form of P_{qq} of Eq. (242)

44. Show that for the energy flow $\frac{d\Sigma}{d\Omega} = \frac{d\Sigma}{d\Omega_{parton}}$

45. Show that the kinematical limits for the Thrust variable T are $1/2 \leq T \leq 1$, where T = 1/2 is achieved for totally isotropic events.

REFERENCES

1. For a review, see for example W.J. Marciano and H. Pagels, Phys. Rept. 36C (1978) 137 or the monograph by Quigg; C. Quigg , Gauge Theories of the Strong Weak and Electromagnetic Interactions (Benjamin, Reading Mass, 1983)

2. S.L. Glashow, Nucl. Phys. 22 (1961) 579;
A. Salam, in Elementary Particle Theory, ed. by N. Svartholm (Almquist and Wiksells, Stockholm 1969);
S. Weinberg, Phys. Rev. Lett. 19 (1967) 1264

3. P.W. Higgs, Phys. Rev. Lett. 12 (1964) 132;
F. Englert and R. Brout, Phys. Rev. Lett. 13 (1964) 321;
G.S. Guralnik, C.R. Hagen and T.W. Kibble, Phys. Rev. Lett. 13 (1964) 585

4. S.L. Glashow, J. Iliopoulos and L. Maiani, Phys. Rev. D2 (1970) 1285

5. N. Cabibbo, Phys. Rev. Lett. 10 (1963) 531

6. M. Kobayashi and T. Maskawa, Prog. Theor. Phys. 49 (1973) 652

7. E.A. Uehling, Phys. Rev. 48 (1935) 55

8. H.D. Politzer, Phys. Rev. Lett. 30 (1973) 1346;
D.J. Gross and F. Wilczek, Phys. Rev. Lett. 30 (1973) 1343;
Phys. Rev. D8 (1973) 3633

9. W.E. Caswell, Phys. Rev. Lett. 33 (1974) 244;
D.R.T. Jones, Nucl. Phys. B75 (1974) 531

10. For a general discussion see the monograph of Feynman;
R.P. Feynman, Photon Hadron Interactions (Benjamin, Reading, Mass. 1972)

11. D.H. Perkins, Proceedings of the 1981 CERN-JINR School of Physics, Hanko, Finland, June 1981

12. M. Jonker et al., Phys. Lett. 99B (1931) 265

13. D.C. Cundy, Proceedings of the XVII International Conference on High Energy Physics, London 1974

14. C. Geweniger, Proceedings of the International Europhysics Conference on High Energy Physics, Brighton, July 1983

15. M. Jonker et al., Phys. Lett. 102B (1981) 67

16. J.E. Kim et al., Rev. Mod. Phys. 53 (1980) 211

17. C.H. Llewellyn Smith, Phys. Rept. 3C (1972) 163

18. C. Callan and D.J. Gross, Phys. Rev. Lett. 22 (1969) 156

19. D.J. Gross and C.H. Llewellyn Smith, Mod. Phys. B14 (1969) 337

20. D. Haidt, Proceedings of the XXI International Conference on High Energy Physics, Paris 1982

21. W.A. Bardeen, A.J. Buras, W. Duke and T. Muta, Phys. Rev. D18 (1978) 3998;
G. Altarelli, R.K. Ellis and G. Martinelli, Nucl. Phys. B143 (1978) 521; ibid., B146 (1978) 544 (E)

22. F. Eisele, Proceedings of the XXI International Conference on High Energy Physics, Paris, 1982

23. C. Prescott et al., Phys. Lett. 77B (1978) 347; ibid, 84B (1979) 524

24. A. Argento et al., Phys. Lett. 140B (1984) 142

25. F. Bergsma et al., Phys. Lett. 147B (1984) 481

26. L. A. Ahrens et al., Phys. Rev. Lett. 51 (1983) 1516; 54 (1984) 18

27. B. Naroska, Proceedings of the 1983 International Symposium on Lepton-Photon Interactions, Ithaca, N.Y.

28. H.U. Martyn, Proceedings of the Symposium on High Energy e^+e^- Interactions, Vanderbilt Univ., Nashville, April 1984

29. G.Arnison et al., Phys. Lett. 122B (1983) 103;
M. Banner et al., Phys. Lett. 122B (1983) 496

30. G. Arnison et al., Phys. Lett. 126B (1983) 393;
P. Bagnaia et al., Phys. Lett. 129B (1983) 130

31. F.E. Paige, Brookhaven report BNL-27066 (1979). For a more up to date calculation see G. Altarelli et al., Nucl. Phys. B246 (1984) 12 and CERN preprint CERN-TH-4015

32. G. Arnison et al., Phys. Lett. 129B (1983) 273

33. P. Bagnaia et al., Zeit. für Physik C24 (1984) 1

34. H.D. Wahl, Proceedings of the 15th International Symposium on Multiparticle Dynamics, Lund, June 1984

35. C.H. Llewellyn Smith and J. Wheater, Phys. Lett. 105B (1981) 486

36. A. Sirlin and W. Marciano, Nucl. Phys. B189 (1981) 442

37. For a review see W. Marciano in Proceedings of the Fourth Topical Workshop on Proton Antiproton Collider Physics, Berne, March 1984

38. A. Sirlin, Phys. Rev. D29 (1984) 89

39. G. Altarelli and G. Parisi, Nucl. Phys. B126 (1977) 298;

40. A.H. Mueller, Phys. Rev. D18 (1978) 3705;
R.K. Ellis, H. Georgi, M. Machacek, H.D. Politzer and G.G. Ross, Nucl. Phys. B156 (1979) 285; Yu Dokshitser, D.I. Dyakonov and S.I. Troyan, Phys. Rept. 58C (1970) 270

41. See Dokshitser et al., Ref. 40

42. H. Georgi and H.D. Politzer, Phys. Rev. D9 (1974) 416

43. D.J. Gross and F. Wilczek, Phys. Rev. D9 (1974) 980

44. A.J. Buras, Rev. Mod. Phys. 52 (1980) 199

45. G. Altarelli, Phys. Rept. 81 (1982) 1

46. F. Dydak, Proceedings of the 1983 International Symposium on Lepton-Photon Interactions, Ithaca, N.Y.

47. A.J. Buras, Proceedings of the 1981 International Symposium on Lepton-Photon Interactions, Bonn

48. B. Cox, Proceedings of the XXI International Conference on High Energy Physics, Paris, 1982

49. T. Appelquist and H. Georgi, Phys. Rev. D8 (1973) 4000;
A. Zee, Phys. Rev. D8 (1973) 4038

50. S.L. Wu, DESY 84-020, to appear in Physics Reports

51. C.L. Basham, L.S. Brown, S.D. Ellis and S.T. Love, Phys. Rev. D17 (1978) 2298

52. C.L. Basham, L.S. Brown, S.D. Ellis and S.T. Love, Phys. Rev. D19 (1979) 2018

53. D. Schlatter et al., Phys. Rev. Lett. 59 (1982) 521

54. E. Fahri, Phys. Rev. Lett. 39 (1977) 1587;
A. De Rujula, J. Ellis, E.G. Floratos and M.K. Gaillard, Nucl. Phys. B138 (1978) 387

55. M. Althoff et al., Zeit. für Phys. C22 (1984) 307

56. P. Hoyer et al., Nucl. Phys. B161 (1979) 349

57. For a review, see for example I. Halliday, Proceedings of the International Europhysics Conference on High Energy Physics, Brighton, July 1983

58. For a review, see for example R.D. Peccei, Proceedings of the Fourth Kyoto Summer Institute on Grand Unified Theories and Related Topics, Kyoto, Japan 1981

59. R.D. Peccei and H.R. Quinn, Phys. Rev. Lett. 38 (1977) 1440;
Phys. Rev. D16 (1977) 1795

60. K. Wilson as quoted in L. Susskind, Phys. Rev. D20 (1979) 2619

61. M. Aizenmann, Phys. Rev. Lett. 47 (1981) 1;
J. Fröhlich, Nucl. Phys. B200 FS4 (1984) 281

62. M. Beg, C. Panagiotakopoulos and A. Sirlin, Phys. Rev. Lett. 52 (1984) 883;
D.J. Callaway, Nucl. Phys. B233 (1984) 189

63. L. Susskind, Phys. Rev. D20 (1979);
S. Weinberg, Phys. Rev. D16 (1976) 974; Phys. Rev. D19 (1979) 1277

64. For a recent review, see for example R.D. Peccei, Proceedings of the International Europhysics Conference on High Energy Physics, Brighton, July 1983

65. J.J. Aubert et al., Phys. Lett. 123B (1983) 275

66. For a review, see C.H. Llewellyn Smith in the Proceedings of the Workshop on Experimentation at HERA, Amsterdam, June 1983

67. R. Jaffe, Phys. Rev. Lett. 50 (1983) 228;
F.E. Close, R.G. Roberts and G.G. Ross, Phys. Lett. 129B (1983) 346

68. F.E. Close, R. Jaffe, R.G. Roberts and G.G. Ross, Phys. Lett. 134B (1984) 449

69. C.T. Hill, Phys. Lett. 97B (1980) 275;
J.S. Hagelin, Nucl. Phys. B193 (1981) 123

70. N.S. Lockeyer et al., Phys. Rev. Lett. 51 (1983) 1316;
E. Fernandez et al., Phys. Rev. Lett. 51 (1983) 1022

71. G. Arnison et al., Phys. Lett. 147B (1984) 493

72. B. Winstein, Proceedings of the XI International Conference on Neutrino Physics and Astrophysics, Nordkirchen, Germany, June 1984

73. R.K. Adair, Proceedings of the XXII International Conference of High Energy Physics, Leipzig 1984

74. For a review, see A.J. Buras, Proceedings of the Workshop on the Future of Medium Energy Physics in Europe, Freiburg, April 1984

75. G. Arnison et al., Phys. Lett. 126B (1983) 398; ibid. 135B (1984) 250;
P. Bagnaia et al., Phys. Lett. 129B (1983) 130

76. G. Arnison et al., Phys. Lett. 139B (1984) 115

77. P. Bagnaia et al., Phys. Lett. 139B (1984) 105

78. R.D. Peccei, Proceedings of the VII European Symposium of Antiproton Interactions, Durham, U.K., July 1984

79. J. Ellis and H. Kowalski, Phys. Lett. 142B (1984) 441 and DESY 84-045;
V. Barger, K. Hagiwara, J. Woodside and W.Y. Keung, Phys. Rev. Lett. 53 (1984) 641;
E. Reya and D.P. Roy, Phys. Rev. Lett. 53 (1984) 881

UNIFICATION AND SUPERSYMMETRY

John Iliopoulos

Laboratoire de Physique Theorique de l'Ecole Normale Superieure
24 rue Lhomond, 75231 Paris cedex 05, France

ABSTRACT

A short review of the various attempts to unify all forces of nature is presented. The concept of Grand Unification is introduced and its dynamical effects are analyzed. A particular emphasis is put on supersymmetry and its phenomenological consequences.

I. THE STANDARD MODEL AND ITS SHORTCOMINGS

The strong, electromagnetic and weak interactions are described by a gauge theory based on the group $U(1) \times SU(2) \times SU(3)$ which is spontaneously broken into $U(1)_{e.m.} \times SU(3)$. The fundamental Lagrangian contains the three following types of fields:

(i) Gauge fields: We have eight gluons associated with the strong interactions and four electroweak vector bosons. After spontaneous breaking, the gluons and the photon remain massless while the three remaining ones ($W^{\pm}$, Z^{o}) acquire a mass.

(ii) Fermion matter fields: The basic unit is a "family" consisting of fifteen two-component spinor fields which, under $SU(2) \times SU(3)$, form the representation

$$(2, 1) + (2, 3) + (1, 1) + (1, 3)$$

The prototype is the electron family

$$\binom{\nu_e}{e^-}_L \quad , \quad \binom{u_i}{d_i}_L \qquad\qquad i = \text{blue, white, red}$$

$$e_R \quad , \quad u_{i_R} \quad , \quad d_{i_R}$$

with the quarks transforming as triplets of SU(3).

For the muon family, $\nu_e \to \nu_\mu$, $e \to \mu$, $u_i \to c_i$ and $d_i \to s_i$ and, similarly, for the tau family, $\nu_e \to \nu_\tau$, $e \to \tau$, $u_i \to t_i$, $d_i \to b_i$. A remarkable property is that the sum of the electric charges in each family vanishes. This turns out to be necessary for the cancellation of triangle anomalies in the Ward identities of axial currents, and, hence, for the construction of a renormalizable theory.

This family structure, which according to the latest news has been once more brilliantly verified with the discovery of top, is an as yet unexplained feature of the theory. Furthermore, the total number of families is not restricted. In fact, we know of no good reason why any, beyond the first one, should exist.

(iii) Higgs scalar fields: We would be very happy if we could live with only the first two kinds of fields, but in fact we need a third one, the scalar Higgs fields. Their non-zero vacuum expectation values break the gauge symmetry spontaneously, thus providing masses to $W^\pm$, Z^o as well as the fermions. In the standard model this is accomplished with a complex doublet of scalar fields. At the end, one neutral spin-zero boson survives as a physical particle. There are no severe restrictions on its mass. In the absence of any concrete experimental evidence, one is left to speculate on the number of physical Higgs particles as well as on their elementary or composite nature. Whichever the ultimate answer to these questions may be, we can say that the Higgs sector is at present the least understood and perhaps the most interesting sector of gauge theories.

The above uncertainties notwithstanding, our confidence in this model is amply justified on the basis of its ability to accurately describe the bulk of our present day data and, especially, of its enormous success in predicting new phenomena. Nevertheless, there are several reasons to suspect that the gauge theory based on U(1) × SU(2) × SU(3) cannot be considered as the final theory. First, it is not a unified theory at all. We know of three fundamental interactions and we have three independent coupling strengths. Each group factor introduces its own. Second, and even worse, is the presence of U(1), because an abelian gauge symmetry allows for arbitrary coupling constants. Indeed, we can easily verify that we can write a consistent theory of quantum electrodynamics with two charged fields, one with charge e and the other with charge πe. However, if we repeat the exercise with a non-abelian gauge theory, say one based on SU(2), we can show that this arbitrariness disappears; all matter fields couple to the gauge bosons with the same coupling strength, up to a possible Clebsch-Gordon coefficient depending on the representation. It follows that the standard model, because of its U(1) factor, cannot explain the observed electric charge quantization. In such a model, the observed very precise equality (up to one part in 10^{20}) of the electric charges of the positron and the proton is accidental.

II. GRAND UNIFICATION

For all these reasons, and others that I skip, several theorists tried to go beyond the standard model. The hypothesis of grand unification states that U(1) × SU(2) × SU(3) is the remnant of a larger, simple or semi-simple group G, which is spontaneously broken at very high energies. The scheme looks like:

$$G \xrightarrow[M]{\;\cdots\cdots\;} U(1) \times SU(2) \times SU(3) \xrightarrow{m_W \sim 10^2 \text{ GeV}} U(1)_{e.m.} \times SU(3) \qquad (1)$$

where the breaking of G may be a multistage and one, and M is one (or several) characteristic mass scale(s). Two questions immediately arise concerning this idea:

(a) Is it possible? In other words are there groups which contain U(1) × SU(2) × SU(3) as a subgroup and which can accommodate the observed particles?

(b) Does it work? That is, is the observed dynamics compatible with this grand-unification idea?

We shall try to answer each of these questions separately.

II.A - Candidates for G.U.T.'s

In this section we shall answer the first question by giving some explicit examples of groups G which satisfy our requirements. We first observe that G must contain electromagnetism, i.e., the photon must be one of the gauge bosons of G. This is contained in the requirement that G contains the group of the standard model U(1) × SU(2) × SU(3). Another way to say the same thing, is to say that the electric charge operator Q must be one of the generators of the algebra of G. Since G is semi-simple all its generators are represented by traceless matrices. It follows that, in any irreducible representation of G, we must have:

$$\mathrm{Tr}\,(Q) = 0 \qquad (2)$$

in other words, the sum of the electric charges of all particles in a given irreducible representation vanishes.

In order to proceed we shall make an important assumption: The fifteen two-component spinors of a family fill a representation of G, i.e., we assume that there are no other, as yet unobserved, particles which sit in the same representation. Property (2), together with the above assumption, have a very important consequence: as we have remarked, the fifteen members of a family satisfy (2) because the sum of their charges vanishes. This, however, is not true if we consider leptons or quarks separately. Therefore each irreducible representation of G will contain both leptons and quarks. This means that there exist gauge bosons of G which can change a lepton into a quark, or vice versa. We conclude that a grand unified theory that satisfies our assumption cannot conserve baryon and lepton numbers separately. This sounds disastrous, since it raises the spectrum of proton decay. The amplitude for such a decay is given by the exchange of the corresponding gauge boson and therefore, it is of order M^{-2}, where M is the gauge boson's mass. The resulting proton life-time τ_p will be of order

$$\tau_p \sim \frac{M^4}{m_p^5} \tag{3}$$

Using the experimental limit of 10^{30} - 10^{32} years we can put a lower limit on M:

$$M \geq 10^{14} - 10^{15} \text{ GeV} \tag{4}$$

Grand unification is not a low-energy phenomenon!

After these general remarks, let us try to find some examples.

II.A.1. The Simplest G.U.T. - SU(5)

$U(1) \times SU(2) \times SU(3)$ is of rank 4 (i.e. there are four generators which commute: one of U(1), one of SU(2) and two of SU(3)). Therefore, let us first look for a grand unification group of rank 4. I list all possible candidates:

$$[SU(2)]^4 \;,\; [SO(5)]^2$$

$$[G_2]^2 \;,\; SO(8) \;,\; SO(9) \;,\; Sp(8) \;,\; F_4$$

$$[SU(3)]^2 \;,\; SU(5)$$

The first two are excluded because they have no SU(3) subgroup. The next five admit no complex representations, therefore, they cannot accommodate the observed families where, as we already saw, the right- and left-handed particles do not transform the same way. (I again assume that no unobserved fermions will complete a given representation.) Finally, in $SU(3) \times SU(3)$, quarks and leptons must live in separate representations because the leptons have no color. But $\Sigma Q_{\text{quarks}} \neq 0$, and the same is true for leptons. This leaves us with SU(5) as the only candidate of a G.U.T. group of rank 4. It is the simplest and, in some sense, the standard model of grand unification.

The gauge bosons belong to the 24-dimensional adjoint representation. It is useful to decompose it into its $SU(2) \times SU(3)$ content. We find:

$$24 = \underbrace{(2,\, 3) \;+\; (2,\, \bar{3})}_{\begin{pmatrix} X \\ Y \end{pmatrix} \begin{matrix} Q = 4/3 \\ Q = 1/3 \end{matrix}} \;+\; \underbrace{(1,\, 8)}_{\text{gluons}} \;+\; \underbrace{(3,\, 1) \;+\; (1,\, 1)}_{W^{\pm} ,\; Z^{\circ} ,\; \gamma} \tag{5}$$

where the first number denotes the SU(2) and the second the SU(3) representation. The eight gluons of Q.C.D. can be identified with the (1, 8) piece (a singlet of SU(2) and an octet of SU(3)), while the electroweak gauge bosons $W^{\pm}$, Z° and γ can be associated with the (3, 1) + (1, 1) piece. We are left with twelve new vector bosons, called X and Y, with electric charges 4/3 and 1/3 respectively, which transform as a doublet of SU(2) and a triplet and anti-triplet of SU(3). They must be heavy, according to the limit (4).

Let us now come to the matter-field assignment. We shall try to put the fifteen two-component spinors of a family in a representation (not necessarily irreducible) of SU(5). But before doing so, we observe that all gauge couplings, being vectorial, conserve helicity. Therefore, we cannot put right- and left-handed spinors in the same representation. We go around this problem by replacing all right-handed spinors by the corresponding left-handed charge conjugate ones. A quick glance at the representation table of SU(5) suggests to use each family in order to fill two distinct representations: the $\overline{5}$ and the 10. Their SU(2) × SU(3) content is:

$$\overline{5} = (1, \overline{3}) + (2, 1) \qquad (6)$$

$$\begin{array}{cc} \downarrow & \downarrow \\ d^c_{i_L} & \begin{pmatrix} \nu \\ e \end{pmatrix}_L \end{array} \qquad \Sigma Q = 0.$$

It contains a singlet of SU(2) and an anti-triplet of SU(3), i.e., three anti-quarks, or, equivalently, a triplet of right-handed quarks, as well as a doublet of SU(2) and a singlet of SU(3), i.e., a doublet of left-handed leptons. Since the sum of the electric charges must vanish, we see that the anti-triplet must contain the charge conjugate of the d-quarks. For the 10-dimensional representation we have:

$$10 = (2, 3) + (1, \overline{3}) + (1, 1) \qquad (7)$$

$$\begin{array}{ccc} \downarrow & \downarrow & \downarrow \\ \begin{pmatrix} u_{i_L} \\ d_{i_L} \end{pmatrix} & u^c_i & e^c_L \end{array} \qquad \Sigma Q = 0$$

and the identification is given by the same reasoning. We often write the representation as:

$$\overline{5} = \begin{pmatrix} d^c_1 \\ d^c_2 \\ d^c_3 \\ \nu \\ e- \end{pmatrix}_L \qquad 10 = \begin{pmatrix} 0 & u^c_3 & -u^c_2 & -u_1 & -d_1 \\ & 0 & u^c_1 & -u_2 & -d_2 \\ & & 0 & -u_3 & -d_3 \\ & & & 0 & -e^c \\ & & & & 0 \end{pmatrix}_L \qquad (8)$$

with the matrix of the 10 being antisymmetric. A technical remark: it is important to notice that the sum of these two representations is anomaly-free. A second physical remark is that, unless one introduces an SU(5) singlet, there is no room for a right-handed neutrino.

Let us finally study the Higgs system. The first symmetry breaking goes through a 24-plet of scalars. It is convenient to represent the 24 as a 5 × 5 traceless matrix. The vacuum expectation value which breaks SU(5) down to U(1) × SU(2) × SU(3) is proportional to the diagonal matrix:

$$\lambda_{24} = \begin{pmatrix} 1 & & & & \\ & 1 & & \bigcirc & \\ & & 1 & & \\ & \bigcirc & & -3/2 & \\ & & & & -3/2 \end{pmatrix} \qquad (9)$$

Can we use the same 24-plet of Higgs in order to obtain the second breaking of the standard model? The answer is no for two reasons: First, the 24 does not contain any (2, 1) piece (see eq. (5)) which is the one needed for the $U(1) \times SU(2) \rightarrow U(1)_{e.m.}$ breaking. Second, the 24 does not have the required Yukawa couplings to the fermions. Indeed, with the $\overline{5}$ and 10 assignment, the fermions can acquire masses through Yukawa couplings with scalars belonging to one of the representations in the products:

$$\overline{5} \times 10 = 5 + 45 \qquad (10)$$

$$10 \times 10 = \overline{5} + \overline{45} + 50 \qquad (11)$$

We see that the 24 is inoperative while the 5 looks promising. If we restrict ourselves to this simplest choice, we have two independent Yukawa couplings. Looking back at the assignment (8), we see that the up-quarks take their masses through (11), while the down-quarks and the leptons through (10).

This discussion answers the first question, namely it shows that there exist groups which have the required representations to be used as groups of grand unification. Before turning to the next question and extracting the dynamical consequences of such a scheme, let us give a second example of a possible G.U.T., which presents some different and interesting features.

II.A.2 A Rank-5 G.U.T. - SO(10)

Following the same method as before, we list all possible rank-5 candidates together with the reasons that exlude them:

$[SU(2)]^5$: no SU(3) subgroup.

SO(11) , Sp(10) : no complex representations, no 15- or 16-dimensional representations.

SU(6) : It has a 15-dimensional representation but the decomposition in $SU(2) \times SU(3)$ representations shows that it cannot accommodate the members of a family. One finds:

$$15 = (2,3) + (1,\overline{3}) + (1,3) + (2,1) + (1,1) \qquad (12)$$

The troublesome piece is the (1,3), which is a singlet of SU(2) and a triplet of color, rather than an anti-triplet.

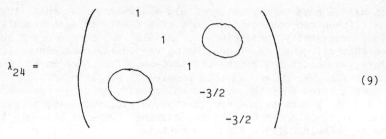

The final candidate for a rank-5 G.U.T. is SO(10), which has a 16-dimensional representation. SO(10) contains SU(5) as a subgroup and the 16-plet decomposes into:

$$16 = 10 + \overline{5} + 1 \tag{13}$$

i.e., under SU(5), we find the $\overline{5}$ and 10 we used before, and a singlet. The obvious interpretation of this last one is a right-handed neutrino (or ν_L^c).

In the simple SU(5) scheme, parity violation, which is observed at present energies, is a fundamental law of nature. Present experimental evidence notwithstanding, it is attractive to speculate that this violation is a low-energy accident and that the underlying theory is ambidextrous. This leads us to extending SU(2) of the standard model to $SU(2)_L \times SU(2)_R$. For phenomenological purposes, it is sufficient that the gauge bosons of $SU(2)_R$ are a few times heavier than $W^{\pm}$. The simplest grandparent of this model is, precisely, SO(10).

The salient features of this G.U.T. are the following: it has 45 gauge bosons which, under the SU(5) subgroup, transform as:

$$45 = 24 + 10 + \overline{10} + 1 \tag{14}$$

As mentioned above, the fermions of each family, together with a corresponding right-handed neutrino, form a 16-dimensional representation. In the long journey from SO(10) down to U(1) × SU(2) × SU(3), nature may chose various paths. She can take the direct road (just one big break), or she may decide to go through one of the intermediate subgroups:

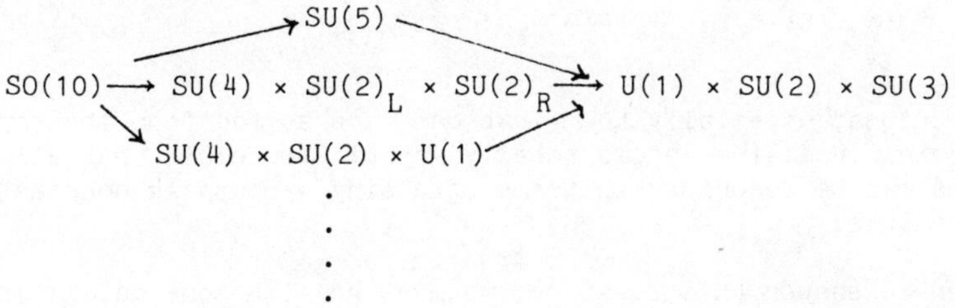

The Higgs system depends on the breaking pattern we choose, but, in any case, it is more complex than that of SU(5). Several representations are necessary.

II.B - Compatibility with Experiment

II.B.1. Dynamics of G.U.T.'s: The Coupling Constants

Let us now turn to the second question, namely, are the dynamical consequences of grand-unified theories compatible with experiment? This is not a trivial question, and, at first sight, the answer seems to be negative. A grand-unified theory is based on a single simple group (or a direct product of simple factors with discreet symmetries which interchange them) and hence it has only one coupling constant. On the other hand, in nature we observe three distinct coupling constants, g_1, g_2, g_3, corresponding to the groups of the standard model U(1) × SU(2) × SU(3). They are often parametrized as g_3 (strong interactions), g_2 (weak interactions) and $\sin\theta_W$, given by

$$\sin^2\theta_W = \frac{g_1^2}{g_1^2 + g_2^2} \tag{15}$$

where the U(1) generator Y is related to the electric charge Q and to the third component of weak isospin T_3 by:

$$Q = T_3 - Y \tag{16}$$

For the embedding of U(1) × SU(2) × SU(3) in the G.U.T. group G, all generators must be normalized the same way. Let us put

$$\mathrm{Tr}(J_i J_j) = R\delta_{ij} \tag{17}$$

where R is a constant that may depend on the representation we use to compute the trace, but it is independent of i and j. Let us now compute $\mathrm{Tr}(T_3^2)$ using, for example, the electron family. (We assume once more that the observed members of a family completely fill a representation, possibly reducible, of G.) We find $\mathrm{Tr}(T_3^2) = 2$. Similarly we find $\mathrm{Tr}(Y^2) = 10/3$. Therefore, we see that, for the embedding, the U(1) generator must be rescaled by $Y \to cY$ with $c^2 = 5/3$. Therefore, (15) gives

$$\sin^2\theta = \frac{g^2/c^2}{g^2/c^2 + g^2} = \frac{3}{8} \tag{18}$$

We conclude that the naive G.U.T. "prediction" is

$$g_3 = g_2 \quad ; \quad \sin^2\theta_W = \frac{3}{8} \tag{19}$$

These predictions, especially the first one, are so far from the truth that the whole G.U.T. idea seems totally wrong! The claim that all interactions can be described in terms of a single coupling constant sounds incredible.

In order to answer this question, we must go with some detail into the renormalization program of a spontaneously broken theory. (The reader who is not interested in technicalities can skip this part and go directly to the conclusions.) For pedagogical purposes let me explain the case of G = SU(5). The changes for any other group are straightforward. Furthermore, let me ignore the second breaking of $U(1) \times SU(2) \to U(1)_{e.m.}$. This means that we shall have all fermions massless and we shall include only the 24-plet of Higgs scalars. We start from the Lagrangian:

$$\mathcal{L} = -\frac{1}{4}\mathrm{Tr}\, G_{\mu\nu}G^{\mu\nu} + \bar{\psi} i \not{D} \psi + \frac{1}{2}\mathrm{Tr}\,(D_\mu \Phi D^\mu \Phi)$$

$$- \frac{1}{2} m^2 \mathrm{Tr}(\Phi^2) - \frac{h_1}{4}\left[\mathrm{Tr}(\Phi^2)\right]^2 - \frac{h_2}{2}\mathrm{Tr}\,(\Phi^4) \tag{20}$$

where $G_{\mu\nu}$ is the usual Yang-Mills field strength of SU(5), which, in terms of the 24 gauge bosons G_μ, is given by:

$$G_{\mu\nu} = \partial_\mu G_\nu - \partial_\nu G_\mu - ig\,[G_\mu, G_\nu] \tag{21}$$

ψ is a column with all fermion fields and Φ is the 24-plet of Higgs fields. $G_{\mu\nu}$, G_μ and Φ are written as 5 × 5 traceless hermitian matrices. $\not{D}$ is the SU(5) covariant derivative for the fermions and Higgs fields. The Lagrangian (20) contains four independent parameters, namely m^2, h_1, h_2 and g. In order to obtain a spontaneous symmetry breaking we choose $m^2 < 0$ and we write

$$\Phi \to \Phi + \frac{V}{\sqrt{2}}\,\lambda_{24} \tag{22}$$

with λ_{24} given by eq. (9). We then find the following mass spectrum:

$$\text{(i)}\quad M^2 = \frac{5}{3}\,V^2 g^2 \text{ for the vector gauge bosons } G_9,..,G_{20} \tag{23a}$$

$$\text{(ii)}\quad M_3^2 = \frac{1}{3}\,V^2 h_2 \text{ for the scalars } \Phi_1,..,\Phi_8 \tag{23b}$$

$$\text{(iii)}\quad M_2^2 = \frac{4}{3}\,V^2 h_2 \text{ for the scalars } \Phi_{21},\Phi_{22},\Phi_{23} \tag{23c}$$

$$\text{(iv)}\quad M_1^2 = 2V^2(h_1 + \frac{7}{15}\,h_2) \text{ for } \Phi_{24} \tag{23d}$$

In terms of the parameters of (20), V^2 is given, in the tree approximation, by:

$$V^2 = -m^2\,[h_1 + \frac{7}{15}\,h_2]^{-1} \tag{24}$$

Notice that, as expected, the gauge bosons $G_1,..,G_8$ (gluons) as well as $G_{21},....,G_{24}$ (electroweak gauge bosons) remain massless, while the scalars $\Phi_9,..,\Phi_{20}$ disappear via the Higgs mechanism. The next step is to renormalize the theory. This means that we have to assign to some quantities prescribed values taken from experiment. In our example of eq. (20) we have four independent parameters, and a convenient choice is to use three masses M^2, M_1^2 and M_2^2, and the gauge coupling constant g. In addition, we need the usual wave-function renormalization conditions for the fermion and boson fields of the theory. The mass renormalizations are done by specifying M^2, M_1^2 and M_2^2 as the poles of the corresponding propagators. The coupling constant g can be chosen to be the value of the fermion-gauge boson three point function at a certain scale of the external momenta $p_i^2 = -\mu^2$.

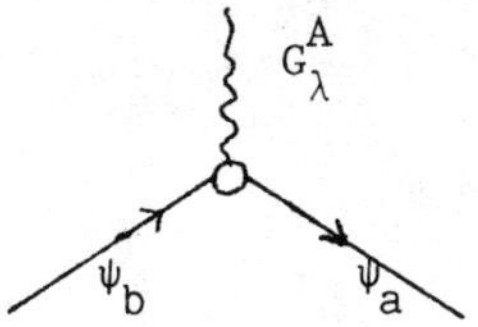

$$\Big|_{p_i^2 = -\mu^2} = i g_A(\mu^2)\,\gamma_\lambda\,(T^A)_{ab} \tag{25}$$

However, since SU(5) is broken, we must specify which particular gauge boson we are using in the definition (25). This is done with the index A, which runs from 1 to 24. The conservation of U(1) × SU(2) × SU(3) implies that we only need to distinguish four distinct cases: A=1,...,8; A=9,...,20; A=21, 22, 23; A=24. Once the condition is imposed using any of these, all the others become finite and calculable. The same remark applies to the wave function renormalization conditions for which again one must specify the component of the vector and scalar fields one is using.

I want to emphasize that, once these three kinds of renormalization conditions (wave functions, masses and coupling constant) have been imposed, the perturbation theory is completely defined and all Green's functions are finite and calculable as formal power series in g_A. Furthermore, it is not possible to impose any further conditions. Notice, in particular, that there exists only one gauge coupling constant, as one should expect from a grand unified theory. Does this mean that we could compute a Q.C.D. process, such as a deep inelastic structure function, as a power series in the weak interaction coupling constant? Formally the answer is yes, but in practice this is not so. Formal perturbation theory guarantees that, if we choose, in (25), A to denote one of the SU(2) gauge bosons, all other three point functions are finite and calculable. In particular, if $\Gamma_3^{(3)}$ is the three-point function with an SU(3) external gauge boson, we can write: (I shall use the following notation: For A = 1,..,8, $g_A = g_3$; for A = 21,...,23, $g_A = g_2$; $g_{A=24} = g_1$):

$$\Gamma_3^{(3)}\left(p_i^2 = -\mu^2\right) \equiv g_3 = g_2 + R_1 g_2^3 + R_2 g_2^5 + \ldots \qquad (26)$$

where the R_n's are finite and calculable functions of μ^2 and of the masses of the theory. However, it is easy to check that R_n is of the form:

$$R_n \sim \left(\ln \frac{M^2}{\mu^2}\right)^n \qquad (27)$$

which, for $M/\mu \sim 10^{14}$ gives $[65]^n$. In other words, although the series (26) is well defined, it is useless for practical computations.

The remedy to this difficulty is simple. We shall renormalize the same broken SU(5) theory in three different ways, where the index A in the condition (25) denotes the bosons of U(1), SU(2) or SU(3). This gives us three perturbation expansions in powers of g_1, g_2, or g_3, always of the same theory, but now each one is suited to particular processes. The values of the g_i's will be fixed by experiment, but we must always remember that we are talking about one and the same theory, so we write the analog of eq. (26):

$$g_i = F_{ij}\left(g_j, \frac{M^2}{\mu^2}, \alpha\right) \qquad (28)$$

where by α we mean the ratios M_1^2/M^2 and M_2^2/M^2. In the limit of exact SU(5) symmetry, i.e. when $M^2/\mu^2 \to 0$, all coupling constants must be equal. This only happens at infinite energy

$$F_{ij}\ (g_j,\ 0,\ \alpha) = g_j \tag{29}$$

By taking $\mu^2 d/d\mu^2$ on both sides of (28) we obtain

$$\beta_i(F_{ij},\ \lambda,\ \alpha) = \left[-\lambda\frac{\partial}{\partial\lambda} + \beta_j(g_j,\ \lambda,\ \alpha)\frac{\partial}{\partial g_j}\ \right] F_{ij} \tag{30}$$

where

$$\beta_k(g_k,\ \lambda,\ \alpha) = \mu^2\frac{d}{d\mu^2}\ g_k \tag{31}$$

and $\lambda = M^2/\mu^2$.

The differential equation (30) with the boundary condition (29) is our basic equation. The β-functions are calculable at any given order of perturbation theory and they are of the form:

$$\beta_k(g_k,\ \lambda,\ \alpha) = b_k^o(\lambda,\alpha)g_k^3 \ + \ldots. \tag{32}$$

Notice that the b coefficients, unlike those of F_{ij} itself, do not contain large logarithms. This can be easily understood since the β-functions, as defined by eqs. (31) and (25), possess well-defined limits both for $M^2/\mu^2 \to \infty$ (when they become the β-functions of SU(3), SU(2) or U(1)) and $M^2/\mu^2 \to 0$ (when they all become equal to that of SU(5)). This is a consequence of the decoupling theorem. On the contrary, F_{ij} has no limit when $M^2/\mu^2 \to \infty$ with g_j kept fixed.

Equation (30) can be solved by the standard methods of characteristics. The solution expresses any coupling constant in terms of any other:

$$g_i = F_{ij} = \eta(g_j,\ \lambda,\ \alpha) \tag{33}$$

with η a given function. For example, using the one-loop β-functions of eq. (32) we find:

$$\frac{1}{g_i^2} = \frac{1}{g_j^2} + 2\int_0^\lambda \frac{dx}{x}\ [b_i^o(x,\alpha) - b_j^o(x,\alpha)] \tag{34}$$

We now must use, as input, the experimentally measured effective strengths of strong, electromagnetic and weak interactions at moderate (say $p^2 \sim 10\text{-}100\ GeV^2$) energies. Using the renormalization group for $\mu^2 = -p^2$, we write $g_i = \bar{g}_i(p^2)$ and $\lambda = -\mu^2/p^2$. The two independent equations given by the relations (33) (i, j = 1, 2, 3) contain three unknown parameters, namely λ and the Higgs masses M_1^2/M^2 and M_2^2/M^2, denoted by α. However, it turns out that the dependence on α is very weak. If we ignore it for the moment, then the two equations can be used to determine λ and to predict the value of $\sin\theta_W$. A precise calculation must take into account the breaking of U(1) × SU(2) as well. In fact, it turns out that the value of $\sin\theta_W$ is quite sensitive to this last breaking. The result, including the two-loop effects, is:

$$M = 3.1 \times 10^{14 \pm 0.3 - 0.2(n_H - 1) + 0.1(F-3)} \left(\frac{\Lambda_{\overline{MS}}}{0.2}\right)^{1.03} \text{ GeV} \tag{35a}$$

$$\tau_p = 10^{29 \pm 2.1 - 0.8(n_H - 1) + 0.4(F-3)} \left(\frac{\Lambda_{\overline{MS}}}{0.2}\right)^{4.1} \text{ years} \tag{35b}$$

$$(\sin^2\theta_W)^{th} \; (\mu^2 = 20 \text{ GeV}^2) = 0.2163 \pm 0.0071 + 0.004(n_H - 1) - 0.0014(F-3) - 0.0056 \ln \left(\frac{\Lambda_{\overline{MS}}}{0.2}\right) \tag{35c}$$

where all uncertainties have been included. By far the largest one is due to the uncertainty in the Q.C.D. coupling constant, or equivalently, the parameter Λ. This has been included in the form of the ratio of its value in the $\overline{MS}$ scheme and 200 MeV. n_H and F are the numbers of the Higgs doublets and fermion families respectively. The remaining uncertainty reflects our ignorance of the heavy Higgs masses (the parameters α in (28)), the precise value of the t-quark mass, etc. τ_p is the predicted proton life-time in years, and an estimation of the uncertainty due to the proton structure is included. We see that, unless the number of families is large, the model is already in trouble with the latest data. The calculated value of $\sin^2\theta_W$ (35c) should be compared with the experimental average:

$$(\sin^2\theta_W)^{exp} \; (\mu^2 = 20 \text{ GeV}^2) = 0.216 \pm 0.01 \tag{36}$$

A new series of precision measurements is already in progress in order to reduce the experimental error on this very fundamental parameter. But we can already say that the agreement between theory and experiment is spectacular.

In Fig. 1a we show the variation of the effective coupling constants, defined by (25), with respect to the scale μ^2. Figure 1b gives, at a larger scale, the region around M. We see that for $\mu \gtrsim 10M$ all the three coupling constants essentially coincide and follow the variation given by the β-function of SU(5). Also, for $\mu \lesssim 10^{-1}M$, the three coupling constants are approximately decoupled and each one varies according to the renormalization group equations of U(1), SU(2) or SU(3). This is the justification of the "step function approximation" which consists in putting, for each scale μ, all the masses larger than μ equal to infinity, and all the masses smaller than μ equal to zero. In fact, Figs. 1a and 1b give a nice illustration of the decoupling theorem: Let us consider a renormalizable theory given by a Lagrangian density $\mathcal{L}(\partial_\mu \phi_i, \phi_i)$ in terms of a set of fields $\phi_i(x) = 1,\ldots,n$. We are interested in the limit in which some of the fields, let us say $\phi_j(x)$ for $j = k + 1,\ldots,n$, become very massive $m_j \to \infty$. The decoupling theorem states that, in this limit, the resulting theory is given by a Lagrangian density $\mathcal{L}'(\partial_\mu \phi_i, \phi_i)$, $i = 1,\ldots,k$, which is obtained from the initial $\mathcal{L}$ by setting all the "superheavy" fields ϕ_j, with $j = k + 1,\ldots,n$, equal to zero, provided the resulting $\mathcal{L}'$ still defines a renormalizable theory. Figures 2a and 2b give the corresponding variation of $\sin^2\theta_W$. The nucleon decay branching ratios can be computed using any specific constituent model. The results are roughly model independent, and the two-body $\pi^0 e^+$ decay mode is predicted to be dominant.

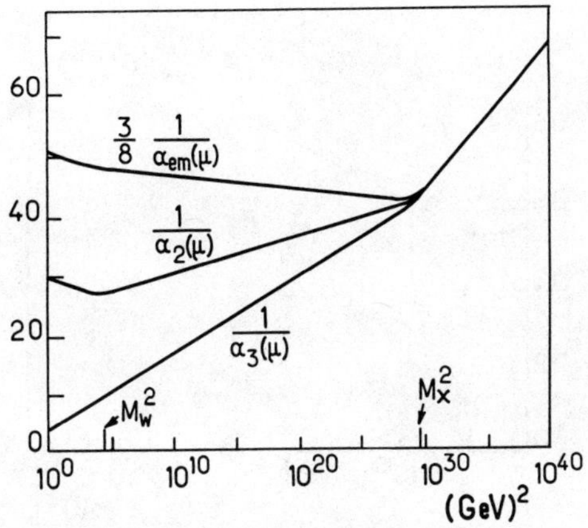

Fig. 1a - The variation of the effective coupling constants with the energy scale.

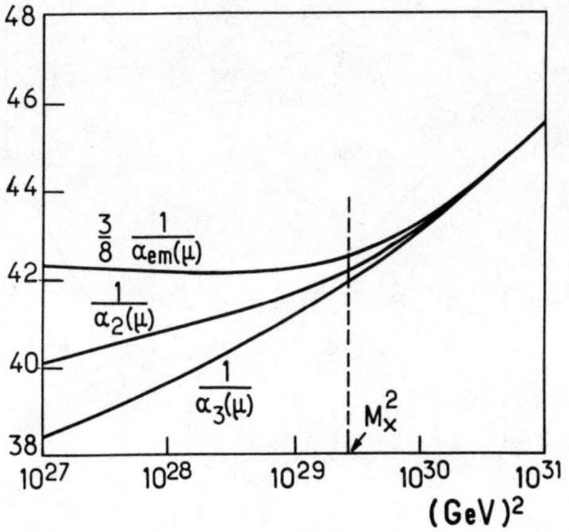

Fig. 1b - Expanded view of the approach to grand unification.

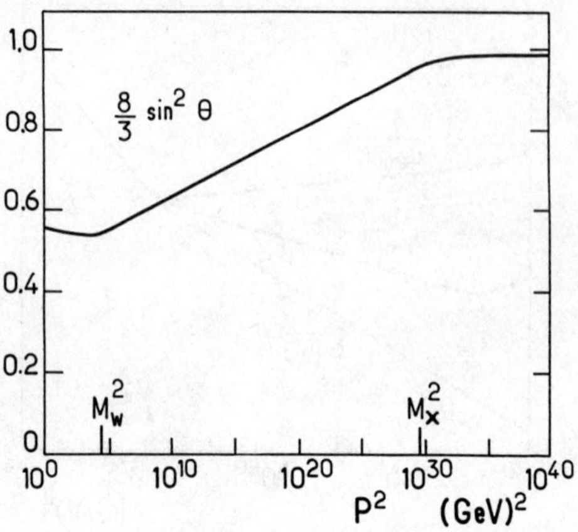

Fig. 2a - The variation of $\sin^2\theta_W$ with the energy scale.

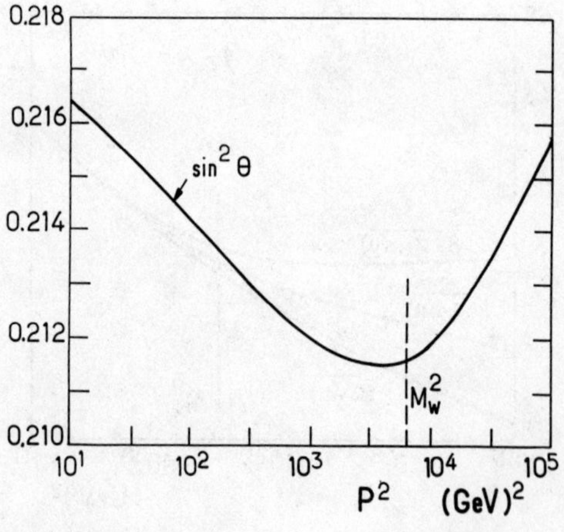

Fig. 2b - Expanded view near M_W^2

This analysis applies to other G.U.T. models as well. For our SO(10) example, the results depend on the symmetry breaking pattern. If we choose a double stage one, we must introduce two mass scales, and we lose the prediction for $\sin^2\theta_W$ which now must be obtained from experiment in order to determine the value of the additional parameter of the model. On the contrary, the nucleon life-time and decay modes are not changed substantially.

II.B.2 Fermion Masses

Fermion masses are generated in G.U.T.'s through the same mechanism as in the standard model, i.e., through Yukawa couplings with Higgs scalars. Therefore, the detailed spectrum is model dependent. In the minimal SU(5) model with a 5-plet of Higgs scalars, we saw, in eqs. (10) and (11), that we have two independent coupling constants for each family. We thus obtain the relations

$$m_d = m_e \; ; \quad m_s = m_\mu \; ; \quad m_b = m_\tau \tag{37}$$

These relations are valid at $\sim 10^{14}$ GeV. We can follow them down, using the renormalization group equations, and we obtain:

$$\frac{m_b}{m_\tau} = 2.7\text{-}3 \quad ; \quad \frac{m_d}{m_s} = \frac{m_e}{m_\mu} \cong \frac{1}{200} \tag{38}$$

The first one is welcome and can be qualitatively understood because the quarks, having strong interactions, become heavier. In fact one finds that:

$$\frac{m_b}{m_\tau} = \left[\frac{\alpha_s(Q = 2m_b)}{\alpha_s(Q = M)}\right]^{\frac{12}{33-F}} \tag{39}$$

where α_s is the strong interaction coupling constant, evaluated at $Q = 2m_b \sim m_T$ (the upsilon mass), and at $M \sim 10^{14} - 10^{15}$ GeV, respectively, and F is the number of families. For F = 3, we find very good agreement with experiment ($m_\tau \sim 1.8$ GeV, $m_T \simeq 2m_b \sim 9.5$ GeV), while the agreement gets much worse with increasing F.

Unfortunately, the second of the relations (38) is in very poor shape. Any estimation of m_d/m_s based on chiral dynamics gives a result which is ten times larger. We can improve the agreement by using a second Higgs multiplet, for example the 45 appearing in (10), but the number of parameters increases accordingly.

II.B.3. B-L and Neutrino Masses

My final remark on G.U.T.'s is devoted to baryon and lepton number violation. let us concentrate on the minimal SU(5) model, involving a 5 and a 24 of Higg's. In this case, the Lagrangian is invariant under a group of U(1) phase transformations (global). If ψ_{10} and $\psi_{\bar{5}}$ are the fermion multiplets and Φ_{24} and ϕ_5 the Higgs scalars, the transformations are:

$$\psi_{10} \to e^{i\theta}\psi_{10} \; ; \; \psi_{\bar{5}} \to e^{-i\theta}\psi_{\bar{5}} \; ; \quad \phi_5 \to e^{-2i\theta}\phi_5 \tag{40}$$

with all other fields invariant. The non-zero vacuum expectation value of ϕ_5 breaks this symmetry spontaneously. This sounds disastrous since it, normally, leads to the apearance of a truly massless Goldstone boson. However, we are saved because the symmetry is not really broken, it is simply changed. We can check immediately that, even after the translation of the Higgs's, the linear combination f_1 + 4Y remains a global symmetry, where f_1 is the generator of (40), and Y is the U(1) part of SU(5) given by (16). The conserved charge of this symmetry is the difference B-L of baryon and lepton numbers. This conservation has some very important consequences. First, it gives precise predictions for the decay properties of the proton. For example, $p \to e^+ + \pi^o$ is allowed, but $n \not\to e^+ + \pi^-$. This property remains true (or very nearly true) for essentially all grand unified models, so its experimental verification (assuming of course that baryon decay is observed) is of the utmost importance. A second consequence of B-L conservation is that the neutrino cannot acquire a Majorana mass. Since, on the other hand, a Dirac mass is impossible (there is no right-hand neutrino component in the standard SU(5)), we conclude that neutrino oscillation experiments can be used to test SU(5). Notice that even if we break B-L by introducing extra Higgs scalars, the expected neutrino Majorana mass is very tiny ($\sim 10^{-5}$ eV). Therefore, if the recent measurement of non-vanishing neutrino mass in tritium β-decay is confirmed, the SU(5) model, at least in its present form, should be abandoned. The only way to save it would be to introduce a ν_L^c as singlet, but then all the simplicity of the model disappears and one would be forced to go to the SO(10) model where, as we already said, a right-handed neutrino is naturally present. In fact, the main experimental prediction of SO(10) that differs substantially from that of SU(5) involves the neutrino mass. In SO(10), B-L is a gauge generator and it is spontaneously violated when SO(10) is broken. (Since it is a gauge generator it has to be violated otherwise there would be a massless gauge boson coupled to it.) The presence of ν_R allows for a neutrino Dirac mass of the form $\bar{\nu}_R \nu_L$ and the violation of B-L allows for a Majorana one. Therefore we naturally expect massive neutrinos. Let me further note that the above-mentioned violation of B-L does not lead to measurable effects in nucleon decay because the branching ratio of forbidden to allowed nucleon decays is predicted to be very small.

$$\frac{n \to e^- + \pi^+}{p \to e^+ + \pi^o} \sim \frac{m_W^2}{M^2} \sim 10^{-24} \tag{41}$$

Proton decay is not sensitive to the different grand-unified models.

III. SUPERSYMMETRY

"Οὐκ ἦλθον καταλῦσαι ἀλλά πληρῶσαι"
"I am not come to destroy, but to fulfil" Matth. E 17

III.A.1. The Trial of Scalars

The purpose of this chapter is not to destroy, but to fulfill. It is our firm belief, shared by most physicists, that gauge theories have come to stay. "Beyond" here does not mean that we propose to replace gauge theories by something else, but rather to embed them into a larger scheme with a tighter structure and higher predictive power. There are several reasons for such a search.

As we said in chapter I, gauge theories contain two and possibly three independent worlds. The world of radiation with the gauge bosons, the world of matter with the fermions and, finally, in our present understanding, the world of Higgs scalars. In the framework of gauge theories these worlds are essentially unrelated to each other. Given a group G, the world of radiation is completely determined, but we have no way to know a priori which and how many fermion representations should be introduced; the world of matter is, to a great extent, arbitrary.

This arbitrariness is even more disturbing if one considers the world of Higgs scalars. Not only their number and their representations are undetermined, but their mere presence introduces a large number of arbitrary parameters into the theory. Just compare Q.C.D. with massless quarks to the $U(1) \times SU(2)$ model. The first contains, according to our present understanding, two arbitrary parameters: a mass scale Λ and a vacuum angle θ. $U(1) \times SU(2)$ on the other hand, with three families contains eighteen parameters. Notice that this is independent of our computational ability, since these are parameters which appear in our fundamental Lagrangian. What makes things worse, is that these arbitrary parameters appear with a wild range of values. For example, in the standard model, the ratio of Yukawa couplings for different fermions equals the ratio of fermion masses. But $m_t/m_e \gtrsim 10^5$, and it is hard to admit that such a number is a fundamental parameter.

The situation becomes even more dramatic in grand unified theories where one may have to adjust parameters to as many as twenty-six significant figures. This is the problem of gauge hierarchy which is connected to the two enormously different mass scales at which spontaneous symmetry breaking occurs. The breaking of G into $U(1) \times SU(2) \times SU(3)$ happens at $M \sim 10^{14}$ GeV. This means that a certain Higgs field Φ acquires a non-zero vacuum expectation value $V = \langle\Phi\rangle_0 \sim 10^{14}$ GeV. The second breaking, that of $U(1) \times SU(2)$, occurs at $\sim 10^2$ GeV, i.e., we must have a second scalar field ϕ with $v = \langle\phi\rangle_0 \sim 10^2$ GeV. But the combined Higgs potential will contain a term of the form $\lambda\phi^2\Phi^2$. Therefore, after the first breaking, the ϕ-mass will be given by:

$$m_\phi^2 = \mu^2 + 2\lambda V^2 \tag{42}$$

where μ is the mass appearing in the symmetric Lagrangian. On the other hand, I want to remind you that $v^2 \sim m_\phi^2$, so unless there is a very precise cancellation between μ^2 and $2\lambda V^2$, a cancellation which should extend to twenty-six decimal figures, v^2 will turn out to be of order V^2 and the two breakings will come together, in other words the theory is not able to sustain naturally a gauge hierarchy. This grand-fine tuning of parameters must be repeated order by order in perturbation theory because, unlike fermions, scalar field masses require quadratically divergent counterterms. The whole structure looks extremely unlikely. The problem is similar to that of the induced cosmological constant in any theory with spontaneous symmetry breaking. I believe that, in spite of its rather technical aspect, the problem is sufficiently important so that some new insight will be gained when it is eventually solved.

One possible remedy is to throw away the scalars as fundamental elementary particles. After all, their sole purpose was to provoke the spontaneous symmetry breaking through their non-vanishing vacuum expectation values. In non-relativistic physics this phenomenon is known to occur, but the role of Higgs fields is played by fermion pairs (ex. the Cooper pairs in superconductivity). Let me also remind you that the spontaneous breaking of chiral symmetry, which is supposed to be a fundamental property of Q.C.D., does show the same feature, namely the "vacuum" is formed by quark-antiquark pair condensates and the resulting Goldstone boson (the pion) is again a $q\bar{q}$ bound state. This idea of dynamical symmetry breaking has been studied extensively, especially under the name of "Technicolor". In spite of its many attractive features, it suffers, up to now, from two main difficulties. First, the available field-theory technology does not allow for any precise quantitative computation of bound-state effects, and everything has to be based on analogy with the chiral symmetry breaking in Q.C.D. Second, nobody has succeeded in producing a satisfactory phenomenological model. Nevertheless, there is still hope that these difficulties may be overcome and, independently, the scheme has some precise predictions which can be tested experimentally in the near future.

III.A.2. The Defense of Scalars

The best defense of scalars is the remark that they are not the only ones to reduce the predictive power of a gauge theory. As we have already seen, going through the chain radiation-fermion matter fields-Higgs scalars we encounter an increasing degree of arbitrariness. One possiblility which presents itself is to connect the three worlds with some sort of symmetry principle. Then, the knowledge of the vector bosons will determine the fermions and the scalars, and the absence of quadratically divergent counterterms from the fermion masses will forbid their appearance in the scalar masses. Let me point out that no such symmetry between fermions and bosons is manifest in the particle spectrum, with the exception of a possible unexplained degeneracy between the photon and the neutrinos.

Is it possible to construct such a symmetry? A general form of an infinitesimal transformation acting on a set of fields $\phi^i(x)$, $i = 1,\ldots, m$ can be written as:

$$\delta\phi^i(x) = \varepsilon^\alpha (T_\alpha)^i_j \, \phi^j(x) \qquad (43)$$

where $\alpha = 1,\ldots,n$, the ε's are infinitesimal parameters and T_α is the matrix of the representation of the fields. Usually the ε's are taken to be c-numbers, in which case transformation (43) mixes only fields with the same spin and obeying the same statistics, fermions with fermions and bosons with bosons. It is clear that if we want to change the spin of the fields with a transformation (43), the corresponding ε's must transform non-trivially under rotations. If they have non-zero integer spin they can mix scalars with vectors or spin-1/2 with spin-3/2 fields. This was the case with the old attempts to construct a relativistic SU(6) theory with its well known shortcomings. If, on the other hand, the ε's are anti-commuting parameters, they will mix fermions with bosons. If they have zero spin, the transformations (43) will change the statistics of the fields without changing their spin, i.e., they will turn a physical field into a ghost. This is the case with the B.R.S. transformation which is so useful in the quantization of non-abelian gauge theories. Here, however, we want to connect physical bosons with physical fermions, therefore the infinitesimal parameters must be anti-commuting spinors. We call such transformations "supersymmetry transformations" and we see that a given irreducible representation will contain both fermions and bosons. It is not a priori obvious that such supersymmetries can be implemented consistently, but in fact they can. In the following I shall give a very brief description of their properties as well as their possible applications to gauge theories.

III.B. - Supersymmetry Transformations

III.B.1. The Algebra

Rather than going through mathematical preliminaries, I shall directly give the algebraic scheme which turns out to be interesting for physics. It is based on the well-known Poincaré algebra whose generators are P_μ, $\mu = 0, 1, 2, 3$ for the four translations and $M_{\mu\nu}$ for rotations and Lorentz transformations. In addition, we shall introduce four new generators Q_α forming a four-component Majorana spinor. This means that the commutator of Q_α with $M_{\mu\nu}$ is given by:

$$[Q_\alpha, M^{\mu\nu}] = i(\gamma^{\mu\nu})_{\alpha\beta}\, Q_\beta \tag{44}$$

with $\gamma^{\mu\nu} = 1/4\, [\gamma^\mu, \gamma^\nu]$. The relation (44) only expresses the fact that Q is a spinor. Furthermore we shall assume that Q is translationally invariant and anti-commutes with itself:

$$[P_\mu, Q_\alpha] = 0 = [Q_\alpha, Q_\beta]_+ = [\overline{Q}_\alpha, \overline{Q}_\beta]_+ \tag{45}$$

where $\overline{Q} = Q^T\gamma^0$ and $[\ ]_+$ denotes the anti-commutator. The last relation required to close the algebra is the anti-commutator of Q with $\overline{Q}$. We postulate:

$$[Q_\alpha, \overline{Q}_\beta]_+ = -2(\gamma^\mu)_{\alpha\beta}\, P_\mu \tag{46}$$

The relations (44)-(46) form the supersymmetry algebra we are going to use. We shall not attempt to justify its particular form on mathematical grounds but we shall derive its physical consequences. Before doing so, let me write it in an alternative form, using two component complex Weyl spinors instead of four component real Majorana ones. We can rewrite (45) and (46) as:

$$[P_\mu, Q] = [P_\mu, \overline{Q}] = [Q,Q]_+ = [\overline{Q},\overline{Q}]_+ = 0 \tag{45'}$$

$$[Q_\alpha, \overline{Q}_{\dot\beta}] = 2(\sigma_\mu)_{\alpha\dot\beta}\, P^\mu \tag{46'}$$

with $\sigma_\mu = (1, \vec{\sigma})$.

An obvious generalization consists in starting from the Poincaré algebra × a compact internal symmetry G with generators A_i. If the Q's belong to a certain representation of the internal symmetry, we write Q^m_α, where α is the spinor index and m that of the internal symmetry, $m = 1,\ldots,N$. We then have:

$$[A_i, A_j] = if_{ijk}\, A_k \qquad i,j,k = 1,\ldots,\ell \tag{47}$$

$$[A_i, Q^m_\alpha] = iS_i^{\ mn}\, Q^n_\alpha \qquad n,m = 1,\ldots, N \tag{48}$$

$$[Q^n_\alpha, Q^m_\beta]_+ = [Q^m_\alpha, P_\mu] = 0 \tag{49}$$

$$[Q^m_\alpha, \bar{Q}^n_{\dot\alpha}]_+ = 2\delta^{mn}\, (\sigma_\mu)_{\alpha\dot\alpha}\, P^\mu \tag{50}$$

The meaning of these relations is clear: The first one defines the internal symmetry, with f_{ijk} the structure constants of the group G. The second gives the N-dimensional representation of G in which the Q's belong and $S_i^{\ nm}$ are the corresponding constants. Finally (49) and (50) generalize in a particular way the basic N = 1 supersymmetry relations (45) and (46) or (45') and (46').

III.B.2. All Possible Supersymmetries of the S-matrix

The reader may feel uneasy with the very particular and seemingly arbitrary form of the algebra we introduced in the previous section. So this is the right moment to state, without proof, a very powerful theorem. It started in the sixties, when people tried unsuccessfully to combine Poincaré invariance with an internal symmetry (at that time it was SU(3) flavor symmetry) into a single larger group. Eventually a no-go theorem was proven showing that the only such combination which may be a symmetry of a unitary S-matrix is the trivial one given by the direct product of Poincaré and internal symmetry. The algebra (47)-(50) seems to contradict this result. In fact it does not, because one of the assumptions of the theorem was that the algebraic scheme used only commutators among the different generators. Now we can go back to the proof of the no-go theorem and relax this assumption allowing for anti-commutators as well as commutators. The remarkable result is that the supersymmetries we considered are essentially (apart from trivial generalizations) the only admissible ones. No other scheme would lead to a unitary S-matrix.

III.B.3. Representations in Terms of One-particle States

In order to extract the possible physical consequences of supersymmetry we must construct the representations of the algebra in terms of one-particle states, i.e., the one-particle "supermultiplets". We start by observing that the spinorial charges commute with P_μ and therefore they do not change the momentum of the one-particle state. Furthermore, the operator P^2 comutes with all the operators of the algebra, which implies that all the members of a supermultiplet will have the same mass. We can distinguish two cases $P^2 \neq 0$ or $P^2 = 0$.

(i) Massive case. We can go to the rest frame in which the r.h.s. of (46') or of (50) becomes a number. Let us first forget about a possible internal symmetry and consider the case N = 1. Eq. (46') gives:

$$[Q_\alpha, \bar{Q}_{\dot\beta}]_+ = 2M\delta_{\alpha\dot\beta} \tag{51}$$

where $P^2 = M^2$. Eq. (51) implies that the operators $Q/\sqrt{2M}$ and $\bar{Q}/\sqrt{2M}$ satisfy the anti-commutation relation for creation and annihilation operators of a system of free fermions. Since the index α can take two values, 1 and 2, and $Q_1^2 = Q_2^2 = 0$, starting from any one-particle state with spin S and projection S_z, we can build a four-dimensional Fock space with states:

$$|S,S_z;\, n_1,n_2\rangle = Q_2^{n_2}\, Q_1^{n_1}\, |\, S,S_z\rangle \qquad n_1,n_2 = 0,1 \tag{52}$$

Some specific examples:

$S = 0 \rightarrow$ one spin-1/2 particle, two spin-zero particles

$S = 1/2,\ S_z = \pm 1/2 \rightarrow$ one spin-zero, two spin-1/2, one spin-one

$S = 1,\ S_z = 0, \pm 1 \rightarrow$ two spin-one, one spin-1/2, one spin-3/2

etc...

The generalization to include internal symmetries is straightforward. The difference is that now we have more creation operators and the corresponding Fock space has 2^{2N} independent states, where N is the number of spinorial charges.

(ii) Massless case. Here we choose the frame $P_\mu = (E,0,0,E)$. The relation (46') yields:

$$[Q_\alpha, \overline{Q}_{\dot\beta}]_+ = 2E\,(1 - \sigma_z) = 4E\,\delta_{\alpha 2}\,\delta_{\dot\beta 2} \tag{53}$$

Only Q_2 and $\overline{Q}_{\dot 2}$ can be considered as creation and anihilation operators. Starting from a one-particle state with helicity $\pm\lambda$, we obtain the state with helicity $\pm(\lambda+1/2)$. Some interesting examples:

$\lambda = 1/2 \rightarrow$ one spin-1/2 and one spin-1 (both massless)

$\lambda = 3/2 \rightarrow$ one spin-3/2 and one spin-2 (both massless)

etc...

If we have more than one spinorial charge, i.e., $N > 1$, we obtain N creation and annihilation operators. A well-established theoretical prejudice is that, if one excludes gravitation, there exist no elementary particles with spin higher than one. This prejudice is based om the great difficulties one encounters if one wants to write consistent field theories with high spin particles. The consequence of such a prejudice is that $N = 4$ is the largest supersymmetry which may be interesting for particle physics without gravitation. The reason is that $N = 4$ contains four creation operators and allows us to go from a helicity state $\lambda = -1$ to that of $\lambda = +1$. Any increase in the number of spinorial charges will automatically yield representations containing higher helicities. Finally, if we include gravitation, the same prejudice tells us that we must allow for elementary particles with helicities $|\lambda| \leq 2$. The previous counting argument now gives $N = 8$ as the maximum allowed supersymmetry.

A concluding remark: All representations contain equal number of bosonic and fermionic states. All states in an irreducible representation have the same mass.

III.B.4. Representations in Terms of Field Operators

In order to use the machinery of Quantum Field Theory we must look for linear representations of supersymmetry in terms of local fields. We remind the reader that our aim was to realize supersymmetry as transformations of the general form (43). Representations of this kind were first obtained by trial and error, but now we have more powerful methods which give, at least for N = 1 supersymmetry, a complete classification of representations. We shall not use them in these lectures and we shall restrict ourselves to presenting some simple examples which turn out to be the most interesting in physics. Two cases should be distinguished: The case of "global" supersymmetry corresponds to transformations (43) with infinitesimal parameters ε which are independent of the space-time point x. The opposite case of x-dependent anticommuting parameters $\varepsilon(x)$ is called "local" supersymmetry or "supergravity". We shall examine first the case of global supersymmetry and we shall postpone the study of supergravity until the last chapter. Some interesting examples of field supermultiplets which transform linearly under global supersymmetry transformations are the following:

(i) The "chiral" multiplet. It consists of a complex Weyl spinor $\psi(x)$ and two complex scalars $A(x)$ and $F(x)$. The latter will turn out to be an auxiliary field. Under an infinitesimal supresymmetry transformation with parameter ξ these fields transform as:

$$\delta A(x) = \xi\psi(x)$$

$$\delta\psi(x) = 2i\sigma_\mu\bar{\xi}\partial^\mu A(x) + 2\xi F(x) \tag{54}$$

$$\delta F(x) = i\partial^\mu\psi(x)\sigma_\mu\bar{\xi}$$

This multiplet will be used to describe "matter" fields, i.e., quarks and leptons.

(ii) The massless vector multiplet which will be used for the gauge bosons before spontaneous symmetry breaking. It contains the free vector boson field strength $V_{\mu\nu}(x)$, a Weyl spinor $\lambda(x)$ and an auxiliary scalar field $D(x)$. Its transformation properties are:

$$\delta V_{\mu\nu}(x) = i\xi\sigma_\nu\partial_\mu\bar{\lambda}(x) - i\xi\sigma_\mu\partial_\nu\bar{\lambda}(x) + \text{h.c.}$$

$$\delta\lambda(x) = \xi\sigma^{\mu\nu}V_{\mu\nu}(x) + \xi D(x) \tag{55}$$

$$\delta D(x) = -\xi\sigma^\mu\partial_\mu\bar{\lambda}(x) + \text{h.c.}$$

with $V_{\mu\nu} = \partial_\mu V_\nu - \partial_\nu V_\mu$ and $\sigma^{\mu\nu} = 1/4\ (\sigma^\mu\sigma'^\nu - \sigma^\nu\sigma'^\mu)$; $(\sigma'_\mu)_{\dot{\alpha}\beta} = [(\sigma_\mu)_{\alpha\dot{\beta}}]^*$.

It is straightforward to verify that (54) or (55) are indeed representations, i.e., if we apply two successive supersymmetry transformations we obtain a third one.

III.C. Simple Field-theoretic Models

In this section we shall construct some simple field-theoretic models with interactions which are invariant under supersymmetry. These models are not realistic but they will reveal to us the mathematical properties of supersymmetry which we shall use in the phenomenological models of later sections.

III.C.1. The Self-interacting Chiral Multiplet

We discuss here the simplest supersymmetric invariant field theory model in four dimensions, that of a self-interacting chiral multiplet. The Langrangian density reads:

$$\mathcal{L} = \mathcal{L}_k + \mathcal{L}_m + \mathcal{L}_I \tag{56}$$

where $\mathcal{L}_k$, $\mathcal{L}_m$ and $\mathcal{L}_I$ are the kinetic energy, the mass and interaction term, respectively. We shall not derive the form of these terms in terms of the fields of eqs. (54), but we shall directly give the result.

$$\mathcal{L}_k = -i\bar{\psi}\partial^\mu\sigma_\mu\psi - \partial_\mu A\partial^\mu A^+ + F^+F \tag{57}$$

$$\mathcal{L}_m = m(AF - 1/2\ \psi\psi) + \text{h.c.} \tag{58}$$

$$\mathcal{L}_I = g(A^2F - \psi\psi A) + \text{h.c.} \tag{59}$$

Some remarks: (i) The Lagrangian (56) is not invariant under supersymmetry transformations. When the fields transform as in (54) we find, for the different terms in $\mathcal{L}$, $\delta\mathcal{L} = \partial_\mu R^\mu$, where R^μ is some vector field constructed out of the basic fields A, ψ, F and their derivatives. In other words, the Lagrangian density is not invariant but the resulting action is. This is not surprising because supersymmetry contains space-time translations. We know that no non-trivial Lagrangian density is invariant under translations. Only actions are. (ii) As announced, the field F is auxiliary because its derivatives do not appear in the kinetic energy term. Using the equations of motion we can eliminate it and we find:

$$F = - mA^+ - gA^+ \tag{60}$$

$$\mathcal{L}_m = - 1/2\ (m^2A^+A + m\psi\psi) + \text{h.c.} \tag{61}$$

$$\mathcal{L}_I = -mgA^2A^+ - g\psi\psi A - \frac{1}{2}g^2(A^+A)^2 + \text{h.c.} \tag{62}$$

In this form, the Lagrangian describes an ordinary renormalizable theory with a Yukawa, a ϕ^3 and a ϕ^4 coupling. Supersymmetry manifests itself in two ways: The masses of the complex scalar field and of the two-component Weyl spinor are equal, and the different coupling strengths are not independent but are all given in terms of a single one g. Notice also that the presence of the auxilairy field F is necessary in order to ensure linear transformation properties for all fields. Indeed, if we replace F in (54) by its equation (60), we find that ψ transforms non-linearly.

This very simple model has some remarkable renormalization properties. First, we can show that all vacuum-to-vacuum diagrams vanish, i.e., no normal ordering is required. This is a consequence of exact supersymmetry and it is valid for every supersymmetric theory. The surprising result, which could not be guessed by supersymmetry considerations alone, is that in this model, mass and coupling constant renormalizations are absent. All Green functions, to every order in perturbation theory, become finite if one introduces a single, common wave-function renormalization counterterm. In other words, in spite of the presence of scalar fields, not only do we not have any quadratically divergent mass counterterms, but we have none whatsoever. What happens is that the divergences due to boson loops are cancelled by those from fermion loops of opposite sign. The equality of masses and coupling constants is essential for this cancellation. We shall use this result extensively later on.

III.C.2. The Supersymetric Extension of Q.E.D.

A combination of supersymmetry with gauge invariance is clearly necessary for the application of these ideas to the real world. We shall first examine an abelian gauge theory, and construct the supersymmetric extension of quantum electrodynamics.

If V_μ is the photon field and ϕ_1 and ϕ_2 the real and imaginary parts of a charged field, an infinitesimal gauge transformation is given by

$$\delta V_\mu = \partial_\mu \Lambda; \quad \delta\phi_1 = e\Lambda\phi_2; \quad \delta\phi_2 = -e\phi_1\Lambda \tag{63}$$

where $\Lambda(x)$ is a scalar function. In order to extend (63) to supersymmetry, we must replace V_μ by a whole vector multiplet. Let us also assume that the matter fields are given by a charged chiral multiplet. We expect, therefore, to describe the interaction of photons with charged scalars and spinors simultaneously. It is obvious that if $\Lambda(x)$ is a scalar function, transformation (63) is not preserved by supersymmetry. The gauge transformation must be generalized so that $\partial_\mu\Lambda(x)$ is a member of a vector multiplet. This can be achieved if $\Lambda(x)$ becomes an entire chiral multiplet. The construction of the Lagrangian requires some kind of tensor calculus of supermultiplets, i.e., the rules of combining supermultiplets in order to obtain new ones. The final result is:

$$\begin{aligned}\mathcal{L} = & - 1/4\, V^2_{\mu\nu} - i\bar{\lambda}\partial^\mu\sigma_\mu\lambda + 1/2\, D^2 \\ & - i\bar{\psi}\partial^\mu\sigma_\mu\psi - \partial_\mu A\partial^\mu A^+ + F^+F + \text{mass terms} \\ & + eV^\mu[\tfrac{1}{2}\bar{\psi}\sigma_\mu\psi + \tfrac{i}{2}A^+\partial_\mu A - \tfrac{i}{2}A\partial_\mu A^+] - \tfrac{1}{4}e^2V_\mu V^\mu A^+A \\ & - \frac{ie}{\sqrt{2}}(A\bar{\lambda}\,\bar{\psi} - A^+\lambda\psi) + \frac{e}{2}DA^+A \end{aligned} \tag{64}$$

It is easy to understand the origin of the various terms in (64). The first line contains the usual kinetic energy terms of the vector multiplet members, the photon V_μ and its spin-1/2 partner λ. The field $D(x)$ is again an auxiliary field. The second line is the kinetic energy of the chiral matter multiplet of Eq. (57) together with a possible mass term given by Eq. (58). In the third line, the photon field V_μ is coupled to the charged current in the usual way with coupling constant e. The well-known "sea-gull" term is also present. The last line contains the new couplings dictated by supersymmetry. They are transformations of those in the previous line and they describe a Yukawa-type interaction between the spinor λ (the partner of the photon) and the scalar and spinor members of the matter multiplet. The last term, after elimination of the auxiliary field D, is the normal $(A^+A)^2$ term in scalar electrodynamics. Notice however that here it does not introduce any new coupling constant. Supersymmetry, once more, forces all couplings in (64) to be given in terms of the electric charge e.

III.C.3. Yang-Mills and Supersymmetry

We can generalize the above results to non-abelian Yang-Mills theories. The gauge bosons belong to vector superfields tranforming according to the adjoint representation of a group G. V_μ is the vector field, $V_{\mu\nu}$ its field-strength given by the analog of Eq. (21), the Weyl spinor λ is its spin-1/2 partner and D the associate auxiliary field. Following the notation of Chapter II (see eq. (21)) we write all these fields as square, traceless, hermitian matrices. The Yang-Mills Lagrangian now reads:

$$\mathcal{L} = -1/4 \text{ Tr } V_{\mu\nu}V^{\mu\nu} - i/2 \text{ Tr } \bar{\lambda} \not{\mathcal{D}} \lambda + 1/2 \text{ Tr } D^2 \qquad (65)$$

where the covariant derivative $\mathcal{D}_\mu$ acting on λ is given by:

$$\mathcal{D}_\mu \lambda = \partial_\mu \lambda + ig[V_\mu, \lambda] \qquad (66)$$

The auxiliary field equation gives, in this case, D = 0. The Lagrangian (65) shows that a Yang-Mills interaction of a massless Weyl fermion in the adjoint representation of the gauge group is automatically supersymmetric. The corresponding spin-3/2 conserved current is:

$$J^\mu \sim \text{Tr } [V_{\nu\rho}\sigma^\nu\sigma^\rho\sigma^\mu\lambda] \qquad (67)$$

where the trace is only taken in the internal symmetry indices and not the σ-matrices.

The introduction of additional matter multiplets in the form of chiral superfields belonging to any desired representation of the gauge group presents no difficulties. An interesting result is obtained if one studies the asymptotic properties of these theories. The one-loop β-function for an SU(m) Yang-Mills supersymmetric theory with n chiral multiplets belonging to the adjoint representation of SU(m) is:

$$\beta(g) = \frac{m(n-3)}{16n^2} g^3 \qquad (68)$$

which means that, for $n < 3$, the theory is asymptotically free, although it contains scalar particles.

Before closing this section, I want to mention a surprising and not yet fully understood result. Until now we have been considering supersymmetric theories with only one spinorial generator. We explained erlier that the generalization to N such generators is straight-forward, and N = 4 is the maximum number which does not introduce spins higher than one. The astonishing result is that the N = 4 super-symmetric Yang-Mills theory based on any group SU(m) seems to have no divergences in perturbation theory. This has been verified up to, and including, three loops. What is the significance of this result? For a theory with only massive particles one can show that the absence of divergences implies either tirviality (free fields) or the presence of states with negative metric. No analogous theorem is known for unbroken gauge theories.

III.D. Breaking of Supersymmetry

The spectrum of elementary particles shows no sign, to any conceivable approximation, of a degeneracy between fermions and bosons. Supersymmetry, if at all relevant, must be broken. Under these circumstances, an entirely explicit breaking is meaningless because the symmetry breaking part of the Hamiltonian could not be considered as a perturbation to the symmetric part. This tells us that the main source of supersymmetry breaking must be spontaneous. In this section we shall study some properties of such a theory.

The usual mechanism for spontaneous symmetry breaking is the introduction of some spin-zero field with negative square mass. This option is not available for supersymmetry because it would imply an imaginary mass for the corresponding fermion. In fact, we can make this argument more general: Taking the trace of the basic anticommutation relation (46'), we find immediately that the Hamiltonian of a supersymmetric system is positive definite. If there exists a state $|a\rangle$ such that $H|a\rangle = 0$, this must be the ground state. But (46') tells us that for such a state we have $Q|a\rangle = 0$, i.e., the state $|a\rangle$ is supersymmetric. We have just proven an important theorem which says that there is no state with energy lower than that of a supersymmetric state. It follows that the only way to break supersymmetry spontaneously is to manage so that there is no supersymmetric state. Supersymmetry is hard to break!

Today we know of three ways to achieve spontaneous supersymmetry breaking in perturbation theory.

(i) Through a U(1) gauge group. This mechanism is not available for G.U.T.'s which are based on simple groups.

(ii) Through "special" chiral multiplets. Although technically possible, this mechanism is subject to even more severe criticism than the ordinary Higgs mechanism regarding its arbitrariness.

(iii) Through supergravity which escapes the consequences of the aforementioned theorem. We shall consider this mechanism in Chapter V.

In the rest of this section we want to present some model-independent consequences of spontaneous supersymmetry breaking.

The first one is a direct application of Goldstone's theorem. Whenever a continuous symmetry is broken spontaneously there appears a massless particle with quantum numbers given by the divergence of the corresponding current. The supersymmetry current J_μ has spin-3/2, therefore, its divergence has spin-1/2. It follows that the associated Goldstone particle is a spin-1/2 fermion of zero mass, the "Goldstino". Where is the Golstino? There are three possible answers to this question.

(i) Our first reaction to the appearance of a massless spin-1/2 fermion is to rejoice because we hope to associate it with one of the neutrinos. Alas, appearances are deceptive! There is a low-energy theorem, satisfied by any Goldstone particle. The field $\psi(x)$ of the Goldstino is proportional to the divergence of the supersymmetry current

$$\psi(x) \sim \partial^{\mu} J_{\mu}(x) \tag{69}$$

Eq.(69) seems strange because the r.h.s. vanishes for a conserved current. Striclty speaking we should introduce a small explicit breaking, in which case the constant of proportionality contains a factor m^{-1}, where m is the Goldstino mass. In the limit when the explicit breaking goes to zero, the ratio of the matrix elements of $\partial^{\mu} J_{\mu}$ divided by m is well defined. Eq.(69) implies that the amplitude $M(a^{\mu} \to b + \psi)$ of the emission (or absorption) of a Goldstino with momentum k_{μ} satisfies the low energy theorem:

$$\lim_{k_{\mu} \to 0} M(k) = 0$$

This is a very powerful prediction and can be checked by studying the end-point spectrum in nuclear β-decay. Unfortunately, it is wrong! Experiments show no such suppression, which means that the electron neutrino cannot be the Golstino.

(ii) Can the Goldstino be a new, as yet unobserved, massless, spin-1/2 fermion? The answer is yes, and I know of at least two ways to implement this idea. One suggestion is to identify it with the right-hand component of the physical neutrino. The low-energy theorem would thus explain why it is not coupled. A second one is to make the Goldstino hard to detect by endowing it with a new, conserved quantum number. This possibility is the only one which has provided some semi-realistic models based on low-energy global supersymmetry. Their predictions have not been ruled out by experiment and can be tested with available technology. We shall study them in the next chapter.

(iii) The third answer to the Goldstino puzzle is a "super-Higgs" mechanism. In the normal Higgs phenomenon we have:

$$(m = 0,\ \text{spin} = 1) + (m = 0,\ \text{spin} = 0) = (m \neq 0,\ \text{spin} = 1)$$

In a "super-Higgs" mechanism we get:

$$(m = 0,\ \text{spin} = 3/2) + (m = 0,\ \text{spin} = 1/2) = (m \neq 0,\ \text{spin} = 3/2)$$

i.e., we need to start with a gauge spin-3/2 field which will absorb the massless Goldstino to give a massive spin-3/2 particle. This mechanism can be applied in the framework of supergravity theories.

The second model-independent consequence of spontaneous breaking of global supersymmetry concerns the particle spectrum. In exact supersymmetry, fermions and bosons are degenerate. After the breaking, the masses are split but a certain pattern remains. The masses squared of the boson fields are equally spaced above and below those of the fermions. More precisely we obtain the mass formula

$$\sum_J (-)^{2J} (2J + 1)\, m_J^2 = 0 \tag{70}$$

where m_J is the mass of the particle of spin J. This formula plays an important role in model building.

IV. PHENOMENOLOGY OF GLOBAL SUPERSYMMETRY

Let us now try to apply these ideas to the real world. We want to build a supersymmetric model which describes the low-energy phenomenology. There may be several answers to this questions but, to my knowledge, there is only one class of models which comes close to being realistic. They assume a superalgebra with only one spinorial generator, consequently all particles of a given supermultiplet must belong to the same representation of the gauge group. In the following we shall try to keep the discussion as general as possible, so that our conclusions will be valid in essentially all models.

IV.A. Building Blocks of Supersymmetric Models

All models based on global supersymmetry use three types of multiplets:

(i) Chiral multiplets. As we said already, they contain one Weyl (or Majorana) fermion and two scalars. Chiral multiplets are used to represent the matter (leptons and quarks) fields as well as the Higgs fields of the standard model.

(ii) Massless vector multiplets. They contain one vector and one Weyl (or Majorana) fermion, both in the adjoint representation of the gauge group. They are the obvious candidates to generalize the gauge bosons.

(iii) Massive vector multiplets. They are the result of ordinary Higgs mechanism in the presence of supersymmetry. A massive vector multiplet is formed by a vector field, a Dirac spinor and a scalar. These degrees of freedom are the combination of those of a massless vector multiplet and of a chiral multiplet.

IV.B. Supersymmetric Extension of the Standard Model

IV.B.1. The Particle Content

In the standard model we have:

Number of bosonic degrees of freedom = 28

Number of fermionic degrees of freedom = 90

It follows that a supersymmetric extension of the standard model will necessarily introduce new particles. Let us go one step further: In N = 1 supersymmetry all the particles of a given supermultiplet must belong to the same representation of the gauge group. For the various particles of the standard model this yields:

(i) The gauge bosons are one color octet (gluons), one SU(2) triplet and one singlet ($W^{\pm}$, $Z°$, γ). No known fermions have these quantum numbers.

(ii) The Higgs scalars transform as SU(2) doublets but they receive a non-zero vacuum expectation value, consequently they cannot be the partners of leptons or quarks otherwise we would have induced a spontaneous violation of lepton or baryon number. Furthermore, we must enlarge the Higgs sector by introducing two complex chiral supermultiplets. This is necessary for several technical reasons that are related to the fact that, in supersymmetry, the Higgs scalars must have their own spin-1/2 partners. This in turn creates new problems such as, for example, new triangle anomalies which must be cancelled. Furthermore, now the operation of complex conjugation on the scalars induces a helicity change of the corresponding spinors. Therefore, we cannot use the same Higgs doublet to give masses to both up and down quarks. Finally, with just one Higgs supermultiplet, we cannot give masses to the charged partners of the W's. The net result of this operation is a much richer spectrum of physical "Higgs" scalars. Since we start with eight scalars (rather than four) we end up having five physical ones (rather than one). They are the scalar partners of the massive vector bosons $W^{\pm}$, Z^{o} and two separate neutral ones.

The conclusion, is that, in the standard model, supersymmetry associates known bosons with unknown fermions and known fermions with unknown bosons. We are far from obtaining a connection between the three independent worlds. For this reason this model should not be considered as a fundamental theory but as an intermediate step towards the higher N supergravity theories which we shall present in the last chapter. Nevertheless, the phenomenological conclusions we shall derive are sufficiently general to be valid, unless otherwise stated, in every theory based on supersymmetry.

We close this section with a table of the particle content in the supersymmetric standard model. Although the spectrum of these particles, as we shall see shortly, is model dependent, their very existence is a crucial test of the whole supersymmetry idea. We shall argue in the following chapters that its experimental verification is within the reach of present machines.

IV.B.2. R-parity

The motivation for introducing a new quantum number for the supersymmetric partners of known particles was the problem of the Goldstino. However, such a quantum number appears naturally in the framework of supersymmetric theories and it is present even in models in which the above motivation is absent. This number is called "R-parity" and one possible definition is:

$$(-)^{R} = (-)^{2S} \quad (-)^{3(B-L)} \tag{71}$$

where S is the spin of the particle and B and L the baryon and lepton numbers, respectively. It is easy to check that Eq. (71) gives R = 0 for all known particles, fermions as well as bosons, while it gives R = ±1 for their supersymmetric partners. Since R is conserved, the R-particles are produced in pairs and the lightest one is stable. In a spontaneously broken global supersymmetry this is the Goldstino, which is massless, but in the supergravity theories the latter is absorbed by the super-Higgs mechanism. The general consensus is that, in this case, the lightest, stable R-particle is the partner of the photon, the "photino".

Table 1

The particle content of the supersymmetric standard model

SPIN-1	SPIN-1/2	SPIN-0	
Gluons	Gluinos		
Photon	Photino		
$W^{\pm}$	2 Dirac Winos	$W^{\pm}$	Higgs Bosons
Z^{o}	2 Major. Zinos	Z	
	1 Major. Higgsino	standard ϕ^{o} pseudosc. $\phi^{o\prime}$	
	Leptons	Spin-0 leptons	
	Quarks	Spin-0 quarks	

IV.B.3. The Goldstino

Looking back at Table 1 we see that an important question - before going into the details of any model - is the indentity of the Goldstino; in particular, can it be identical, say to the photino? As we said earlier, the mechanism of spontaneous symmetry breaking, which is at the origin of the existence of the Goldstino, allows us to find some properties of the latter, independently of the details of a particular model. In a spontaneously broken theory the spin-3/2 conserved current is given by:

$$J^{\mu} = d\gamma^{\mu}\gamma_{5}\psi_{g} + \hat{J}^{\mu} \qquad (72)$$

where d is a parameter with dimensions (mass)², ψ_g is the Goldstino field and $\hat{J}^{\mu}$ is the usual part of the current which is at least bilinear in the fields. In other words, the field of a Goldstone particle can be identified with the linear piece in the current. The conservation of J^{μ} gives:

$$d\gamma_5 \not{\partial}\psi_g = \partial_\mu \hat{J}^\mu \qquad (73)$$

This is the equations of motion of the Goldstino. In the absence of spontaneous breaking d = 0 and $\partial_\mu \hat{J}^\mu = 0$. In fact, to lowest order, the contribution of a given multiplet to $\partial_\mu \hat{J}^\mu$ is proportional to the square mass-splitting Δm^2. Thus, the coupling constant of the Goldstino to a spin-0-spin-1/2 pair is given by

$$f_g = \pm \frac{\Delta m^2}{d} \qquad (74$$

where the sign depends on the chirality of the fermion. It follows that, if the Goldstino were the photino, $f_g \sim e$ and the $(\text{mass})^2$-splittings would have been proportional to the electric charge. For example, if S_e and t_e were the charged spin-zero partners of the electron, we would have:

$$m^2(S_e) + m^2(t_e) = 2m^2(e) \qquad (75)$$

Eq.(75), which is clearly unacceptable, is a particular example of mass-formula (70) and here we see how restrictive the latter is. The conclusion is that the photon cannot be the bosonic partner of the Goldstino. With a similar argument we prove that the same is true for the Z boson, the Higgs's or any linear combination of them. This is a model-independent result. Not only can we not identify the Goldstino with the neutrino, but we also cannot pair it with any of the known neutral particles. Therefore, strictly speaking, there is no acceptable supersymmetric extension of the standard model. The one that comes closest to it assumes an enlargement of the gauge group to U(1) × U(1) × SU(2) × SU(3) thus involving a new neutral gauge boson. We shall not study it in detail here but we shall rather extract those features which are model independent and are likely to be present in any supersymmetric theory.

IV.B.4. Particle Spectrum and Decay Modes

As we said earlier, supersymmetry predicts a rich spectroscopy of new particles whose existence is an important test of the theory. Such a test, however, is only meaningful if the masses of the new particles are also predicted, at least to an order of magnitude. Let me remind you of the situation when the charmed particles were predicted. The motivation was the need to suppress unwanted processes like strangeness changing neutral current transitions. Such a suppression was effective only if the charmed particles were not too heavy. No precise value could be given but the prediction was powerful enough to be testable. We have a similar situation with supersymmetry. The need for introducing supersymmetry was to control the bad behavior of elementary scalar fields, and this can be achieved only if supersymmetry is not too badly broken which, in practice, means that the masses of the partners of the known particles cannot exceed the mass of the W too much. The most convincing argument comes from grand-unified theories, and it will be presented in the next chapter. The conclusion is that supersymmetry can be tested with present machines, existing or under construction (CERN and FNAL $\bar{p}p$ colliders and LEP).

Let us now briefly discuss some results on masses and decay properties. In the absence of any concrete experimental evidence, I can only quote limits and expected signatures. The mass spectrum is very model dependent but some general features can be expected.

(i) Scalar partners of quarks and leptons (squarks, sleptons). The best limits on the masses of the charged ones come from PETRA. With small variations they are of the order of 20 GeV. Theoretical arguments almost always predict squarks heavier than sleptons. The reason is that in most models the masses are set equal at the grand unification scale and the differences are due to the strong interactions of the squarks (see eq. (39)). For the same reason the masses of sneutrinos are predicted to be of the same order as those of the corresponding charged sleptons. The recent collider results, which at the moment I am writing these lines still need confirmation, suggest that squarks are at least as heavy as 40 GeV.

(ii) Gluinos. From beam dump results we can extract a model-dependent limit of the order of 2 GeV. From the same theoretical arguments, the gluinos are always heavier than photinos. The same UA1 results indicate again a limit of the order of 40 GeV.

(iii) Charged gauginos, the partners of $W^{\pm}$. We have the same PETRA bounds.

(iv) Neutral gauginos and Higgsinos. They mix among themselves, so they should be treated simultaneously. The lightest among them, which for the purposes of this discussion we shall call the photino, is likely to be the lighest particle with $R \neq 0$ and therefore stable. I remind you that the Goldstino, which, if it exists, is massless, is absent from theories derived from supergravity (see last chapter). No precise prediction for the photino mass exists but we shall assume $m_{\tilde{\gamma}} \sim$ 1-10 GeV. Supergravity-type models have a tendency to give masses close to the upper value while cosmological arguments exclude values much lighter than 1 GeV.

The picture that emerges is that supersymmetry particles may be spread all over from 10 to 100 GeV with squarks and gluinos heavier than sleptons and photino.

Let us now come to possible decay modes and signatures. They obviously depend on the detailed mass spectrum but two considerations must be kept in mind: Supersymmetric particles do not introduce new coupling constants. Thus squarks and gluinos are coupled with α_s and sleptons and gauginos with α. Secondly, we must remember that R is conserved, therefore all new particles will eventually end up giving photinos whose interactions are comparable to those of the neutrinos and they leave undetected. Hence the great importance of a precise determination of missing transverse momentum as a handle in the search of supersymmetric particles. After these preliminaries we list some specific examples:

Squarks are produced in hadron collsions either in pairs or in association with gluinos (R must be conserved). Their decay modes are

$$\tilde{q} \to q + \tilde{\gamma} \quad \text{(quark + photino)} \tag{76}$$

or, if phase space permits,

$$\tilde{q} \to q + \tilde{g} \quad \text{(quark + gluino)} \tag{77}$$

The gluino in turn decays either as

$$\tilde{g} \to q + \bar{q} + \tilde{\gamma} \tag{78}$$

or as

$$\tilde{g} \to g + \tilde{\gamma} \quad \text{(gluon + photino)} \tag{79}$$

In most models,(78) dominates over (79). We see that the signature for squarks or gluinos is missing P_T + jets. It was from the study of the small sample of the UA1 monojet events that the limits of 40 GeV for squarks and gluinos were obtained.

Sleptons behave similarly and give

$$\tilde{\ell} \to \ell + \tilde{\gamma} \quad \text{(lepton + photino)} \tag{80}$$

Thus the signal at PETRA or LEP will, again, be missing energy and acoplanar and acollinear events. The jump in R (the total hadronic e+e- cross section divided by the $\mu+\mu-$ one) due to a single charged scalar particle is too small to be detectable, but, if LEP is above several thresholds, it will be seen.

An important source of information is provided by the W and Z decays. A precise measurement of the Z° width, like R in e^+e^- collisions, counts the number of particle species that are produced. In the most favorable case, i.e., several decay channels available, the total Z° width could be increased by a factor of two. If the charged gauginos are substantially lighter than the W, one could have, for example:

$$\begin{array}{l} W \to \text{wino + photino} \\ \quad\quad\;\; \hookrightarrow \text{photino} + e + \nu_e \end{array} \tag{81}$$

which could be detected by a precise measurement of the electron spectrum. Another possible decay mode of the W, if $m_W > m_{\tilde{e}} + m_{\tilde{\nu}}$, is:

$$W \to \text{selectron + sneutrino} \tag{82}$$

The sneutrino, if it is relatively light, decays into neutrino + photino. If it is heavier than the other gauginos, it can decay into them. In all cases,(82) will result in a final state which, like that of (81), looks like an ordinary one and can be identified only through detailed measurement of the electron spectrum.

A final interesting possibility for LEP is the study of quarkonium decay. This has already been considered for ψ in order to put limits on the photino mass. The process was $\psi \to$ Goldstino + photino. At present, our prejudices exclude such a decay, but toponium is still interesting. If $m_t > m_{\tilde{t}} + m_{\tilde{\gamma}}$ one could have $T \to \bar{t} + \tilde{t} + \tilde{\gamma}$. If $m_t < (m_{\tilde{t}} + m_{\tilde{\gamma}})$, but $m_{\tilde{t}} < m_t$, we could have $T \to \tilde{t} + \bar{\tilde{t}}$. Depending on the gluino mass, we could have $T \to g + \tilde{g} + \tilde{g}$, or $T \to g + \tilde{g} + \tilde{\gamma}$ or even $T \to \tilde{g} + \tilde{g}$.

I believe that looking for supersymmetric particles will dominate experimental research in the second half of this decade. I hope that it is going to be both exciting and rewarding and, in any case, by the early nineties we shall know for sure whether supersymmetry is a fundamental symmetry of particle forces.

IV.C. Supersymmetry and Grand-Unified Theories

Let me remind you that one of the reasons why we decided to study supersymmetry in connection with gauge theories was the presence of the gauge hierarchy problem that plagues all known G.U.T.'s. It is now time to study this question. The problem has two aspects: The first one, the physical aspect, is to find a natural way to create these too largely separated mass scales. Supersymmetry offers no new insight to this very fundamental question. The second is the technical aspect. In the notation of Section III.A.1, the 24-plet Φ takes a vacuum expectation value V. As before, in order for the model to be able to sustain a gauge hierarchy, we must impose a very precise relation among the parameters of the potential, of the form $m \sim \lambda V$. It is this relation which is destroyed by renormalization effects and has to be enforced artificially order by order in perturbation theory. This is the technical aspect. Supersymmetry can eliminate this part of the problem. The key is the non-renormalization theorems we mentioned in Section III.C.1. If supersymmetry is exact, the parameters of the potential do not get renormalized. What happens is that the infinities coming from fermion loops cancel those coming from boson loops. When supersymmetry is spontaneously broken, the cancellation is not exact, but the corrections are finite and calculable. They are of the order $\sim \Delta m^2$, where Δm is the mass-splitting in the supermultiplet. Here comes the estimation we used in the previous section. For the gauge hierarchy to remain, Δm^2 should not be much larger than the small mass scale, namely m_W^2. A badly broken supersymmetry is not effective in protecting the small mass scale. Let me make the logic of this argument clear: Supersymmetry does not solve the gauge hierarchy problem in the sense of providing an explanation for the existence of two very different mass scales. It only provides a framework to stabilize the hierarchy, once it is imposed. For this last function the breaking cannot be arbitrarily large. Hence, the upper limit on the masses of supersymmetric particles.

After these remarks on the gauge hierarchy problem, one can proceed in supersymmetrizing one's favorite G.U.T. model. The construction parallels that of the low-energy standard model, with similar conclusions. Again, no known particle can be the superpartner of another known particle. Furthermore, assuming a spontaneous symmetry breaking, we can repeat the analysis which led us to conclude that $U(1) \times SU(2) \times SU(3)$ was too small. The corresponding conclusion here will be that SU(5) is too small, since SU(5) does not contain anything larger than the group of the standard model.

Finally, we can repeat the renormalization group estimation of the grand-unification scale and the proton life-time. We found in Chapter II that at low energies the effective coupling constants evolve following, approximately, the renormalization group equations of U(1), SU(2) or SU(3). The same remains true in a supersymmetric theory, but now the values of the β-functions are different. The number of Yang-Mills gauge bosons is the same as before. They are the ones which give rise to negative β-functions. On the other hand, supersymmetric theories have a larger number of "matter" fields, spinors and scalars, which give positive contributions. The net result is a smaller β-function (in absolute value) and, therefore, a slower variation of the asymptotically-free coupling constants. We expect larger values of the grand-unification scale M and, indeed, we find $M \sim 10^{16} - 10^{17}$ GeV. If nothing else contributes to proton decay, it will be invisible! Fortunately, there are other contributions, which, although of higher order, turn out to be dominant. An example is shown in figure 3. The important point is that the superheavy particle is now a fermion and the amplitude is of order M^{-1} rather than M^{-2}. The resulting proton life-time is $\tau_p \sim 10^{30}$ years, with large uncertainties. The absolute value has not changed significantly with the introduction of supersymmetry, but the dominant decay modes are now different. Since the intermediate particle is a Higgsino, heavy flavors are favored.

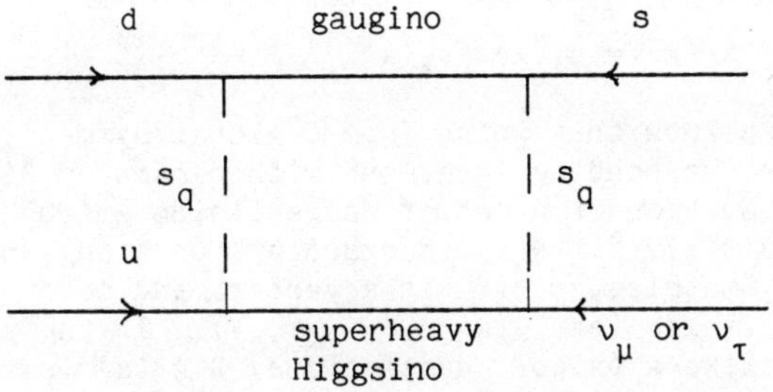

Fig. 3

We expect to find

$$\begin{aligned} p &\to \bar{\nu}_\mu + K^+ \\ &\to \bar{\nu}_\tau + K^+ \end{aligned} \tag{83}$$

which considerably complicates the task of experimental detection. One cannot rely on a two-body back-to-back decay mode like the $e^+\pi^0$ for background rejection. Maybe the last word has not yet been said in baryon decay.

V. SUPERGRAVITY

Supergravity is the theory of local supersymmetry, i.e., supersymmetry transformations whose infinitesimal parameters - which are anticommuting spinors - are also functions of the space-time point x. There are several reasons to go from global to local supersymmetry:

(i) We have learned in the past few years that all fundamental symmetries in nature are local (or gauge) symmetries.

(ii) The supersymmetry algebra contains translations, and we know that invariance under local translations leads to general relativity which, at least at the classical level, gives a perfect description of gravitational interactions.

(iii) As we already noticed, local supersymmetry provided the most attractive explanation for the absence of a physical Goldstino.

(iv) In the last chapter we saw that in a supersymmetric grand-unified theory the unification scale approaches the Planck mass (10^{19} GeV), at which gravitational interactions can no longer be neglected.

The gauge fields of local supersymmetry can be easily deduced. Let us introduce an anticommuting spinor ε for every spinorial charge Q, and write the basic relation (50) as a commutator:

$$[\varepsilon^m Q^m, \bar{Q}^n \bar{\varepsilon}^n] = 2\delta^{mn} \; \varepsilon^m \sigma_\mu \bar{\varepsilon}^n P^\mu \quad ; \quad n, m=1,\dots,N \tag{84}$$

where no summation over m and n is implied. In a local supersymmetry transformation, ε becomes a function $\varepsilon(x)$. Eq.(84) implies that the product of two supersymmetry transformations, with parameters $\varepsilon_1(x)$ and $\varepsilon_2(x)$, is a local translation with parameter

$$A_\nu(x) = \varepsilon_1(x)\sigma_\nu \bar{\varepsilon}_2(x) \tag{85}$$

On the other hand, we know that going from a global symmetry with parameter θ to the corresponding local one with parameter $\theta(x)$, results in the introduction of a set of gauge fields which have the quantum numbers of $\partial_\mu\theta(x)$. If $\theta(x)$ is a scalar function, which is the case for internal symmetries, $\partial_\mu\theta(x)$ is a vector, and so are the corresponding gauge fields (ex. gluons, $W^\pm$, Z, γ). If the parameter is itself a vector, like $A_\nu(x)$ of translations, $\partial_\mu A_\nu(x)$ is a two-index tensor, and the associated gauge field has spin two. In super-symmetry, the parameters $\varepsilon^m(x)$ have spin one-half, so the gauge fields will have spin three-half. We conclude that the gauge fields of local supersymmetry, otherwise called supergravity, are one spin-two field and N spin-three-half ones. To these we have to add the ordinary vector gauge fields of whichever internal symmetry we are considering.

V.A. - N=1 Supergravity

"Οὐκ ἦν ἐκεῖνος τὸ φῶς, ἀλλ' ἵνα μαρτυρήσῃ περὶ τοῦ φωτός"

"He was not that Light, but was sent to bear witness of that Light" John A 8

This is the simplest supergravity theory. As I shall explain in the next section, I do not consider it as the fundamental theory of particle physics, but I believe that it provides for a good basis for a phenomenological analysis. The gauge fields are the metric tensor $g_{\mu\nu}(x)$, which represents the graviton and a spin-three-half Majorana "gravitino" $\psi_\mu(x)$. We can start by writing the Lagrangian of "pure" supergravity, i.e., without any matter fields. The Lagrangian of general relativity can be written as:

$$\mathcal{L}_G = -\frac{1}{2\kappa^2}\sqrt{-g}\,R = \frac{1}{2\kappa^2}\;eR \tag{86}$$

where $g_{\mu\nu}$ is the metric tensor and $g=\det g_{\mu\nu}$. R is the curvature constructed out of $g_{\mu\nu}$ and its derivatives. We have also introduced the vierbein field e^m_μ in terms of which $g_{\mu\nu}$ is given as $g_{\mu\nu} = e^m_\mu e^n_\nu \eta_{mn}$, with the Minkowski space metric. It is well known that if one wants to study spinor fields in general relativity the vierbien, or tetrad, formalism is more convenient. e equals $-\sqrt{-g}$; κ^2 is the gravitational coupling constant. Eq.(86) is the Lagrangian of the gravitational field in empty space. We add to it the Rarita-Schwinger Lagrangian of a spin-three-half massless field in interaction with gravitation:

$$\mathcal{L}_{RS} = -\frac{1}{2}\,\varepsilon^{\mu\nu\rho\sigma}\,\bar{\psi}_\mu\,\gamma_5\gamma_\nu\,\mathcal{D}_\rho\psi_\sigma \tag{87}$$

where $\varepsilon^{\mu\nu\rho\sigma}$ is the completely antisymmetric tensor, which equals one (minus one) if its four indices form an even (odd) permutation of 1, 2, 3, 4, and zero otherwise. $\mathcal{D}_\rho$ is the covariant derivative

$$\mathcal{D}_\rho = \partial_\rho + 1/2\;\omega_\rho^{mn}\,\gamma_{mn} \quad ; \quad \gamma_{mn} = 1/4\;[\gamma_m, \gamma_n] \tag{88}$$

and ω_ρ^{mn} is the spin connection. Although ω_ρ^{mn} can be treated as an independent field, its equation of motion expresses it in terms of the vierbein and its derivatives.

The remarkable result is that the sum of (86) and (87)

$$\mathcal{L} = \mathcal{L}_G + \mathcal{L}_{RS} \tag{89}$$

gives a theory invariant under local supersymmetry transformations with parameter $\varepsilon(x)$

$$\delta\, e^m_\mu = \frac{\kappa}{2}\;\bar{\varepsilon}(x)\;\gamma^m\;\psi_\mu \tag{90a}$$

$$\delta\,\omega^{mn}_\mu = 0 \tag{90b}$$

$$\delta\psi_\mu = \frac{1}{\kappa}\;\mathcal{D}_\mu\;\varepsilon(x) = \frac{1}{\kappa}\left(\partial_\mu + \frac{1}{2}\,\omega^{mn}_\mu\;\gamma_{mn}\right)\varepsilon(x) \tag{90c}$$

Two remarks are in order here: First, the invariance of (89) reminds us of the similar result obtained in global supersymmetry, where we found that the sum of a Yang-Mills Lagrangian and that of a set of Majorana spinors belonging to the adjoint representation, was automatically supersymmetric. Second, we must point out that the transformations (90) close an algebra only if one uses the equations of motion derived from (89). We can avoid this inconvenience by introducing a set of auxiliary fields. In fact, we have partly done so, because the spin connection is already an auxiliary field.

The next step is to couple the N = 1 supergravity fields with matter in the form of chiral or vector multiplets. The resulting

Lagrangian is quite complicated and will not be given explicitly here. Let me only mention that, in the most general case, it involves two arbitary functions. If I call z the set of complex scalar fields, the two functions are:

$G(z,z^*)$: a real function, invariant under whichever gauge group we have used.

$f_{ij}(z)$: an analytic function which transforms as a symmetric product of two adjoint representations of the gauge group.

One may wonder why we have obtained arbitrary functions of the fields, but we must remember that, in the absence of gravity, we impose to our theories the requirement of renormalizability, which restricts the possible terms in a Lagrangian to monomials of low degree. In the presence of gravity, however, renormalizability is lost anyway, so no such restriction exists. In view of this, it is quite remarkable that only the two aforementioned functions occur.

As in ordinary gauge theories, the spontaneous breaking of local supersymmetry results in a super-Higgs mechanism. The gravitino, which is the massless gauge field of local supersymmetry, absorbs the massless Goldstino and becomes a massive spin-three-half field. At ordinary energies we can take the limit of the Planck mass going to infinity. In this case, gravitational interactions decouple and the spontaneously broken supergravity behaves like an explicitly but softly broken global supersymmetry. The details of the final theory, such as particle spectra, depend on the initial choice of the functions G and f_{ij}, but the general phenomenology is the same as the one presented in the previous chapter.

Before closing this section let me mention a famous unsolved problem for which supergravity offers a new line of approach. The Einstein Lagrangian (86) is not the most general one. We could add a constant Λ with dimensions $|\text{mass}|^4$ and write:

$$\mathcal{L}_G = -\frac{1}{2\kappa^2}\sqrt{-g}\,(R + \Lambda) \qquad (91)$$

Λ is called "the cosmological constant", and represents the energy density of empty space, but in the presence of the gravitational field this is no longer an unphysical quantity which one can set equal to zero. In fact, any matter field gives an infinite contribution to Λ. Experimentally, Λ is very small, $\Lambda < 10^{-48}[\text{GeV}]^4$. If we have exact supersymmetry, Λ vanishes identically because the infinite vacuum energy of the bosons cancels that of the fermions. However, in a spontaneously broken global supersymmetry the vacuum energy is always positive, as we explained in Section III.D, and this yields a positive cosmological constant. In a spontaneously broken supergravity this is no longer true, and one can arrange to have $E_{vac} = 0$ and hence $\Lambda = 0$. In a realistic theory this must be the consequence of a certain symmetry and, indeed, such models have been constructed and are under study. I believe that ultimately this problem will be connected to the way one obtains N = 1 supergravity, as an intermediate step between low-energy phenomenology and the fundamental theory to which I shall now turn.

V.B. - N = 8 Supergravity

"... ἵνα ὦσι τετελειωμένοι εἰς ἕν..."

"..., that they may be made perfect in one;..." John IZ 23

Let me remind you that one of the arguments to introduce supersymmetry was the desire to obtain a connection among the three independent worlds of gauge theories, the worlds of radiation, matter and Higgs fields. None of the models presented so far have achieved this goal. They all enlarged each world separately into a whole supermultiplet, but they did not put them together. N = 8 supergravity is the only one which attempts such unification. It is the largest super-symmetry we can consider, if we do not want to introduce states with spin higher than two. Following the method of Section III.B.3, we construct the irreducible representation of one-particle states which contains:

1 spin-2 graviton

8 spin-3/2 Majorana gravitini

28 spin-1 vector bosons (92)

56 spin-1/2 Majorana fermions

70 spin-0 scalars

We shall not write down the Lagrangian that involves all these fields, and is invariant under eight local supersymmetry transformations, but we shall mention some of its properties. Contrary to the N = 1 case, there is no known system of auxiliary fields. Since we have 28 vector bosons, we expect the natural gauge symmetry to be SO(8). This is bad news because SO(8) does not contain U(1) × SU(2) × SU(3) as a subgroup. The remarkable property of the theory, which raised N = 8 to the status of a candidate for a truly fundamental theory, is the fact that the final Lagrangian has unexpected symmetries: (i) a global non-compact E_7 symmetry, and (ii) a gauge SU(8) symmetry whose gauge bosons are not elementary fields. They are composites made out of the 70 scalars. SU(8) is large enough to contain the symmetries of the standard model, but this implies that all known gauge fields (gluons, $W^{\pm}$, Z°, γ) are in fact composite states. The elementary fields are only the members of the fundamental multiplet (92). We believe that none of the particles we know is among them, they should all be obtained as bound states.

N = 8 supergravity promises to give us a truly unified theory of all interactions, including gravitation, and a description of the world in terms of a single fundamental multiplet. We still have many problems to solve before reaching this final step, and I shall list some of them: (i) Is it a consistent field theory? The Lagrangian is certainly non renormalizable, but the large number of supersymmetries and the good convergence properties that often accompany them may make the problem tractable. (ii) Can we solve the bound-state problem and obtain the known particles out of the fields of the fundamental supermultiplet? (iii) It seems that only N = 1 supersymmetry may be relevant for low-energy phenomenology. Is it possible to break N = 8 into N = 1 spontaneously? (iv) Can N = 8 accommodate chiral fermions? The problems are formidable but the expectations are great!

COSMOLOGY AND PARTICLE PHYSICS

Michael S. Turner

NASA/Fermilab Astrophysics Center
Fermi National Accelerator Laboratory
Batavia, Illinois 60510

and

Departments of Physics and Astronomy and Astrophysics
Enrico Fermi Institute
The University of Chicago
Chicago, Illinois 60637

INTRODUCTION

In the past five years or so progress in both elementary particle physics and in cosmology has become increasingly dependent upon the interplay between the two disciplines. On the particle physics side, the $SU(3)_C \times SU(2)_L \times U(1)_Y$ model seems to very accurately describe the interactions of quarks and leptons at energies below, say, 10^3 GeV. At the very least, the so-called standard model is a satisfactory, effective low energy theory. The frontiers of particle physics now involve energies of much greater than 10^3 GeV--energies which are not now available in terrestrial accelerators, nor are ever likely to be available in terrestrial accelerators. For this reason particle physicists have turned both to the early Universe with its essentially unlimited energy budget (up to 10^{19} GeV) and high particle fluxes (up to 10^{107} cm^{-2} s^{-1}), and to various unique, contemporary astrophysical environments (centers of main sequence stars where temperatures reach 10^8 K, neutron stars where densities reach 10^{14}-10^{15} g cm^{-3}, our galaxy whose magnetic field can impart 10^{11} GeV to a Dirac magnetic charge, etc.) as non-traditional laboratories for studying physics at very high energies and very short distances.

On the cosmological side, the hot big bang model, the so called standard model of cosmology, seems to provide an accurate accounting of the history of the Universe from about 10^{-2} s after 'the bang' when the temperature was about 10 MeV, until today, some 10-20 billion years after 'the bang' and temperature of about 3 K ($\simeq 3 \times 10^{-13}$ GeV). Extending our understanding further back, to earlier times and higher temperatures, requires knowledge about the fundamental particles (presumably quarks and leptons) and their interactions at very high energies. For this reason, progress in cosmology has become linked to progress in elementary particle physics.

In these 4 lectures I will try to illustrate the two-way nature of the interplay between these fields by focusing on a few selected topics. In Lecture 1 I will review the standard cosmology, especially concentrating on primordial nucleosynthesis, and discuss how the standard cosmology has been used to place constraints on the properties of various particles. Grand Unification makes two striking predictions: (1) B non-conservation; (2) the existence of stable, superheavy magnetic monopoles. Both have had great cosmological impact. In Lecture 2 I will discuss baryogenesis, the very attractive scenario in which the B, C, CP violating interactions in GUTs provide a dynamical explanation for the predominance of matter over antimatter, and the present baryon-to-photon ratio. Baryogenesis is so cosmologically attractive, that in the absence of observed proton decay it has been called 'the best evidence for some kind of unification.' Monopoles are a cosmological disaster, and an astrophysicist's delight. In Lecture 3 I will discuss monopoles, cosmology, and astrophysics. To date, the most important 'cosmological payoff' of the Inner Space/Outer Space connection is the inflationary Universe scenario. In Lecture 4 I will discuss how a very early ($t \lesssim 10^{-34}$ sec) phase transition associated with spontaneous symmetry breaking (SSB) has the potential to explain a handful of very fundamental cosmological facts, facts which can be accommodated by the standard cosmology, but which are not 'explained' by it. The 5th Lecture will be devoted to a discussion of structure formation in the Universe. For at least a decade cosmologists have had a general view of how structure developed, but have been unable to fill in details because of the lack of knowledge of the initial data for the problem (quantity and composition of the matter in the Universe and the nature of the initial density perturbations). The study of the very early Universe has provided us with important hints as to the initial data, and this has led to significant progress in our understanding of how structure formed in the Universe.

By selecting just 5 topics I have left out some other very important and interesting topics -- supersymmetry/supergravity and cosmology, superstrings and cosmology in extra dimensions, and axions, astrophysics, and cosmology -- to mention just a few. I refer the interested reader to references 1-3.

LECTURE 1 -- THE STANDARD COSMOLOGY

The hot big bang model nicely accounts for the universal (Hubble) expansion, the 2.7 K cosmic microwave background radiation, and through primordial nucleosynthesis, the abundances of D, ^{4}He and perhaps also ^{3}He and ^{7}Li. Light received from the most distant objects observed (QSOs at redshifts $\simeq 3.5$) left these objects when the Universe was only a few billion years old, and so observations of QSOs allow us to directly probe the history of the Universe to within a few billion years of 'the bang'. The surface of last scattering for the microwave background is the Universe about 100,000 yrs after the bang when the temperature was about 1/3 eV. The microwave background is a fossil record of the Universe at that very early epoch. In the standard cosmology an epoch of nucleosynthesis takes place from $t \simeq 10^{-2}$ s - 10^2 s when the temperature was $\simeq$ 10 MeV - 0.1 MeV. The light elements synthesized, primarily D, ^{3}He, ^{4}He, and ^{7}Li, are relics from this early epoch, and comparing their predicted big bang abundances with their inferred primordial abundances is the most stringent test of the standard cosmology we have at present. [Note that I must say inferred primordial abundance because contemporary astrophysical processes can affect the abundance of these light isotopes, e.g., stars very efficiently burn D, and produce ^{4}He.] At present the standard cosmology passes this test with flying colors (as we shall see shortly).

On the large scale (>> 100 Mpc), the Universe is isotropic and homogenous, and so it can accurately be described by the

Robertson-Walker line element

$$ds^2=-dt^2+R(t)^2[dr^2/(1-kr^2)+r^2\,d\theta^2+r^2\sin^2\theta d\phi^2], \quad (1.1)$$

where ds^2 is the proper separation between two spacetime events, $k = 1$, 0, or -1 is the curvature signature, and $R(t)$ is the cosmic scale factor. The expansion of the Universe is embodied in $R(t)$--as $R(t)$ increases all proper (i.e., measured by meter sticks) distances scale with $R(t)$, e.g., the distance between two galaxies comoving with the expansion (i.e., fixed r, θ, ϕ), or the wavelength of a freely-propagating photon ($\lambda \propto R(t)$). The $k > 0$ spacetime has positive spatial curvature and is finite in extent; the $k < 0$ spacetime has negative spatial curvature and is infinite in extent; the $k = 0$ spacetime is spatially flat and is also infinite in extent.

The evolution of the cosmic scale factor is determined by the Friedmann equations:

$$H^2 \equiv (\dot{R}/R)^2 = 8\pi G\rho/3 - k/R^2, \quad (1.2)$$

$$d(\rho R^3) = -p\, d(R^3), \quad (1.3)$$

where ρ is the total energy density and p is the pressure. The expansion rate H (also called the Hubble parameter) sets the characteristic time for the growth of $R(t)$; $H^{-1} \approx$ e-folding time for R. The present value of H is $100\ h\ \mathrm{kms}^{-1}\ \mathrm{Mpc}^{-1} \approx h\ (10^{10}\ \mathrm{yr})^{-1}$; the observational data strongly suggest that $1 \gtrsim h \gtrsim 1/2$ (ref. 4). As it is apparent from Eqn. 1.2 model Universes with $k \lesssim 0$ expand forever, while a model Universe with $k > 0$ must eventually recollapse. The sign of k (and hence the geometry of spacetime) can be determined from measurements of ρ and H:

$$k/H^2R^2 = \rho/(3H^2/8\pi G) - 1, \quad (1.4)$$

$$\equiv \Omega - 1,$$

where $\Omega = \rho/\rho_{crit}$ and $\rho_{crit} = 3H^2/8\pi G \approx 1.88\ h^2 \times 10^{-29}\ \mathrm{gcm}^{-3}$. The cosmic surveying required to directly determine ρ is far beyond our capabilities (i.e., weigh a cube of cosmic material 10^{25} cm on a side!). However, based upon the amount of luminous matter (i.e., baryons in stars) we can set a lower limit to Ω: $\Omega \gtrsim \Omega_{lum} \approx 0.01$. The best upper limit to Ω follows by considering the age of the Universe:

$$t_U = 10^{10}\ \mathrm{yr}\ h^{-1}\ f(\Omega), \quad (1.5)$$

where $f(\Omega) \lesssim 1$ and is monotonically decreasing (e.g., $f(0) = 1$ and $f(1) = 2/3$). The ages of the oldest stars (in globular clusters) strongly suggest that $t_U \gtrsim 10^{10}$ yr; combining this with Eqn. 1.5 implies that: $\Omega f^2(\Omega) \gtrsim \Omega h^2$. The function Ωf^2 is monotonically increasing and asymptotically approaches $(\pi/2)^2$, implying that independent of h, $\Omega h^2 \lesssim 2.5$. Restricting h to the interval (1/2, 1) it follows that: $\Omega h^2 \lesssim 0.8$ and $\Omega \lesssim 3.2$.

The energy density contributed by nonrelativistic matter varies as $R(t)^{-3}$--due to the fact that the number density of particles is diluted by the increase in the proper (or physical) volume of the Universe as it expands. For relativistic particles the energy density varies as $R(t)^{-4}$, the extra factor of R due to the redshifting of the particle's momentum (recall $\lambda \propto R(t)$). The energy density contributed by a relativistic species ($T \gg m$) at temperature T is

$$\rho = g_{eff}\pi^2T^4/30, \quad (1.6)$$

where g_{eff} is the number of degrees of freedom for a bosonic species, and 7/8 that number for a fermionic species. Note that $T \propto R(t)^{-1}$. Here and throughout I have taken $\hbar = c = k_B = 1$, so that 1 GeV = $(1.97 \times 10^{-14}\ \mathrm{cm})^{-1}$ = $(1.16 \times 10^{13}\ \mathrm{K})$ = $(6.57 \times 10^{-25}\ \mathrm{s})^{-1}$, $G = m_{pl}^{-2}$ ($m_{pl} = 1.22 \times 10^{19}$ GeV), and 1 $\mathrm{GeV}^4 = 2.32 \times 10^{17}\ \mathrm{g\ cm^{-3}}$. By the way, 1 light year $\approx 10^{18}$ cm; 1 pc $\approx$ 3 light year; and 1 Mpc $\approx 3 \times 10^{24}$ cm $\approx 1.6 \times 10^{38}\ \mathrm{GeV}^{-1}$.

Today, the energy density contributed by relativistic particles (photons and 3 neutrino species) is negligible: $\Omega_{rel} \approx 4 \times 10^{-5} h^{-2} (T/2.7\ \mathrm{K})^4$. However, since $\rho_{rel} \propto R^{-4}$, while $\rho_{nonrel} \propto R^{-3}$, early on relativistic species dominated the energy density. For $R/R(\mathrm{today}) < 4 \times 10^{-5} (\Omega h^2)^{-1} (T/2.7\ \mathrm{K})^4$, which corresponds to $t < 4 \times 10^{10}\ \mathrm{s}\ (\Omega h^2)^{-2} (T/2.7\ \mathrm{K})^6$ and $T > 6\ \mathrm{eV}\ (\Omega h^2)(2.7\ \mathrm{K}/T)^3$, the energy density of the Universe was dominated by relativistic particles. Since the curvature term varies as $R(t)^{-2}$, it too was small compared to the energy density contributed by relativistic particles early on, and so Eqn. 1.2 simplifies to:

$$H \equiv (\dot{R}/R) \approx (4\pi^3 g_*/45)^{1/2}\, T^2/m_{pl}, \tag{1.7}$$

$$\approx 1.66\, g_*^{1/2}\, T^2/m_{pl},$$

$$(\text{valid for } t \lesssim 10^{10}\ \mathrm{s},\ T \gtrsim 10\ \mathrm{eV}).$$

Here g_* counts the total number of effective degrees of freedom of all the relativistic particles (i.e., those species with mass $\ll T$):

$$g_* = \sum_{\mathrm{Bose}} g_i (T_i/T)^4 + 7/8 \sum_{\mathrm{Fermi}} g_i (T_i/T)^4 \quad , \tag{1.8}$$

where T_i is the temperature of species i, and T is the photon temperature. For example: $g_*(3\ \mathrm{K}) \approx 3.36$ (γ, 3 $\nu\bar{\nu}$); $g_*(\text{few MeV}) \approx 10.75$ (γ, $e^\pm$, 3 $\bar{\nu}\nu$); $g_*(\text{few 100 GeV}) \approx 110$ (γ, $W^\pm$ Z^0, 8 gluons, 3 families of quarks and leptons, and 1 Higgs doublet).

If thermal equilibrium is maintained, then the second Friedmann equation, Eqn. 1.3 - conservation of energy, implies that the entropy per comoving volume (a volume with fixed r, θ, ϕ coordinates) $S \propto sR^3$ remains constant. Here s is the entropy density, which is dominated by the contribution from relativistic particles, and is given by:

$$s = (\rho + p)/T \approx 2\pi^2 g_* T^3/45. \tag{1.9}$$

The entropy density s itself is proportional to the number density of relativistic particles. So long as the expansion is adiabatic (i.e., in the absence of entropy production) S (and s) will prove to be useful fiducials. For example, at low energies ($E \ll 10^{14}$ GeV) baryon number is effectively conserved, and so the net baryon number per comoving volume $N_B \propto n_B (\equiv n_b - n_{\bar{b}}) R^3$ remains constant, implying that the ratio n_B/s is a constant of the expansion. Today $s \approx 7 n_\gamma$, so that $n_B/s \approx \eta/7$, where $\eta \equiv n_b/n_\gamma$ is the baryon-to-photon ratio, which as we shall soon see, is known from primordial nucleosynthesis to be in the range: $4 \times 10^{-10} \lesssim \eta \lesssim 7 \times 10^{-10}$. The fraction of the critical density contributed by baryons (Ω_b) is related to η by:

$$\Omega_b \approx 3.53 \times 10^{-3} (\eta/10^{-10}) h^{-2} (T/2.7\ \mathrm{K})^3. \tag{1.10}$$

Whenever $g_* \approx$ constant, the constancy of the entropy per comoving volume implies that $T \propto R^{-1}$; together with Eqn. 1.7 this gives

$$R(t) = R(t_o)(t/t_o)^{1/2}, \tag{1.11}$$

$$t \simeq 0.3\, g_*^{-1/2}\, m_{pl}/T^2,$$
$$\simeq 2.4 \times 10^{-6}\ \text{s}\ g_*^{-1/2}\, (T/\text{GeV})^{-2}, \tag{1.12}$$

valid for $t \lesssim 10^{10}$ s and $T \gtrsim 10$ eV.

Finally, let me mention one more important feature of the standard cosmology, the existence of particle horizons. The distance that a light signal could have propagated since the bang is finite, and easy to compute. Photons travel on paths characterized by $ds^2 = 0$; for simplicity (and without loss of generality) consider a trajectory with $d\theta = d\phi = 0$. The coordinate distance covered by this photon since 'the bang' is just $\int_0^t dt'/R(t')$, corresponding to a physical distance (measured at time t) of

$$d_H(t) = R(t) \int_0^t dt'/R(t') \tag{1.13}$$
$$= t/(1-n) \qquad [\text{for } R \propto t^n,\ n < 1].$$

If $R \propto t^n$ $(n < 1)$, then the horizon distance is finite and $\simeq t \simeq H^{-1}$. Note that even if $d_H(t)$ diverges (e.g., if $R \propto t^n$, $n \geq 1$), the Hubble radius H^{-1} still sets the scale for the 'physics horizon'. Since all physical lengths scale with $R(t)$, they e-fold in a time of $O(H^{-1})$. Thus a coherent microphysical process can only operate over a time interval $\lesssim O(H^{-1})$, implying that a causally-coherent microphysical process can only operate over distances $\lesssim O(H^{-1})$.

During the radiation-dominated epoch $n = 1/2$ and $d_H = 2t$; the baryon number and entropy within the horizon at time t are easily computed:

$$S_{HOR} = (4\pi/3)t^3\, s,$$
$$\simeq 0.05\, g_*^{-1/2}\, (m_{pl}/T)^3; \tag{1.14}$$

$$N_{B-HOR} = (n_B/s) \times S_{HOR},$$
$$\simeq 10^{-12}\, (m_{pl}/T)^3; \tag{1.15a}$$
$$\simeq 10^{-2}\, M_\odot (T/\text{MeV})^{-3}; \tag{1.15b}$$

where I have assumed that n_B/s has remained constant and has the value $\simeq 10^{-10}$. A solar mass ($M_\odot$) of baryons is $\simeq 1.2 \times 10^{57}$ baryons (or 2×10^{33} g).

Although our verifiable knowledge of the early history of the Universe only takes us back to $t \simeq 10^{-2}$ s and $T \simeq 10$ MeV (the epoch of primordial nucleosynthesis), nothing in our present understanding of the laws of physics suggests that it is unreasonable to extrapolate back to times as early as $\simeq 10^{-43}$ s and temperatures as high as $\simeq 10^{19}$ GeV. At high energies the interactions of quarks and leptons are asymptotically free (and/or weak) justifying the dilute gas approximation made in Eqn. 1.6. At energies below 10^{19} GeV quantum corrections to General Relativity are expected to be small. I hardly need to remind the reader that 'reasonable' does not necessarily mean 'correct'. Making this extrapolation, I have summarized 'The Complete History of the Universe' in Fig. 1.1. [For more complete reviews of the standard cosmology I refer the interested reader to refs. 5 and 6.]

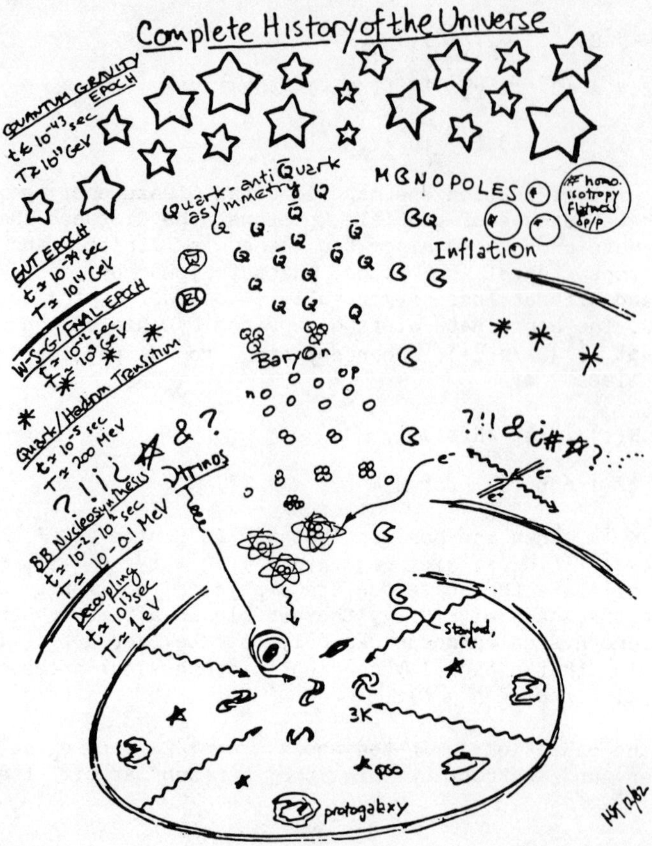

Fig. 1.1 'The Complete History of the Universe'. Highlights include: decoupling ($t \simeq 10^{13}$ s, $T \simeq 1/3$ eV) - the surface of last scattering for the cosmic microwave background, epoch after which matter and radiation cease to interact and matter 'recombines' into neutral atoms (D, ^{3}He, ^{4}He, ^{7}Li); also marks the beginning of the formation of structure; primordial nucleosynthesis ($t \simeq 10^{-2}$ s, $T \simeq 10$ MeV) - epoch during which all of the free neutrons and some of the free protons are synthesized into D, ^{3}He, ^{4}He, and ^{7}Li, and the surface of last scattering for the cosmic neutrino backgrounds; quark/hadron transition ($t \simeq 10^{-5}$ s, $T \simeq$ few 100 MeV) - epoch of 'quark enslavement' [confinement transition in SU(3)]; W-S-G epoch associated with electroweak breaking, SU(2) x U(1) → U(1); GUT epoch (?? $t \simeq 10^{-34}$ s, $T \simeq 10^{14}$ GeV??) - SSB of the GUT, during which the baryon asymmetry of the Universe evolves, monopoles are produced, and 'inflation' may occur; the Quantum Gravity Wall ($t \simeq 10^{-43}$ s, $T \simeq 10^{19}$ GeV).

Primordial Nucleosynthesis

At present the most stringent test of the standard cosmology is big bang nucleosynthesis. Here I will briefly review primordial nucleosynthesis, discuss the concordance of the predictions with the observations, and mention one example of how primordial nucleosynthesis has been used as a probe of particle physics--counting the number of light neutrino species.

The two fundamental assumptions which underlie big bang nucleosynthesis are: the validity of General Relativity and that the Universe was once hotter than a few MeV. An additional assumption (which, however, is not necessary) is that the lepton number, $n_L/n_\gamma = (n_{e^-} - n_{e^+})/n_\gamma + (n_\nu - n_{\bar\nu})/n_\gamma \approx \eta + (n_\nu - n_{\bar\nu})/n_\gamma$, like the baryon number ($\approx \eta$) is small. Having swallowed these assumptions, the rest follows like 1-2-3.

Frame 1: $t \approx 10^{-2}$ sec, $T \approx 10$ MeV. The energy density of the Universe is dominated by relativistic species: γ, e^+e^-, $\nu_i\bar\nu_i$ (i = e, μ, τ,...); $g_* \approx 10.75$ (assuming 3 neutrino species). Thermal equilibrium is maintained by weak interactions ($e^+ + e^- \leftrightarrow \nu_i + \bar\nu_i$, $e^+ + n \leftrightarrow p + \bar\nu_e$, $e^- + p \leftrightarrow n + \nu_e$) as well as electromagnetic interactions ($e^+ + e^- \leftrightarrow \gamma + \gamma$, $\gamma + p \leftrightarrow \gamma + p$, etc.), both of which are occurring rapidly compared to the expansion rate $H = \dot R/R$. Thermal equilibrium implies that $T_\nu = T_\gamma$ and that $n/p = \exp(-\Delta m/T)$; here n/p is the neutron to proton ratio and $\Delta m = m_n - m_p$. No nucleosynthesis is occurring yet because of the tiny equilibrium abundance of D: $n_D/n_b \approx \eta \exp(2.2 \text{ MeV}/T) \approx 10^{-10}$, where n_b, n_D, and n_γ are the baryon, deuterium, and photon number densities, and 2.2 MeV is the binding energy of the deuteron. This is the so-called deuterium bottleneck.

Frame 2: $t \approx 1$ sec, $T \approx 1$ MeV. At about this temperature the weak interaction rates become slower than the expansion rate and thus weak interactions effectively cease occurring. The neutrinos decouple and thereafter expand adiabatically ($T_\nu \propto R^{-1}$). This epoch is the surface of last scattering for the neutrinos; detection of the cosmic neutrino seas would allow us to directly view the Universe as it was 1 sec after 'the bang'. From this time forward the neutron to proton ratio no longer 'tracks' its equilibrium value, but instead 'freezes out' a value $\approx 1/6$, very slowly decreasing, due to occasional free neutron decays. A little bit later ($T \approx m_e/3$), the $e^\pm$ pairs annihilate and transfer their entropy to the photons, heating the photons relative to the neutrinos, so that from this point on $T_\nu \approx (4/11)^{1/3}T_\gamma$. The 'deuterium bottleneck' continues to operate, preventing nucleosynthesis.

Frame 3: $t \approx 200$ sec, $T \approx 0.1$ MeV. At about this temperature the 'deuterium bottleneck' breaks [$n_D/n_b \approx \eta \exp(2.2 \text{ MeV}/T) \approx 1$], and nucleosynthesis begins in earnest. Essentially all the neutrons present ($n/p \approx 1/7$) are quickly incorporated first into D, and then into ^{4}He nuclei. Trace amounts of D and ^{3}He remain unburned; substantial nucleosynthesis beyond ^{4}He is prevented by the lack of stable isotopes with A = 5 and 8, and by coulomb barriers. A small amount of ^{7}Li is synthesized by ^{4}He(t, γ)^{7}Li (for $\eta \lesssim 3 \times 10^{-10}$) and by ^{4}He(^{3}He, γ)^{7}Be followed by the eventual β-decay of ^{7}Be to ^{7}Li (for $\eta \gtrsim 3 \times 10^{-10}$).

The nucleosynthetic yields depend upon η, N_ν (which I will use to parameterize the number of light ($\lesssim 1$ MeV) species present, other than γ and $e^\pm$), and in principle all the nuclear reaction rates which go into the reaction network. In practice, most of the rates are known to sufficient precision that the yields only depend upon a few rates. ^{4}He production depends only upon η, N_ν, and $\tau_{1/2}$, the neutron half-life, which determines the rates for all the weak processes which interconvert neutrons and protons. The mass fraction Y_p of ^{4}He produced increases monotonically with increasing values of η, N_ν, and $\tau_{1/2}$ - a fact which is simple to understand. Larger η means that the 'deuterium bottleneck' breaks earlier, when the value of n/p is larger. More light species (i.e., larger value of N_ν) increases the expansion rate (since $H \propto (G\rho)^{1/2}$), while a larger value of $\tau_{1/2}$ means slower weak interaction rates ($\propto \tau_{1/2}^{-1}$) - both effects cause the weak interactions to freeze out earlier, when n/p is larger. The yield of ^{4}He is determined by the n/p ratio when nucleosynthesis commences, $Y_p \approx 2(n/p)/(1 + n/p)$, so that a higher n/p ratio means more ^{4}He is synthesized. At present the value of the neutron half-life is only known to an accuracy of about 2%: $\tau_{1/2}$ = 10.6 min ± 0.2 min. Since ν_e and ν_μ are known (from laboratory measurements) to be light, $N_\nu \gtrsim 2$. Based upon the luminous matter in galaxies, η is known to be $\gtrsim 0.3 \times 10^{-10}$. If all the mass in binary galaxies and small groups of galaxies (as inferred by dynamical measurements) is baryonic, then η must be $\gtrsim 2 \times 10^{-10}$.

To an accuracy of about 10%, the yields of D and ^{3}He only depend upon η, and decrease rapidly with increasing η. Larger η corresponds to a higher nucleon density and earlier nucleosynthesis, which in turn results in less D and ^{3}He remaining unprocessed. Because of large uncertainties in the rates of some reactions which create and destroy ^{7}Li, the predicted primordial abundance of ^{7}Li is only accurate to within about a factor of 2.

In 1946 Gamow[7] suggested the idea of primordial nucleosynthesis. In 1953, Alpher, Follin, and Herman[8] all but wrote a code to determine the primordial production of ^{4}He. Peebles[9] (in 1966) and Wagoner, Fowler, and Hoyle[10] (in 1967) wrote codes to calculate the primordial abundances. Yahil and Beaudet[11] (in 1976) independently developed a nucleosynthesis code and also extensively explored the effect of large lepton number ($n_\nu - n_{\bar\nu} \approx O(n_\gamma)$) on primordial nucleosynthesis. Wagoner's 1973 code[12] has become the 'standard code' for the standard model. In 1981 the reaction rates were updated by Olive et al.[13], the only significant change which resulted was an increase in the predicted ^{7}Li abundance by a factor of O(3). In 1982 Dicus et al.[14] corrected the weak rates in Wagoner's 1973 code for finite temperature effects and radiative/coulomb corrections, which led to a systematic decrease in Y_p of about 0.003. Figs. 1.2, 1.3 show the predicted abundances of D, ^{3}He, ^{4}He, and ^{7}Li, as calculated by the most up to date version of Wagoner's 1973 code.[15] The numerical accuracy of the predicted abundances is about 1%. Now let me discuss how the predicted abundances compare with the observational data. [This discussion is a summary of the collaborative work in ref. 15.]

The abundance of D has been determined in solar system studies and in UV absorption studies of the local interstellar medium (ISM). The solar system determinations are based upon measuring the abundances of deuterated molecules in the atmosphere of Jupiter and inferring the pre-solar (i.e., at the time of the formation of the solar system) D/H ratio from meteoritic and solar data on the abundance of ^{3}He. These determinations are consistent with a pre-solar value of (D/H) $\approx$ (2 ± 1/2) x 10^{-5}. An average ISM value for (D/H) $\approx 2 \times 10^{-5}$ has been derived from UV absorption studies of the local ISM ($\lesssim$ few 100 pc), with individual measurements spanning the range $(1 - 4) \times 10^{-5}$. Note that these measurements are consistent with the solar system determinations of D/H.

The deuteron being very weakly-bound is easily destroyed and hard to produce, and to date, it has been difficult to find an astrophysical site where D can be produced in its observed abundance.[16] Thus, it is generally accepted that the presently-observed deuterium abundance provides a lower bound to the primordial abundance. Using $(D/H)_p \geq 1 \times 10^{-5}$ it follows that η must be less than about 10^{-9} in order for the predictions of primordial nucleosynthesis to be concordant with the observed abundance of D. [Note: because of the rapid variation of $(D/H)_p$ with η, this upper bound to η is rather insensitive to the precise lower bound to $(D/H)_p$ used.] Using Eqn. 1.10 to relate η to Ω_b, this implies an upper bound to Ω_b: $\Omega_b \lesssim 0.035 h^{-2} (T/2.7K)^3 \lesssim 0.19$ -- baryons alone cannot close the Universe. One would like to also exploit the sensitive dependence of $(D/H)_p$ upon η to derive a lower bound to η for concordance; this is not possible because D is so easily destroyed. However, as we shall soon see, this end can be accomplished instead by using both D and ^3He.

The abundance of ^3He has been measured in solar system studies and by observations of the $^3He^+$ hyperfine line in galactic HII regions (the analog of the 21 cm line of H). The abundance of ^3He in the solar wind has been determined by analyzing gas-rich meteorites, lunar soil, and the foil placed upon the surface of the moon by the Apollo astronauts. Since D is burned to ^3He during the sun's approach to the main sequence, these measurements represent the pre-solar sum of D and ^3He. These determinations of D + ^3He are all consistent with a pre-solar $[(D + {}^3He)/H] \simeq (4.0 \pm 0.3) \times 10^{-5}$. Earlier measurements of the $^3He^+$ hyperfine line in galactic HII regions and very recent measurements lead to derived present abundances of ^3He: $^3He/H \approx (3\text{-}20) \times 10^{-5}$. The fact that these values are higher than the pre-solar abundance is consistent with the idea that the abundance of ^3He should increase with time due to the stellar production of ^3He by low mass stars.

^3He is much more difficult to destroy than D. It is very hard to efficiently dispose of ^3He without also producing heavy elements or large amounts of ^4He (environments hot enough to burn ^3He are usually hot enough to burn protons to ^4He). In ref. 15 we have argued that in the absence of a Pop III generation of very exotic stars which process essentially all the material in the Universe and in so doing destroy most of the ^3He without overproducing ^4He or heavy elements, ^3He can have been astrated (i.e. reduced by stellar burning) by a factor of no more than $f_a \approx 2$. [The youngest stars, e.g. our sun, are called Pop I; the oldest observed stars are called Pop II. Pop III refers to a yet to be discovered, hypothetical first generation of stars.] Using this argument and the inequality

$$[(D+{}^3He)/H]_p \lesssim \text{pre-solar}(D/H)+f_a\ \text{pre-solar}({}^3He/H) \qquad (1.16)$$
$$\lesssim (1-f_a)\text{pre-solar}(D/H)+f_a\text{pre-solar}(D+{}^3He)/H;$$

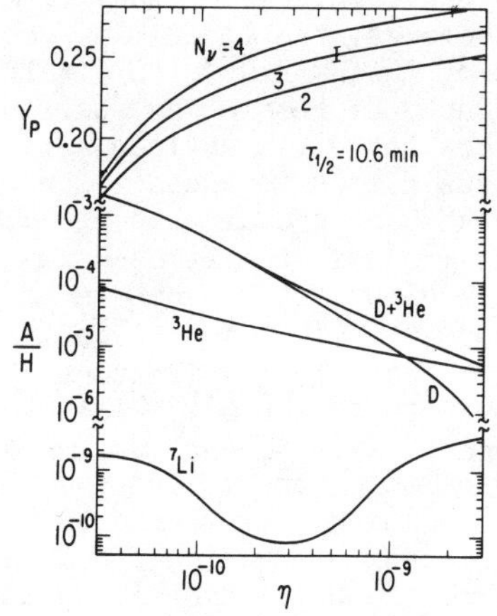

Fig. 1.2 The predicted primordial abundances of D, ^3He, ^4He, and ^7Li. [Note $\tau_{1/2}$ = 10.6 min was used; error bar shows $\Delta\tau_{1/2} = \pm 0.2$ min; Y_p = mass of ^4He; N_ν = equivalent number of light neutrino species.] Inferred primordial abundances: $Y \simeq 0.23\text{-}0.25$; $(D/H) > 1 \times 10^{-5}$; $(D + {}^3He)/H \lesssim 10^{-4}$; $^7Li/H \simeq (1.1 \pm 0.4) \times 10^{-10}$. Concordance requires: $\tilde{\eta} \simeq (4\text{-}7) \times 10^{-10}$ and $N_\nu \lesssim 4$.

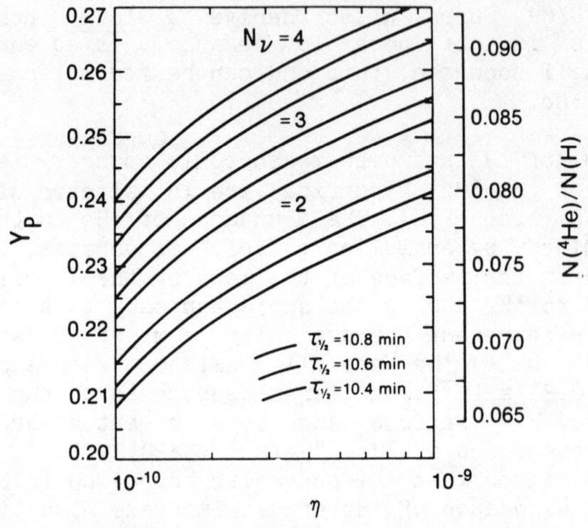

Fig. 1.3 The predicted primordial abundance of ^4He. Note that Y_p increases with increasing values of $\tau_{1/2}$, η, and N_ν. Hence lower bounds to η and $\tau_{1/2}$ and an upper bound to Y_p imply an upper bound to N_ν. Taking $\tau_{1/2} \approx 10.4$ min, $\eta \gtrsim 4 \times 10^{-10}$ (based on D + ^3He production), and $Y_p \lesssim 0.25$, it follows that N_ν must be $\lesssim 4$.

the presolar abundances of D and D + ^3He can be used to derive an upper bound to the primordial abundance of D + ^3He: $[(D + {}^3He)/H]_p \lesssim 8 \times 10^{-5}$. [For a very conservative astration factor, $f_a \approx 4$, the upper limit becomes 13×10^{-5}.] Using 8×10^{-5} as an upper bound on the primordial D + ^3He production implies that for concordance, η must be greater than 4×10^{-10} (for the upper bound of 13×10^{-5}, η must be greater than 3×10^{-10}). To summarize, consistency between the predicted big bang abundances of D and ^3He, and the derived abundances observed today requires η to lie in the range $\approx (4 - 10) \times 10^{-10}$.

Until very recently, our knowledge of the ^7Li abundance was limited to observations of meteorites, the local ISM, and Pop I stars, with a derived present abundance of $^7Li/H \approx 10^{-9}$ (to within a factor of 2). Given that ^7Li is produced by cosmic ray spallation and some stellar processes, and is easily destroyed (in environments where $T \gtrsim 2 \times 10^6$K), there is not the slightest reason to suspect (or even hope!) that this value accurately reflects the primordial abundance. Recently, Spite and Spite[17] have observed ^7Li lines in the atmospheres of 13 unevolved halo and old disk stars with very low metal abundances ($Z_\odot/12 - Z_\odot/250$), whose masses span the range of $\approx (0.6 - 1.1)M_\odot$. Stars less massive than about 0.7 $M_\odot$ are expected to astrate (by factors $\gtrsim$ O(10)) their ^7Li abundance during their approach to the MS, while stars more massive than about 1 $M_\odot$ are not expected to significantly astrate ^7Li in their outer layers. Indeed, they see this trend in their data, and deduce a primordial ^7Li abundance of: $^7Li/H \approx (1.12 \pm 0.38) \times 10^{-10}$. Remarkably, this is the predicted big bang production for η in the range $(2 - 5) \times 10^{-10}$. If we take this to be the primordial ^7Li abundance, and allow for a possible factor of 2 uncertainty in the predicted abundance of Li (due to estimated uncertainties in the reaction rates which affect ^7Li), then concordance for ^7Li restricts η to the range $(1 - 7) \times 10^{-10}$. Note, of course, that their derived ^7Li abundance is the pre-Pop II abundance, and may not necessarily reflect the true primordial abundance (e.g., if a Pop III generation of stars processed significant amounts of material).

In sum, the concordance of big bang nucleosynthesis predictions with the derived abundancess of D and ^{3}He requires $\eta \simeq (4 - 10) \times 10^{-10}$; moreover, concordance for D, ^{3}He, and ^{7}Li further restricts η: $\eta \simeq (4 - 7) \times 10^{-10}$.

In the past few years the quality and quantity of ^{4}He observations has increased markedly. In Fig. 1.4 all the ^{4}He abundance determinations derived from observations of recombination lines in HII regions (galactic and extragalactic) are shown as a function of metalicity Z (more precisely, 2.2 times the mass fraction of ^{16}O).

Since ^{4}He is also synthesized in stars, some of the observed ^{4}He is <u>not</u> primordial. Since stars also produce metals, one would expect some correlation between Y and Z, or at least a trend: lower Y where Z is lower. Such a trend is apparent in Fig. 1.4. From Fig. 1.4 it is also clear that there is a large primordial component to ^{4}He: $Y_p \simeq 0.22 - 0.26$. Is it possible to pin down the value of Y_p more precisely?

There are many steps in going from the line strengths (what the observer actually measures), to a mass fraction of ^{4}He (e.g., corrections for neutral ^{4}He, reddening, etc.). In galactic HII regions, where abundances can be determined for various positions within a given HII region, variations are seen within a given HII region. Observations of extragalactic HII regions are actually observations of a superposition of <u>several</u> HII regions. Although observers have quoted statistical uncertainties of $\Delta Y \simeq \pm 0.01$ (or lower), from the scatter in Fig. 1.4 it is clear that the systematic uncertainties must be larger. For example, different observers have derived ^{4}He abundances of between 0.22 and 0.25 for I Zw18, an extremely metal-poor dwarf emission line galaxy.

Perhaps the safest way to estimate Y_p is to concentrate on the ^{4}He determinations for metal-poor objects. From Fig. 1.4 $Y_p \simeq 0.23 - 0.25$ appears to be consistent with all the data (although Y_p as low as 0.22 or high as 0.26 could not be ruled out). Recently Kunth and Sargent[18] have studied 13 metal-poor ($Z \lesssim Z_\odot/5$) Blue Compact galaxies. From a weighted average for their sample they derive a <u>primordial abundance</u> $Y_p \simeq 0.245 \pm 0.003$; allowing for a 3σ variation this suggests $0.236 \lesssim Y_p \lesssim 0.254$.

For the concordance range deduced from D, ^{3}He, and ^{7}Li ($\eta \gtrsim 4 \times 10^{-10}$) and $\tau_{1/2} \gtrsim 10.4$ min, the predicted ^{4}He abundance is

$$Y_p \gtrsim \begin{cases} 0.230 & N_\nu = 2. \\ 0.244 & = 3. \\ 0.256 & = 4. \end{cases}$$

[Note, that $N_\nu = 2$ is permitted only if the τ-neutrino is heavy ($\gtrsim$ few MeV) and unstable; the present experimental upper limit on its mass is 160 MeV.] Thus, since $Y_p \simeq 0.23 - 0.25$ (0.22 - 0.26?) there are values of η, N_ν, and $\tau_{1/2}$ for which there is agreement between the abundances predicted by big bang nucleosynthesis and the primordial abundances of D, ^{3}He, ^{4}He, and ^{7}Li derived from observational data.

To summarize, the only isotopes which are predicted to be produced in significant amounts during the epoch of primordial nucleosynthesis are: D, ^{3}He, ^{4}He, and ^{7}Li. At present there is concordance between the predicted primordial abundances of all 4 of these elements and their observed abundances for values of N_ν, $\tau_{1/2}$, and η in the following intervals: $2 \lesssim N_\nu \lesssim 4$; 10.4 min $\lesssim \tau_{1/2} \lesssim$ 10.8 min; and $4 \times 10^{-10} \lesssim \eta \lesssim 7 \times 10^{-10}$ (or 10×10^{-10} if the ^{7}Li abundance is not used). This is a

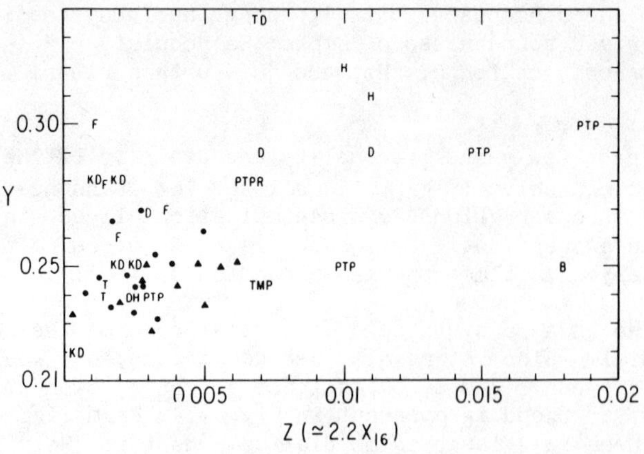

Fig. 1.4 Summary of ^4He abundance determinations (galactic and extragalactic) from recombination lines in HII regions vs. mass fraction of heavy ($A \gtrsim 12$) elements Z ($\simeq$ 2.2 mass fraction of ^{16}O). Note, observers do not usually quote errors for individual objects--scatter is probably indicative of the uncertainties. The triangles and filled circles represent two data sets of note: circles - 13 very metal poor emission line galaxies (Kunth and Sargent[18]); triangles - 9 metal poor, compact galaxies (Lequeux etal.[18]).

truly remarkable achievement, and strong evidence that the standard model is valid back as early as 10^{-2} sec after 'the bang'.

The standard model will be in serious straights if the primordial mass fraction of ^4He is unambiguously determined to be less than 0.22. What alternatives exist if $Y_p \lesssim 0.22$? If a generation of Pop III stars which efficiently destroyed ^3He and ^7Li existed, then the lower bound to η based upon D, ^3He, (and ^7Li) no longer exists. The only solid lower bound to η would then be that based upon the amount of luminous matter in galaxies (i.e., the matter inside the Holmberg radius): $\eta \gtrsim 0.3 \times 10^{-10}$. In this case the predicted Y_p could be as low as 0.15 or 0.16. Although small amounts of anisotropy increase[19] the primordial production of ^4He, recent work[20] suggests that larger amounts could <u>decrease</u> the primordial production of ^4He. Another possibility is neutrino degeneracy; a large lepton number ($n_\nu - n_{\bar{\nu}} \simeq O(n_\gamma)$) drastically modifies the predictions of big bang nucleosynthesis.[21] Finally, one might have to discard the standard cosmology altogether.

<u>Primordial Nucleosynthesis as a Probe</u>

If, based upon its apparent success, we accept the validity of the standard model, we can use primordial nucleosynthesis as a probe of cosmology and particle physics. For example, concordance requires: $4 \times 10^{-10} \lesssim \eta \lesssim 7 \times 10^{-10}$ and $N_\nu \lesssim 4$. This is the most precise determination we have of η and implies that

$$0.014h^{-2}(T/2.7K)^3 \lesssim \Omega_b \lesssim 0.024h^{-2}(T/2.7K)^3 \tag{1.17}$$

$$0.014 \lesssim \Omega_b \lesssim 0.14,$$

$$n_B/s \simeq \eta/7 \simeq (6 - 10) \times 10^{-11}. \tag{1.18}$$

If, as some dynamical studies suggest, $\Omega > 0.14$, then some other non-baryonic form of matter must account for the difference between Ω

and Ω_b. [For a recent review of the measurements of Ω, see refs. 22, 23.] Numerous candidates have been proposed for the dark matter, including primordial black holes, axions, quark nuggets, photinos, gravitinos, relativistic debris, massive neutrinos, sneutrinos, monopoles, pyrgons, maximons, etc. [A discussion of some of these candidates is given in refs. 3, 24.]

With regard to the limit on N_ν, Schvartsman[25] first emphasized the dependence of the yield of ^{4}He on the expansion rate of the Universe during nucleosynthesis, which in turn is determined by g_*, the effective number of massless degrees of freedom. As mentioned above the crucial temperature for ^{4}He synthesis is $\simeq$ 1 MeV -- the freeze out temperature for the n/p ratio. At this epoch the massless degrees of freedom include: γ, $\nu\bar{\nu}$, $e^\pm$ pairs, and any other light particles present, and so

$$g_*=g_\gamma+7/8(g_{e^\pm} + N_\nu g_{\nu\bar{\nu}}) + \sum_{Bose} g_i\ (T_i/T)^4+7/8 \sum_{Fermi} g_i(T_i/T)^4$$
$$=5.5 + 1.75N_\nu+ \sum_{Bose} g_i(T_i/T)^4+ 7/8 \sum_{Fermi} g_i(T_i/T)^4. \quad (1.19)$$

Here T_i is the temperature of species i, T is the photon temperature, and the total energy density of relativistic species is: $\rho = g_*\pi^2T^4/30$. The limit $N_\nu \lesssim 4$ is obtained by assuming that the only species present are: γ, $e^\pm$, and N_ν neutrinos species, and follows because for $\eta \gtrsim 4 \times 10^{-10}$, $\tau_{1/2} \gtrsim 10.4$ min, and $N_\nu \gtrsim 4$, the mass fraction of ^{4}He produced is $\gtrsim 0.25$ (which is greater than the observed abundance). More precisely, $N_\nu \lesssim 4$ implies

$$g_* \lesssim 12.5 \quad (1.20)$$

or

$$1.75 \gtrsim 1.75(N_\nu-3) + \sum_{Bose} g_i(T_i/T)^4 + \sum_{Fermi} g_i(T_i/T)^4. \quad (1.21)$$

At most 1 additional light ($\lesssim$ MeV) neutrino species can be tolerated; many more additional species can be tolerated if their temperatures T_i are < T. [Big bang nucleosynthesis limits on the number of light ($\lesssim$ MeV) species have been derived and/or discussed in refs. 26.]

The number of neutrino species can also be determined by measuring the width of the Z^0 boson: each neutrino flavor less massive than $O(m_Z/2)$ contributes $\simeq$ 190 MeV to the width of the Z^0. Preliminary results on the width of the Z^0 imply that $N_\nu \lesssim O(20)$[27]. Note that while big bang nucleosynthesis and the width of the Z^0 both provide information about the number of neutrino flavors, they 'measure' slightly different quantities. Big bang nucleosynthesis is sensitive to the number of light ($\lesssim$ MeV) neutrino species, and all other light degrees of freedom, while the width of the Z^0 is determined by the number of particles less massive than about 50 GeV which couple to the Z^0 (neutrinos among them). This issue has been recently discussed in ref. 28.

Given the important role occupied by big bang nucleosynthesis, it is clear that continued scrutiny is in order. The importance of new observational data cannot be overemphasized: extragalactic D abundance determinations (Is the D abundance universal? What is its value?); more measurements of the ^{3}He abundance (What is its primordial value?); continued improvement in the accuracy of ^{4}He abundances in very metal poor HII regions (Recall, the difference between $Y_p = 0.22$ and $Y_p = 0.23$ is crucial); and further study of the ^{7}Li abundance in very old stellar populations (Has the primordial abundance of ^{7}Li already been measured?). Data from particle physics will prove useful too: a high precision determination of $\tau_{1/2}$ (i.e., $\Delta\tau_{1/2} \lesssim \pm 0.05$ min) will all but eliminate the uncertainty in the predicted ^{4}He primordial abundance; an accurate measurement of the width of the recently-found Z^0 vector boson will determine the total number of neutrino species (less massive than about 50 GeV) and thereby bound the total number of light neutrino species. All these data will not only make primordial nucleosynthesis a more stringent test of the standard cosmology, but they will also make primordial nucleosynthesis a more powerful probe of the early Universe.

'Freeze-out' and the Making of a Relic Species

In Eqns. 1.19, 1.21 I allowed for a species to have a temperature T_i which is less than the photon temperature. What could lead to this happening? As the Universe expands it cools ($T \propto R^{-1}$), and a particle species can only remain in 'good thermal contact' if the reactions which are important for keeping it in thermal equilibrium are occurring rapidly compared to the rate at which T is decreasing (which is set by the expansion rate $-\dot{T}/T = \dot{R}/R = H$). Roughly-speaking the criterion is

$$\Gamma \gtrsim H, \tag{1.22}$$

where $\Gamma = n\langle\sigma v\rangle$ is the interaction rate per particle, n is the number density of target particles and $\langle\sigma v\rangle$ is the thermally-averaged cross section. When Γ drops below H, that reaction is said to 'freeze-out' or 'decouple'. The temperature T_f (or T_d) at which $H = \Gamma$ is called the freeze-out or decoupling temperature. [Note that if $\Gamma = aT^n$ and the Universe is radiation-dominated so that $H = (2t)^{-1} \simeq 1.67\ g_*^{1/2}T^2/m_{pl}$, then the number of interactions which occur for $T \lesssim T_f$ is just: $\int_{T_f}^{0} \Gamma dt \simeq (\Gamma/H)|_{T_f}/(n-2) \simeq (n-2)^{-1}$]. If the species in question is relativistic ($T_f >> m_i$) when it decouples, then its phase space distribution (in momentum space) remains thermal (i.e., Bose-Einstein or Fermi-Dirac) with a temperature $T_i \propto R^{-1}$. [It is a simple exercise to show this.] So long as the photon temperature also decreases as R^{-1}, $T_i = T$, as if the species were still in good thermal contact.

However, due to the entropy release when various massive species annihilate (e.g., $e^\pm$ pairs when $T \simeq 0.1$ MeV), the photon temperature does not always decrease as R^{-1}. Entropy conservation ($S \propto g_*T^3$=constant) can, however, be used to calculate its evolution; if g_* is decreasing, then T will decrease less rapidly than R^{-1}. As an example consider neutrino freeze-out. The cross section for processes like $e^+e^- \leftrightarrow \nu\bar{\nu}$ is: $\langle\sigma v\rangle \simeq 0.2G_F^2T^2$, and the number density of targets $n \simeq T^3$, so that $\Gamma \simeq 0.2\ G_F^2T^5$. Equating this to H it follows that

$$T_f \simeq (30\ m_{pl}^{-1}G_F^{-2})^{1/3} \tag{1.23}$$
$$\simeq \text{few MeV},$$

i.e., neutrinos freeze out before $e^\pm$ annihilations and do not share in subsequent entropy transfer. For $T \lesssim$ few MeV, neutrinos are decoupled and $T_\nu \propto R^{-1}$, while the entropy density in $e^\pm$ pairs and γs $s \propto R^{-3}$. Using the fact that before $e^\pm$ annihilation the entropy density of the $e^\pm$ pairs and γs is: $s \propto (7/8g_{e^\pm} + g_\gamma)T^3 = 5.5\ T^3$ and that after $e^\pm$ annihilation $s \propto g_\gamma T^3 = 2T^3$, it follows that after the $e^\pm$ annihilations

$$T_\nu/T = [g_\gamma/(g_\gamma + 7/8\ g_{e^\pm})]^{1/3}$$
$$= (4/11)^{1/3}. \tag{1.24}$$

Similarly, the temperature at the time of primordial nucleosynthesis T_i of a species which decouples at an arbitrary temperature T_d can be calculated:

$$T_i/T = [(g_\gamma + 7/8(g_{e^\pm} + N_\nu g_{\nu\bar{\nu}}))/g_{*d}]^{1/3}$$
$$\approx (10.75/g_{*d})^{1/3} \qquad (\text{for } N_\nu = 3). \qquad (1.25)$$

Here $g_{*d} = g_*(T_d)$ is the number of species in equilibrium when the species in question decouples. Species which decouple at a temperature 30 MeV $\approx m_\mu/3 \lesssim T \lesssim$ few 100 MeV do not share in the entropy release from $\mu^\pm$ annihilations, and $T_i/T \approx 0.91$; the important factor for limits based upon primordial nucleosynthesis $(T_i/T)^4 \approx 0.69$. Species which decouple at temperatures $T_d \gtrsim$ the temperature of the quark/hadron transition $\approx$ few 100 MeV, do not share in the entropy transfer when the quark-gluon plasma $[g_* \approx g_\gamma + g_{Gluon} + 7/8(g_{e^\pm} + g_{\mu^\pm} + g_{\nu\bar{\nu}} + g_{u\bar{u}} + g_{d\bar{d}} + g_{s\bar{s}} + ..) \gtrsim 62]$ hadronizes, and $T_i/T \approx 0.56$; $(T_i/T)^4 \approx 0.10$.

<u>'Hot' relics</u>- Consider a stable particle species X which decouples at a temperature $T_f >> m_X$. For $T < T_f$ the number density of Xs n_X just decreases as R^{-3} as the Universe expands. In the absence of entropy production the entropy density s also decreases as R^{-3}, and hence the ratio n_X/s remains constant. At freeze-out

$$n_X/s = (g_{Xeff}\zeta(3)/\pi^2)/(2\pi^2 g_{*d}/45),$$
$$\approx 0.278 g_{Xeff}/g_{*d}, \qquad (1.26)$$

where $g_{Xeff} = g_X$ for a boson or $3/4\ g_X$ for a fermion, $g_{*d} = g_*(T_d)$, and $\zeta(3) = 1.20206...$. Today $s \approx 7.1\ n_\gamma$, so that the number density and mass density of Xs are

$$n_X \approx (2g_{Xeff}/g_{*d})n_\gamma, \qquad (1.27)$$

$$\Omega_X = \rho_X/\rho_c \approx 7.6(m_X/100\text{eV})(g_{Xeff}/g_{*d})h^{-2}(T/2.7\text{K})^3. \qquad (1.28)$$

Note, that if the entropy per comoving volume S has increased since the X decoupled, e.g., due to entropy production in a phase transition, then these values are decreased by the same factor that the entropy increased. As discussed earlier, Ωh^2 must be $\lesssim O(1)$, implying that for a <u>stable</u> particle species

$$m_X/100\ \text{eV} \lesssim 0.13\ g_{*d}/g_{Xeff}; \qquad (1.29)$$

for a neutrino species: $T_d \approx$ few MeV, $g_{*d} \approx 10.75$, $g_{Xeff} = 2 \times (3/4)$, so that $n_{\nu\bar{\nu}}/n_\gamma \approx 3/11$ and m_ν must be $\lesssim 96$ eV. Note that for a species which decouples very early (say $g_{*d} = 200$), the mass limit (1.7 keV for $g_{Xeff} = 1.5$) which $\propto g_{*d}$ is much less stringent.

Constraint (1.29) obviously does not apply to an unstable particle with $\tau <$ 10-15 billion yrs. However, any species which decays radiatively is subject to other very stringent constraints, as the photons from its decays can have various unpleasant astrophysical consequences, e.g., dissociating D, distorting the microwave background, 'polluting' various diffuse photon backgrounds, etc. The astrophysical/cosmological constraints on the mass/lifetime of an unstable neutrino species and the photon spectrum of the Universe are shown in Figs. 1.5, 1.6.

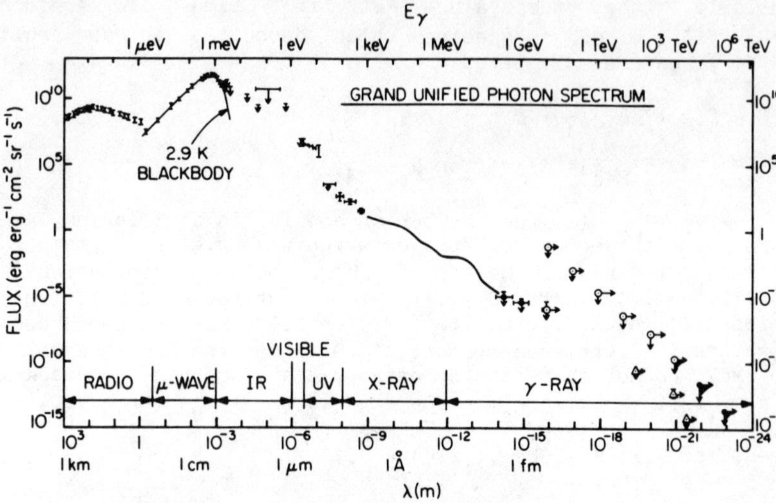

Fig. 1.5 The diffuse photon spectrum of the Universe from λ = 1 km to 10^{-24} m. Vertical arrows indicate upper limits.

<u>'Cold' relics</u>- Consider a stable particle species which is still coupled to the primordial plasma ($\Gamma > H$) when $T \simeq m_X$. As the temperature falls below m_X, its equilibrium abundance is given by

$$n_X/n_\gamma \simeq (g_{Xeff}/2)(\pi/8)^{1/2}(m_X/T)^{3/2}\exp(-m_X/T), \tag{1.30}$$

$$n_X/s \simeq 0.17(g_{Xeff}/g_*)(m_X/T)^{3/2}\exp(-m_X/T), \tag{1.31}$$

and in order to maintain an equilibrium abundance Xs must diminish in number (by annihilations since by assumption the X is stable). So long as $\Gamma_{ann} \simeq n_X(\sigma v)_{ann} \gtrsim H$ the equilibrium abundance of Xs is maintained. When $\Gamma_{ann} \simeq H$, when $T=T_f$, the Xs 'freeze-out' and their number density

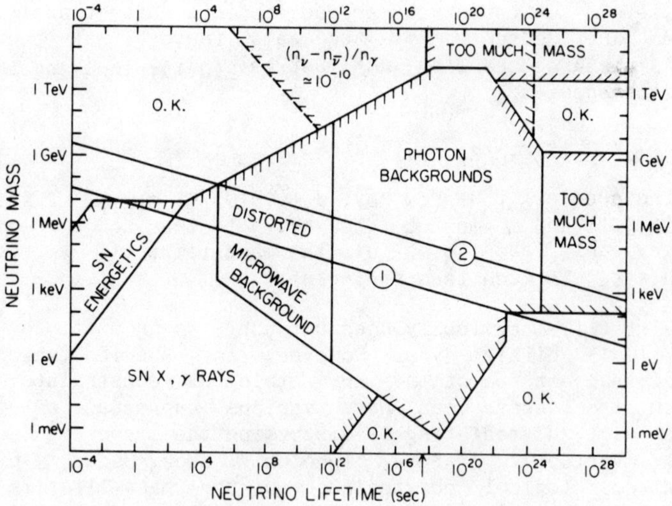

Fig. 1.6 Summary of astrophysical/cosmological constraints on neutrino masses/lifetimes. Lines 1 and 2 represent mass/lifetime relationships: τ = a x 10^{-6} sec $(m_\mu/m_\nu)^5$, for a = 1, 10^{12}.

n_X decreases only due to the volume increase of the Universe, so that for $T \lesssim T_f$:

$$n_X/s \simeq (n_X/s)|_{T_f}. \qquad (1.32)$$

The equation for freeze-out ($\Gamma_{ann} \simeq H$) can be solved approximately, giving

$$m_X/T_f \simeq \ln[0.04(\sigma v)_o m_X m_{pl} g_X g_*^{-1/2}]$$
$$+ (1/2 - n) \ln\{\ln[0.04(\sigma v)_o m_X m_{pl} g_X g_*^{-1/2}]\},$$
$$\simeq 39+\ln[(\sigma v)_o m_X]+(1/2 - n)\ln[39+\ln[(\sigma v)_o m_X]], \qquad (1.33)$$

$$n_X/s \simeq 5\{\ln[0.04(\sigma v)_o m_X m_{pl} g_X g_*^{-1/2}]\}^{1+n}/[(\sigma v)_o m_X m_{pl} g_*^{1/2}],$$
$$\simeq 4 \times 10^{-19}\{39+\ln[(\sigma v)_o m_X]\}^{1+n}/[(\sigma v)_o m_X g_*^{1/2}] \qquad (1.34)$$

where $(\sigma v)_{ann}$ is taken to be $(\sigma v)_o(T/m_X)^n$, and in the second form of each equation $g_X = 2$, $g_* \simeq 100$, and all dimensional quantities are to be measured in GeV units.

[The 'correct way' to solve for n_X/s is to integrate the Boltzmann equation which governs the X abundance, $d/dt\ (n_X/s) = -(\sigma v)s[(n_X/s)^2-(n_{Xeq}/s)^2]$. This has been done in ref. 29, and the 'freeze-out' approximation used in Eqns. 1.33, 1.34 is found to be an excellent one.]

As an example, consider a heavy neutrino species (m_X >> MeV), for which $(\sigma v) \simeq O(1)\ m_X^2 G_F^2$. In the absence of annihilations this species would decouple at T ≃ few MeV which is << m_X, and so the X will become a 'cold relic'. Using Eqns. 1.33, 1.34, we find that today:

$$n_X/s \simeq 5 \times 10^{-9}/(m_X/\mathrm{GeV})^3, \qquad (1.35)$$

$$\Omega_X h^2 \simeq 2(m_X/\mathrm{GeV})^{-2}, \qquad (1.36)$$

implying that a stable, heavy neutrino species must be <u>more</u> massive than a few GeV. [This calculation was first done by Lee and Weinberg,[30] and independently by Kolb.[31]] Note that $\rho_X \propto n_X m_X \propto (\sigma v)_o^{-1}$ -- implying that the more weakly-interacting a particle is, the more 'dangerous' it is cosmologically. If a particle species is to saturate the mass density bound and provide most of the mass density today ($\Omega_X h^2 \simeq 1$) then its mass and annihilation cross section must satisfy the relation:

$$(\sigma v)_o g_*^{1/2} \simeq 10^{-10}\{39+\ln[m_X(\sigma v)_o]\}^{1+n} \qquad (1.37)$$

where as usual all dimensional quantities are in GeV units.

LECTURE 2 - BARYOGENESIS

I'll begin by briefly summarizing the evidence for the baryon asymmetry of the Universe and the seemingly insurmountable problems that render baryon symmetric cosmologies untenable. For a more detailed discussion of these I refer the reader to Steigman's review of the subject[33]. For a review of recent attempts to reconcile a symmetric Universe with both baryogenesis and the observational constraints, I refer the reader to Stecker[34].

Evidence for a Baryon Asymmetry

Within the solar system we can be very confident that there are no concentrations of antimatter (e.g., antiplanets). If there were, solar wind particles striking such objects would be the strongest γ-ray sources in the sky. Also, NASA has yet to lose a space probe because it annihilated with antimatter in the solar system.

Cosmic rays more energetic than O(0.1 GeV) are generally believed to be of "extrasolar" origin, and thereby provide us with samples of material from throughout the galaxy (and possibly beyond). The ratio of antiprotons to protons in the cosmic rays is about 3×10^{-4}, and the ratio of anti-^{4}He to ^{4}He is less than 10^{-5} (ref. 35). Antiprotons are expected to be produced as cosmic-ray secondaries (e.g. $p + p \to 3p + \bar{p}$) at about the 10^{-4} level. At present both the spectrum and total flux of cosmic-ray antiprotons are at variance with the simplest model of their production as secondaries. A number of alternative scenarios for their origin have been proposed including the possibility that the detected $\bar{p}$s are cosmic rays from distant antimatter galaxies. Although the origin of these $\bar{p}$s remains to be resolved, it is clear that they do not provide evidence for an appreciable quantity of antimatter in our galaxy. [For a recent review of antimatter in the cosmic rays we refer the reader to ref. 35.]

The existence of both matter and antimatter galaxies in a cluster of galaxies containing intracluster gas would lead to a significant γ-ray flux from decays of π^0s produced by nucleon-antinucleon annihilations. Using the observed γ-ray background flux as a constraint, Steigman[33] argues that clusters like Virgo, which is at a distance $\approx$20 Mpc ($\approx 10^{26}$ cm) and contains several hundred galaxies, must not contain both matter and antimatter galaxies.

Based upon the above-mentioned arguments, we can say that if there exist equal quantities of matter and antimatter in the Universe, then we can be absolutely certain they are separated on mass scales greater than 1 $M_\odot$, and reasonably certain they are separated on scales greater than (1-100) $M_{galaxy} \approx 10^{12}$-$10^{14} M_\odot$. As discussed below, this fact is virtually impossible to reconcile with a symmetric cosmology.

It has often been pointed out that we drive most of our direct knowledge of the large-scale Universe from photons, and since the photon is a self-conjugate particle we obtain no clue as to whether the source is made of matter or antimatter. Neutrinos, on the other hand, can in principle reveal information about the matter-antimatter composition of their source. Large neutrino detectors such as DUMAND may someday provide direct information about the matter-antimatter composition of the Universe on the largest scales.

Baryons account for only a tiny fraction of the particles in the Universe, the 3K-microwave photons being the most abundant species (yet detected). The number density of 3K photons is $n_\gamma = 399(T/2.7K)^3$ cm^{-3}. The baryon density is not nearly as well determined. Luminous matter (baryons in stars) contribute at least 0.01 of closure density ($\Omega_{lum} >$ 0.01), and as discussed in Lecture 1 the age of the Universe requires that Ω_{tot} (and Ω_b) must be $<$ O(2). These direct determinations place the baryon-to-photon ratio $\eta \equiv n_b/n_\gamma$ in the range 3×10^{-11} to 6×10^{-8}. As I also discussed in Lecture 1 the yields of big-bang nucleosynthesis depend directly on η, and the production of amounts of D, ^{3}He, ^{4}He, and ^{7}Li that are consistent with their present measured abundances restricts η to the narrow range (4-7) $\times 10^{-10}$.

Since today it appears that $n_b >> n_{\bar{b}}$, η is also the ratio of net baryon number to photons. The number of photons in the Universe has not remained constant, but has increased at various epochs when particle species have annihilated (e.g. $e^{\pm}$ pairs at $T \approx 0.5$ MeV). Assuming the expansion has been isentropic (i.e. no significant entropy production), the entropy per comoving volume ($\propto sR^3$) has remained constant. The "known entropy" is presently about equally divided between the 3K photons and the three cosmic neutrino backgrounds (e, μ, τ). Taking this to be the present entropy, the ratio of baryon number to entropy is

$$n_B/s \approx (1/7)\eta \approx (6\text{-}10) \times 10^{-11}, \qquad (2.1)$$

where $n_B \equiv n_b - n_{\bar{b}}$ and η is taken to be in the range $(4\text{-}7) \times 10^{-10}$. So long as the expansion is isentropic and baryon number is at least effectively conserved this ratio remains constant and is what I will refer to as the baryon number of the Universe.

Although the matter-antimatter asymmetry appears to be "large" today (in the sense that $n_B \approx n_b >> n_{\bar{b}}$), the fact that $n_B/s \simeq 10^{-10}$ implies that at very early times the asymmetry was "tiny" ($n_B << n_b$). To see this, let us assume for simplicity that nucleons are the fundamental baryons. Earlier than 10^{-6} s after the bang the temperature was greater than the mass of a nucleon. Thus nucleons and antinucleons should have been about as abundant as photons, $n_N \approx n_{\bar{N}} \approx n_\gamma$. The entropy density s is $\approx g_* n_\gamma \approx g_* n_N \approx O(10^2) n_N$. The constancy of $n_B/s \approx O(10^{-10})$ requires that for $t < 10^{-6}$s, $(n_N - n_{\bar{N}})/n_N (\approx 10^2 n_B/s) \approx O(10^{-8})$. During its earliest epoch, the Universe was nearly (but not quite) baryon symmetric.

The Tragedy of a Symmetric Cosmology

Suppose that the Universe were initially locally baryon symmetric. Earlier than 10^{-6} s after the bang nucleons and antinucleons were about as abundant as photons. For $T < 1$ GeV the equilibrium abundance of nucleons and antinucleons is $(n_N/n_\gamma)_{EQ} \approx (m_N/T)^{3/2} \exp(-m_N/T)$, and as the Universe cooled the number of nucleons and antinucleons would decrease tracking the equilibrium abundance as long as the annihilation rate $\Gamma_{ann} \approx n_N(\sigma v)_{ann} \approx n_N m_\pi^{-2}$ was greater than the expansion rate H. At a temperature T_f annihilations freeze out ($\Gamma_{ann} \approx H$), nucleons and antinucleons being so rare they can no longer find each other to annihilate. Using Eqn. 1.33 we can compute T_f: $T_f \simeq O(20$ MeV). Because of the incompleteness of the annihilations, residual nucleon and antinucleon to photon ratios (given by Eqn. 1.34) $n_{\bar{N}}/n_\gamma = n_N/n_\gamma \simeq 10^{-18}$ are "frozen in." Even if the matter and antimatter could subsequently be separated, n_N/n_γ is a factor of 10^8 too small. To avoid 'the annihilation catastrophe', matter and antimatter must be separated on large scales before $t \approx 3 \times 10^{-3}$ s($T \approx 20$ MeV).

Statistical fluctuations: One possible mechanism for doing this is statistical (Poisson) fluctuations. The co-moving volume that encompasses our galaxy today contains $\simeq 10^{12}$ $M_\odot \approx 10^{69}$ baryons and $\simeq 10^{79}$ photons. Earlier than 10^{-6} s after the bang this same comoving volume contained $\simeq 10^{79}$ photons and $\simeq 10^{79}$ baryons and antibaryons. In order to avoid the annihilation catastrophe, this volume would need an excess of baryons over antibaryons of $\simeq 10^{69}$, but from statistical fluctuations one would expect $N_b - N_{\bar{b}} \approx O(N_b^{1/2}) \simeq 3 \times 10^{39}$ - a mere 29 1/2 orders of magnitude too small!

Causality constraints: Clearly, statistical fluctuations are of no help, so consider a hypothetical interaction that separates matter and antimatter. In the standard cosmology the distance over which light signals (and hence causal effects) could have propagated since the bang (the horizon distance) is finite and $\simeq 2t$. When $T \approx 20$ MeV (t A≈ s V 10^{-3} s) causally coherent regions contained only about $10^{-5}\ M_\odot$. Thus, in the standard cosmology causal processes could have only separated matter and antimatter into lumps of mass $\lesssim 10^{-5}\ M_\odot \ll M_{galaxy} \approx 10^{12}\ M_\odot$. [In Lecture 4 I will discuss inflationary scenarios; in these scenarios it is possible that the Universe is globally symmetric, while asymmetric locally (within our observable region of the Universe). This is possible because inflation removes the causality constraint.]

It should be clear that the two observations, $n_b \gg n_{\bar{b}}$ on scales at least as large as $10^{12}\ M_\odot$ and $n_b/n_\gamma \approx (4\text{-}7) \times 10^{-10}$, effectively render all baryon-symmetric cosmologies untenable. A viable pre-GUT cosmology needed to have as an initial condition a tiny baryon number, $n_B/s \approx (6\text{-}10) \times 10^{-11}$--a very curious initial condition at that!

The Ingredients Necessary for Baryogenesis

More than a decade ago Sakharov[36] suggested that an initially baryon-symmetric Universe might dynamically evolve a baryon excess of $O(10^{-10})$, which after baryon-antibaryon annihilations destroyed essentially all of the antibaryons, would leave the one baryon per 10^{10} photons that we observe today. In his 1967 paper Sakharov outlined the three ingredients necessary for baryogenesis: (a) B-nonconserving interactions; (b) a violation of both C and CP; (c) a departure from thermal equilibrium.

It is clear that B(baryon number) must be violated if the Universe begins baryon symmetric and then evolves a net B. In 1967 there was no motivation for B nonconservation. After all, the proton lifetime is more than 35 orders of magnitude longer than that of any unstable elementary particle--pretty good evidence for B conservation. Of course, grand unification provides just such motivation, and proton decay experiments are likely to detect B nonconservation in the next decade if the proton lifetime is $\lesssim 10^{33}$ years.

Under C (charge conjugation) and CP (charge conjugation combined with parity), the B of a state changes sign. Thus a state that is either C or CP invariant must have B = 0. If the Universe begins with equal amounts of matter and antimatter, and without a preferred direction (as in the standard cosmology), then its initial state is both C and CP invariant. Unless both C and CP are violated, the Universe will remain C and CP invariant as it evolves, and thus cannot develop a net baryon number even if B is not conserved. Both C and CP violations are needed to provide an arrow to specify that an excess of matter be produced. C is maximally violated in the weak interactions, and both C and CP are violated in the K^0-$\bar{K}^0$ system. Although a fundamental understanding of CP violation is still lacking at present, GUTs can accommodate CP violation. It would be very surprising if CP violation only occurred in the K^0-$\bar{K}^0$ system and not elsewhere in the theory also (including the B-nonconserving sector). In fact, without miraculous cancellations the CP violation in the neutral kaon system will give rise to CP violation in the B-nonconserving sector at some level.

The necessity of a departure from thermal equilibrium is a bit more subtle. It has been shown that CPT and unitary alone are sufficient to guarantee that equilibrium particle phase space distributions are given by: $f(p) = [\exp(\mu/T+E/T)\pm 1]^{-1}$. In equilibrium, processes like $\gamma + \gamma \leftrightarrow b + \bar{b}$ imply that $\mu_b = -\mu_{\bar{b}}$, while processes like (but not literally) $\gamma + \gamma \leftrightarrow b + b$ require that $\mu_b = 0$. Since $E^2 = p^2 + m^2$ and $m_b = m_{\bar{b}}$ by CPT, it

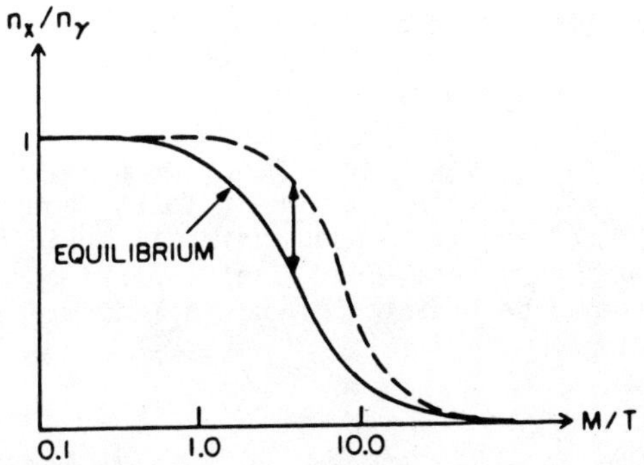

Fig. 2.1 The abundance of X bosons relative to photons. The broken curve shows the actual abundance, while the solid curve shows the equilibrium abundance.

follows that in thermal equilibrium, $n_b \equiv n_{\bar{b}}$. [Note, $n = \int d^3p f(p)/(2\pi)^3$.]

Because the temperature of the Universe is changing on a characteristic timescale H^{-1}, thermal equilibrium can only be maintained if the rates for reactions that drive the Universe to equilibrium are much greater than H. Departures from equilibrium have occurred often during the history of the Universe. For example, because the rate for γ + matter $\rightarrow \gamma'$ + matter' is << H today, matter and radiation are not in equilibrium, and nucleons do not all reside in ^{56}Fe nuclei (thank God!).

<u>The Standard Scenario: Out-of-Equilibrium Decay</u>

The basic idea of baryogenesis has been discussed by many authors.[37-42] The model that incorporates the three ingredients discussed above and that has become the "standard scenario" is the

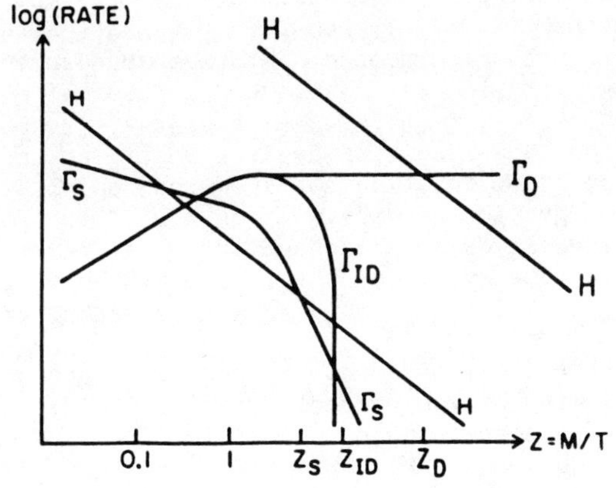

Fig. 2.2 Important rates as a function of z = M/T. H is the expansion rate, Γ_D the decay rate, Γ_{ID} the inverse decay rate, and Γ_S the $2 \leftrightarrow 2$ $\not{B}$ scattering rate. Upper line marked H corresponds to case where K << 1; lower line the case where K > 1. For K << 1, Xs decay when $z = z_D$; for K > 1, freeze out of IDs and S occur at $z = z_{ID}$ and z_S.

so-called out-of-equilibrium decay scenario. I now describe the scenario in some detail.

Denote by "X" a superheavy ($\gtrsim 10^{14}$ GeV) boson whose interactions violate B conservation. X might be a gauge or a Higgs boson (e.g., the XY gauge bosons in SU(5), or the color triplet component of the 5 dimensional Higgs). [Scenarios in which the X particle is a superheavy fermion have also been suggested.] Let its coupling strength to fermions be $\alpha^{1/2}$, and its mass be M. From dimensional considerations its decay rate $\Gamma_D = \tau^{-1}$ should be

$$\Gamma_D \approx \alpha M. \tag{2.2}$$

At the Planck time ($\simeq 10^{-43}$ s) assume that the Universe is baryon symmetric ($n_B/s = 0$), with all fundamental particle species (fermions, gauge and Higgs bosons) present with equilibrium distributions. At this epoch $T \simeq g_*^{-1/4} m_{pl} \simeq 3 \times 10^{18}$ GeV >> M. (Here I have taken $g_* \simeq O(100)$; in minimal SU(5) $g_* \approx 160$.) So at the Planck time X, $\bar{X}$ bosons are very relativistic and up to statistical factors as abundant as photons: $n_X = n_{\bar{X}} \approx n_\gamma$. Nothing of importance occurs until $T \approx M$.

For T < M the equilibrium abundance of X, $\bar{X}$ bosons relative to photons is

$$X_{EQ} \approx (M/T)^{3/2} \exp(-M/T),$$

where $X \equiv n_X/n_\gamma$ is just the number of X, $\bar{X}$ bosons per comoving volume. In order for X, $\bar{X}$ bosons to maintain an equilibrium abundance as T falls below M, they must be able to diminish in number rapidly compared to $H = |\dot{T}/T|$. The most important process in this regard is decay; other processes (e.g. annihilation) are higher order in α. If Γ_D >> H for T = M, then X, $\bar{X}$ bosons can adjust their abundance (by decay) rapidly enough so that X "tracks" the equilibrium value. In this case thermal equilibrium is maintained and no asymmetry is expected to evolve.

More interesting is the case where $\Gamma_D < H \approx 1.66\, g_*^{1/2} T^2/m_{pl}$ when T = M, or equivalently $M > g_*^{-1/2} \alpha 10^{19}$ GeV. In this case, X, $\bar{X}$ bosons are not decaying on the expansion timescale (τ > t) and so remain as abundant as photons (X = 1) for $T \lesssim M$; hence they are overabundant relative to their equilibrium number. This overabundance (indicated with an arrow in Fig. 2.1) is the departure from thermal equilibrium. Much later, when T << M, $\Gamma_D \approx H$ (i.e. $t \approx \tau$), and X, $\bar{X}$ bosons begin to decrease in number as a result of decays. To a good approximation they decay freely since the fraction of fermion pairs with sufficient center-of-mass energy to produce an X or $\bar{X}$ is $\simeq \exp(-M/T) << 1$, which greatly spresses inverse decay processes ($\Gamma_{ID} \approx \exp(-M/T)\Gamma_D << H$). Fig. 2.1 summarizes the time evolution of X; Fig. 2.2 shows the relationship of the various rates (Γ_D, Γ_{ID}, and H) as a function of $M/T (\propto t^{1/2})$.

Now consider the decay of X and $\bar{X}$ bosons: suppose X decays to channels 1 and 2 with baryon numbers B_1 and B_2, and branching ratios r and (1-r). Denote the corresponding quantities for $\bar{X}$ by $-B_1$, $-B_2$, $\bar{r}$, and $(1-\bar{r})$ [e.g. $1 = (\bar{q}\bar{q})$, $2 = (q\ell)$, $B_1 = -2/3$, and $B_2 = 1/3$]. The mean net baryon number of the decay products of the X and $\bar{X}$ are, respectively, $B_X = rB_1 + (1-r)B_2$ and $B_{\bar{X}} = -\bar{r}B_1 - (1-\bar{r})B_2$. Hence the decay of an X, $\bar{X}$ pair on average produces a baryon number ε,

$$\varepsilon \equiv B_X + B_{\bar{X}} = (r-\bar{r})(B_1 - B_2). \tag{2.3}$$

If $B_1 = B_2$, $\varepsilon = 0$. In this case X could have been assigned a baryon number B_1, and B would not be violated by X, $\bar{X}$ bosons.

It is simple to show that $r = \bar{r}$ unless both C and CP are violated. Let $\bar{X}$ = the charge conjugate of X, and $r_\uparrow$, $r_\downarrow$, $\bar{r}_\uparrow$, $\bar{r}_\downarrow$ denote the respective branching ratios in the upward and downward directions. [For simplicity, I have reduced the angular degree of freedom to up and down.] The quantities r and $\bar{r}$ are branching ratios averaged over angle: $r = (r_\uparrow + r_\downarrow)/2$, $\bar{r} = (\bar{r}_\uparrow + \bar{r}_\downarrow)/2$ and $\varepsilon = (r_\uparrow - \bar{r}_\uparrow + r_\downarrow - \bar{r}_\downarrow)/2$. If C is conserved, $r_\uparrow = \bar{r}_\uparrow$ and $r_\downarrow = \bar{r}_\downarrow$, and $\varepsilon = 0$. If CP is conserved $r_\uparrow = \bar{r}_\downarrow$ and $r_\downarrow = \bar{r}_\uparrow$, and once again $\varepsilon = 0$.

When the X, $\bar{X}$ bosons decay ($T << M$, $t \approx \tau$) $n_X = n_{\bar{X}} \approx n_\gamma$. Therefore, the net baryon number density produced is $n_B \approx \varepsilon n_\gamma$. The entropy density $s \approx g_* n_\gamma$, and so the baryon asymmetry produced is $n_B/s \approx \varepsilon/g_* \approx 10^{-2}\ \varepsilon$.

Recall that the condition for a departure from equilibrium to occur is $K \equiv (\Gamma_D/H)|_{T=M} < 1$ or $M > g_*^{-1/2}\ \alpha m_{pl}$. If X is a gauge boson then $\alpha \approx 1/45$, and so M must be $\gtrsim 10^{16}$ GeV. If X is a Higgs boson, then α is essentially arbitrary, although $\alpha \approx (m_f/M_W)^2\ \alpha_{gauge} \approx 10^{-3} - 10^{-6}$ if the X is in the same representation as the light Higgs bosons responsible for giving mass to the fermions (here m_f = fermion mass, M_W = mass of the W boson ≈ 83 GeV). It is apparently easier for Higgs bosons to satisfy this mass condition than it is for gauge bosons. If $M > g_*^{-1/2}\ \alpha m_{pl}$, then only a modest C, CP-violation ($\varepsilon \approx 10^{-8}$) is necessary to explain $n_B/s \approx (6-10) \times 10^{-11}$. As I will discuss below ε is expected to be larger for a Higgs boson than for a gauge boson. For both these reasons a Higgs boson is the more likely candidate for producing the baryon asymmetry.

Numerical Results

Boltzmann equations for the evolution of n_B/s have been derived and solved numerically in refs. 43, 44. They basically confirm the correctness of the qualitative picture discussed above, albeit, with some important differences. The results can best be discussed in terms of

$$K \equiv \Gamma_D/2H(M) \simeq \alpha m_{pl}/3g_*^{1/2}M, \qquad (2.4)$$

$$\simeq 3 \times 10^{17}\ \alpha\ \text{GeV}/M$$

K measures the effectiveness of decays, i.e., rate relative to the expansion rate. K measures the effectiveness of B-nonconserving processes in general because the decay rate characterizes the rates in general for B nonconserving processes, for $T \lesssim M$ (when all the action happens):

$$\Gamma_{ID} \simeq (M/T)^{3/2} \exp(-M/T)\ \Gamma_D, \qquad (2.5)$$

$$\Gamma_S \simeq A\alpha (T/M)^5\ \Gamma_D, \qquad (2.6)$$

where Γ_{ID} is the rate for inverse decays (ID), and Γ_S is the rate for 2 ↔ 2 B nonconserving scatterings (S) mediated by X. [A is a numerical factor which depends upon the number of scattering channels, etc, and is typically O(100-1000).]

[It is simple to see why $\Gamma_S \propto \alpha (T/M)^5 \Gamma_D \propto \alpha^2 T^5/M^4$. $\Gamma_S \simeq n(\sigma v)$; $n \simeq T^3$ and for $T < M$, $(\sigma v) \propto \alpha^2 T^2/M^4$. Note, in some supersymmetric GUTs, there exist fermionic partners of superheavy Higgs which mediate B̸ (and also lead to dim-5 B̸ operators). In this case $(\sigma v) \propto \alpha^2/M^2$ and $\Gamma_S \simeq A\alpha (T/M)^3 \Gamma_D$, and 2 ↔ 2 B̸ scatterings are much more important.]

The time evolution of the baryon asymmetry (n_B/s vs $z = M/T \propto t^{1/2}$) and the final value of the asymmetry which evolves are shown in Figs.

2.3 and 2.4 respectively. For $K < 1$ all B nonconserving processes are ineffective (rate < H) and the asymmetry which evolves is just ε/g_* (as predicted in the qualitative picture). For $K_c > K > 1$, where K_c is determined by

$$K_c\ (\ell n\ K_c)^{-2.4} \simeq 300/A\alpha, \tag{2.7}$$

S 'freeze out' before IDs and can be ignored. Equilibrium is maintained to some degree (by Ds and IDs), however a sizeable asymmetry still evolves

$$n_B/s \simeq (\varepsilon/g_*)\ 0.3\ K^{-1}(\ell n K)^{-0.6}. \tag{2.8}$$

This is the surprising result: for $K_c > K >> 1$, equilibrium is not well maintained and a significant n_B/s evolves, whereas the qualitative picture would suggest that for $K >> 1$ no asymmetry should evolve. For $K > K_c$, S are very important, and the n_B/s which evolves becomes exponentially small:

$$n_B/s \simeq (\varepsilon/g_*)(AK\alpha)^{1/2}\ \exp[-4/3\ (AK\alpha)^{1/4}]. \tag{2.9}$$

[In supersymmetric models which have dim-5 $\not{B}$ operators, $K_c(\ell n K_c)^{-1.2} \simeq 18/A\alpha$ and the analog of Eqn. 2.9 for $K > K_c$ is: $n_B/s \simeq (\varepsilon/g_*)\ A\alpha K \exp[-2(A\alpha K)^{1/2}]$.]

For the XY gauge bosons of SU(5) $\alpha \simeq 1/45$, $A \simeq$ few x 10^3, and $M \simeq$ few x 10^{14} GeV, so that $K_{XY} \simeq 0(30)$ and $K_c \simeq 100$. The asymmetry which

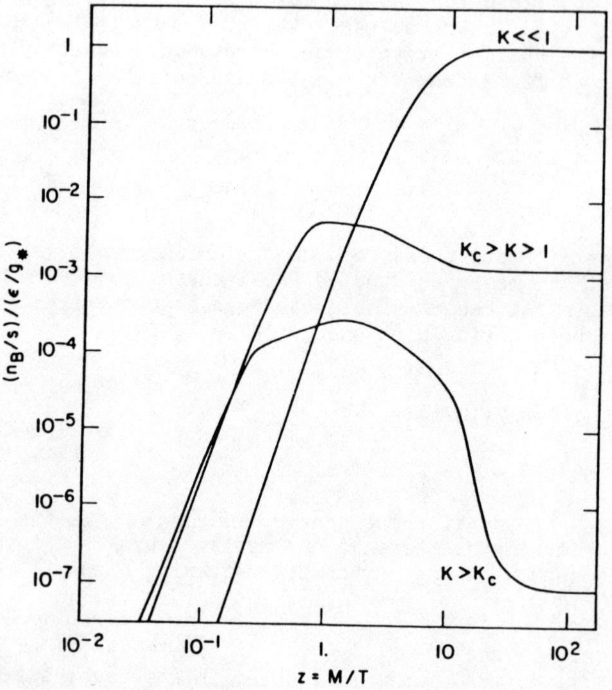

Fig. 2.3 Evolution of n_B/s as a function of $z = M/T$ ($\sim t^{1/2}$). For $K << 1$, n_B/s is produced when Xs decay out-of-equilibrium ($z >> 1$). For $K_c > K > 1$, $n_B/s \propto z^{-1}$ (due to IDs) until the IDs freeze out ($z \simeq 10$). For $K > K_c$ $2 \leftrightarrow 2$ scatterings are important, and n_B/s decreases very rapidly until they freeze out.

could evolve due to these bosons is $\approx 10^{-2}$ (ε_{XY}/g_*). For a color triplet Higgs $\alpha_H \approx 10^{-3}$ (for a top quark mass of 40 GeV) and $A \approx$ few x 10^3, leading to $K_H \approx 3 \times 10^{14}$ GeV/M_H and $K_c \approx$ few x 10^3. For $M_H \lesssim 3 \times 10^{14}$ GeV, $K_H < 1$ and the asymmetry which could evolve is $\approx \varepsilon_H/g_*$.

Very Out-of-Equilibrium Decay

If the X boson decays very late, when $M \gg T$ and $\rho_X > \rho_{rad}$, the additional entropy released in its decays must be taken into account. This is very easy to do. Before the Xs decay, $\rho = \rho_X + \rho_{rad} \approx \rho_X = Mn_X$. After they decay $\rho_X \approx \rho_{rad} = (\pi^2/30)g_* T_{RH}^4 = (3/4)sT_{RH}$ (s, T_{RH} = entropy density and temperature after the X decays). As usual assume that on average each decay produces a mean net baryon number ε. Then the resulting n_B/s produced is

$$n_B/s = \varepsilon n_X/s,$$

$$\approx (3/4)\varepsilon\, T_{RH}/M \tag{2.10}$$

[Note, I have assumed that when the Xs decay $\rho_X \gg \rho_{rad}$ so that the initial entropy can be ignored compared to entropy produced by the decays; this assumption guarantees that $T_{RH} \lesssim M$. I have also assumed that $T \ll M$ so that IDs and S processes can be ignored. Finally, note that how the Xs produce a baryon number of ε per X is irrelevant; it could be by $X \to q$'s ℓ's, or equally well by $X \to \phi s \to q$'s ℓ's (ϕ = any other particle species).]

Note that the asymmetry produced depends upon the ratio T_{RH}/M and not T_{RH} itself--this is of some interest in inflationary scenarios in which the Universe does not reheat to a high enough temperature for baryogenesis to proceed in the standard way (out-of- equilibrium decays). For reference T_{RH} can be calculated in terms of $\tau_X \approx \Gamma^{-1}$; when the Xs decay ($t \approx \tau_X$, $H \approx t^{-1} \approx \Gamma$): $\Gamma^2 = H^2 = 8\pi\rho_X/3m_{pl}^2$. Using the fact that $\rho_X \approx g_*(\pi^2/30)T_{RH}^4$ it follows that

$$T_{RH} \approx g_*^{-1/4}\,(\Gamma m_{pl})^{1/2} \tag{2.11}$$

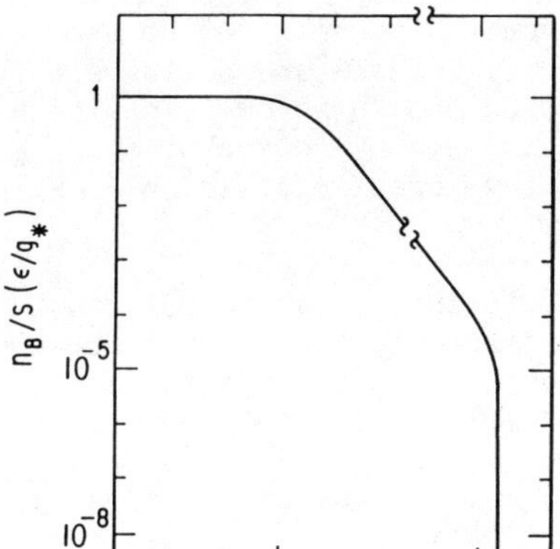

Fig. 2.4 The final baryon asymmetry (in units of ε/g_*) as a function of $K \approx 3 \times 10^{17}\, \alpha$ GeV/M. For $K \lesssim 1$, n_B/s is independent of K and $\approx \varepsilon/g_*$. For $K_c > K > 1$, n_B/s decreases slowly, $\propto 1/(K(\ell n K)^{0.6})$. For $K > K_c$ (when $2 \leftrightarrow 2$ scatterings are important), n_B/s decreases exponentially with $K^{1/4}$.

The C,CP Violation ε

The crucial quantity for determining n_B/s is ε--the C, CP violation in the superheavy boson system. Lacking 'The GUT', ε cannot be calculated precisely, and hence n_B/s cannot be predicted, as, for example, the ^{4}He abundance can be.

The quantity $\varepsilon \propto (r-\bar{r})$; at the tree graph (i.e., Born approximation) level $r-\bar{r}$ must vanish. Non-zero contributions to $(r-\bar{r})$ arise from higher order loop corrections due to Higgs couplings which are complex.[41,45,46] For these reasons, it is generally true that:

$$\varepsilon_{Higgs} \lesssim O(\alpha^N) \sin\delta, \tag{2.12}$$

$$\varepsilon_{gauge} \lesssim O(\alpha^{N+1}) \sin\delta, \tag{2.13}$$

where α is the coupling of the particle exchanged in loop (i.e., $\alpha = g^2/4\pi$), $N \gtrsim 1$ is the number of loops in the diagrams which make the lowest order, non-zero contributions to $(r-\bar{r})$, and δ is the phase of some complex coupling. The C, CP violation in the gauge boson system occurs at 1 loop higher order than in the Higgs because gauge couplings are necessarily real. Since $\alpha \lesssim \alpha_{gauge}$, ε is at most $O(10^{-2})$--which is plenty large enough to explain $n_B/s \approx 10^{-10}$. Because K for a Higgs is likely to be smaller, and because C, CP violation occurs at lower order in the Higgs boson system, the out-of-equilibrium decay of a Higgs is the more likely mechanism for producing n_B/s. [No additional cancellations occur when calculating $(r-\bar{r})$ in supersymmetric theories, so these generalities also hold for supersymmetric GUTs.]

In minimal SU(5)--one $\underset{\sim}{5}$ and one $\underset{\sim}{24}$ of Higgs, and three families of fermions, N = 3. This together with the smallness of the relevant Higgs couplings implies that $\varepsilon_H \lesssim 10^{-15}$ which is not nearly enough.[41,45,46] With 4 families the relevant couplings can be large enough to obtain $\varepsilon_H \approx 10^{-8}$--if the top quark and fourth generation quark/lepton masses are $O(m_W)$ (ref. 47). By enlarging the Higgs sector (e.g., by adding a second $\underset{\sim}{5}$ or a $\underset{\sim}{45}$), $(r-\bar{r})$ can be made non-zero at the 1-loop level, making $\varepsilon_H \approx 10^{-8}$ easy to achieve.

In more complicated theories, e.g., E6, S(10), etc., $\varepsilon \approx 10^{-8}$ can also easily be achieved. However, to do so restricts the possible symmetry breaking patterns. Both E6 and SO(10) are C-symmetric, and of course C-symmetry must be broken before ε can be non-zero. In general, in these models ε is suppressed by powers of M_C/M_G where M_C (M_G) is the scale of C(GUT) symmetry breaking, and so M_C cannot be significantly smaller than M_G.

It seems very unlikely that ε can be related to the parameters of the K^0-$\bar{K}^0$ system, the difficulty being that not enough C, CP violation can be fed up to the superheavy boson system. It has been suggested that ε could be related to the electric dipole moment of the neutron.[48]

Although baryogenesis is nowhere near being on the same firm footing as primordial nucleosynthesis, we now at least have for the first time a very attractive framework for understanding the origin of $n_B/s \approx 10^{-10}$. A framework which is so attractive, that in the absence of observed proton decay, the baryon asymmetry of the Universe is probably the best evidence for some kind of quark/lepton unification. [In writing up this lecture I have borrowed freely and heavily from the review on baryogenesis written by myself and E. W. Kolb (ref.49) and refer the interested reader there for a more thorough discussion of the details of baryogenesis.]

LECTURE 3: MONOPOLES, COSMOLOGY, AND ASTROPHYSICS

Birth: Glut or Famine

In 1931 Dirac[50] showed that if magnetic monopoles exist, then the single-valuedness of quantum mechanical wavefunctions require the magnetic charge of a monopole to satisfy the quantization condition

$$g = ng_D,\ n = 0,\ \pm 1,\ \pm 2\ \ldots$$

$$g_D = 1/2e \approx 69e.$$

However, one is not required to have Dirac monopoles in the theory--you can take 'em or leave 'em! In 1974 't Hooft[51] and Polyakov[52] independently made a remarkable discovery. They showed that monopoles are obligatory in the low-energy theory whenever a semi-simple group G, e.g., SU(5), breaks down to a group G' x U(1) which contains a U(1) factor [e.g., SU(3) x SU(2) x U(1)]; this, of course, is the goal of unification. These monopoles are associated with nontrivial topology in the Higgs field responsible for SSB, topological knots if you will, have a mass $m_M \approx O(M/\alpha)$ [$\approx 10^{16}$ GeV in SU(5); M = scale of SSB], and have a magnetic charge which is a multiple of the Dirac charge.

Since there exist no contemporary sites for producing particles of mass even approaching 10^{16} GeV, the only plausible production site is the early Universe, about 10^{-34} s after 'the bang' when the temperature was $\approx O(10^{14}$ GeV). There are two ways in which monopoles can be produced: (1) as topological defects during the SSB of the unified group G; (2) in monopole-antimonopole pairs by energetic particle collisions. The first process has been studied by Kibble[53], Preskill[54], and Zel'dovich and Khlopov[55], and I will review their important conclusions here.

The magnitude of the Higgs field responsible for the SSB of the unified group G is determined by the minimization of the free energy. However, this does not uniquely specify the direction of the Higgs field in group space. A monopole corresponds to a configuration in which the direction of the Higgs field in group space at different points in physical space is topologically distinct from the configuration in which the Higgs field points in the same direction (in group space) everywhere in physical space (which corresponds to no monopole):

→ = direction of Higgs field in group space

↑ ↑ ↑ ↑ ↑ ↑	↑ ← → ↓
no monopole	monopole

Clearly monopole configurations cannot exist until the SSB [G → G' x U(1)] transition takes place. When spontaneous symmetry breaking occurs, the Higgs field can only be smoothly oriented (i.e., the no monopole configuration) on scales smaller than some characteristic correlation length ξ. On the microphysical side, the inverse Higgs mass at the Ginzburg temperature (T_G) sets such a scale: $\xi \approx m_H^{-1}(T_G)$ (in a second-order phase transition)[56]. [The Ginzburg temperature is the temperature below which it becomes improbable for the Higgs field to fluctuate between the SSB minimum and $\phi = 0$.] Cosmological considerations set an absolute upper bound: $\xi \lesssim d_H$($\approx$ t in the standard cosmology). [Note, even if the horizon distance $d_H(t)$ diverges, e.g., because $R \propto t^n$ $(n > 1)$ for $t \lesssim t_{pl}$, the physics horizon H^{-1} sets an absolute upper bound on ξ, which is numerically identical.] On scales larger than ξ the Higgs field must be uncorrelated, and thus we expect of order 1 monopole per correlation volume ($\simeq \xi^3$) to be produced as a topological defect when the Higgs field freezes out.

Let's focus on the case where the phase transition is either second order or weakly-first order. Denote the critical temperature for the transition by T_c ($\simeq O(M)$), and as before the monopole mass by $m_M \simeq O(M/\alpha)$. The age of the Universe when $T \simeq T_c$ is given in the standard cosmology by: $t_c \simeq 0.3\, g_*^{-1/2} m_{pl}/T_c^2$, cf. Eqn. 1.12. For SU(5): $T_c \simeq 10^{14}$ GeV, $m_M \simeq 10^{16}$ GeV and $t_c \simeq 10^{-34}$ s. Due to the fact that the freezing of the Higgs field must be uncorrelated on scales $\gtrsim \xi$, we expect an initial monopole abundance of O(1) per correlation volume; using $d_H(t_c)$ as an absolute upper bound on ξ this leads to: $(n_M)_i \simeq O(1)\, t_c^{-3}$. Comparing this to our fiducials S_{HOR} and N_{B-HOR}, we find that the initial monopole-to-entropy and monopole-to-baryon number ratios are:

$$n_M/s \gtrsim 10^2\, (T_c/m_{pl})^3, \tag{3.1a}$$

$$n_M/n_B \gtrsim 10^{12}\, (T_c/m_{pl})^3. \tag{3.1b}$$

[Note: $\langle F_M \rangle$, the average monopole flux in the Universe, and Ω_M, the fraction of critical density contributed by monopoles, are related to n_M/s and n_M/n_B by:

$$\langle F_M \rangle \simeq 10^{10} (n_M/s)\ \mathrm{cm^{-2}\ sr^{-1}\ sec^{-1}}, \tag{3.2a}$$

$$\simeq (n_M/n_B)\ \mathrm{cm^{-2}\ sr^{-1}\ sec^{-1}}, \tag{3.2b}$$

$$\Omega_M h^2 \simeq 10^{24} (n_M/s)(m_M/10^{16}\mathrm{GeV}), \tag{3.3a}$$

$$\simeq 10^{14} \langle F_M \rangle (m_M/10^{16}\mathrm{GeV}), \tag{3.3b}$$

where the monopole velocity has been assumed to be $\simeq 10^{-3}c$ (this assumption will be discussed in detail later).

Preskill[54] has shown that unless n_M/s is $> 10^{-10}$ monopole-antimonopole annihilations do not significantly reduce the initial monopole abundance. If $n_M/s > 10^{-10}$, he finds that n_M/s is reduced to $\simeq 10^{-10}$ by annihilations. For $T_c < 10^{15}$ GeV our estimate for n_M/s is $< 10^{-10}$, and we will find that in the standard cosmology T_c must be $\ll 10^{15}$ GeV to have an acceptable monopole abundance, so for our purposes we can ignore annihilations. Assuming that the expansion has been adiabatic since $T \simeq T_c$, this estimate for n_M/s translates into:

$$\langle F_M \rangle \simeq 10^{-3}\, (T_c/10^{14}\ \mathrm{GeV})^3\ \mathrm{cm^{-2}\ sr^{-1}\ s^{-1}}, \tag{3.4a}$$

$$\Omega_M \simeq 10^{11}\, (T_c/10^{14}\ \mathrm{GeV})^3 (m_M/10^{16}\ \mathrm{GeV}) \tag{3.4b}$$

--a flux that would make any monopole hunter/huntress ecstatic, and an Ω_M that is unacceptably large (except for $T_c \ll 10^{14}$ GeV). As was discussed previously, Ω can be at most O(few), so we have a very big problem with the simplest GUTs (in which $T_c \simeq 10^{14}$ GeV). This is the so-called 'Monopole Problem'. The statement that $\Omega_M \simeq 10^{11}$ for $T_c \simeq 10^{14}$ GeV is a bit imprecise; clearly if $k < 0$ (corresponding to $\Omega < 1$) monopole production cannot close the Universe (and in the process change the geometry from being infinite in extent and negatively-curved, to being finite in extent and positively-curved). More precisely, a large monopole abundance would result in the Universe becoming matter-dominated much earlier, at $T \simeq 10^3$ GeV $(T_c/10^{14}\ \mathrm{GeV})^3\ (m_M/10^{16}$ GeV), and eventually reaching a temperature of 3 K at the young age of t

$\approx 10^4$ yrs$(T_c/10^{14}$ GeV$)^{-3/2}$ $(m_M/10^{16}$ GeV$)^{-1/2}$. The requirement that $\Omega_M \lesssim$ O(few) implies that

$$T_c \lesssim 10^{11} \text{ GeV} \qquad (\Omega_M \lesssim \text{few})$$

where I have taken m_M to be O(100 T_c). Note, given the generous estimate for ξ, even this is probably not safe; if one had a GUT in which $T_c \approx 10^{11}$ GeV a more careful estimate for ξ would be called for.

The Parker bound (to be discussed below) on the average monopole flux in the galaxy, $\langle F_M \rangle \lesssim 10^{-15}$ cm^{-2} sr^{-1} s^{-1}, results in a slightly more stringent constraint:

$$T_c \lesssim 10^{10} \text{ GeV} \qquad (\text{Parker bound})$$

The most restrictive constraints on T_c follow from the neutron star catalysis bounds on the monopole flux (also to be discussed below) and the most restrictive of those, $\langle F_M \rangle \lesssim 10^{-27}$ cm^{-2} sr^{-1} s^{-1}, implies that

$$T_c \lesssim 10^{6} \text{ GeV} \qquad (\text{Neutron star catalysis bound})$$

Note, to obtain these bounds I have compared my estimate for the average monopole flux in the Universe, Eqn. 3.4a, with the astrophysical bounds on the average flux of monopoles in our galaxy. If monopoles cluster in galaxies (which I will later argue is unlikely), then the average galactic flux of monopoles is greater than the average flux of monopoles in the Universe, making the above bounds on T_c more restrictive.

If the GUT transition is strongly first order (I am excluding inflationary Universe scenarios for the moment), then the transition will proceed by bubble nucleation at a temperature T_n ($\ll T_c$), when the nucleation rate becomes comparable to the expansion rate H. Within each bubble the Higgs field is correlated; however, the Higgs field in different bubbles should be uncorrelated. Thus one would expect O(1) monopole per bubble to be produced. When the Universe supercools to a temperature T_n, bubbles nucleate, expand, and rapidly fill all of space; if r_b is the typical size of a bubble when this occurs, then one expects n_M to be $\approx r_b^{-3}$. After the bubbles coalesce, and the Universe reheats, the entropy density is once again $s \approx g_* T_c^3$, so that the resulting monopole to entropy ratio is: $n_M/s \approx (g_* r_b^3 T_c^3)^{-1}$. Guth and Weinberg[57] have calculated r_b and find that $r_b \approx (m_{pl}/T_c^2)/\ln(m_{pl}^4/T_c^4)$, leading to a relatively accurate estimate for the monopole abundance:

$$n_M/s \approx [\ln(m_{pl}^4/T_c^4)(T_c/m_{pl})]^3 \quad , \tag{3.5}$$

which is even more disasterous than the estimate for a second order phase transition [recall, however, estimate 3.1 was an absolute lower bound].

The bottom line is that we have a serious problem here--the standard cosmology extrapolated back to $T \approx T_c$ and the simplest GUTs are incompatible (to say the least). One (or both) must be modified. Although this result is discouraging (especially when viewed in the light of the great success of baryogenesis), it does provide a valuable piece of information about physics at very high energies and/or the earliest moments of the Universe, in that regard a 'window' to energies $\gtrsim 10^{14}$ GeV and times $\lesssim 10^{-34}$ sec.

A number of possible solutions have been suggested. To date the most attractive is the new inflationary Universe scenario (which will be the subject of Lecture 4). In this scenario, a small region (size < the horizon) within which the Higgs field could be correlated, grows to a size which encompasses all of the presently observed Universe, due to the exponential expansion which occurs during the phase transition. This results in less than one monopole in the entire observable Universe (due to Kibble production).

Let me very briefly review some of the other attempts to solve the monopole problem. Several people have pointed out that if there is no complete unification [e.g., if G = H x U(1)], or if the full symmetry of the GUT is not restored in the very early Universe (e.g., if the maximum temperature the Universe reached was $< T_c$, or if a large lepton number[58], $n_L/n_\gamma > 1$, prevented symmetry restoration at high temperature), then there would be no monopole problem. However, none of these possibilities seems particularly attractive.

Several authors[59-62] have studied the possibility that monopole-antimonopole annihilation could be enhanced over Preskill's estimate, due to 3-body annihilations or the gravitational clumping of monopoles (or both). Thus far, this approach has not solved the problem.

Bais and Rudaz[63] have suggested that large fluctuations in the Higgs field at temperatures near T_c could allow the monopole density to relax to an acceptably small value. They do not explain how this mechanism can produce the acausal correlations needed to do this.

Scenarios have been suggested in which monopoles and antimonopoles form bound pairs connected by flux tubes, leading to rapid monopole-antimonopole annihilation. For example, Linde[64] proposed that at high temperatures color magnetic charge is confined, and Lazarides and Shafi[65] proposed that monopoles and antimonopoles become connected by Z^0 flux tubes after the SU(2) x U(1) SSB phase transition. In both cases, however, the proposed flux tubes are not topologically stable, nor has their existence even been demonstrated.

Langacker and Pi[66] have suggested a solution which does seem to work. It is based upon an unusual (although perhaps contrived) symmetry breaking pattern for SU(5):

$$SU(5) \xrightarrow[T_c \simeq 10^{14}\,\text{GeV}]{} SU(3) \times SU(2) \times U(1) \xrightarrow[T_1]{} SU(3) \xrightarrow[T_2]{} SU(3) \times U(1)$$

$$\underbrace{T_1 \longleftrightarrow T_2}_{\text{superconducting phase}}$$

(note T_1 could be equal to T_c). The key feature of their scenario is the existence of the epoch ($T \simeq T_1 \rightarrow T_2$) in which the U(1) of electromagnetism is spontaneously broken (a superconducting phase); during this epoch magnetic flux must be confined to flux tubes, leading to the annihilation of the monopoles and antimonopoles which were produced earlier on, at the GUT transition. Although somewhat contrived, their scenario appears to be viable (however, I'll have more to say about it shortly).

Finally, one could invoke the Tooth Fairy (in the guise of a perfect annihilation scheme). E. Weinberg[67] has recently made a very interesting point regarding 'perfect annihilation schemes', which applies to the Langacker-Pi scenario[66], and even to a Tooth Fairy which operates causally. Although the Kibble mechanism results in equal numbers of monopoles and antimonopoles being produced, E. Weinberg points out that in a finite volume there can be magnetic charge fluctuations. He shows that if the Higgs field 'freezes out' at $T \simeq T_c$ and is uncorrelated on scales larger than the horizon at that time, then the expected net RMS magnetic charge in a volume V which is much bigger than the horizon is

$$\Delta n_M \simeq (V/t_c^3)^{1/3}. \tag{3.6}$$

He then considers a perfect, causal annihilation mechanism which operates from $T = T_1 \rightarrow T_2$ (e.g., formation of flux tubes between monopoles and antimonopoles). At best, this mechanism could reduce the monopole abundance down to the net RMS magnetic charge contained in the horizon at $T = T_2$, leaving a final monopole abundance of

$$n_M/s \simeq 10^2\, T_c T_2^2/m_{pl}^3, \tag{3.7}$$

resulting in

$$\Omega_M \gtrsim 0.1(T_c/10^{14}\text{ GeV})(m_M/10^{16}\text{ GeV})(T_2/10^8\text{ GeV})^2, \tag{3.8a}$$

$$\langle F_M \rangle \gtrsim 10^{-15}(T_c/10^{14}\text{GeV})(T_2/10^8\text{GeV})^2\text{cm}^{-2}\text{sr}^{-1}\text{s}^{-1}. \tag{3.8b}$$

It is difficult to imagine a perfect annihilation mechanism which could operate at temperatures $\lesssim 10^3$ GeV, without having to modify the standard SU(2) x U(1) electroweak theory; for $T_c \simeq 10^{14}$ GeV and $T_2 \simeq 10^3$ GeV, E. Weinberg's argument[67] implies that $\langle F_M \rangle$ must be $\gtrsim 10^{-25}$ cm^{-2} sr^{-1} s^{-1}, which would be in conflict with the most stringent neutron star catalysis bound, $F_M < 10^{-27}$ cm^{-2} sr^{-1} s^{-1}.

Finally, I should emphasize that the estimate of n_M/s based upon $\xi \lesssim d_H(t)$ is an absolute (and very generous) lower bound to n_M/s. Should a model be found which succeeds in suppressing the monopole abundance to an acceptable level (e.g., by having $T_c \ll 10^{14}$ GeV or by a perfect annihilation epoch), then the estimate for ξ must be refined and scrutinized.

If the glut of monopoles produced as topological defects in the standard cosmology can be avoided, then the only production mechanism is pair production in very energetic particle collisions, e.g., particle(s) + antiparticle(s) $\rightarrow$ monopole + antimonopole. [Of course, the 'Kibble production' of monopoles might be consistent with the standard cosmology (and other limits to the monopole flux) if the SSB transition occurred at a low enough temperature, say $\ll O(10^{10}$ GeV).] The numbers produced are intrinsically small because monopole configurations do not exist in the theory until SSB occurs ($T_c \simeq M$ = scale of SSB), and have a mass $O(M/\alpha) \simeq 100$ M $\simeq 100\, T_c$. For this reason they are never present in equilibrium numbers; however, some are produced due to the rare collisions of particles with sufficient energy. This results in a present monopole abundance of [68-70]

$$n_M/s \simeq 10^2\, (m_M/T_{max})^3 \exp(-2m_M/T_{max}), \tag{3.9a}$$

$$\Omega_M \simeq 10^{26}(m_M/10^{16}\text{GeV})(m_M/T_{max})^3\exp(-2m_M/T_{max}), \tag{3.9b}$$

$$\langle F_M \rangle \simeq 10^{12}\text{cm}^{-2}\text{sr}^{-1}\text{s}^{-1}(m_M/T_{max})^3 \exp(-2m_M/T_{max}), \tag{3.9c}$$

where T_{max} is the highest temperature reached after SSB.

In general, $m_M/T_{max} \simeq O(100)$ so that $\Omega_M \simeq O(10^{-40})$ and $\langle F_M \rangle \simeq O(10^{-32}$ cm^{-2} sr^{-1} s^{-1})--a negligible number of monopoles. However, the number produced is exponentially sensitive to m_M/T_{max}, so that a factor of 3-5 uncertainty in m_M/T_{max} introduces an enormous uncertainty in the predicted production. For example, in the new inflationary Universe, the monopole mass can be $\propto$ the Higgs field responsible for SSB, and as that field oscillates about the SSB minimum during the reheating process m_M also oscillates, leading to enhanced monopole production [m_M/T_{max} in Eqns. 3.9a,b,c is replaced by fm_M/T_{max}, where $f < 1$ depends upon the details of reheating; see refs. 71, 72].

Cosmology seems to leave the poor monopole hunter/huntress with two firm predictions: that there should be equal numbers of north and south poles; and that either far too few to detect, or far too many to be consistent with the standard cosmology should have been produced. The detection of any superheavy monopoles would necessarily send theorists back to their chalkboards!

From Birth Through Adolescence ($t \approx 10^{-34}$sec to $t \approx 3x10^{17}$sec)

As mentioned in the previous section, monopoles and antimonopoles do not annihilate in significant numbers; however, they do interact with the ambient charged particles (e.g., monopole + $e^- \leftrightarrow$ monopole + e^-) and thereby stay in kinetic equilibrium (KE $\approx$ 3T/2) until the epoch of $e^\pm$ annihilations (T $\approx$ 1/2 MeV, t $\approx$ 10 s). At the time of $e^\pm$ annihilations monopoles and antimonopoles should have internal velocity dispersions of:

$$\langle v_M^2 \rangle^{1/2} \approx 30 \text{ cm s}^{-1} \, (10^{16} \text{ GeV}/m_M)^{1/2}.$$

After this monopoles are effectively collisionless, and their velocity dispersion decays $\propto R(t)^{-1}$, so that if we neglect gravitational and magnetic effects, today they should have an internal velocity dispersion of

$$\langle v_M^2 \rangle^{1/2} \approx 10^{-8} \text{ cm s}^{-1} \, (10^{16} \text{ GeV}/m_M)^{1/2}.$$

Since they are collisionless, only their velocity dispersion can support them against gravitational collapse. With such a small velocity dispersion to support them they are gravitationally unstable on all scales of astrophysical interest ($\lambda_{Jeans} \approx 10^{-10}$ LY).

After decoupling (T $\approx$ 1/3 eV, t $\approx 10^{13}$ s) [or the epoch of matter domination in scenarios where the mass of the Universe is dominated by a nonbaryonic component], matter can begin to clump, and structure can start to form. Monopoles, too, should clump and participate in the formation of structure. However, since they cannot dissipate their gravitational energy, they cannot collapse into the more condensed objects (such as stars, planets, the disk of the galaxy, etc.) whose formation clearly must have involved the dissipation of gravitational energy. Thus, one would only expect to find monopoles in structures whose formation did not require dissipation (such as clusters of galaxies, and galactic haloes). However, galactic haloes are not likely to be a safe haven for monopoles in galaxies with magnetic fields; monopoles less massive than about 10^{20} GeV will, in less than 10^{10} yrs, gain sufficient KE from a magnetic field of strength a few x 10^{-6} G to reach escape velocity[73]. We are led to the conclusion that initially monopoles should either be uniformly distributed through the cosmos, or clumped in clusters of galaxies or in the haloes of galaxies with weak or non-existent magnetic fields. Since our own galaxy has a magnetic field of strength $\approx$ few x 10^{-6} G, and is not a member of a cluster of galaxies, we would expect the local flux of monopoles to be not too different from the average monopole flux in the Universe.

Although monopoles initially have a very small internal velocity dispersion, there are many mechanisms for increasing their velocities. First, typical peculiar velocities (i.e., velocities relative to the Hubble flux) are $O(10^{-3}$ c), leading to a typical monopole-galaxy velocity of 10^{-3}c. Monopoles will be accelerated by the gravitational fields of galaxies (to $\approx 10^{-3}$ c $\approx$ orbital velocity in the galaxy), and if they encounter them, clusters of galaxies (to $\approx 3 \times 10^{-3}$ c). A typical monopole, however, will never encounter a galaxy or a cluster of galaxies, the respective mean free paths being: L_{gal} ($\approx 10^{26}$ cm $\approx 10^{-2}$ c x age of the Universe) and $L_{cluster} \approx 3 \times 10^{28}$ cm.

Monopoles will also be accelerated by magnetic fields. The intragalactic magnetic field strength is $\lesssim 3 \times 10^{-11}$ G (ref. 74), and results in a monopole velocity of

$$v_M \approx 3 \times 10^{-4}\ c\ (B/10^{-11}\ G)(10^{16}\ GeV/m_M).$$

The galactic magnetic field will accelerate monopoles in our galaxy to velocities of[73]

$$v_M \approx 3 \times 10^{-3}\ c\ (10^{16}\ GeV/m_M)^{1/2}.$$

Taking all of these 'sources of velocity' into account, we can make an educated estimate of the typical monopole-detector relative velocity (see Table 3.1). From Table 3.1 below it should be clear that the typical monopole should be moving with a velocity of <u>at least</u> a few x 10^{-3} c with respect to an earth-based detector. It goes without saying that 'this fact' is an important consideration for detector design.

Although planets, stars, etc. should be monopole-free at the time of their formation, they will accumulate monopoles during their lifetimes. The number captured by an object is

$$N_M = (4\pi R^2)(\pi\text{-sr})(1 + 2GM/Rv_M{}^2)\langle F_M\rangle\varepsilon\tau, \qquad (3.10)$$

where M, R and τ are the mass, radius and age of the object, v_M is the monopole velocity, and ε is the efficiency with which the object stops monopoles which strikes its surface. The efficiency of capture ε depends

Table 3.1 Typical Monopole-Detector Relative Velocities

DETECTOR VELOCITY		MONOPOLE VELOCITY	
orbit in galaxy	$2/3 \times 10^{-3}$ c	galactic B-field	3×10^{-3} c $(10^{16}GeV/m_M)^{1/2}$
orbit in solar system	10^{-4} c	grav. acceleration by galaxy	10^{-3} c
		grav. acceleration by sun	10^{-4} c
		monopole-galaxy relative velocity	10^{-3} c

upon the mass and velocity of the monopole, and its rate of energy loss in the object. The quantity $(1 + 2GM/R\, v_M^2)$ is just the ratio of the capture cross section to the geometric cross section. Main sequence stars of mass $(0.6 - 30)M_\odot$ will capture monopoles less massive than about 10^{18} GeV with velocities $\lesssim 10^{-3}$ c with good efficiency ($\varepsilon \simeq 1$); in its main sequence lifetime a star will capture approximately 10^{24} F_{-16} monopoles[75] (essentially independent of its mass). Here $\langle F_M \rangle = F_{-16}$ 10^{-16} cm^{-2} sr^{-1} s^{-1}. Neutron stars will capture monopoles less massive than about 10^{20} GeV with velocities $\lesssim 10^{-3}$ c with unit efficiency, capturing about 10^{21} F_{-16} monopoles in 10^{10} yrs. Planets like Jupiter can stop monopoles less massive than about 10^{16} GeV with velocities < 10^{-3}, accumulating about 10^{22} F_{-16} monopoles in 10^{10} yrs.[76]. A planet like the earth can only stop light or slowly-moving monopoles[76] (for $m_M = 10^{16}$ GeV, v_M must be $\lesssim 3 \times 10^{-5}$ c). Once inside, monopoles can do interesting things, like catalyze nucleon decay (to be discussed below), which keeps the object hot (and leads to a potentially observable photon flux), and eventually depletes the object of all its nucleons. A monopole flux of F_{-21} 10^{-21} cm^{-2} sr^{-1} s^{-1} will cause a neutron star to evaporate in 10^{11} $F_{-21}^{\ -1/2}$ yrs, a Jupiter-like planet to evaporate in 5×10^{15} $F_{-21}^{\ -1/2}$ yrs, and an Earth-like planet to evaporate in 10^{18} $F_{-21}^{\ -1/2}$ yrs[77]. Accretion of monopoles by astrophysical objects, however, does not significantly reduce the monopole flux; the mean free path of a monopole in the galaxy is $\simeq 10^{42}$ cm.

What are Monopoles Doing Today?--Astrophysical Constraints

The three most conspicuous properties of a GUT monopole are: (i) macroscopic mass ($\simeq M/\alpha$--10^{16} GeV $\simeq 10^{-8}$ g for SU(5)); (ii) hefty magnetic charge $h \simeq n\, 69e$ ($n = \pm 1, \pm 2, \ldots$); (iii) the ability to catalyze nucleon decay. Because of these properties, monopoles, if present, should be doing very astrophysically interesting things today--so interesting and so conspicuous that very stringent astrophysical bounds can be placed upon their flux (summarized in Fig. 3.1).

Theoretical prejudice strongly favors the flat cosmological model (i.e., $\Omega \simeq 1$). As I discussed in Lecture 1 big bang nucleosynthesis strongly suggests that baryons contribute $\Omega_b \lesssim 0.15$. In addition, the flat rotation curves of galaxies provide strong evidence that most of the mass associated with a galaxy is dark and exists in an extended structure (most likely a spherical halo). Monopoles are certainly a candidate for the dark matter in galaxies and for providing the closure density.

As I discussed in the first lecture the age of the Universe implies that $\Omega h^2 \lesssim O(1)$; if monopoles are uniformly distributed in the cosmos, then this constrains their average flux to be

$$\langle F_M \rangle \lesssim 10^{-14}\ cm^{-2} sr^{-1} s^{-1} (m_M/10^{16}\ GeV)^{-1}, \qquad (3.11)$$

cf. Eqn. 3.3b. For comparison 10^{-14} cm^{-2} sr^{-1} s^{-1} $\simeq$ 30 monopoles $(\text{soccerfield})^{-1}$ $yr.^{-1}$

If monopoles are clustered in galaxies the local galactic flux can be significantly higher. The mass density in the neighborhood of the sun is about 10^{-23} gcm^{-3}; of this about 1/2 is accounted for (stars, gas, dust, etc.). Monopoles can at most provide the other 1/2, resulting in the flux bound

$$F_M \lesssim 5 \times 10^{-10}\ cm^{-2}\ sr^{-1}\ s^{-1}\ (m_M/10^{16}\ GeV)^{-1}. \qquad (3.12)$$

Actually the bound is probably at least a factor of 10-30 more

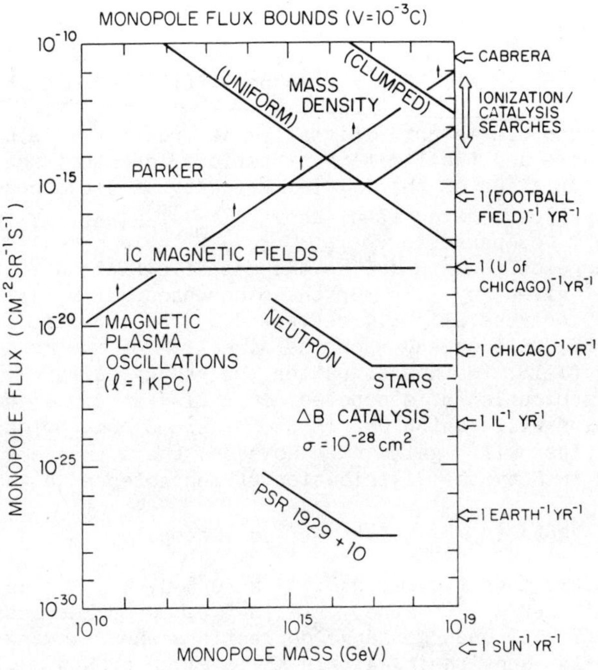

Fig. 3.1 Summary of the astrophysical/cosmological limits to the monopole flux as a function of monopole mass. Wherever necessary the monopole velocity is taken to be 10^{-3} c. The monopole catalysis bound based upon white dwarfs (ref. 93) is: $F_M \lesssim 2 \times 10^{-18}$ $(\sigma v)^{-1}_{-28}$ cm^{-2} sr^{-1} s^{-1} (not shown here). The line labeled 'magnetic plasma oscillations' is the lower bound to the flux predicted in scenarios which evade the 'Parker bound' by having monopoles participate in the maintenance of the galactic B field.

stringent. The unseen material has a column density (= $\int \rho dz$) of no more than about (30 kpc)(10^{-25} gcm^{-3}) (as determined by studying the motions of stars in the stellar neighborhood[78]). Since monopoles are effectively collisionless, if present, they would be distributed in an extended spherical halo. Flat rotation curves indicate that the scale of galactic halos is O(30 kpc), so that the local column density of halo material is ρ_{halo} x 30 kpc. Comparing this to the bound on the local column density of unseen material it follows that locally $\rho_{halo} \lesssim 10^{-25}$ g cm^{-3}. Using this as the limit to the density contributed by monopoles the flux bound 3.12 becomes

$$F_M \lesssim 10^{-11} cm^{-2} sr^{-1} s^{-1} (m_M/10^{16} GeV)^{-1}. \tag{3.13}$$

A monopole by virtue of its magnetic charge will be accelerated by magnetic fields, and in the process can gain KE. Of course, any KE gained must come from somewhere. Any gain in KE is exactly compensated for by a loss in field energy: $\Delta KE = -\Delta[(B^2/8\pi) \times Vol]$. Consider a monopole which is initially at rest in a region of uniform magnetic field. It will be accelerated along the field and after moving a distance ℓ the monopole will have

$$KE \approx hB\ell \approx 10^{11} GeV (B/3x10^{-6}G)(\ell/300pc), \tag{3.14}$$

$$v_{mag} \approx (2hB\ell/m_M)^{1/2}$$

$$\approx 3\times10^{-3}c(B/3\times10^{-6}G)^{1/2}(\ell/300pc)^{1/2}(10^{16}GeV/m_M)^{1/2}. \qquad (3.15)$$

If the monopole is not initially at rest the story is a bit different. There are two limiting situations, and they are characterized by the relative sizes of the initial velocity of the monopole, v_o, and the velocity just calculated above, v_{mag}. First, if the monopole is moving slowly compared to v_{mag}, $v_o << v_{mag} \approx (2hB\ell/m)^{1/2}$, then it will undergo a large deflection due to the magnetic field and its change in KE will be given by 3.14. On the other hand, if $v_o >> v_{mag}$, then the monopole will only be slightly deflected by the magnetic field, and its change in KE will depend upon the direction of its motion relative to the magnetic field. In this situation the energy gained by a spatially isotropic distribution of monopoles, or a flux of equal numbers of north and south poles will vanish at first order in B-some poles will lose KE and some poles will gain KE. However, there is a net gain in KE at second order in B by the distribution of monopoles as a whole:

$$\langle\Delta KE\rangle \approx (hB\ell)\,(v_o/v_{mag})^2/4 \qquad \text{(per monopole).} \qquad (3.16)$$

For the galactic magnetic field $B \approx 3 \times 10^{-6}$ G, $\ell \approx 300$ pc, and $v_{mag} \approx 3 \times 10^{-3}c\,(10^{16}\ \text{GeV/m})^{1/2}$. Since $v_o \approx 10^{-3}$ c, monopoles less massive than about 10^{17} GeV will undergo large deflections when moving through the galactic field and their gain in KE is given by Eqn. 3.14. Because of this energy gain, monopoles less massive than 10^{17} GeV will be ejected from galaxies in a very short time, and thus are unlikely to cluster in the haloes of galaxies. In fact the second order gain in KE will "evaporate" monopoles as massive as $O(10^{20}$GeV) in a time less than the age of the galaxy[73]. Although consideration of galaxy formation would suggest that monopoles should cluster in galactic haloes, galactic magnetic fields should prevent monopoles less massive than $O(10^{20}$ GeV) from clumping in galactic haloes. [These conclusions are not valid if the magnetic field of the galaxy is in part produced by monopoles, a point to which I will return.]

The "no free-lunch principle" ($\Delta KE = -\Delta$ Magnetic Field Energy) and formulae 3.15 and 3.16 can be used to place a limit on the average flux of monopoles in the galaxy.[73,79-80] If, as it is commonly believed, the origin of the galactic magnetic field is due to dynamo action, then the time required to generate/regenerate the field is of the order of a galactic rotation time $\approx O(10^8$ yr). Demanding that monopoles not drain the field energy in a time shorter than this results in the following constraints:

$m_M \lesssim 10^{17}$ GeV:

$$F \lesssim 10^{-15}cm^{-2}sr^{-1}s^{-1}(B/3\times10^{-6}\ G)\ (3\times10^7\ yr/\tau)\ \times$$
$$(r/30\ kpc)^{1/2}\ (300\ pc/\ell)^{1/2}, \qquad (3.17)$$

$m_M \gtrsim 10^{17}$ GeV:

$$F \lesssim 10^{-16}cm^{-2}sr^{-1}s^{-1}(m_M/10^{16}\ GeV)(3\times10^7yr/\tau)(300pc/\ell), \qquad (3.18)$$

where v_o has been assumed to be 10^{-3} c, τ is the regeneration time of the field, ℓ is the coherence length of the field, and r is the size of the magnetic field region in the galaxy. Constraint 3.17 which applies to 10^{16} GeV monopoles is very stringent (less than 3 monopoles soccer $field^{-1}$ yr^{-1}) and is known as the "Parker bound." For more massive monopoles ($\gtrsim 10^{17}$ GeV) the "Parker bound" becomes less restrictive[70,73] (because the KE gain is a second order effect); however, the mass density constraint becomes more restrictive (cf. Fig. 3.1). These two bounds together restrict the flux to be $\lesssim 10^{-13}$ cm^{-2} sr^{-1} s^{-1} (which is allowed for monopoles of mass $\approx 3 \times 10^{19}$ GeV).

Analogous arguments can be applied to other astrophysical magnetic fields. Rephaeli and Turner[81] have analyzed intracluster (IC) magnetic fields and derived a flux bound of $O(10^{-18}$ cm^{-2} sr^{-1} $s^{-1})$ for monopoles less massive than $O(10^{18}$ GeV). Although the presence of such fields has been inferred from diffuse radio observations for a number of clusters (including Coma), the existence of IC fields is not on the same firm footing as galactic fields. It is also interesting to note that the IC magnetic fields are sufficiently weak so that only monopoles lighter than $O(10^{16}$ GeV) should be ejected, and thus it is very likely that monopoles more massive than 10^{16} GeV will cluster in rich clusters of galaxies, where the local mass density is $O(10^2-10^3)$ higher than the mean density of the Universe. Unfortunately, our galaxy is not a member of a rich cluster.

Several groups have pointed out that the 'Parker bound' can be evaded if the monopoles themselves participate in the maintenance of the galactic magnetic field.[73,82-83] In such a scenario a monopole magnetic plasma mode is excited, and monopoles only 'borrow the KE' they gain from the magnetic field, returning it to the magnetic field a half cycle later. In order for this to work the monopole oscillations must maintain coherence; if they do not 'phase-mixing' (Landau damping) will cause the oscillations to rapidly damp. The criterion for coherence to be maintained is that the phase velocity of the oscillations $v_{ph} \approx \omega_{pl}(\ell/2\pi)$ be greater than the gravitational velocity dispersion of the monopoles ($\approx 10^{-3}c$); $\ell \approx$ wavelength of the relevant mode $\approx$ coherence length of the galactic field $\lesssim 1$ kpc. The monopole plasma frequency is given by

$$\omega_{pl} = (4\pi h^2 n_M/m_M)^{1/2}, \tag{3.19}$$

where n_M is the monopole number density. The condition that v_{ph} be $\gtrsim 10^{-3}c$ implies a <u>lower</u> bound to flux of

$$F_M \gtrsim 1/4\ m_M\ v^3_{grav}\ (h\ell)^{-2},$$
$$\gtrsim 10^{-14}(m_M/10^{16}\text{GeV})(1\text{kpc}/\ell)^2\text{cm}^{-2}\text{sr}^{-1}\text{s}^{-1}. \tag{3.20}$$

Incidently, this also implies an upper bound to the oscillation period: $\tau = 2\pi/\omega_{pl} \lesssim \ell/v_{grav} \approx 3 \times 10^6$ yr $(\ell/1\text{kpc})$--a very short time compared to other galactic timescales.

While it is <u>possible</u> that such scenarios could allow one to beat the 'Parker bound', a number of hurdles remain to be cleared before these scenarios can be called realistic or even viable. To mention a few, monopole oscillations can always be damped on sufficiently small scales (recall $v_{ph} \approx (\omega_{pl}/2\pi)\ell$), and nonlinear effects in this very complicated system--coupled electric and magnetic plasmas in a self-gravitating fluid, tend to feed power from large scales down to small scales. Can the coherence of the oscillations which is so crucial be maintained both spatially and temporally in the presence of inhomogeneities (after all the galaxy is not a homogeneous fluid)?

Finally, as the observational limits continue to improve, the large monopole flux predicted in these models will be the ultimate test. Already, the oscillation scenario for $m_M = 10^{16}$ GeV is probably observationally excluded.

Perhaps the most intriguing property of the monopole is its ability to catalyze nucleon decay with a strong interaction cross section: $(\sigma v) \simeq 10^{-28}$ cm^2. Since the symmetry of the GUT is restored at the monopole core, one would expect, on geometric grounds, that monopoles would catalyze nucleon decay with a cross section $\simeq M^{-2} \simeq 10^{-56}$ cm^2 ($M^{-1} \simeq$ size of monopole core)--which of course is utterly negligible. Rubakov[84] and independently Callan[85] showed that due to the singular nature of the potential between the s-wave of a fermion and a monopole, the fermion wave function is literally sucked into the core (technically, one might call this 's-wave sucking'), with the cross section saturating the unitarity bound: $(\sigma v) \simeq$ (fermion energy)$^{-2}$, or for low energies $(\sigma v) \simeq$ (fermion mass)$^{-2}$.

Needless to say, monopole catalysis has great astrophysical potential! For comparison, the nuclear reaction $4p \rightarrow {}^4He + 2e^+ + 2\nu_e$ which powers most stars proceeds at a weak interaction rate (first step: $p + p \rightarrow D$) and releases only about 0.7% of the rest mass involved, while monopole catalysis proceeds at a strong interaction rate and releases 100% of the rest mass of the nucleon (e.g., $M + n \rightarrow M + \pi^- + e^+$). The energy released by monopole catalysis is 3×10^3 erg s^{-1} $(\sigma v)_{-28}(\rho/1\mathrm{gcm}^{-3})$ per monopole; only about 10^{30} monopoles in the sun ($\simeq 10^{57}$ nucleons) are needed to produce the solar luminosity ($\simeq 4 \times 10^{33}$ erg s^{-1}). Here and throughout I will parameterize (σv) by:

$$(\sigma v) = (\sigma v)_{-28}\ c\ 10^{28}\ \mathrm{cm}^2.$$

Because of their awesome power to release energy via catalysis, there can't be too many monopoles in astrophysical objects like stars, planets, etc., otherwise the sky would be aglow in all wavebands from the energy released by monopoles. [This energy released in catalysis would be thermalized and radiated from the surface of the object.] The measured luminosities of neutron stars (some as low as 3×10^{30} erg s^{-1}); white dwarfs (some as low as 10^{29} erg s^{-1}); Jupiter (10^{25} erg s^{-1}); and the Earth (3×10^{20} erg s^{-1}) imply upper limits to the number of monopoles in these objects: some neutron stars ($\lesssim 10^{12}\ (\sigma v)^{-1}_{-28}$ monopoles); some white dwarfs ($\lesssim 10^{18}\ (\sigma v)^{-1}_{-28}$ monopoles); Jupiter ($\lesssim 10^{20}\ (\sigma v)^{-1}_{-28}$ monopoles); and the Earth ($\lesssim 3 \times 10^{15}\ (\sigma v)^{-1}_{-28}$ monopoles). In order to translate these limits into bounds on the monopole flux and abundance we need to know how many monopoles would be expected in each of these objects. As I discussed earlier, ab initio we would expect very few; those present must have been captured since the formation of the object. The number is $\propto F_M$ and is given by Eqn. 3.10; hence the limits above can be used to constrain the monopole flux.

The most stringent limit on F_M follows from considering neutron stars. A variety of techniques have been used to obtain limits to the luminosities of neutron stars [recall the limit to the number of monopoles is: $N_M \lesssim$ luminosity/(10^{18}erg s^{-1} $(\sigma v)_{-28}$ $(\rho/3\times10^{14}$g cm^{-3}))]. I will just discuss one. The other techniques lead to similar bounds on F_M and are reviewed in ref.86.

PSR 1929 + 10 is an old ($\simeq 3\times10^6$ yr), radio pulsar whose distance from the earth is about 60 pc. The Einstein x-ray observatory was used to measure the luminosity of this pulsar, and it was determined to be $L \simeq 3 \times 10^{30}$ erg s^{-1} corresponding to a surface temperature of about 30 eV, making it the coolest neutron star yet observed. In its tenure as a neutron star it should have captured $10^{17}\ F_{-16}$ monopoles. The measured luminosity sets a limit to the number of monopoles in PSR 1929 + 10, $N_M \lesssim 10^{12}\ (\sigma v)^{-1}_{-28}$, which in turn can be used to bound $\langle F_M \rangle$:

$$\langle F_M \rangle \lesssim 10^{-21} (\sigma v)^{-1}_{-28} \mathrm{cm}^{-2} \mathrm{sr}^{-1} \mathrm{s}^{-1} \qquad (3.21)$$

--which is less than one monopole Munich^{-1} yr^{-1}!

The progenitors of neutron stars are main sequence (MS) stars of mass $(1-30)M_\odot$ which were either too massive to become white dwarfs (WDs), or evolved to the WD state and were pushed over the Chandrasekhar limit by accretion from a companion star. Freese etal.[75] have calculated that MS stars in the mass range $(1-30)M_\odot$ will during their MS lifetime capture $(10^{23}-10^{25})F_{-16}$ monopoles (for $v_M \simeq 10^{-3}c$ and $m_M \lesssim 10^{18}$ GeV, and depending on the star's mass). The progenitor of PSR 1929 + 10 should have captured at least 10^6 times more monopoles than the neutron star, and Freese etal.[75] argue that it is likely that a fair fraction of them should be retained in the neutron star. If we include these monopoles, the bound improves significantly, to

$$\langle F_M \rangle \lesssim 10^{-27}(\sigma v)^{-1}_{-28}\,cm^{-2}sr^{-1}s^{-1} \tag{3.22}$$

--less than one monopole earth^{-1} yr^{-1}!

How reliable are these astrophysical bounds? The most stringent, Eqn. 3.22, relies upon an additional assumption, that the monopoles captured by the progenitor MS star make their way into the neutron star. Both bounds (and all catalysis bounds) are $\propto (\sigma v)^{-1}$. If the cross section for catalysis is not large, e.g., because the physics at the core of the monopole does not violate B conservation (such is the case for the Z_2 monopoles in SU(10))[87,88], or because the Callan-Rubakov calculation is incorrect, then the catalysis limits are not stringent.

In addition there are astrophysical uncertainties. Hot neutron stars radiate both γs and $\nu\bar{\nu}$s, but only the photons can be detected. The ratio of these luminosities has been calculated for various neutron star equations of state and was taken into account in deriving the catalysis bounds. [For $L_\gamma \lesssim 10^{32}$ erg s^{-1}, L_ν is typically $\lesssim L_\gamma$; while for $L_\gamma \gtrsim 10^{32}$ erg s^{-1} L_ν can be $(10^3-10^6)\,L_\gamma$, see Fig. 3.2.] Monopoles less massive than about 10^{14} GeV may be deflected away from neutron stars with B fields $\gtrsim 10^{12}$ G; monopoles inside neutron stars which have pion condensates in their cores may be ejected by the so-called 'pion-slingshot effect'.[89]

The strength of the neutron star catalysis bounds lies in the number of different techniques which have been used. Individual objects have been studied[90] (PSR 1929 + 10 and 10 or so other old radio pulsars); searches for bright, nearby x-ray point sources have been made with negative results[91] [the number density of old ($\simeq 10^{10}$ yrs) neutron stars in our neighborhood should be $\gtrsim 10^{-4}$ pc^{-3}, implying that there should be O(100) or so within 100 pc of the solar system - if due to 'monopole heating' their luminosities were $\gtrsim 10^{31}$ erg s^{-1} they would surely have been detected]; the integrated contribution of old neutron stars to the diffuse soft x-ray background has been used to limit the average luminosity of an old neutron star ($\lesssim 10^{32}$ erg s^{-1}) and in turn the monopole flux.[86,91,92] The three techniques just mentioned involve different astrophysical assumptions and uncertainties, but all result in comparable bounds to $\langle F_M \rangle$: $\langle F_M \rangle \lesssim 10^{-21}\,(\sigma v)^{-1}_{-28}$ cm^{-2} sr^{-1} s^{-1}. Although I will not discuss it here, the same analysis has been applied to WDs,[93] and results in a less stringent bound, $\langle F_M \rangle \lesssim 2 \times 10^{-18}(\sigma v)^{-1}_{-28}cm^{-2}sr^{-1}s^{-1}$, but more importantly one which involves a different astrophysical system.

If monopoles catalyze nucleon decay with a large cross section, $(\sigma v)_{-28}$ not too much less than order unity, then, based upon the astrophysical arguments, it seems certain that the monopole flux must be small ($\ll 10^{-18}cm^{-2}sr^{-1}sec^{-1}$). On the other hand, if the monopoles of interest do not catalyze nucleon decay at a significant rate (for whatever reason), then the 'Parker bound' is the relevant (and I believe

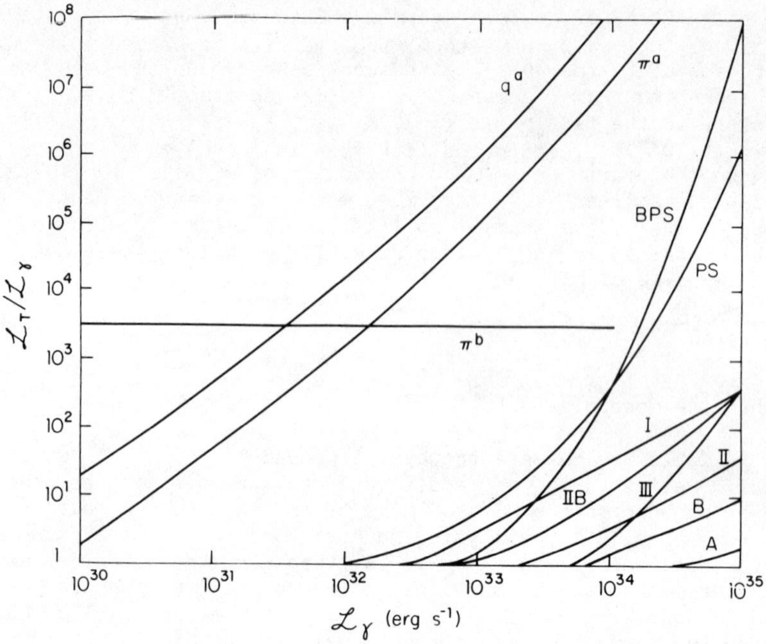

Fig. 3.2 The ratio of the total luminosity (= $L_\gamma + L_\nu$) of a hot neutron star to its photon luminosity as a function of L_γ. The different curves represent different neutron star equations of state: q (quark matter); π^a, π^b (pion condensate); the rest are more conventional equations of state (from ref. 86).

reliable) constraint, with the outside possibility that it could be exceeded due to monopole plasma oscillations (--a scenario which is very astrophysically interesting!).

Monopole Hunting

There are two basic techniques for detecting a monopole: (1) inductive - a monopole which passes through a loop will induce a persistent current $\propto h/L$ (L = inductance of the loop $\propto$ radius, for a circular loop); (2) energy deposition - a monopole can deposit energy due to ionization [$dE/dx \simeq (10\ \mathrm{MeV/cm})(v/10^{-3}c)(\rho/1\mathrm{gcm}^{-3})$], or indirectly by any nucleon decays it catalyzes. Method (1) has the advantage that the signal only depends upon the monopole's magnetic charge (and can be calculated by any first year graduate student who knows Maxwell's equations), and furthermore because of its unique signature (step function in the current) has the potential for clean identification. However, because the induced current $\propto L^{-1} \propto \mathrm{Area}^{-1/2}$, the simplest loop detectors are limited in size to $\lesssim 1\mathrm{m}^2$ ($1\mathrm{m}^2 \times 2\pi$ - sr x 1yr $\simeq 10^{12}$ cm^2 sr sec). In method (2) the detection signal depends upon other properties of the monopole (e.g. velocity, ability to catalyze nucleon decay), and the calculation of the energy loss is not so straightforward, as it involves the physics of the detector material. However, it is very straightforward to fabricate very large detectors of this type.

On 14 February 1982 using a superconducting loop Blas Cabrera detected a jump in current of the correct amplitude for a Dirac magnetic charge.[94] His exposure at that time was about 2×10^9 cm^2 sr s--which naively corresponds to an enormous flux ($\simeq 6 \times 10^{-10}$ cm^{-2} sr^{-1} s^{-1}), especially when compared to the astrophysical bounds discussed above. Sadly, since then his exposure has increased more than 100-fold with no additional candidates.[95] Ionization type searches with exposures upto 10^{14} cm sr s, sensitivities to monopole velocities $3 \times 10^{-4} - 3 \times 10^{-3}$ c, and no candidates have been reported. Searches which employ large proton decay detectors to search for multiple, colinear proton decays caused by a passing monopole with similar exposures (although these searches are only sensitive to specific windows in the $(\sigma v) - v_M$ space) have seen no candidate events. [There is a bit of a Catch 22 here; if σv is large enough so that a monopole would catalyze a string of proton decays in a proton decay detector ($(\sigma v)_{-28} \approx O(1)$), then the astrophysical bounds strongly suggest that $\langle F_M \rangle \lesssim 10^{-21}$ cm^{-2} sr^{-1} s^{-1}.] The most intriguing search done to date involves the etching of a 1/2 Byr old piece of mica of size a few cm^2 (exposure $\approx 10^{18}$ cm^2 sr s).[96] A monopole passing through mica leaves no etchable track; however, a monopole with a nucleus with $Z \gtrsim 10$ (e.g. Al) attached to it leaves an etchable track. Unfortunately, the negative results of searches of this type imply flux limits $\propto$ (probability of a monopole picking a nucleus and holding on to it)$^{-1}$. However exposures of up to 10^{22} cm^2 sr s can possibly be achieved, and if a track is seen, it would be a strong candidate for a monopole. [Very thorough and excellent reviews of monopole searches and searching techniques can be found in refs.97, 98.]

Concluding Remarks

What have we learned about GUT monopoles? (1) They are exceedingly interesting objects, which, if they exist, must be relics of the earliest moments of the Universe. (2) They are one of the very few predictions of GUTs that we can attempt to verify and study in our low energy environment. (3) Because of the glut of monopoles that should have been produced as topological defects in the very early Universe, the simplest GUTs and the standard cosmology (extrapolated back to times as early as $\approx 10^{-34}$ s) are not compatible. This is a very important piece of information about physics at very high energies and/or the earliest moments of the Universe. (4) There is no believable prediction for the flux of relic, superheavy magnetic monopoles. (5) Based upon astrophysical considerations, we can be reasonably certain that the flux of relic monopoles is small. Since it is not obligatory that monopoles catalyze nucleon decay at a prodigious rate, a firm upper limit to the flux is provided by the Parker bound[73], $\langle F_M \rangle \lesssim 10^{-15}$ cm^{-2} sr^{-1} s^{-1}. Note, this is not a predicted flux, it is only a firm upper bound to the flux. It is very likely that flux has to be even smaller, say $\lesssim 10^{-18}$ cm^{-2} sr^{-1} s^{-1} or even 10^{-21} cm^{-2} sr^{-1} s^{-1}. (6) There is every reason to believe that typical monopoles are moving with velocities (relative to us) of at least a few x 10^{-3} c. [Although it is possible that the largest contribution to the local monopole flux is due to a cloud of monopoles orbiting the sun with velocities $\approx (1 - 2) \times 10^{-4}$ c, I think that it is very unlikely.[99,100]]

LECTURE 4 - INFLATION

As I have discussed in Lecture 1 the hot big bang model seems to provide a reliable accounting of the Universe at least as far back as 10^{-2} sec after 'the bang' ($T \lesssim 10$ MeV). There are, however, a number of very fundamental 'cosmological facts' which the hot big bang model by itself does not elucidate (although it can easily accomodate them). The inflationary Universe paradigm, as originally proposed by Guth,[101] and modified by Linde,[102] and Albrecht and Steinhardt,[103] provides for the first time a framework for understanding the origin of these cosmological facts in terms of dynamics rather than just as particular initial data. As we shall see the underlying mechanism of their solution is rather generic--the temporary abolition of particle horizons and the production of entropy, and while inflation is the first realization of this mechanism which is based upon relatively well-known physics (spontaneous symmetry breaking (SSB) phase transitions), it may not prove to be the only such framework. I will begin by reviewing the cosmological puzzles, and then will go on to discuss the new inflationary Universe scenario.

Large-Scale Homogeneity and Isotropy

The observable Universe ($d \approx H^{-1} \approx 10^{28}$ cm $\approx$ 3000 Mpc) is to a high degree of precision isotropic and homogenous on the largest scales (> 100 Mpc). The best evidence for this is provided by the uniformity of the cosmic background temperature: $\Delta T/T \lesssim 10^{-3}$ (10^{-4} if the dipole anisotropy is interpreted as being due to our peculiar motion through the cosmic rest frame; see Fig. 4.1). Large-scale density inhomogeneities or an anisotropic expansion would result in fluctuations in the microwave background temperature of a comparable size (see, e.g., refs. 104, 105). The smoothness of the observable Universe is puzzling if one wishes to understand it as a result of microphysical processes operating in the early Universe. As I mentioned in Lecture 1 the standard cosmology has particle horizons, and when matter and radiation last vigorously interacted (decoupling: $t \approx 10^{13}$ s, $T \approx 1/3$ eV) what was to become the presently observable Universe was comprised of $\approx 10^6$ causally-distinct regions. Put slightly differently, the particle horizon at decoupling only subtends an angle of about 1/2° on the sky today; how is it that the microwave background temperature is so uniform on angular scales >> 1/2°?

Small-Scale Inhomogeneity

As any astronomer will gladly tell you on small scales ($\lesssim$ 100 Mpc) the Universe is very lumpy (stars, galaxies, clusters of galaxies, etc.). [Note, today $\delta\rho/\rho \approx 10^5$ on the scale of a galaxy.] The uniformity of the microwave background on very small angular scales (<<1°) indicates that the Universe was smooth, even on these scales at the time of decoupling (see Fig. 4.1). [The relationship between angle subtended on the sky and mass contained within the corresponding length scale at decoupling is: $\theta \approx 1' \, h(M/10^{12}M_\odot)^{1/3}$.] Whence came the structure that is so conspicuous today? Once matter decouples from the radiation and is free of the pressure support provided by the radiation, small inhomogeneities will grow via the Jeans (gravitational) instability: $\delta\rho/\rho \propto t^{2/3} \propto R$ (in the linear regime). [If the mass density of the Universe is dominated by a collisionless particle species, e.g., a light relic neutrino species, or axions, density perturbations in these particles can begin to grow when the Universe becomes matter-dominated, $R \approx 3 \times 10^{-5} R_{today}$ for $\Omega h^2 = 1$.] Density perturbations of amplitude $\delta\rho/\rho \approx 10^{-3}$ or so, on the scale of a galaxy ($\approx 10^{12} M_\odot$) at the time of decoupling seem to be required to account for the small-scale structure observed today. Their origin, their spectrum (certainly perturbations should exist on scales other than $10^{12}M_\odot$), their nature (adiabatic or isothermal), and the composition of the dark matter (see ref. 3) are all crucial questions for understanding the formation of structure, which to date remain unanswered.

Flatness

The quantity $\Omega \equiv \rho/\rho_c$ measures the ratio of the energy density of the Universe to the critical energy density ($\rho_c \equiv 3H^2/8\pi G$). Although Ω is not known with great precision, from Lecture 1 we know that $0.01 \lesssim \Omega \lesssim$ few. Using Eqn. 1.5 Ω can be written as

$$\Omega = 1/(1 - x(t)), \tag{4.1a}$$

$$x(t) = (k/R^2)/(8\pi G\rho/3). \tag{4.1b}$$

Note that Ω is not constant, but varies with time since $x(t) \propto R(t)^n$ (n = 1 - matter-dominated, or 2 - radiation-dominated). Since $\Omega \approx O(1)$ today, x_{today} must be at most O(1). This implies that at the epoch of nucleosynthesis: $x_{BBN} \lesssim 10^{-16}$ and $\Omega_{BBN} = 1 \pm O(\lesssim 10^{-16})$, and that at the Planck epoch: $x_{pl} \lesssim 10^{-60}$ and $\Omega_{pl} = 1 \pm O(\lesssim 10^{-60})$. That is, very early on the ratio of the curvature term to the density term was extremely small, or equivalently, the expansion of the Universe proceeded at the critical rate ($H^2_{crit} = 8\pi G\rho/3$) to a very high degree of precision. Since x(t) has apparently always been $\lesssim 1$, our Universe is today and has been in the past closely-described by the k = 0 flat model. Were the ratio x not exceedingly small early on, the Universe would have either recollapsed long ago (k > 0), or began its coasting phase (k < 0) where $R \propto t$. [If k < 0 and $x_{BBN} = 1$, then T = 3K for t $\approx$ 300 yrs!] The smallness of the ratio x required as an 'initial condition' for our Universe is puzzling. [The flatness puzzle has been emphasized in refs. 101,106.]

Predominance of Matter Over Antimatter

The puzzle involving the baryon number of the Universe, and its attractive explanation by B, C, CP violating interactions predicted by GUTs has been discussed at length in Lecture 2.

The Monopole Problem

The glut of monopoles predicted in the standard cosmology ('the monopole problem') and the lack of a compelling solution (other than inflation) has been discussed in Lecture 3.

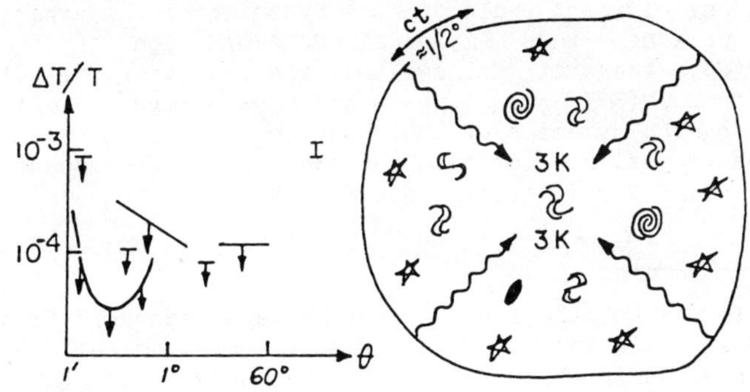

Fig. 4.1 Summary of measurements of the anisotropy of the 3K background on angular scales > 1' (from refs. 112, 113).

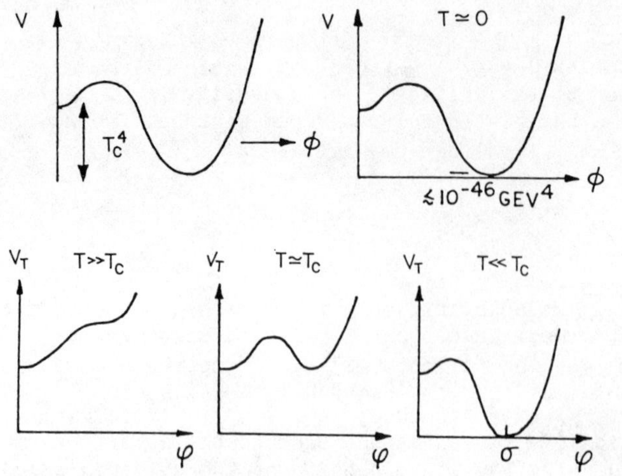

Fig. 4.2 The finite temperature effective potential V_T, for $T > T_c$; $T \approx T_c$; and $T \ll T_c$; here $\phi = \sigma$ is the SSB minimum of V.

The Smallness of the Cosmological Constant

With the possible exception of supersymmetry and supergravity theories, the absolute scale of the effective potential $V(\phi)$ is not determined in gauge theories (ϕ = one or more Higgs field). At low temperatures $V(\phi)$ is equivalent to a cosmological term (i.e., contributes $Vg_{\mu\nu}$ to the stress energy of the Universe). The observed expansion rate of the Universe today ($H \approx 50 - 100$ km s^{-1} Mpc^{-1}) limits the total energy density of the Universe to be $\lesssim O(10^{-29}$ g cm$^{-3}) \approx 10^{-46}$ GeV^4. Thus empirically the vacuum energy of our $T \approx 0$ SU(3) x U(1) vacuum (= $V(\phi)$ at the SSB minimum) must be $\lesssim 10^{-46}$ GeV^4. Compare this to the difference in energy density between the false ($\phi = 0$) and true vacua, which is $O(T_c^4)$ ($T_c \approx$ symmetry restoration temperature): for $T_c \approx 10^{14}$ GeV, $V_{SSB}/V(\phi = 0) \lesssim 10^{-102}$! At present there is no satisfactory explanation for the vanishingly small value of the $T \approx 0$ vacuum energy density (equivalently, the cosmological term).

Today, the vacuum energy is apparently negligibly small and seems to play no significant role in the dynamics of the expansion of the Universe. If we accept this empirical determination of the absolute scale of $V(\phi)$, then it follows that the energy of the false ($\phi = 0$) vacuum is enormous ($\approx T_c^4$), and thus could have played a significant role in determining the dynamics of the expansion of the Universe. Accepting this very non-trivial assumption about the zero of the vacuum energy is the starting point for inflation (see Fig. 4.2).

Generic New Inflation

The basic idea of the inflationary Universe scenario is that there was an epoch when the vacuum energy density dominated the energy density of the Universe. During this epoch $\rho \approx V \approx$ constant, and thus $R(t)$ grows exponentially ($\propto \exp(Ht)$), allowing a small, causally-coherent region (initial size $\lesssim H^{-1}$) to grow to a size which encompasses the region which eventually becomes our presently-observable Universe. In Guth's original scenario[101], this epoch occurred while the Universe was trapped in the false ($\phi = 0$) vacuum during a strongly first-order phase transition. Unfortunately, in models which inflated enough (i.e., underwent sufficient exponential expansion) the Universe never made a 'graceful return' to the usual radiation-dominated FRW cosmology.[57,107] Rather than discussing the original model and its shortcomings in detail, I will instead focus on the variant, dubbed 'new inflation', proposed independently by Linde[102] and Albrecht and Steinhardt[103]. In this scenario, the vacuum-dominated epoch occurs while the region of the Universe in question is slowly, but inevitably, evolving toward the true, SSB vacuum. Rather than considering specific models in this section, I will discuss new inflation for a generic model.

Consider a SSB phase transition which occurs at an energy scale M_G. For $T \gtrsim T_c \approx M_G$ the symmetric ($\phi = 0$) vacuum is favored, i.e., $\phi = 0$ is the global minimum of the finite temperature effective potential $V_T(\phi)$ (= free energy density). As T approaches T_c a second minimum develops at $\phi \neq 0$, and at $T = T_c$ the two minima are degenerate. [I am assuming that this SSB transition is a first-order phase transition.] At temperatures below T_c the SSB ($\phi = \sigma$) minimum is the global minimum of $V_T(\phi)$ (see Fig. 4.2). However, the Universe does not instantly make the transition from $\phi = 0$ to $\phi = \sigma$; the details and time required are a question of dynamics. [The scalar field ϕ is the order parameter for the SSB transition under discussion; in the spirit of generality ϕ might be a gauge singlet field or might have nontrivial transformation properties under the gauge group, possibly even responsible for the SSB of the GUT.]

Assuming a barrier exists between the false and true vacua, thermal fluctuations and/or quantum tunneling must be responsible for taking ϕ across the barrier. The dynamics of this process determine when and how the process occurs (bubble formation, spinodal decomposition, etc.) and the value of ϕ after the barrier is penetrated. For definiteness suppose that the barrier is overcome when the temperature is T_{MS} and the value of ϕ is ϕ_o. From this point the journey to the true vacuum is downhill (literally) and the evolution of ϕ should be adequately described by the semi-classical equations of motion for ϕ:

$$\ddot{\phi} + 3H\dot{\phi} + \Gamma\dot{\phi} + V' = 0, \tag{4.2}$$

where ϕ has been normalized so that its kinetic term in the Lagrangian is $1/2\ \partial_\mu\phi\partial^\mu\phi$, and prime indicates a derivative with respect to ϕ. The subscript T on V has been dropped; for $T \ll T_c$ the temperature dependence of V_T can be neglected and the zero temperature potential ($\equiv$ V) can be used. The $3H\dot{\phi}$ term acts like a frictional force, and arises because the expansion of the Universe 'redshifts away' the kinetic energy of $\phi(\propto R^{-3})$. The $\Gamma\dot{\phi}$ term accounts for particle creation due to the time-variation of ϕ[refs. 108-110]. The quantity Γ is determined by the particles which couple to ϕ and the strength with which they couple ($\Gamma^{-1} \simeq$ lifetime of a ϕ particle). As usual, the expansion rate H is determined by the energy density of the Universe: ($H^2 = 8\pi G\rho/3$), with

$$\rho \approx 1/2\ \dot{\phi}^2 + V(\phi) + \rho_r, \tag{4.3}$$

where ρ_r represents the energy density in radiation produced by the time variation of ϕ. For $T_{MS} \ll T_c$ the original thermal component makes a negligible contribution to ρ. The evolution of ρ_r is given by

$$\dot{\rho}_r + 4H\rho_r = \Gamma\dot{\phi}^2, \tag{4.4}$$

where the $\Gamma\dot{\phi}^2$ term accounts for particle creation by ϕ.

In writing Eqns. 4.2-4.4 I have implicitly assumed that ϕ is spatially homogeneous. In some small region (inside a bubble or a fluctuation region) this will be a good approximation. The size of this smooth region will be unimportant; take it to be of order the 'physics horizon', H^{-1}. Now follow the evolution of ϕ within the small, smooth patch of size H^{-1}.

If V is sufficiently flat somewhere between $\phi = \phi_o$ and $\phi = \sigma$, then ϕ will evolve very slowly in that region, and the motion of ϕ will be 'friction-dominated' so that $3H\dot{\phi} \simeq -V'$ (in the slow growth phase particle creation is not important[110]). If V is sufficiently flat, then the time required for ϕ to transverse the flat region can be long compared to the expansion timescale H^{-1}, say for definiteness, $\tau_\phi = 100\ H^{-1}$. During this slow growth phase $\rho \simeq V(\phi) \simeq V(\phi = 0)$; both ρ_r and $1/2\ \dot{\phi}^2$ are << $V(\phi)$. The expansion rate H is then just

$$H \simeq (8\pi V(0)/3m_{pl}{}^2)^{1/2} \qquad (4.5)$$

$$\simeq M_G^2/m_{pl},$$

where V(0) is assumed to be of order M_G^4. While $H \simeq$ constant R grows exponentially: $R \propto \exp(Ht)$; for $\tau_\phi = 100\ H^{-1}$ R expands by a factor of e^{100} during the slow rolling period, and the physical size of the smooth region increases to $e^{100}H^{-1}$. This exponential growth phase is called a deSitter phase.

As the potential steepens, the evolution of ϕ quickens. Near $\phi = \sigma$, ϕ oscillates around the SSB minimum with frequency ω: $\omega^2 \simeq V''(\sigma) \simeq M_G{}^2 >> H^2 \simeq M_G{}^4/m_{pl}{}^2$. As ϕ oscillates about $\phi = \sigma$ its motion is damped by particle creation and the expansion of the Universe. If $\Gamma^{-1} << H^{-1}$, the coherent field energy density $(V + 1/2\ \dot{\phi}^2)$ is converted into radiation in less than an expansion time ($\Delta t_{RH} \simeq \Gamma^{-1}$), and the patch is reheated to a temperature $T \simeq O(M_G)$ - the vacuum energy is efficiently converted into radiation ('good reheating'). On the other hand, if $\Gamma^{-1} >> H^{-1}$, then ϕ continues to oscillate and the coherent field energy redshifts away with the expansion: $(V + 1/2\ \dot{\phi}^2) \propto R^{-3}$. [The coherent field energy behaves like nonrelativistic matter; see ref. 111 for more details.] Eventually, when $t \simeq \Gamma^{-1}$ the energy in radiation begins to dominate that in coherent field oscillations, and the patch is reheated to a temperature $T \simeq (\Gamma/H)^{1/2}M_G \simeq (\Gamma m_{pl})^{1/2} << M_G$ ('poor reheating'). The evolution of ϕ is summarized in Fig. 4.3.

For the following discussion let us assume 'good reheating' ($\Gamma >> H$). After reheating the patch has a physical size $e^{100}H^{-1}$ ($\simeq 10^{17}$cm for $M_G \simeq 10^{14}$ GeV), is at a temperature of order M_G, and in the approximation that ϕ was initially constant throughout the patch, the patch is exactly smooth. From this point forward the region evolves like a radiation-dominated FRW model. How have the cosmological conundrums been 'explained'? First, the homogeneity and isotropy; our observable Universe today ($\simeq 10^{28}$cm) had a physical size of about 10 cm ($= 10^{28}$cm x 3K/10^{14} GeV) when T was 10^{14} GeV. Thus it lies well within one of the smooth regions produced by the inflationary epoch. At this point the inhomogeneity puzzle has not been solved, since the patch is precisely uniform. Due to deSitter space produced quantum fluctuations in ϕ, ϕ is not exactly uniform even in a small patch. Later, I will discuss the density inhomogeneities that result from the quantum fluctuations in ϕ. The flatness puzzle involves the smallness of the ratio of the curvature term to the energy density term. This ratio is exponentially smaller after inflation: x(after) $\simeq e^{-200}$ x(before) since the energy density before and after inflation is $O(M_G^4)$, while k/R^2 has decreased exponentially (by e^{200}). Since the ratio x is reset to an exponentially small value, the inflationary scenario predicts that today Ω should be $1 \pm O(10^{-BIG})$. If the Universe is reheated to a temperature of order M_G, a baryon asymmetry can evolve in the usual way, although the quantitative details may be slightly different[49,110]. If the Universe is not efficiently reheated ($T_{RH} << M_G$), it may be possible for n_B/s to be produced directly in the decay of the coherent field oscillations (which behave just like NR ϕ particles). This is an example of very out-of-equilibrium decay (discussed in Lecture 2), in which case the n_B/s produced is $\propto T_{RH}/(m_\phi \approx \omega)$ and does not depend upon T_{RH} being of order 10^{14} GeV or so. In any case, it is absolutely necessary to have baryogenesis occur after reheating since any baryon number (or any other quantum number) present before inflation is diluted by a factor $(M_G/T_{MS})^3 \exp(3H\tau_\phi)$ - the factor by which the total entropy increases. Note that if C, CP are violated spontaneously, then ε (and n_B/s) could have a different sign in different patches--leading to a Universe which on the very largest scales ($>> e^{100}H^{-1}$) is baryon symmetric.

Since the patch that our observable Universe lies within was once (at the beginning of inflation) causally-coherent, the Higgs field could have been aligned throughout the patch (indeed, this is the lowest energy configuration), and thus there is likely to be $\lesssim 1$ monopole within the entire patch which was produced as a topological defect. The glut of monopoles which occurs in the standard cosmology does not occur. [The production of other topological defects (such as domain walls, etc.) is avoided for similar reasons.] As discussed in Lecture 3, some monopoles will be produced after reheating in rare, very energetic particle collisions. The number produced is exponentially small and exponentially uncertain. [In discussing the resolution of the monopole problem I am tacitly assuming that the SSB of the GUT is occurring during the SSB transition in question, or that it has already occurred in an earlier SSB transition; if not then one has to worry about the monopoles produced in the subsequent GUT transition.]

The key point is that although monopole production is intrinsically small in inflationary models, the uncertainties in the number of monopoles produced are exponential. Of course, it is also possible that monopoles might be produced as topological defects in a subsequent phase transition[114], although it may be difficult to arrange that they not be overproduced.

Finally, the inflationary scenario sheds no light upon the cosmological constant puzzle. Although it can potentially successfully

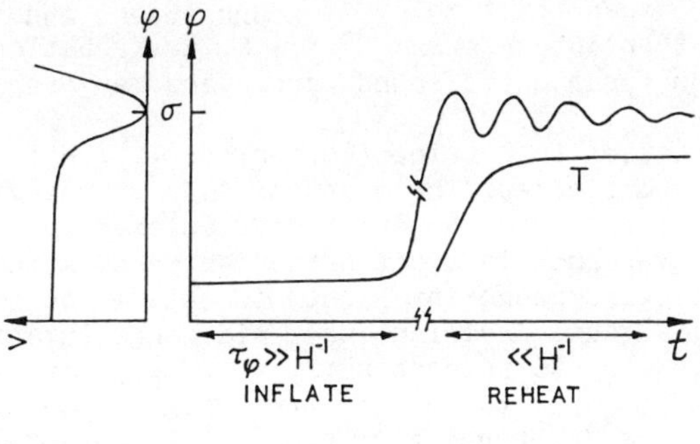

Fig. 4.3 The time evolution of ϕ. During the slow growth phase the time required for ϕ to change appreciably is $>> H^{-1}$. As the potential steepens ϕ evolves more rapidly (timescale $<< H^{-1}$), eventually oscillating about the SSB minimum. Particle creation damps the oscillations in a time $\approx \Gamma^{-1}$ ($<<H^{-1}$, if $\Gamma>>H$ as shown here) reheating the patch to $T \approx \min[M_G, (\Gamma m_{pl})^{1/2}]$.

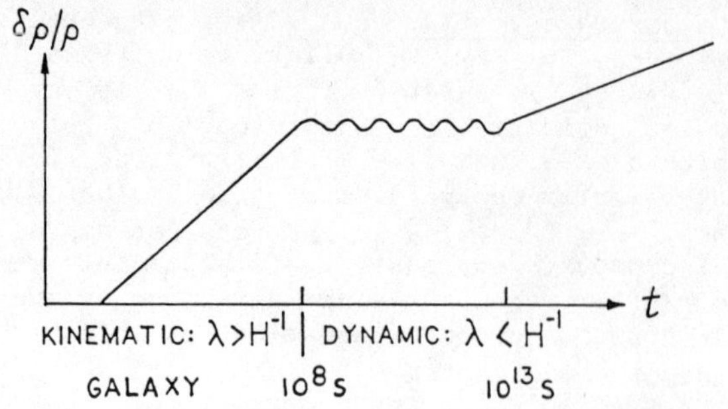

Fig. 4.4 Evolution of a galactic mass adiabatic density perturbation.

resolve all of the other puzzles in my list, inflation is, in some sense, a house of cards built upon the cosmological constant puzzle.

Density Inhomogeneities

Before I discuss the production of density inhomogeneities during the inflationary transition I will briefly review some of the 'Standard Lore'. [A more thorough and systematic treatment of the subject can be found in ref. 105, and in Lecture 5.]

A density perturbation is described by its wavelength λ or its wavenumber $k(= 2\pi/\lambda)$, and its amplitude $\delta\rho/\rho$ (ρ = average energy density). As the Universe expands the physical (or proper) wavelength of a given perturbation also expands; it is useful to scale out the expansion so that a particular perturbation is always labeled by the same comoving wavelength $\lambda_c \equiv \lambda/R(t)$ or comoving wavenumber $k_c \equiv kR(t)$. [$R(t)$ is often normalized so that $R_{today} = 1$.] Even more common is to label a perturbation by the comoving baryon mass (or total mass in nonrelativistic particles if $\Omega_b \neq \Omega_{TOT}$) within a half wavelength $M = \pi\lambda^3 n_B m_N/6$ (n_B = net baryon number density, m_N = nucleon mass).

The relative sizes of λ and H^{-1} (= 'physics horizon' and particle horizon also in the standard cosmology) are crucial for determining the evolution of $\delta\rho/\rho$. When $\lambda \lesssim H^{-1}$ (the perturbation is said to be inside the horizon) microphysics can affect the perturbation. If $\lambda > \lambda_J \simeq v_s H^{-1}$ (physically λ_J, the Jeans length, is the distance a pressure wave can propagate in an expansion time; v_s = sound speed) and the Universe is matter-dominated, then $\delta\rho/\rho$ grows $\propto t^{2/3} \propto R$. Perturbations with $\lambda < \lambda_J$ oscillate as pressure-supported sound waves (and may even damp).

When a perturbation is outside the horizon ($\lambda > H^{-1}$) the situation is a bit more complicated. The quantity $\delta\rho/\rho$ is not gauge-invariant; when $\lambda < H^{-1}$ this fact creates no great difficulties. However when $\lambda > H^{-1}$ the gauge-noninvariance is a bit of a nightmare. Although Bardeen[115] has developed an elegant gauge-invariant formalism to handle density perturbations in a gauge-invariant way, his gauge invariant quantities are not intuitively easy to understand. I will try to give a brief, intuitive description in terms of the gauge dependent, but more intuitive quantity $\delta\rho/\rho$. Physically, only real, honest-to-God wrinkles in the geometry (called curvature fluctuations or adiabatic fluctuations) can 'grow'. In the synchronous gauge ($g_{00} = -1$, $g_{0i} = 0$) $\delta\rho/\rho$ for these perturbations grows $\propto t^n$ (n = 1 - radiation dominated, = 2/3 - matter dominated). Geometrically, when $\lambda > H^{-1}$ these perturbations are just wrinkles in the space time which are evolving kinematically (since microphysical processes cannot affect their evolution). Adiabatic perturbations are characterized by $\delta\rho/\rho \neq 0$ and $\delta(n_B/s) = 0$; while isothermal perturbations (which do not grow outside the horizon) are characterized by $\delta\rho/\rho = 0$ and $\delta(n_B/s) \neq 0$. [With greater generality $\delta(n_B/s)$ can be replaced by any spatial perturbation in the equation of state $\delta p/p$, where $p = p(\rho, \ldots)$.] In the standard cosmology $H^{-1} \propto t$ grows monotonically; a perturbation only crosses the horizon once (see Fig. 4.5). Thus it should be clear that microphysical processes cannot create adiabatic perturbations (on scales $\gtrsim H^{-1}$) since microphysics only operates on scales $\lesssim H^{-1}$. In the standard cosmology adiabatic (or curvature) perturbations were either there ab initio or they are not present. Microphysical processes can create isothermal (or pressure perturbations) on scales $\gtrsim H^{-1}$ (of course, they cannot grow until $\lambda \lesssim H^{-1}$). Fig. 4.4 shows the evolution of a galactic mass ($\simeq 10^{12} M_\odot$) adiabatic perturbation: for $t \lesssim 10^8$ s, $\lambda > H^{-1}$ and $\delta\rho/\rho \propto t$; for 10^{13} s $\gtrsim t \gtrsim 10^8$ s, $\lambda < H^{-1}$ and $\delta\rho/\rho$ oscillates as a sound wave since matter and radiation are still coupled ($v_s \simeq c$) and hence $\lambda_J \simeq H^{-1}$; for $t \gtrsim 10^{13}$ s, $\lambda < H^{-1}$ and $\delta\rho/\rho \propto t^{2/3}$ since matter and radiation are decoupled ($v_s \ll c$) and $\lambda_J < \lambda_{Galaxy}$. [Note: in an $\Omega = 1$ Universe the mass inside the horizon $\simeq (t/\mathrm{sec})^{3/2} M_\odot$.]

Finally, at this point it should be clear that a convenient epoch to specify the amplitude of a density perturbation is when it crosses the horizon. It is often supposed (in the absence of knowledge about the origin of perturbations) that the spectrum of fluctuations is a power law (i.e., no preferred scale):

$$(\delta\rho/\rho)_H = \varepsilon M^{-\alpha}.$$

If $\alpha > 0$, then on some small scale perturbations will enter the horizon with amplitude $\gtrsim O(1)$--this leads to black hole formation; if this scale is $\gtrsim 10^{15}$ g (mass of a black hole evaporating today) there will be too many black holes in the Universe today. On the other hand, if $\alpha < 0$ then the Universe becomes more irregular on larger scales (contrary to observation). In the absence of a high or low mass cutoff, the $\alpha = 0$ (so-called Zel'dovich spectrum[116]) of density perturbations seems to be the only 'safe' spectrum. It has the attractive feature that all scales cross the horizon with the same amplitude (i.e., it is scale-free). Such a spectrum is not required by the observations; however, such a spectrum with amplitude of $O(10^{-4})$ probably leads to an acceptable picture of galaxy formation (i.e., consistent with all present observations--microwave background fluctuations, galaxy correlation function, etc.; for a more detailed discussion see ref. 3.)

Origin of Density Inhomogeneities in the New Inflationary Universe

The basic result is that quantum fluctuations in the scalar field ϕ (due to the deSitter space event horizon which exists during the exponential expansion (inflation) phase) give rise to an almost scale-free (Zel'dovich) spectrum of density perturbations of amplitude

$$(\delta\rho/\rho)_H \simeq (4 \text{ or } 2/5) H\, \Delta\phi/\dot{\phi}(t_1), \tag{4.6}$$

where 4 applies if the scale in question reenters the horizon when the Universe is radiation-dominated and $(\delta\rho/\rho)_H$ is then the amplitude of the sound wave; 2/5 applies if the scale in question reenters the horizon when the Universe is matter-dominated and $(\delta\rho/\rho)_H$ is then the amplitude of the growing mode perturbation at horizon crossing; H is the value of the Hubble parameter during inflation; $\dot{\phi}(t_1)$ is the value of $\dot{\phi}$ when the perturbation left the horizon during the deSitter phase; and $\Delta\phi \approx H/2\pi$ is the fluctuation in ϕ. This result was derived independently by the authors of refs. 117-120. Rather than discussing the derivation in detail here, I will attempt to physically motivate the result. This result turns out to be the most stringent constraint on models of new inflation.

The crucial difference between the standard cosmology and the inflationary scenario for the evolution of density perturbations is that H^{-1} (the 'physics horizon') is not strictly monotonic; during the inflationary (deSitter) epoch it is constant. Thus, a perturbation can cross the horizon ($\lambda = H^{-1}$) twice (see Fig. 4.5)! The evolution of two scales (λ_G = galaxy and λ_H = presently observable Universe) is shown in Fig. 4.5. Earlier than t_1 (time when $\lambda_G \approx H^{-1}$) $\lambda_G < H^{-1}$ and microphysics (quantum fluctuations, etc.) can operate on this scale. When $t = t_1$ microphysics 'freezes out' on this scale; the density perturbation which exists on this scale, say $(\delta\rho/\rho)_1$, then evolves 'kinematically' until it reenters the horizon at $t = t_H$ (during the subsequent radiation-dominated FRW phase) with amplitude $(\delta\rho/\rho)_H$.

DeSitter space is exactly time-translationally-invariant; the inflationary epoch is approximately a deSitter phase - ϕ is almost, but not quite constant (see Fig. 4.3). [In deSitter space $p + \rho = 0$; during inflation $p + \rho = \dot{\phi}^2$.] This time-translation invariance is crucial; as each scale leaves the horizon (at $t = t_1$) $\delta\rho/\rho$ on that scale is fixed by microphysics to be some value, say, $(\delta\rho/\rho)_1$. Because of the (approximate) time-translation invariance of the inflationary phase this value $(\delta\rho/\rho)_1$ is (approxmately) the same for all scales. [Recall H, ϕ, $\dot{\phi}$ are all approximately constant during this epoch, and each scale has the same physical size ($= H^{-1}$) when it crosses outside of the horizon.] The precise value of $(\delta\rho/\rho)_1$ is fixed by the amplitude of the quantum fluctuations in ϕ on the scale H^{-1}; for a free scalar field $\Delta\phi = H/2\pi$ (the Hawking temperature). [Recall, during inflation V'' ($\approx$ the effective mass-squared) is very small.]

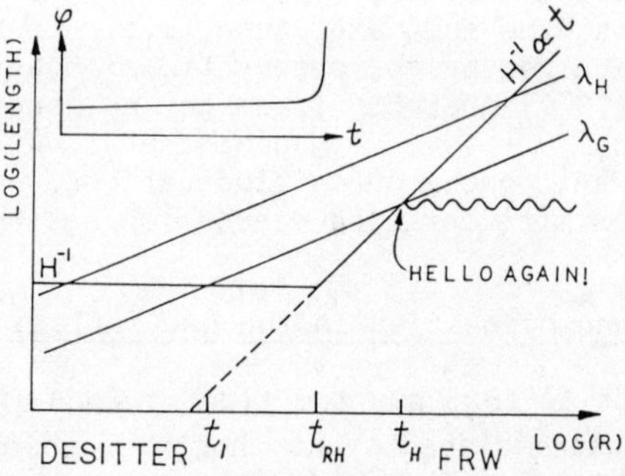

Fig. 4.5 The evolution of the 'physics horizon' ($\approx H^{-1}$) and the physical sizes of perturbations on the scale of a galaxy (λ_G) and on the scale of the present observable Universe (λ_H). Reheating occurs at $t = t_{RH}$. **For reference the evolution of ϕ** is also shown. The broken line shows the evolution of H^{-1} in the standard cosmology. In the inflationary cosmology a perturbation crosses the horizon twice, which makes it possible for causal microphysics (in this case, quantum fluctuaions in ϕ) to produce large-scale density perturbations.

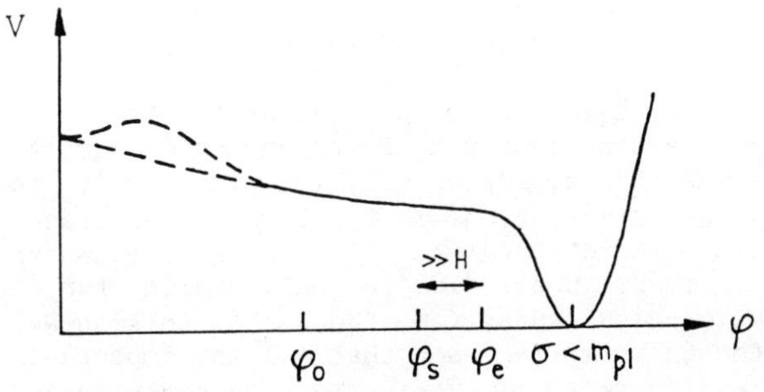

Fig. 4.6 The 'prescribed potential' for successful inflation.

While outside the horizon ($t_1 \lesssim t \lesssim t_H$) a perturbation evolves 'kinematically' (as a wrinkle in the geometry); viewed in some gauges the amplitude changes (e.g., the synchronous gauge), while in others (e.g., the uniform Hubble constant gauge) it remains constant. However, in all gauges the kinematic evolution is independent of scale (intuitively this makes sense since this is the kinematic regime). Given these 'two facts': $(\delta\rho/\rho)_1 \simeq$ scale-independent and the kinematic evolution $\simeq$ scale-independent, it follows that all scales reenter the horizon (at $t = t_H$) with (approximately) the same amplitude, given by Eqn. 4.6. Not only is this a reasonable spectrum (the Zel'dovich spectrum), but this is one of the very few instances that the spectrum of density perturbations has been calculable from first principles. [The fluctuations produced by strings are another such example, see, e.g. ref. 121; however, in a string scenario without inflation the homogeneity of the Universe must be assumed.]

Coleman-Weinberg SU(5) Model

The first model of new inflation[102,103] studied was the Coleman-Weinberg SU(5) model, with T = 0 effective potential

$$V(\phi) = 1/2\, B\sigma^4 + B\phi^4[\ln(\phi^2/\sigma^2) - 1/2], \tag{4.7}$$

$$\simeq 1/2\, B\sigma^4 - \lambda(\phi)\phi^4 \qquad (\phi \ll \sigma)$$

where Φ is the 24 dimensional field responsible for GUT SSB, ϕ is the magnitude of Φ in the SU(3) x SU(2) x U(1) SSB direction, $B = 25g^4/256\pi^2$ (g = gauge coupling constant), $\sigma \simeq 1.2 \times 10^{15}$ GeV, and for $\phi \simeq 10^9$ GeV, $\lambda(\phi) \simeq 0.1$. [V may not look familiar; this is because ϕ is normalized so that its kinetic term is $1/2\ \dot{\phi}^2$ rather than the usual $(15/4)\dot{\phi}^2$.] Albrecht and Steinhardt[15] showed that when $T \simeq 10^8 - 10^9$ GeV the metastability limit is reached, and thermal fluctuations drive ϕ over the T-dependent barrier (height $\simeq T^4$) in the finite temperature effective potential. Naively, one expects that $\phi_0 \simeq T_{MS}$ since for $\phi \ll \sigma$ there is no other scale in the potential (this is a point to which I will return). The potential is sufficiently flat that the approximation $3H\dot{\phi} \simeq -V'$ is valid for $\phi \ll \sigma$, and it follows that

$$(\phi/H)^2 = (3/2\lambda)[H(\tau_\phi - t)]^{-1}, \tag{4.8}$$

where $H\tau_\phi \simeq (3/2\lambda)(H/\phi_0)^2$ (recall τ_ϕ = time it takes ϕ to traverse the flat portion of the potential). Physically, $H\tau_\phi$ is the number of e-folds of R which occur during inflation, which to solve the homogeneity-isotropy and flatness puzzles must be $\gtrsim$ 0(60). For this model $H \simeq 7 \times 10^9$ GeV; setting $\phi_0 \simeq 10^8 - 10^9$ GeV results in $H\tau_\phi \simeq$ 0(500-50000) - seemingly more than sufficient inflation.

There is however, a very basic problem here. Eqn. 4.8 is derived from the semi-classical equation of motion for ϕ [Eqn. 4.1], and thus only makes sense when the evolution of ϕ is 'classical', that is when $\phi >> \Delta\phi_{QM}$ (= quantum fluctuations in ϕ). In deSitter space the scale of quantum fluctuations is set by H: $\Delta\phi_{QM} \simeq H/2\pi$ (on the length scale H^{-1}). Roughly speaking then, Eqn. 4.8 is only valid for $\phi >> H$. However, sufficient inflation requires $\phi_0 \lesssim H$. Thus the Coleman-Weinberg model seems doomed for the simple reason that all the important physics must occur when $\phi \lesssim \Delta\phi_{QM}$. This is basically the conclusion reached by Linde[122] and Vilenkin and Ford[123] who have analyzed these effects carefully. Note that by artificially reducing λ by a factor of 10-100 sufficient inflation can be achieved $\phi_0 >> H$ (i.e., the potential becomes sufficiently flat that the classical part of the evolution, $\phi >> H$, takes a time $\gtrsim 60\ H^{-1}$). In the Coleman-Weinberg model $\Gamma >> H$ and the Universe reheats to $T \simeq M_G \simeq 10^{14}$ GeV.

Let's ignore for the moment the difficulties associated with the need to have $\phi_0 < H$, and examine the question of density fluctuations. Combining Eqns. 4.6 and 4.8 it follows that

$$(\delta\rho/\rho)_H \simeq (4 \text{ or } 2/5)100\lambda^{1/2}[1 + \ln(M/10^{12}M_\odot)/171 + \ln(g\sigma/10^{15}\text{ GeV})/57]^{3/2}, \quad (4.9)$$

where M is the comoving mass within the perturbation. Note that the spectrum is almost, but not quite scale-invariant (varying by less than a factor of 2 from $1M_\odot$ to $10^{22}M_\odot$ = present horizon mass). Blindly plugging in $\lambda \simeq 0.1$, results in $(\delta\rho/\rho)_H \simeq 0(10^2)$ which is clearly a disaster. [On angular scales >> 1° the Zel'dovich spectrum results in temperature fluctuations of[104] $\Delta T/T \simeq 1/2(\delta\rho/\rho)_H$ which must be $\lesssim 10^{-4}$ to be consistent with the observed isotropy.] To obtain perturbations of an acceptable amplitude one must artificially set $\lambda \simeq 10^{-12}$ or so. [In an SU(5) GUT λ is determined by the value of $\alpha_{GUT} = g^2/4\pi \simeq 1/45$, which implies $\lambda \simeq 0.1$.] As mentioned earlier the density fluctuation constraint is a very severe one; recall that $\lambda \simeq 10^{-2} - 10^{-3}$ would solve the difficulties associated with the quantum fluctuations in ϕ. To say the least, the Coleman-Weinberg SU(5) model seems untenable.

Lessons Learned--A Prescription for Successful New Inflation

Other models for new inflation have been studied, including supersymmetric models which employ the inverse hierarchy scheme,[124] supersymmetric/supergravity models[124-126] and just plain GUT models[127] No model has led to a completely satisfactory new inflationary scenario, some failing to reheat sufficiently to produce a baryon asymmetry, others plagued by large density perturbations, etc. Unlike the situation with 'old inflation' a few years ago, the situation does not appear hopeless. The early failures have led to a very precise prescription for a potential which will successfully implement new inflation.[128] Among the necessary conditions are:

(1) A flat region where the motion of ϕ is 'friction-dominated', i.e., $\ddot{\phi}$ term negligible so that $3H\dot{\phi} = -V'$. This i.e., $\ddot{\phi}$ term negligible so that $3H\dot{\phi} = -V'$. This requires an interval where $V'' \lesssim 9H^2$.

(2) Denote the starting and ending values of ϕ in this interval by ϕ_s and ϕ_e respectively (note: ϕ_s must be $\gtrsim \phi_0$). The length of the interval should be much greater than H (which sets the scale of quantum

fluctuations in ϕ): $\phi_e - \phi_s >> H$. This insures that quantum fluctuations will not drive ϕ across the flat region too quickly.

(3) The time required for ϕ to traverse the flat region should be $\gtrsim 60\ H^{-1}$ (to solve the homogeneity-isotropy and flatness problems). This implies that

$$\int H dt \approx -\int_{\phi_s}^{\phi_e} (3H^2\, d\phi/V') \gtrsim 60. \tag{4.10}$$

(4) In order to achieve an acceptable amplitude for density fluctuations, $(\delta\rho/\rho)_H \approx H^2/\dot{\phi}(t_1)$, $\dot{\phi}$ must be $\approx 10^4\ H^2$ when a galactic size perturbation crosses outside the horizon. This occurs about 50 Hubble times before the end of inflation.

(5) Sufficiently high reheat temperature so that the Universe is radiation-dominated at the time of primordial nucleosynthesis ($t \approx 10^{-2} - 10^2$ sec; $T \approx 10$ MeV $- 0.1$ MeV), and so that a baryon-asymmetry of the correct magnitude can evolve. As discussed earlier, the reheat temperature is:

$$T_{RH} = \min\{M_G, (\Gamma m_{pl})^{1/2}\}; \tag{4.11}$$

this must exceed min{10 MeV, T_B}, where T_B is the smallest reheat temperature for which an acceptable baryon asymmetry will evolve.

(6) The potential be part of a 'sensible particle physics' model.

These conditions and a few others which are necessary for a successful implementation of new inflation are discussed in detail in ref.128. Potentials which satisfy all of the constraints tend to be very flat (for a long run in ϕ), and necessarily involve fields which are very weakly coupled (self couplings $\lesssim 10^{-10}$; see Fig. 4.6). To insure that radiative corrections do not spoil the flatness it is almost essential that the field ϕ be a gauge singlet field.

Concluding Remarks

New inflation is an extremely attractive cosmological program. It has the potential to 'free' the present state of the Universe (on scales at least as large as 10^{28} cm) from any dependence on the initial state of the Universe, in that the current state of the observable Universe in these models depends only upon microphysical processes which occurred very early on ($t \lesssim 10^{-34}$s). [I should mention that this conjecture of 'Cosmic Baldness'[129] is still just that; it has not been demonstrated that starting with the most general cosmological solution to Einstein's equations, there exist regions which undergo sufficient inflation. The conjecture however has been addressed perturbatively; pre-inflationary perturbations remain constant in amplitude, but are expanded beyond the present horizon[130] and neither shear nor negative-curvature can prevent inflation from occurring[131].]

At present there exists no completely successful model of new inflation. However, one should not despair, as I have just described, there does exist a clear-cut and straightforward prescription for the desired potential (see Fig. 4.6). Whether one can find a potential which fits the prescription and also predicts sensible particle physics remains to be seen. If such a theory is found, it would truly be a monumental achievement for the Inner Space/Outer Space connection.

Now for some sobering thoughts. The inflationary scenario does not address the issue of the cosmological constant; in fact, the small value of the cosmological constant today is its foundation. If some relaxation mechanism is found to insure that the cosmological constant is always small, the inflationary scenario (in its present form at least) would vanish into the vacuum. It would be fair to point out that inflation is not the only approach to resolving the cosmological puzzles discussed above. The homogeneity, isotropy, and inhomogeneity puzzles all involve the apparent smallness of the horizon. Recall that computing the horizon distance

$$d_H = R(t) \int_0^t dt'/R(t') \qquad (4.12)$$

requires knowledge of R(t) all the way back to t = 0. If during an early epoch ($t \lesssim 10^{-43}$s?) R increased as or more rapidly than t (e.g. $t^{1.1}$), then $d_H \to \infty$, eliminating the 'horizon constraint'. The monopole and flatness problems can be solved by producing large amounts of entropy since both problems involve a ratio to the entropy. Dissipating anisotropy and/or inhomogeneity is one possible mechanism for producing entropy. One alternative to inflation is Planck epoch physics. Quantum gravitational effects could both modify the behaviour of R(t) and through quantum particle creation produce large amounts of entropy [see e.g., the recent review in ref. 132].

Two of the key 'predictions' of the inflationary scenario, $\Omega = 1 \pm O(10^{-BIG})$ and scale-invariant density perturbations, are such natural and compelling features of a reasonable cosmological model, that their ultimate verification (my personal bias here!) as cosmological facts will shed little light on whether or not we live in an inflationary Universe. Although the inflationary Universe scenario is not the only game in town, right now it does seem to be the best game in town.

LECTURE 5: FORMATION OF STRUCTURE IN THE UNIVERSE

Overview

On small scales the Universe today is very lumpy. For example, the average density in a galaxy ($\simeq 10^{-24}$g cm^{-3}) is about 10^5 the average density of the Universe. The average density in a cluster of galaxies is about 100 times the average density in the Universe. Of course, on very large scales, say >> 100 Mpc, the Universe is smooth, as evidenced by the isotropy of the microwave background, number counts of radio sources, and the isotropy of the x-ray background.

The surface of last scattering for the 3K microwave background is the Universe at 200,000 years after 'the bang', when T $\simeq$ 1/3 eV and R $\simeq 10^{-3} R_{today} = 10^{-3}$ (it is convenient to set $R_{today} \equiv 1$). Thus the μ-wave background is a fossil record of the Universe at that very early epoch. The isotropy of the μ-wave background, $\delta T/T \lesssim O(10^{-4})$ on angular scales ranging from 1' to 180° (see Fig. 5.1), implies that the Universe was smooth at that early epoch: $\delta\rho/\rho << 1$. There is a calculable relationship between $\delta T/T$ and $\delta\rho/\rho$ (which depends upon the nature of density perturbations present -- type and spectrum), but typically

$$(\delta\rho/\rho)_{DEC} = \#(\delta T/T) \lesssim O(10^{-2}-10^{-3}), \qquad (5.1)$$

where # is O(10-100); for the detailed calculations I refer the interested reader to refs. 133-136.

So, the Universe was very smooth, and today it is very lumpy -- how did it get to here from there?? For the past decade, or so cosmologists have had a general picture of how this took place: small density inhomogeneities present initially grew via the Jeans, or gravitational instability, into the large inhomogeneities we observe today, i.e., galaxies, clusters of galaxies, etc. After decoupling, when the Universe is matter-dominated and baryons are free of the pressure support provided by photons, density inhomogeneities in the baryons and other components grow as

$$\delta\rho/\rho \propto \begin{cases} R & \delta\rho/\rho \lesssim 1 \\ \gtrsim R^3 & \delta\rho/\rho \gtrsim 1 \end{cases} \tag{5.2}$$

The isotropy of the μ-wave background allows for perturbations as large as 10^{-2} - 10^{-3} at decoupling, and the cosmic scale factor R(t) has grown by slightly more than a factor of 10^3 since decoupling, thus it is possible for the large perturbations we see today to have grown from small perturbations present at decoupling. This is the basic picture which is generally accepted as part of the 'standard cosmology'. [For a detailed discussion of structure formation in the Universe, see ref. 137.]

One would like to fill in the details, so that we can understand the formation of structure in the same detail that we do, say, primordial nucleosynthesis. The formation of structure (or galaxy formation as it is sometimes referred) began in earnest when the Universe became matter-dominated [$t_{eq} \simeq 3\times10^{10}$sec $(\Omega h^2/\theta^3)^{-2}$; $T_{eq} \simeq 6.8$eV $\Omega h^2/\theta^3$]; that is the time when density perturbations in the matter component can begin to grow. In order to fill in the details of structure formation one needs the 'initial data' for that epoch; in this case, they include: the total amount of non-relativistic stuff in the Universe, quantified by Ω; the composition, i.e., fraction Ω_i of the various components [(i = baryons, relic WIMPs (weakly-interacting massive particles), cosmological constant, relic WIRPs (weakly-interacting relativistic particles)]; spectrum and type (i.e., 'adiabatic' or 'isothermal') of density perturbations initially present. Given these 'initial data' one can construct a detailed scenario (e.g., by numerical simulation), which can then be compared to the Universe we observe today.

I want to emphasize the importance of the 'initial data' for this problem; without such, it is clear that a detailed picture of structure formation cannot be put together. As I will discuss, it is in this regard that the Inner Space/Outer Space connection has been so very important in recent years. Events which we believe took place during the earliest moments of the history of the Universe and which we are just now beginning to understand, have given us important hints as to the 'initial data' for this problem. These hints include: $\Omega \simeq 1.0$, adiabatic density perturbations with the Zel'dovich spectrum (from inflation); $\Omega_b \simeq 0.014 - 0.15$ (primordial nucleosynthesis); the possibility that the Universe is dominated by a massive, relic particle species (or WIMP) -- see Table 5.1 for a list of candidates; and other even more exotic possibilities -- topological strings, isothermal axion perturbations, a relic cosmological constant, and relativistic relic particles, to mention a few. Because of these 'hints from the very early Universe', the problem has become much more focused, and at present two detailed scenarios exist -- the 'cold dark matter' picture and the 'hot dark matter' picture. Unfortunately, as I will discuss, neither appears to be completely satisfactory at present.

In the following subsections I will discuss: 'The Observed Universe' -- the final test of any scenario is how well it reproduces the observed Universe; the standard notation and definitions, associated with discussing the evolution of density inhomogeneities in an expanding Universe; the standard lore about the evolution of density perturbations; the hot and cold dark matter scenarios; and finally, I'll mention some of the pressing issues and current ideas.

The Observed Universe

The 'basic building blocks' of the Universe we see are galaxies; a typical galaxy has a mass of $O(10^{11}\, M_\odot)$. For reference, a solar mass ($M_\odot$) is 2×10^{33}g or about 10^{57} baryons. We observe galaxies with redshifts of greater than 1; the redshift z that a photon emitted at time t suffers by the present epoch is just:

$$(1 + z) = R_{today}/R(t) = R(t)^{-1}$$

The fact that we see galaxies with $z \gtrsim O(1)$ implies that galaxies were present and 'lit up' by the time the Universe was about 1/2 its present size. QSOs are the most distant objects we can see, and many QSOs with redshifts in excess of 3 have been observed. [A typical QSO is 1-100 times as luminous as a galaxy, with a size of less than a light-month, or $\lesssim 10^{-5}$ that of a galaxy.] This fact implies that QSOs were present and 'lit up' when the Universe was 1/4 its present size.

To a first approximation galaxies are uniformly distributed in space; however, they do have a tendency to cluster and their clustering has been quantified by the galaxy-galaxy correlation function $\xi(r)$. The probability of finding a galaxy at a distance r from another galaxy is $(1+\xi(r))$ greater than if galaxies were just distributed randomly. The galaxy-galaxy correlation function is well-studied[137,138] and

$$\xi(r) \simeq (r/5h^{-1}\mathrm{Mpc})^{-1.8}$$

Something like 10% of all galaxies are found in clusters of galaxies, a cluster being a bound and sometimes virialized system of O(100) galaxies. Particularly populous clusters are called rich clusters and many of these rich clusters are affectionately known by their Abell numbers, as the astronomer George Abell studied and classified many of these objects. There is some evidence that clusters cluster, and this has been quantified by the cluster-cluster correlation function[139]

$$\xi_{cc}(r) \simeq (r/25h^{-1}\mathrm{Mpc})^{-1.8}$$

Interestingly enough, the cluster-cluster correlation function has the same slope, and a larger amplitude. The larger amplitude is a fact which is not presently understood; a number of explanations have been suggested[140,141], including the possibility that the amplitude of the cluster-cluster correlation function has not yet been correctly determined. One very promising idea[140] is that this just reflects the fact that rich clusters are 'rare events' and that for Gaussian statistics rare events are more highly correlated than typical events (i.e. galaxies not in clusters).

There is some evidence for still larger-scale structure -- superclusters[142], objects which may contain many rich clusters and are not yet virialized; voids[143], large regions of space (perhaps as large as 100 Mpc) which contain far fewer than the expected number of galaxies; and filamentary structures[144], long chains of galaxies. As of yet, there are no unambiguous statistics to quantify these features of the Universe.

How much stuff is there in the Universe? This is usually quantified as the fraction of critical density Ω ($\equiv \rho/\rho_{crit}$), where

$$\rho_{crit} \simeq 1.88h^2 \times 10^{-29} \text{g cm}^{-3} , \quad (5.3)$$

$$\simeq 1.05h^2 \times 10^4 \text{eV cm}^{-3} ,$$

$$\simeq 0.810h^2 \times 10^{-46} \text{ GeV}^4 .$$

From primordial nucleosynthesis we know that the fraction contributed by baryons must be:

$$0.014 \lesssim \Omega_b \lesssim 0.15.$$

What can we say based on more direct observations of the amount of stuff in the Universe? The standard approach is not too different from that of the poor drunk faced with the task of locating his/her lost keys in a large, poorly-lit parking lot on a moonless night. He/she focuses his/her search in the vicinity of the only lamp post in the parking lot -- not because he/she thinks he/she lost his/her keys there, but rather because he/she realizes that this is the only place he/she could find them should they be there.

By determining the average mass per galaxy, one can convert the observed number density of galaxies into a mass density:

$$\langle\rho\rangle \simeq \langle M_{GAL}\rangle\langle n_{GAL}\rangle \quad (5.4)$$

[To be more accurate, what astronomers actually do is to measure the mass to luminosity (or mass-to-light ratio, M/L, for short) for a typical galaxy and then multiply this by the observed luminosity density of the Universe to obtain the mass density; see ref. 145 for the dirty details of this procedure.]

The mass associated with a galaxy can be determined by a number of dynamical techniques, all of which basically involve Kepler's 3rd law in some way, or another. For a system with spherical symmetry:

$$GM = v^2 r , \quad (5.5)$$

where M is the mass interior to the orbit of an object with orbital velocity v and orbital radius r.

By studying the orbital motion of stars at the radius where the light has crapped out (the characteristic radius associated with the fall off in luminosity is called the Holmberg radius), one can measure the mass associated with the luminous material: M_{lum}. Converting this mass into an estimate for Ω one obtains:

$$\Omega_{LUM} \simeq 0.01 ,$$

which is disappointingly distant from $\Omega = 1$, but consistent with baryons being the luminous material (thank God!).

Studies of the orbits of stars (in spiral galaxies) beyond the radius where the light has 'crapped out' have revealed an important and startling result[146] -- their orbital velocities do not decrease (as they would if the luminous mass were the whole story, $v \propto r^{-1/2}$), but rather stay constant (this is the phenomenon referred to as 'flat rotation curves'). This, of course, indicates that mass continues to increase linearly with radius (or $\rho \propto r^{-2}$), whereas the light does not, indicating that the additional mass is dark. Flat rotation curves are the best evidence for the existence of dark matter. As of yet, there is no convincing evidence for a rotation curve that 'turns over' (which would indicate that the total mass in that galaxy has started to converge). Thus the $\langle M_{GAL} \rangle$ obtained this way provides a lower limit to $\langle M_{GAL} \rangle$; this lower limit is <u>at least</u> 3-10 times the luminous mass, implying

$$\Omega_{HALO} \gtrsim 0.03 - 0.10 \ .$$

The dark matter inferred from the rotation curves of spiral galaxies is often referred to as the 'halo material', as it is less condensed, seems to have a spherical distribution, and has a 'density run', $\rho \propto r^{-2}$, characteristic of a self-gravitating isothermal sphere of particles. Dynamical studies of the stars in our galaxy indicate that at our position ($\simeq$ 8 kpc from the center) the disk component (i.e., the component distributed like the stars in the disk) dominates the halo component by about a factor of 30[147]. Incidently, these same studies[147] indicate that not all of the disk component is accounted for by stars, dust, gas, <u>etc.</u> Locally, about half the density, or $\simeq 6{\times}10^{-24}$g cm^{-3}, is unaccounted for. The 'dark matter in the disk' must be material which is capable of undergoing dissipation, since the formation of the disk clearly involved dissipation of gravitational energy.

[As of yet there is no undisputable evidence for dark matter in elliptical galaxies. They are more difficult to study since they lack 'test particles' far from the center of the galaxy. Attempts have been made to use x-ray measurements of the gravitational potential, and these seem to indicate the presence of dark matter.[148]]

By studying the dynamics of galaxies in bound, virialized systems (binary galaxy systems, small groups of galaxies, and galaxy clusters) one can try to infer $\langle M_{GAL} \rangle$. Again one is basically using Kepler's third law (in the guise of the virial theorem). The results indicate

$$\Omega_{BS} \simeq 0.1 - 0.5 \ ,$$

have a higher degree of uncertainty, and are somewhat more ambiguous to interpret. For example, the highest values for Ω come from clusters, but only one galaxy in ten is found in a cluster. Most galaxies are found in binary systems or small groups. There is also the difficulty of identifying which galaxies are members of a given system, and determining whether or not a system has 'settled down' (i.e., is well-virialized) sufficiently so that the virial theorem is applicable.

There are many other techniques -- 'infall arguments': our motion toward the Virgo supercluster allows us to weigh the Virgo cluster, and our motion toward the Andromeda galaxy allows us to weigh our galaxy and Andromeda; the cosmic virial theorem[149], which relates the peculiar velocity field of the Universe[150] to Ω; all seem to point to a value of Ω in the range of $\simeq$ 0.1 - 0.3. Based upon the very non-trivial assumption that light is a good tracer of mass the observations seem to indicate that

$$\Omega_{OBS} \simeq 0.2 \text{ '}\pm 0.1\text{'} \ ,$$

where '±0.1' is not meant as a formal error, but rather is meant to indicate the spread of the observations.

To summarize what we know about the amount of stuff in the Universe: (1) dark matter dominates -- by at least a factor of 3-10, and is less-condensed than the luminous component (which shows clear signs of having undergone dissipation); (2) the dark component could be baryonic -- until Ω_{DARK} is shown to be $\gtrsim 0.15$; (3) baryons cannot provide closure density as primordial nucleosynthesis restricts $\Omega_b \simeq 0.014 - 0.15$; (4) no observation made yet indicates that Ω_{TOT} is close to 1.0! Our knowledge of Ω is summarized in Fig. 5.2.

Finally, there is the 3K μ-wave background radiation, the fossil record of the Universe at 200,000 yrs after the bang when $T \simeq 1/3$ eV and $R \simeq 10^{-3}$. The angular scale viewed on the sky and the corresponding length scale are related by

$$\lambda/\mathrm{Mpc} = 1.8(\phi/1')h^{-1} , \tag{5.6a}$$

$$M/10^{12}M_{\odot} \simeq 0.85(\phi/1')^3\ \Omega h^{-1} . \tag{5.6b}$$

Two additional scales of importance are: the thickness of the surface of last scattering ($\simeq 6'$) and the size of the horizon at decoupling ($\simeq 1°$). On small angular scales ($\lesssim 1°$) the temperature fluctuations in the μ-wave background are dominated by those intrinsic to the photons. For adiabatic perturbations:

$$(\delta T/T) \simeq (1/3)(\delta n_\gamma/n_\gamma) \simeq (1/3)(n_b/n_b); \tag{5.7}$$

in WIMP-dominated model Universes perturbations in the WIMPs (which began to grow as soon as the Universe becomes matter-dominated) have

Table 5.1 - WIMP Candidates for the Dark Matter

Particle	Mass	Place of origin	Abundance*
Invisible Axion	10^{-5}eV	10^{-30}sec, 10^{12}GeV	10^9cm^{-3}
Neutrino	30eV	1 sec, 1 MeV	100cm^{-3}
Photino/Gravitino/ Mirror Neutrino	keV	10^{-4}sec, 100MeV	10cm^{-3}
Photino/Sneutrino/Axino Gravitino/Shadow Matter/ Heavy Neutrino	GeV	10^{-3} sec, 10 MeV	10^{-5}cm^{-3}
Superheavy Magnetic Monopoles	10^{16}GeV	10^{-34}sec, 10^{14}GeV	10^{-21}cm^{-3}
Pyrgons/Maximons/Newtorites	$\gtrsim 10^{19}$GeV	10^{-43}sec, 10^{19}GeV	$\lesssim 10^{-24}$cm^{-3}
Quark Nuggets	$\simeq 10^{15}$g	10^{-5}sec, 300MeV	10^{-44}cm^{-3}
Primordial Black Holes	$\gtrsim 10^{15}$g	$\gtrsim 10^{-12}$sec, $\lesssim 10^3$GeV	$\lesssim 10^{-44}$cm^{-3}

*Abundance required for closure density: $n_{WIMP} \simeq 1.05h^2 \times 10^{-5}$ cm$^{-3}/m_{WIMP}$ (GeV)

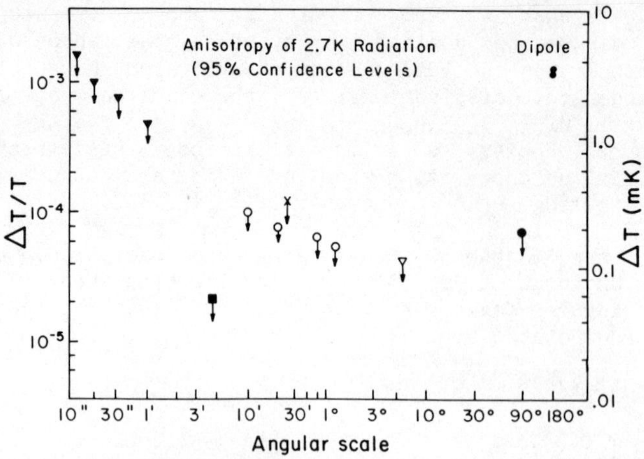

Fig. 5.1 Summary of the measurements of the isotropy of the microwave background (from ref. 113).

grown by a factor of ≃ (10-30) by decoupling, so that

$$(\delta T/T) \simeq (\delta\rho_X/\rho_X)_{dec}/30 \; . \tag{5.8}$$

On very small angular-scales ($\lesssim 6'$) the fluctuations in the μ-wave background are 'washed-out' due to the thickness of the last scattering surface.

On large-angular scales ($\gtrsim 1°$), the temperature fluctuations are primarily due to the Sachs-Wolfe effect (i.e., induced by the fluctuations in the gravitational potential): $\delta T/T \simeq \delta\phi/\phi \simeq 1/2\,(\delta\rho/\rho)_H$, and

$$(\delta T/T) \simeq 0.5(\delta\rho/\rho)_{HOR} \; , \tag{5.9}$$

where $(\delta\rho/\rho)_{HOR}$ is the amplitude of the density perturbation on the scale corresponding to θ when that density perturbation crossed inside the horizon. Note that since angular scales $\gtrsim 1°$ correspond to linear scales which were outside the horizon at decoupling, fluctuations on these scales provide us with information about the 'unprocessed, primordial spectrum of density perturbations.' [In the case that the Universe is reionized again after decoupling, the scale of 'virgin perturbations' may be somewhat larger, see ref. 151.] The main point here is that the 3K background is a fossil record of the early Universe, and density perturbations in particular. For this reason it provides a very stringent test of scenarios of galaxy formation. The interpretation of microwave background fluctuations is discussed in greater detail in refs. 133-136.

Notation and Definitions

It is convenient to discuss density perturbations in terms of the density contrast,

$$\delta \equiv \delta\rho(x)/\bar{\rho} \; , \tag{5.10}$$

and to expand the density contrast δ in a Fourier expansion:

$$\delta(x) = \Sigma\, e^{-ikx}\delta_k \; , \tag{5.11}$$

$$\equiv V/(2\pi)^3 \int \delta_k e^{-ikx} d^3x ,$$

$$\delta_k \equiv V^{-1} \int \delta(x) e^{ikx} d^3k .$$

Here $\bar{\rho}$ is the average density of the Universe, periodic boundary conditions have been imposed, and V is the volume of the fundamental cube. Strictly speaking, this expansion is only valid in spatially-flat models (k=0); however, at early times the effect of spatial curvature is small and can usually be neglected. [An analogous expansion exists for spatially-curved models; see ref. 152.] One other very important warning: $\delta(x)$ is not a gauge invariant quantity; in fact it is always possible to make $\delta(x) = 0$ by a suitable gauge transform: $x^\mu \to x^\mu + \varepsilon^\mu$, $g \to g + O(\varepsilon)$, $\varepsilon^\mu \simeq O(kt)$. J. Bardeen[152] has developed an elegant but non-trivial gauge invariant treatment of this problem. Rather than develop his formalism here, I will instead discuss the standard lore in the synchronous gauge ($g_{oo} = -1$, $g_{oi} = 0$). [For more details concerning the standard lore in the synchronous gauge, see, e.g., ref. 153.] I should emphasize that the gauge non-invariance of $\delta(x)$ only rears its ugly head for perturbations larger than the horizon.

A particular Fourier component is characterized by an amplitude and a wavenumber k. Let x and k be co-ordinate (or comoving) quantities, so that the physical distance

$$dx_{phys} = R(t)\, dx$$

and the physical wavenumber

$$k_{phys} = k/R(t) . \tag{5.12}$$

The wavelength of a perturbation is

$$\lambda \equiv 2\pi/k , \tag{5.13}$$

$$\lambda_{phys} = R(t)\, \lambda . \tag{5.14}$$

In the linear regime ($\delta < 1$), the physical size of a given density perturbation grows with the expansion. The comoving label (k or λ) for that perturbation is quite useful because it does not change with time -- i.e., the same physical perturbation is characterized by the same comoving label, while the physical labels (k(phys) and λ(phys)) do change with the expansion. It is also useful (and conventional) to characterize a density perturbation by the (invariant) rest mass M in a sphere of radius $\lambda/2$:

$$M \simeq 1.5\times10^{11} M_\odot (\Omega h^2) \lambda^3_{Mpc} , \tag{5.15}$$

where as before Ω is the fraction of critical density in NR particles (today) and R(t) has been normalized so that $R_{today} = 1$. A given Fourier component then is labeled by λ (= its physical size today, assuming that $\delta(x)$ were still < 1), k, its comoving wavenumber, and M, the mass contained within a sphere of radius $\lambda/2$. For reference, a typical galaxy mass ($10^{12} M_\odot$ including the dark matter) corresponds to

$$\lambda_{GAL} \simeq 1.9 \text{ Mpc } (\Omega h^2)^{-1/3} . \tag{5.16}$$

Of course the physical size of a galaxy is much less, more like 100 kpc, because galactic-sized perturbations have 'gone non-linear' ($\delta > 1$) and have ceased to continue to grow with the expansion; 1.9 Mpc, then, is the size a perturbation containing $10^{12} M_\odot$ would have today had it not 'gone nonlinear' and pulled away from the general expansion of the Universe.

With these definitions we can proceed to discuss quantities like: $\delta\rho/\rho$, $\delta M/M$, and $\xi(r)$, in terms of $|\delta_k|^2$. First consider $\delta\rho/\rho$:

$$\delta\rho/\rho = \langle\delta(x)\delta(x)\rangle^{1/2}, \tag{5.17}$$

where $\langle\ \rangle$ indicates the average over all space, i.e., $\delta\rho/\rho$ is the RMS value of $\delta(x)$. A bit of Fourier algebra yields:

$$(\delta\rho/\rho)^2 = V/(2\pi)^3\int_0^\infty |\delta_k|^2\ d^3k\ . \tag{5.18}$$

Evidently, the contribution to $(\delta\rho/\rho)^2$ from a given scale (specified by k) is:

$$(\delta\rho/\rho)^2|_k \approx V/(2\pi)^3\ k^3|\delta_k|^2\ . \tag{5.19}$$

Now consider $(\delta M/M)$, the RMS mass fluctuation on a given mass scale. This is what most people mean when they discuss the density contrast on a given mass scale. Mechanically, one would measure $(\delta M)_{RMS}$ by taking a volume V_W, which on average contains mass M, and placing it at all points throughout the space and computing the RMS mass fluctuation. Although it is simplest to choose a spherical volume V_W with a sharp surface, to avoid surfaces effects we must take care to smooth the surface. This is done by using a 'window function' W(r) (see below), which smoothly defines a volume V_W and mass $M = \bar{\rho}\ V_W$, where

$$V_W = 4\pi\int_0^\infty r^2W(r)dr\ . \tag{5.20}$$

A particularly simple window function is a Gaussian:

$$W(r) = \exp(-r^2/2r_o^2)\ , \tag{5.21a}$$

$$V_W = (2\pi)^{3/2}\ r_o^3\ , \tag{5.21b}$$

$$W(k) = (2\pi)^{3/2}r_o^3\ \exp(-k^2r_o^2/2)/V\ . \tag{5.21c}$$

The RMS mass fluctuation on the mass scale $M \equiv \bar{\rho}V_W$ is then:

$$(\delta M/M)^2 = \langle(\delta M/M)^2\rangle \equiv \langle(\int\delta(\vec{x}+\vec{r})W(\vec{r})d^3r)^2/V_W^2\rangle; \tag{5.22}$$

after some simple Fourier algebra it follows that

$$(\delta M/M)^2 = V^3(2\pi)^{-3}V_W^{-2}\int|\delta_k|^2|W(k)|^2d^3k\ . \tag{5.23}$$

For the Gaussian window function:

$$(\delta M/M)^2 = V(2\pi)^{-3}\int|\delta_k|^2e^{-k^2r_o^2}\ d^3k. \tag{5.24}$$

That is, $(\delta M/M)^2$ is equal to the integral of $|\delta_k|^2d^3k$ over all scales larger than r_o (all wavenumbers smaller than r_o^{-1}). If $|\delta_k|^2$ is given by a power law, $|\delta_k|^2 \propto k^n$ with $n > -3$, then the dominant contribution to $(\delta M/M)$ comes from the Fourier component on scale $k \simeq r_o^{-1}$ (as one would have hoped!), and

$$(\delta M/M)^2 \simeq V(2\pi)^{-3}k^3|\delta_k|^2/(n+3)|_{k\simeq r_o^{-1}}. \tag{5.25}$$

When one refers to '$\delta\rho/\rho$ on a given mass scale' what one means more precisely is $\delta M/M$, and as we have shown here that is just the power on that scale:

$$(\delta\rho/\rho)_M \propto k^{3/2}|\delta_k|\ , \tag{5.26}$$

$$\propto k^{3/2+n/2},$$

$$\propto M^{-1/2-n/6},$$

where we have used the fact that the comoving label $M \propto \lambda^3 \propto k^{-3}$, and assumed (as is usually done) that $|\delta_k|^2 \propto k^n$.

Finally, consider the galaxy-galaxy correlation function $\xi(r)$. Let $n(x)$ be the number density of galaxies at position x, and $\bar{n}$ be the average number density of galaxies: $\bar{n} = \langle n(x)\rangle$. The joint probability δP_{12} of finding galaxies in volumes δV_1 and δV_2 (centered at positions 1 and 2) is:

$$\delta P_{12} = \bar{n}^2 \delta V_1 \delta V_2 (1+\delta n(x_1)/\bar{n})(1+\delta n(x_2)/\bar{n}), \tag{5.27}$$

where $\delta n/\bar{n} = (n(x) - \bar{n})/\bar{n}$. If positions 1 and 2 are separated by a distance r, say, then the average probability of finding a galaxy a distance r from another galaxy is:

$$\langle\delta P_{12}\rangle = \bar{n}^2 \delta V_1 \delta V_2 (1+\langle\delta n(x+r)\delta n(x)\rangle/\bar{n}^2),$$

$$\equiv \bar{n}^2 \delta V_1 \delta V_2 (1 + \xi(r)) , \tag{5.28}$$

where $\xi(r)$ is the excess probability (over a random distribution), and is known as the galaxy-galaxy correlation function. Note that $\xi(r)$ is just $\langle\delta n(x+r)\delta n(x)\rangle/\bar{n}^2$. In terms of its Fourier transform $f(k) \equiv V^{-1} \int (\delta n(x)/\bar{n}) e^{ikx} dx$,

$$\xi(r) = V(2\pi)^{-3} \int |f(k)|^2 e^{-ikr} d^3k . \tag{5.29}$$

If, as is often done, we make the assumption that 'light faithfully traces mass', then $n(x) = a\rho(x)$ and $\delta(x) = \delta n(x)/\bar{n}$, so that

$$\xi(r) = \langle(\delta\rho(x+r)/\bar{\rho})(\delta\rho(x)/\bar{\rho})\rangle, \tag{5.30}$$

$$\xi(r) = V/(2\pi)^3 \int |\delta_k|^2 e^{-ikr} d^3k . \tag{5.31}$$

Note, the power spectrum $|\delta_k|^2$ is the Fourier transform of the galaxy-galaxy correlation function. Also note that

$$\xi(r=0) = \langle(\delta\rho/\rho)^2\rangle . \tag{5.32}$$

Perturbations are usually characterized as being adiabatic (more properly, 'curvature perturbations') or isothermal (more properly, 'isocurvature perturbations'). An adiabatic perturbation is an honest-to-God, gauge-invariant, wrinkle in the space-time manifold. That is '$\delta\rho \neq 0$'. By the equivalence principle, all components participate in an adiabatic perturbation:

$$\delta n_b/n_b = \delta n_\gamma/n_\gamma = \delta n_x/n_x , \tag{5.33}$$

where 'x' represents any other species, e.g., a relic WIMP. Note for an adiabatic perturbation $\delta(n_x/n_\gamma) = \delta(n_B/n_\gamma) = 0$.

On the other hand, an isothermal perturbation is not an honest-to-God wrinkle in the space-time manifold. That is '$\delta\rho = 0$'. Rather it is a spatial variation in the equation of state, a pressure perturbation, if you will. The usual example of an isothermal perturbation is a spatial fluctuation in the baryon-to-photon ratio, i.e., $\delta(n_B/n_\gamma) = f(x)$. Such perturbations are referred to as isothermal, because at very early times when $\rho_{TOT} \simeq \rho_\gamma$, $\delta\rho = 0$ is equivalent to $\delta T = 0$. Generalizing, isothermal perturbations at early times are characterized by:

$$\delta n_\gamma/n_\gamma = 3(\delta T/T) = 0$$

$$(\delta n_i/n_i) \neq 0 \text{ (for some species i).}$$

This corresponds to some species (or quantum number) being 'laid down' non-uniformly. Note, that since an isothermal perturbation is really a 'pressure perturbation', pressure gradients can and do, by moving matter about, change a pressure perturbation into a density perturbation on scales $\lesssim v_s t$, (v_s = sound speed). Thus on physical scales $\lesssim v_s t$, a pressure perturbation results in a density perturbation of the same amplitude. [For further discussion of adiabatic and isothermal perturbations, see, e.g., refs. 152-153.]

The Standard Lore

In this subsection I will attempt to briefly summarize the standard results (in the synchronous gauge). For a more detailed discussion of the evolution of density perturbations I refer the reader to refs. 137, 153; for a gauge-invariant discussion I refer the reader to Bardeen[152].

A very important scale when discussing the evolution of density perturbations is H^{-1}, the Hubble radius, or as I like to refer to it, 'the physics horizon'. H^{-1} is the expansion timescale; thus, in a time $\simeq H^{-1}$ all physical distances roughly double. Therefore it is the timescale for 'coherent microphysics' -- physical processes operating on longer timescales will have their effects distorted by the expansion. Therefore, H^{-1} is the distance over causal, coherent microphysics operates -- hence, the phrase 'physics horizon'. If $R \propto t^n$ $(n < 1)$, then H^{-1} is also the particle horizon. The evolution of a given Fourier component of $\delta\rho/\rho$ is naturally divided into two regimes: (1) $\lambda_{phys} \gtrsim H^{-1}$ -- when the perturbations said to be 'outside the horizon'; (2) $\lambda_{phys} \lesssim H^{-1}$ -- when the perturbation is said to be 'inside the horizon'. Since $\lambda_{phys} \propto R(t)$, so long as $R(t)$ increases more slowly than t, a perturbation begins outside the horizon and then at some point enters the horizon (see Fig. 5.3). For reference, during the radiation-dominated epoch, a perturbation on the mass scale M enters the horizon at time $t \simeq (M/M_\odot)^{2/3}$ sec.

(1) $\lambda_{phys} \gtrsim H^{-1}$: While outside the physics horizon, a perturbation cannot be affected by microphysics; its evolution is purely kinematical -- it evolves like a ripple in space-time. In the synchronous gauge,

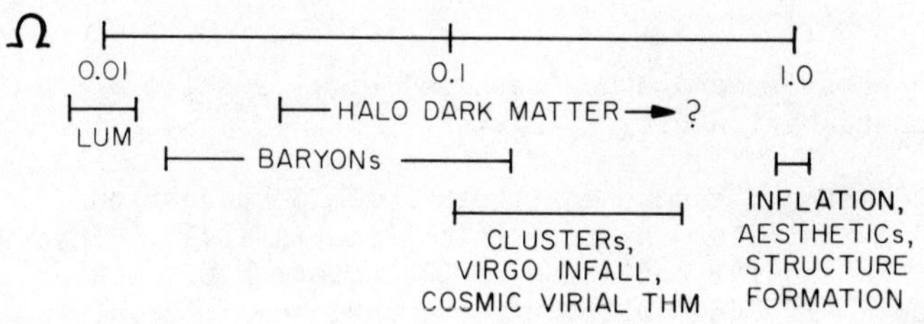

Fig. 5.2 Summary of our knowledge of Ω.

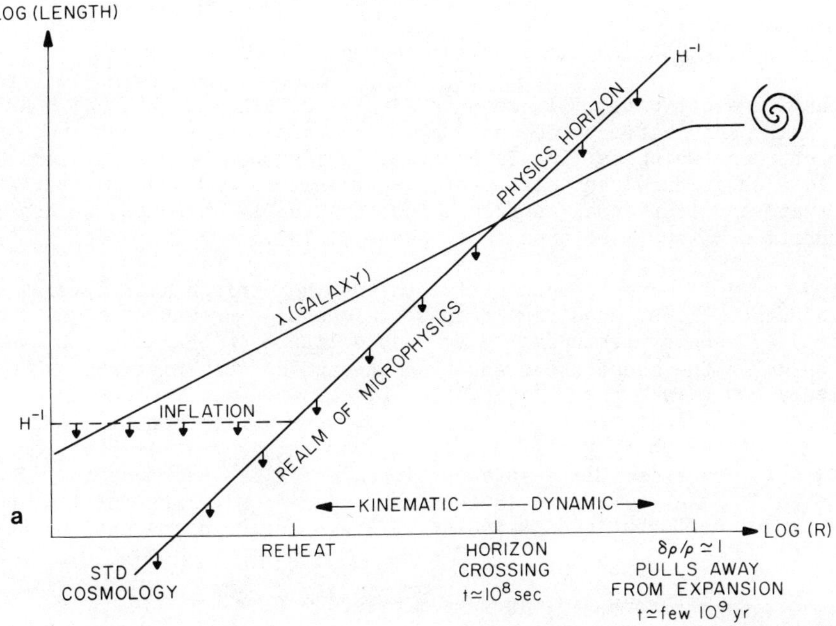

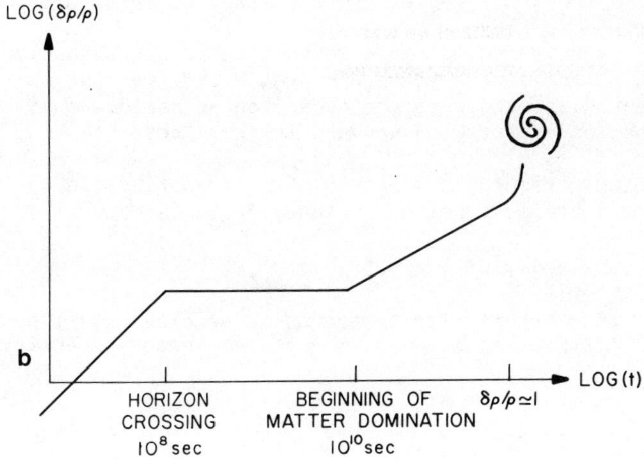

Fig. 5.3a Evolution of the physical size of a galactic-scale perturbation and of the 'physics horizon' (= H^{-1}) as a function of the scale factor R(t). Causal, coherent microphysics only operates on physical scales $\lesssim H^{-1}$. The kinematic ($\lambda \gtrsim H^{-1}$) and dynamical ($\lambda \lesssim H^{-1}$) regimes are indicated. Once $\delta\rho/\rho$ becomes of O(1) the galactic perturbation pulls away from the expansion, and its size remains constant.

Fig. 5.3b Evolution of the amplitude of a galactic-scale perturbation -- amplitude $\delta\rho/\rho \propto k^{3/2}|\delta_k|$.

adiabatic perturbations grow:

$$\delta\rho/\rho \propto t^n \quad n = \begin{cases} 1 & \text{radiation-dominated} \\ 2/3 & \text{matter-dominated} \\ 0 & \text{curvature-dominated} \end{cases}$$

By curvature-dominated, I mean $k/R^2 \gg 8\pi G\rho/3$ so that $H^2 \simeq k/R^2$. Isothermal perturbations do not grow -- after all, they are not honest-to-God wrinkles in space-time. [I should be careful here, as this is a gauge-dependent statement. What is true and gauge-invariant, is that an isothermal perturbation eventually becomes a density perturbation of the same amplitude; see ref. 153.]

(2) $\lambda_{phys} \lesssim H^{-1}$: Once a perturbation enters the horizon microphysics can be important. First consider pressure forces. Pressure forces can support a density perturbation against collapse on scales $\lesssim v_s t$, where $v_s^2 \equiv dp/d\rho$ is the sound speed and t is the age of the Universe. [More, precisely t is the dynamical timescale, i.e., timescale for gravitational collapse: $t_{dyn} \simeq (G\rho)^{-1/2}$. If $\rho = \rho_{TOT}$, then t_{dyn} is also the age of the Universe; if $\rho < \rho_{TOT}$, then $t_{dyn} \gtrsim t$.] Perturbations on scales $\lesssim \lambda_J \simeq v_s t$ oscillate as acoustic waves; λ_J is known as the Jeans length and $M_J = \pi\lambda_J^3\rho_{NR}/6$ is the Jeans mass (ρ_{NR} = mass density in NR particles). Perturbations on scales $\gtrsim \lambda_J$ are unstable against collapse, and grow:

$$\delta\rho/\rho \propto t^n \quad \begin{cases} n = 2/3 & \text{matter-dominated} \\ n = 0 & \text{radiation-dominated, or curvature-dominated} \end{cases}$$

Note that if the Universe is radiation-dominated, perturbations in a NR component even on scales $\lambda \gtrsim \lambda_J$ cannot grow. Simply put, the energy density of the radiation drives the expansion so rapidly that the growth of the perturbation cannot keep up, and it just 'hovers'.

The equation governing the growth of perturbations in species i (i = baryons, exotic particle relic, photons, etc.) is:

$$\ddot{\delta}_i + 2H\dot{\delta}_i + k^2 v_{si}^2\delta_i/R^2 = 4\pi G\rho\Sigma(\rho_i\delta_i/\rho), \tag{5.34}$$

where δ_i is the kth-Fourier component in species i, v_{si}^2 is the sound speed squared in component i, ρ is the total energy density of the

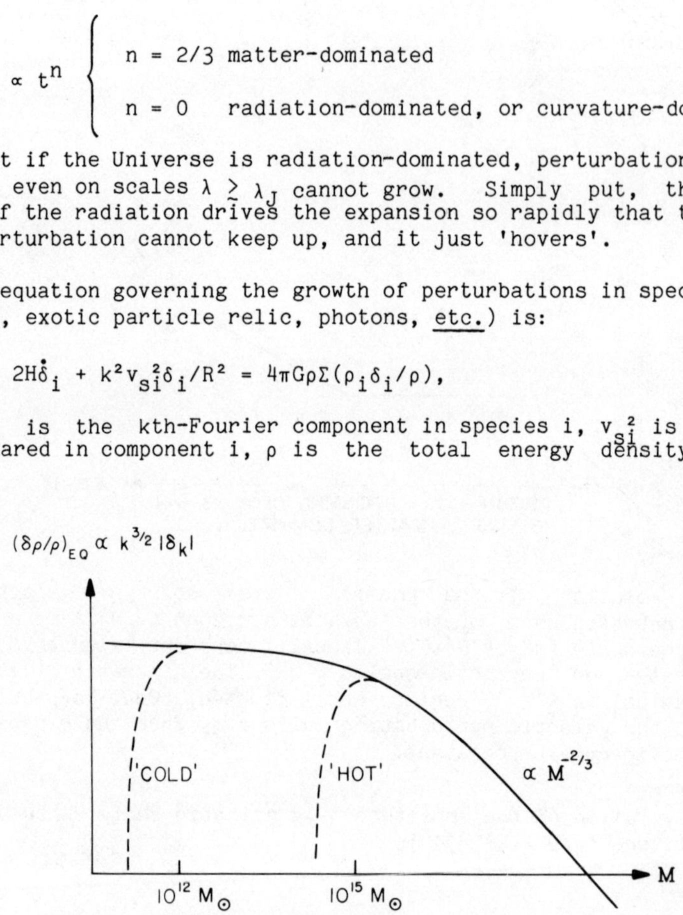

Fig. 5.4 The 'Processed Spectra' at $t \simeq t_{eq}$

Universe, and ρ_i is the energy density in species i. For a one species fluid, it is straightforward to verify the results quoted above.

A few words about Eqn. (5.34). In a non-expanding Universe (i.e., H=0), the Jeans instability is exponential on scales $\lambda \gtrsim \lambda_J$, with characteristic time $\tau \simeq t_{dyn} \simeq (G\rho_i)^{-1/2}$. The expansion of the Universe slows the growth and results in a power law instability. When the Universe is radiation-dominated (i.e., $R \sim t^{1/2}$, $H \sim 1/2t$, and $\rho_i/\rho \ll 1$), the solution to Eqn. (5.34) is $\delta_i = a + b\ln t$, so a perturbation with an initial velocity ($\dot{\delta}_i \neq 0$) can actually grow, albeit logarithmically.

Finally, one last bit of important microphysics. Perturbations in a collisionless component (e.g., neutrinos, axions, etc.) are subject to Landau damping, or 'freestreaming'. Until they become Jeans unstable, they can 'stream out' of overdense regions and into underdense regions, in the process smoothing out density perturbations. [Note, this effect has not been taken into account in Eqn. (5.34); in order to correctly take this effect into account one must integrate the collisionless Boltzmann Equation for the collisionless component.] The comoving freestreaming scale is easily calculated:

$$\lambda_{FS}(t) = \int_0^t v(t')dt'/R(t') , \quad (5.35)$$

where v(t) is the velocity of the species in question. Most of the contribution to the integral comes at or just after the time the species goes NR; after this epoch $v \propto R^{-1}$ (for a collisionless species) and

$$\lambda_{FS} \simeq (t_{NR}/R_{NR}) \begin{cases} 3 & t_{eq} \lesssim t_{NR} \\ \ln(t_{eq}/t_{NR}) & t_{eq} \gtrsim t_{NR} \end{cases} \quad (5.36)$$

where t_{NR} and R_{NR} are the time and the scale factor when the species became NR, and t_{eq} is the time when the Universe became matter-dominated. [Note, in the case that the Universe is still radiation-dominated when the species goes NR, λ_{FS} continues to grow after $t \simeq t_{NR}$ -- albeit logarithmically.] For reference:

$$t_{eq} \simeq 3\times10^{10}\ \mathrm{sec}(\Omega h^2/\theta^3)^{-2} , \quad (5.37a)$$

$$T_{eq} \simeq 6.8\mathrm{eV}\ (\Omega h^2/\theta^3) , \quad (5.37b)$$

$$T_{NR} \simeq (m_X/3)(T_\gamma/T_X) , \quad (5.37c)$$

$$R_{NR} \simeq 7\times10^{-7}\theta(T_X/T_\gamma)(\mathrm{keV}/m_X) , \quad (5.37d)$$

$$t_{NR} \simeq 2.2\times10^{7}\mathrm{sec}(m_X/\mathrm{keV})^{-2}(T_X/T_\gamma)^2 , \quad (5.37e)$$

where T_γ is the photon temperature, T_X/T_γ is the ratio the temperature of species X to the photons, e.g., for neutrinos $T_X/T_\gamma \simeq (4/11)^{1/3}$, and as usual θ is the present photon temperature in units of 2.7K.

Assuming that $t_{NR} \lesssim t_{eq}$, which is almost always the case, it follows that:

$$\lambda_{FS} \simeq 1\ \mathrm{Mpc}(m_X/\mathrm{keV})^{-1}(T_X/T_\gamma)[\ln(t_{eq}/t_{NR}) + 1]. \quad (5.38)$$

For a neutrino species, $t_{NR} \simeq t_{eq}$ and $(T_\nu/T_\gamma) \simeq 0.71$, so that the freestreaming scale is

$$\lambda_\nu \simeq 30\ \mathrm{Mpc}\ (m_\nu/30\mathrm{eV})^{-1}. \quad (5.39)$$

Although photons and baryons are certainly not collisionless (at least until after decoupling), there is a similar effect due to the fact that the mean free path of a photon is finite (and so the two are not a perfect fluid). In this case the operative word is not 'free streaming', but rather 'diffusion', and the effect is known as 'Silk damping'.[154] The scale associated with photon diffusion is

$$\lambda_S \approx 60\text{Mpc}/(1 + 25\Omega_b{}^{3/4}\Omega^{1/4}h^2). \tag{5.40}$$

Because of photon diffusion, or freestreaming of a collisionless species all initial perturbations on scales less than λ_{FS} (and for baryons λ_S) are strongly damped.

The Spectrum of Density Perturbations

We know how to specify the spectrum of density perturbations, but when does one specify it? From our discussion of the evolution of density perturbations, it seems very sensible to specify the amplitude of a density perturbation when it crosses the horizon, before any microphysical processing can occur. The amplitude at horizon crossing also has a Newtonian interpretation -- it is the perturbation in the gravitational potential. Note that by doing such, the amplitude on different scales is specified at different times. It is traditional to suppose that the spectrum is a featureless power law. Until recently this was merely an assumption since one had no fundamental understanding of the origin of density inhomogeneities. As discussed in Lecture 4, the inflationary paradigm has changed that situation. Write the amplitude at horizon crossing $(\delta\rho/\rho)_{HOR}$ as a power law,

$$(\delta\rho/\rho)_{HOR} \propto M^{-\alpha} ,$$

and remember that $(\delta\rho/\rho)$ means $k^{3/2}|\delta_k|$. If $|\delta_k|^2 \propto k^n$, then $\alpha = -1/2 - n/6$. The inflationary Universe scenario predicts $\alpha=0$ or $n=-3$.

What can we say more generally about α. The formation of galaxies requires that $(\delta\rho/\rho)_{HOR} \simeq 10^{-4}$ on the scale of a galaxy, i.e., $10^{12}M_\odot$. The measured isotropy of the μ-wave background implies that $(\delta\rho/\rho)_{HOR}$ on the scale of the present horizon ($\simeq 10^{22}M_\odot$) is less than 10^{-3} (to be very conservative). This means that α must be $\gtrsim -0.1$ (or else the spectrum must have a cutoff).

Consider perturbations on scales $\ll 10^{12}M_\odot$. If a perturbation crosses the horizon with amplitude greater than order unity, black hole formation is inevitable -- regions of space-time will pinch off before pressure forces can respond to prevent black hole formation.[155] Black holes less massive than order 10^{15}g will have evaporated before the present epoch (via the Hawking process[156]), however holes more massive than 10^{15}g will still be with us today. If $(\delta\rho/\rho)_{HOR}$ were $\gtrsim O(1)$ on a scale $\gtrsim 10^{15}$g, there would be far too many black holes with us today. This implies that α must be $\lesssim 0.2$ (or that the spectrum must be cut off on the low mass end). The $\alpha=0$, constant-curvature spectrum is clearly singled out. It is the so-called Harrison-Zel'dovich[157] spectrum.

It is sometimes convenient to specify the spectrum at a fixed time, e.g., at $t \gtrsim t_{eq}$ when structure formation is proceeding:

$$(\delta\rho/\rho)_t \propto M^{-\gamma} ; \tag{5.41}$$

for scales which are still outside the horizon at time t, $\gamma = \alpha + 2/3$. [This fact is straightforward to show.] Note that when specified at a fixed time, the Zel'dovich spectrum (on scales larger than the horizon) corresponds to $\gamma = 2/3$ and $n = 1$.

The Processed Spectrum

Given the initial spectrum, i.e., $(\delta\rho/\rho)_{HOR}$, we can use our knowledge of the microphysics to calculate 'the processed spectrum.' A convenient time to specify the processed spectrum is at $t \simeq t_{eq}$ when structure formation really begins. Let's assume that the initial spectrum is the Zel'dovich spectrum. If not, the slopes of the various regions of the processed spectrum are obtained by changing those shown by α. Scales which cross the horizon before t_{eq}, i.e., those with

$$\lambda \lesssim \lambda_{eq} \simeq 2ct_{eq}/R_{eq} \simeq 13\text{Mpc}(\theta^2/\Omega h^2), \qquad (5.42)$$

grow only logarithmically from horizon-crossing until $t \simeq t_{eq}$ when the Universe becomes matter-dominated (and perturbations start to grow as $R(t)$), by a factor of $O(20)$ for scales $<< \lambda_{eq}$. On scales $\gtrsim \lambda_{eq}$, $(\delta\rho/\rho) \propto M^{-2/3}$, as discussed above. Now add free-streaming and we have the fully-processed spectrum, from which structure formation will proceed (see Fig. 5.4).

From the epoch of matter domination until decoupling ($R \simeq 10^{-3}$, $T \simeq 1/3$ eV, $t_{dec} \simeq 7\times10^{12}\text{sec}(\Omega h^2)^{-1/2}$) perturbations in the WIMPs can grow ($\delta\rho/\rho \propto R \propto t^{2/3}$); however, perturbations in the baryons cannot yet grow since they are still tightly coupled to the photons. After decoupling, the baryons are free of the pressure support provided by the photons, and quickly fall into the potential wells formed by the WIMPs. In a few expansion times the baryon perturbations catch up with the WIMP perturbations (so long as $\Omega_{WIMP} >> \Omega_b$), and then perturbations in both components grow together (see Fig. 5.5).

Starting at $t \simeq t_{eq}$, all scales grow together, $\delta\rho/\rho \propto t^{2/3}$, so that the shape of the spectrum remains the same (while the overall amplitude increases). When $(\delta\rho/\rho)$ becomes unity on a scale, structures of that mass begin to form bound systems whose self-gravitational attraction dominates that of the rest of the Universe. These structures cease to participate in the general expansion of the Universe -- 'they pull away from the expansion'. For the Zel'dovich spectrum (or any spectrum which decreases with increasing mass scale), the first scale on which structures form is set by λ_D (see Fig. 5.4).

There are two limiting cases: (1) $\lambda_D \simeq \lambda_{eq} \simeq 13h^{-2}$ Mpc (for relic WIMPs which go NR at $t \simeq t_{eq}$ -- of the candidates in Table 5.1, this only applies to light neutrinos). In this case the damping scale is much greater than a galactic mass (closer to the mass of a supercluster). This case is known as 'hot dark matter'. (2) $\lambda_D \lesssim 1$ Mpc (i.e., $\lambda_D << \lambda_{eq}$), which occurs for relic WIMPs which go NR long before the epoch of matter-domination. In this case the first bound structures to form are galactic mass (or smaller). There is, of course, an intermediate possibility, $\lambda_D \simeq 1$ Mpc - 10 Mpc, referred to as 'warm dark matter', although I will not discuss that case here. In this case the first objects to form are necessarily of galactic mass. I'll briefly review structure formation in the hot and cold dark matter scenarios.

Two Stories: Hot and Cold Dark Matter

The Hints - Before going on to discuss the hot and cold dark matter scenarios, let me once again emphasize the hints which the early Universe has provided us with, and which has led to these two detailed scenarios. The structure-formation problem is basically an initial data problem, and the study of the very early Universe has helped to focus our thinking in this regard. First, the density perturbations. It has

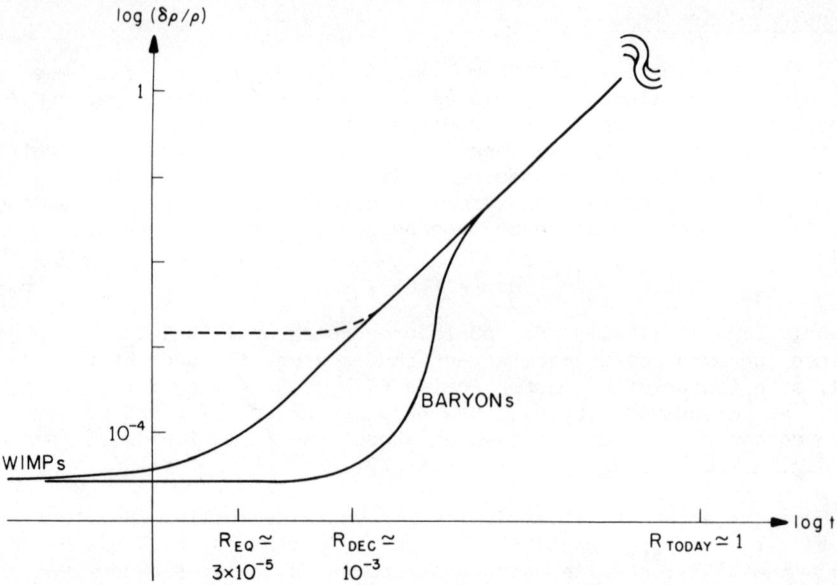

Fig. 5.5 The evolution of $(\delta\rho/\rho)$ in the Baryon and WIMP components. Note that in a baryon-dominated Universe $(\delta\rho/\rho)$ cannot begin to grow until decoupling and must necessarily begin with a larger amplitude.

long been realized that primordial perturbations were necessary, however until very recently their type, spectrum and origin were largely a mystery. Inflation has provided for the very first time a scenario for the origin of perturbations and makes a very definite prediction -- adiabatic perturbations with the Harrison-Zel'dovich spectrum. [Inflation in an axion-dominated Universe also results in isothermal, axion perturbations, see Seckel and Turner, ref. 3.] Next, the composition of the Universe, Ω_i. We know that the dominant component is dark, and since Ω_b must be ≤ 0.15, the dominant component must be non-baryonic if Ω is to be ≥ 0.15. Inflation predicts $\Omega = 1.0$ (more precisely, $k = 0$). Structure formation also favors a large value of Ω, as in a $\Omega < 1$ Universe perturbations have less time to grow -- perturbations cease to grow when the Universe becomes curvature-dominated, at a scale factor $R \approx \Omega$. [Because of this fact larger initial perturbations are needed (implying larger $\delta T/T$); the small isotropy of the microwave background implies that $\Omega h^{4/3} > 0.2$, see ref. 135.] There is no lack of candidate particles which could have sufficient relic abundance to provide $\Omega = 1$ (see Table 5.1). The hint, then, is that the Universe is dominated by relic WIMPs, with $\Omega_b \approx 0.1$ or so and $\Omega_{WIMP} \approx 0.9$. This too aids the growth of density perturbations; while perturbations in the baryons cannot begin to grow until after decoupling $(R \approx 10^{-3})$, perturbations in the dominant WIMP component can begin to grow as soon as the Universe becomes matter-dominated $(R_{eq} \approx 3\times10^{-5}h^{-2})$ -- providing for an additional factor of $30h^2$ growth in $\delta\rho/\rho$ over an $\Omega = 1$ baryon-dominated Universe. [In that regard a $\Omega = \Omega_b \approx 0.1$ Universe is in sad shape; perturbations can grow only from $R \approx 10^{-3}$ to $R \approx \Omega_b \approx 10^{-1}$, for a total growth factor of only 100! In fact, the small-scale microwave anisotropy measurement of Uson and Wilkinson rules out a baryon-dominated Universe with adiabatic perturbations; see,

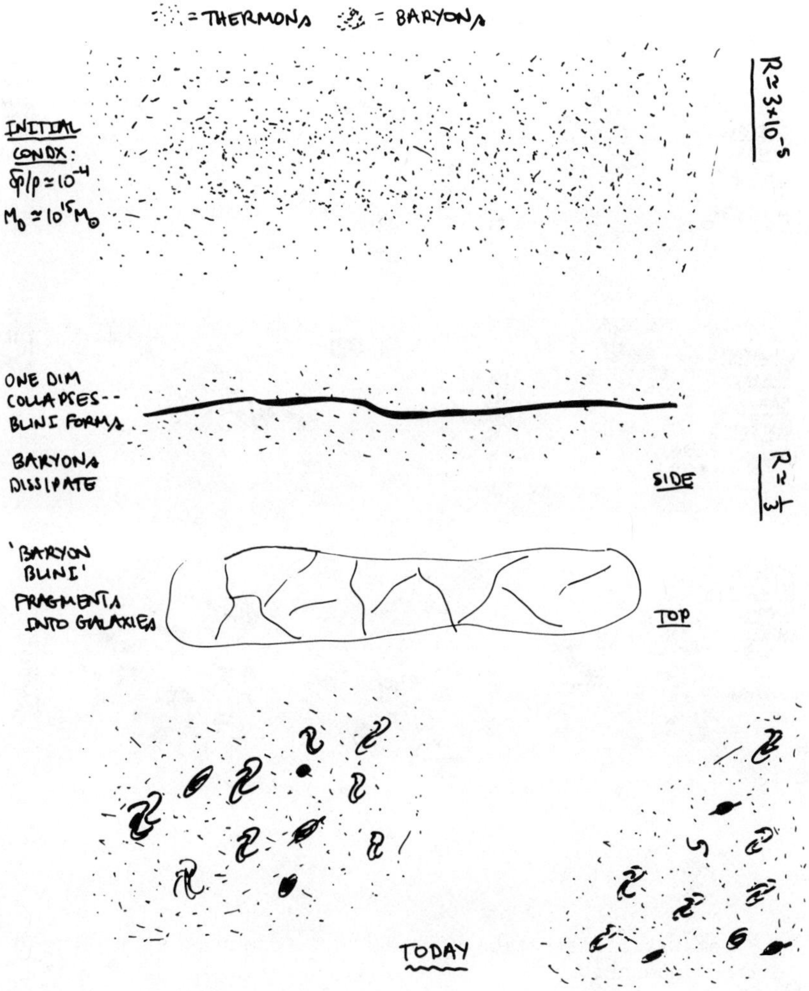

Fig. 5.6 Schematic illustration of structure formation in the 'hot dark matter' scenario.

refs. 134-135.] These hints have led to two detailed scenarios -- hot and cold dark matter.

Hot dark matter - The first structures to form are of supercluster size ($10^{15}M_\odot$ or so), and they do so rather recently ($z \lesssim 3$). Zel'dovich has argued rather convincingly that the collapse of the objects should be very non-spherical, and very nearly 1-dimensional (like a pancake or 'blini'). [For a review of the pancake scenario, see, ref. 158.] Once the pancake forms and goes non-linear in one dimension, the baryons within it can collide with each other and dissipate their gravitational energy. Thereby, the baryons in the pancake can fragment and condense into smaller (say galaxy-sized) objects. The neutrinos, being so weakly interacting, do not collide with each other or the baryons, cannot dissipate their gravitational energy and therefore cannot collapse into more tightly-bound objects. Thus they should remain as a halo. Some slow-moving neutrinos may subsequently be captured by the baryon-dominated galaxies. Structure formation in the hot dark matter

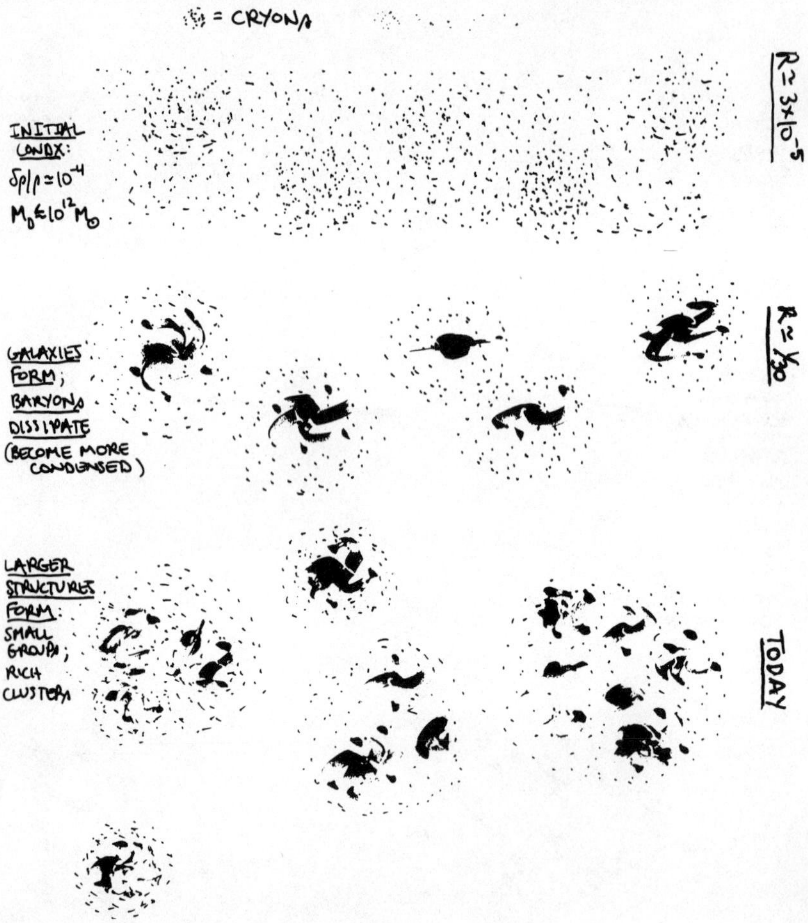

Fig. 5.7 Schematic illustration of structure formation in the 'cold dark matter' scenario.

case is schematically illustrated in Fig. 5.6. In a phrase, the structure in a hot dark matter Universe forms from 'the top down'.

Several groups[159,160] have performed numerical simulations of the neutrino-dominated Universe. The simulations reproduce nicely the voids and filamentary structures which seem to exist in the Universe (see Fig. 5.8). However, in order to reproduce the observed galaxy-galaxy correlation function, the epoch of pancaking must be made to occur yesterday -- that is at a very low redshift ($z \lesssim 1$). This fact is difficult to reconcile with the many galaxies with redshifts greater than 1 and QSO's with redshifts greater than 3 (the current record holder has $z \simeq 3.8$). It should be mentioned that these simulations only simulate the behaviour of the neutrinos, whereas, the galaxy-galaxy correlation function clearly is determined by where the baryons are. The small-scale microwave anisotropy predicted in the neutrino scenario is marginally consistent with the upper limit[112] of Uson and Wilkinson on 4.5': $\delta T/T \lesssim 3\times10^{-5}$, so long as Ω is close to 1. To summarize, hot dark matter is down, but not quite out (yet). [For more details about how the baryons cool and fragment, and the hot dark matter scenario in general, see refs. 158-162.]

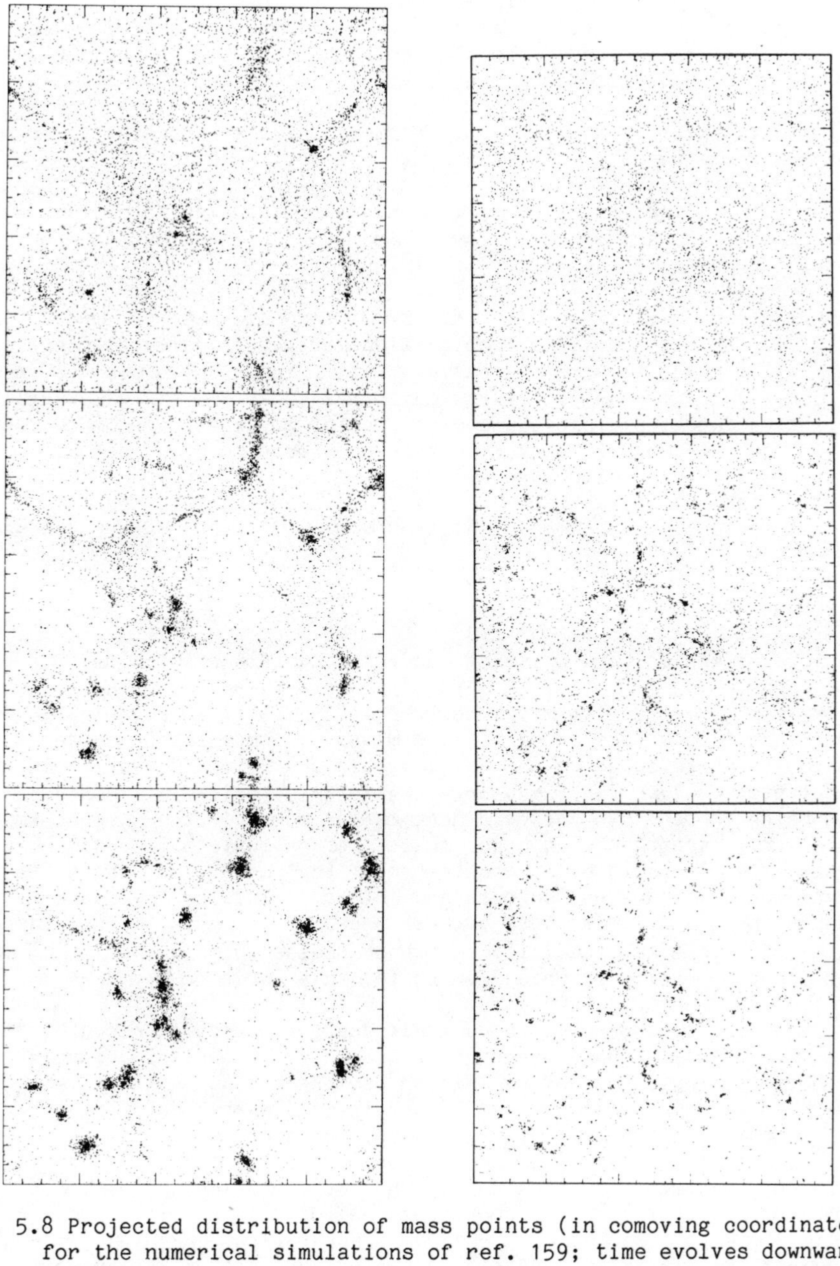

Fig. 5.8 Projected distribution of mass points (in comoving coordinates) for the numerical simulations of ref. 159; time evolves downward. For 'hot dark matter' (left), the panels represent the present Universe (z=0), in models where the onset of galaxy formation occurred at z_{GF} = .4, 1.1, 2.5 . For 'cold dark matter' (right), the last panel corresponds to the present Universe in a model with $\Omega h \simeq 0.2$ -- the model which gives the best fit to the observed Universe.

Cold dark matter -- Since the damping scale is less than a galactic mass in this case, the first structures to form should be galaxies. This should occur at a redshift of 10-20. Once galactic-sized objects have formed and have virialized, some of the baryons will collide and dissipate their gravitational energy, thereby settling into more compact objects at the center of the WIMP halo. These objects include stars, star clusters, and galactic disks.

Once galaxies form, they will tend to cluster together, forming larger aggregates such as small groups of galaxies, clusters of galaxies, and eventually superclusters. This process of 'hierarchical clustering' is helped along by the initial density perturbations which exist on these larger scales. The cold dark matter or 'bottom up' scenario is shown schematically in Fig. 5.7.

Numerical simulations of cold dark matter have been performed (see Fig. 5.8). Here there is no problem with getting galaxies to form early enough. The simulators[163,164] can get their model Universes to match the observed Universe in very many respects -- the galaxy-galaxy correlation function, masses and densities of galaxies, and many other details, but at a price: they find that they must require $\Omega h \simeq 0.2$ to do so. The small-scale microwave anisotropy predicted in the cold dark matter scenario is a factor of 4 or so less than the observed upper limit at present. [For more details of the cold dark matter scenario see refs. 163-165.]

The Ω-Problem

This brings us to a pressing and very significant problem: the fact that $\Omega_{OBS} \simeq 0.2 \pm$ '0.1', while theoretical prejudice would have us believe that $\Omega \simeq 1.0$. As mentioned earlier, Ω_{OBS} is determined assuming that 'light' is a good tracer of the mass. A component of matter which is smoothly distributed on the scales which are being probed would not be detected. Ω_{OBS} only measures the material which is clumped on the scale being probed. In the cold dark matter scenario the Ω-problem is particularly acute as on the scales of galactic haloes and larger, baryons (light) should be a good tracer of the mass, as there is no mechanism to separate baryons and WIMPs. In the hot dark matter scenario the situation is less clear. Because of their large damping length, neutrinos are initially smooth on scales up to 30 Mpc or so, and Ω has yet to be reliably probed on scales this large.

The Ω-problem has received a great deal of attention recently and a number of solutions have been proposed. They all basically involve the same idea: the existence of a smooth component $\Omega_{SMOOTH} = 1-\Omega_{OBS} \simeq 0.8\pm$'0.1' which would provide the additional mass density required to bring Ω to 1. This component, of course, would go undetected because it is more smoothly distributed than the observed component. [It could, of course, be detected by measuring the deceleration parameter q_o.] Suggestions for the smooth component include: (1) A relic cosmological term[166,167] (i.e., $\Lambda = 3\Omega_{SMOOTH}H_o^2$), which by definition is of course absolutely smooth. The possible origin of such a term, at present, has not even the slightest hint of an explanation. (2) Relativistic (or very fast-moving) particles,[166,168] produced by the recent (redshift 3-10) decay of an unstable WIMP (e.g., a 100 eV neutrino which decays to a light neutrino and a massless scalar). Very fast-moving particles cannot, by virtue of their high speeds, cluster. This scenario has received a great deal of attention lately. (3) A network of strings, or very fast-moving strings which too cannot cluster.[169] (4) A more smoothly-distributed component of galaxies which are either too faint to be seen or never lit up.[170]

The idea of 'failed galaxies' is an intriguing one. In this scenario, the very overdense regions ('3σ peaks') which collapse first would correspond to the galaxies we see. The more typical regions ('1 σ peaks' in the density distribution) would be the 'failed galaxies'. It is a well-known property of gaussian statistics that the rare events ('3 σ peaks' here) are more highly correlated. Put another way, the common events ('1 σ peaks') are less correlated, i.e., more smoothly distributed. If this idea is correct, then today we are only seeing the tips of the icebergs so to speak. What is presently lacking in this scenario is a plausible mechanism for 'biasing' galaxy formation (i.e., inhibiting the 1 σ peaks from lighting up).

Epilogue

The hints provided by the very early Universe have helped to focus the efforts of those studying galaxy formation. We have at present two rather detailed stories - hot and cold dark matter. Neither story however provides a totally satisfactory picture. We have at least one major problem -- the Ω problem. I have discussed some possible modifications of the two stories which might help to resolve the Ω problem.

Lest we become over confident, we should realize that the eventual sorting out of the details of structure formation may involve a bold departure from the two stories I discussed. For example, I did not discuss the role that cosmic strings may play. It may be that structure formation was initiated not by adiabatic density perturbations, but rather by strings and loops and the isothermal perturbations induced by them. I refer the interested reader to ref. 171, for further discussion of strings and galaxy formation.

The discussion of structure formation usually focuses on the role of gravitational forces, astrophysical fireworks (energy produced by the galaxies themselves by nuclear and other processes) are usually ignored. During the very early stages of structure formation (before any structures form) this is probably a very good approximation, but once a few galaxies light up, the energy released by astrophysical processes within them may play an important role. This point has been particularly emphasized by Ostriker and his collaborators.[172] All of the processes suggested for biasing galaxy formation involve astrophysical processes.

While I have focused on the role of 'primordial perturbations' (i.e., those produced in the very early Universe), it could be that the relevant perturbations arise due to physical processes which occur rather late, e.g., processes which result in black hole formation (say masses 10^{-6}-$10^{6}M_{\odot}$), with these 'small' black holes then playing the role of seeds for structure formation. Carr[174] has particularly emphasized this possibility.

Although much progress has been made toward understanding structure formation, there may yet be some very interesting surprises in store for us! More than likely observational/experimental data will play an important role. For example, the list of candidate WIMPs may be whittled down by searches for SUSY partners at the CERN Sp$\bar{p}$S and TeV 1 (or one of the candidate WIMPs may be detected!), the situation with regard to the mass of the electron neutrino may become clearer, the 'not so invisible' halo of axions may be detected by the technique advocated by Sikivie[173], further refinement of the measurement of the small-scale anisotropy of the microwave background may provide a signal(!) or even tighter constraints, deeper galaxy surveys may lead to the discovery (or lack of discovery) of galaxies (or QSOs) at very high redshifts ($z > 4$), which would have important implications for whether things were 'top down' or 'bottom up' (or neither), and other results not even dreamed of by simple-minded theorists!

Due to the brevity of this course in particle physics/cosmology there are many important and interesting topics which I have not covered (some of which are discussed in refs. 1-3). I apologize for any omissions and/or errors I may be guilty of. I thank my collaborators who have allowed me to freely incorporate material from co-authored works; they include E. W. Kolb, P. J. Steinhardt, G. Steigman, D. N. Schramm, K. Olive and J. Yang. This work was supported in part by the DOE (at Chicago and Fermilab), NASA (at Fermilab), and an Alfred P. Sloan Fellowship.

REFERENCES

1. M. B. Green and J. H. Schwarz, Nucl. Phys. B181, 502 (1981); B198, 252 (1982); B198, 441 (1982); Phys. Lett. 109B, 444 (1982); M. B. Green, J. H. Schwarz, and L. Brink, Nucl. Phys. B198, 474 (1982); J. H. Schwarz, Phys. Rep. 89, 223 (1982); M. B. Green and J. H. Schwarz, Phys. Lett. 149B, 117 (1984); B 151B, 21 (1985); Nucl. Phys. B243, 475 (1984); L. Alvarez-Gaume and E. Witten, Nucl. Phys. B243, 475 (1984); G. Chapline and N. Manton, Phys. Lett. 120B, 105 (1983); D. Gross, J. Harvey, E. Martinec, and R. Rohm, Phys. Rev. Lett. 54, 502 (1984); P. Candelas, G. Horowitz, A. Strominger, and E. Witten, Nucl. Phys. B, in press (1985); M. B. Green, Nature 314, 409 (1985); E. Kolb, D. Seckel, and M. S. Turner, Nature 314, 415 (1985); K. Huang and S. Weinberg, Phys. Rev. Lett. 25, 895 (1970).
2. M. Srednicki, Nucl. Phys. B202, 327 (1982); D. V. Nanopoulos and K. Tamvakis, Phys. Lett. 110B, 449 (1982); J. Ellis, J. S. Hagelin, D. V.Nanopoulos, K. Olive, and M. Srednicki, Nucl. Phys. B 238, 453 (1984); A. Salam and J. Strathdee, Ann. Phys. (NY) 141, 316 (1982); P. G. O. Freund, Nucl. Phys. B209, 146 (1982); P. G. O. Freund and M. Rubin, Phys. Lett. 37B, 233 (1980); E. Kolb and R. Slansky, Phys. Lett. 135B, 378 (1984); C. Kounnas etal., Grand Unification With and Without Supersymmetry (World Scientific, Singapore, 1984); Q. Shafi and C. Wetterich, Phys. Lett. 129B, 387 (1983); P. Candelas and S. Weinberg, Nucl. Phys. B237, 397 (1984); R. Abbott, S. Barr, and S. Ellis, Phys. Rev. D30, 720 (1984); E. Kolb, D. Lindley, and D. Seckel, Phys. Rev. D30, 1205 (1984).
3. P. Sikivie, "Axions and Cosmology", in Internationale Univeersitätswochen für Kernphysik der Karl-Franzens- Universität XXIth (Schladming, Austria 1982); K. Sato and H. Sato, Prog. Theor. Phys. 54, 1564 (1975); D. Dicus, E. Kolb, V. Teplitz, and R. Wagoner, Phys. Rev. D27, 839 (1980); L. Abbott and P. Sikivie, Phys. Lett. 120B, 133 (1983); M. Dine and W. Fischler, Phys. Lett. 120B, 137 (1983); J. Preskill, M. Wise, and F. Wilczek, Phys. Lett. 120B, 127 (1983); M. S. Turner, F. Wilczek, and A. Zee, Phys. Lett. 125B, 35 and 519 (1983); P. Sikivie, Phys. Rev. Lett. 48, 1156 (1982); P. Sikivie, Phys. Rev. Lett. 51, 1415 (1985); P. J. Steinhardt and M. S. Turner, Phys. Lett. 129B, 51 (1983); M. Fukugita, S. Watamura, and M. Yoshimura, Phys. Rev. D26, 1840 (1982); D. Seckel and M. S. Turner, Phys. Rev. D, in press (1985).
4. R. Buta and G. deVaucouleurs, Astrophys. J. 266, 1 (1983); A. Sandage and G. A. Tammann, Astrophys. J. 256, 339 (1982).
5. S. Weinberg, Gravitation and Cosmology (Wiley: NY, 1972).
6. Physical Cosmology, eds. R. Balian, J. Audouze, and D. N. Schramm (North-Holland: Amsterdam, 1980).
7. G. Gamow, Phys. Rev. 70, 572 (1946).

8. R. A. Alpher, J. W. Follin, and R. C. Herman, Phys. Rev. 92, 1347 (1953).
9. P. J. E. Peebles, Astrophys. J. 146, 542 (1966).
10. R. V. Wagoner, W. A. Fowler, and F. Hoyle, Astrophys. J. 148, 3(1967).
11. A. Yahil and G. Beaudet, Astrophys. J. 206, 26 (1976).
12. R. V. Wagoner, Astrophys. J. 179, 343 (1973).
13. K. Olive, D. N. Schramm, G. Steigman, M. S. Turner, and J. Yang, Astrophys. J. 246, 557 (1981).
14. D. A. Dicus etal., Phys. Rev. D26, 2694 (1982).
15. J. Yang, M. S. Turner, G. Steigman, D. N. Schramm, and K. Olive, Astrophys. J. 281, 493 (1984).
16. R. Epstein, J. Lattimer, and D. N. Schramm, Nature 263, 198 (1976).
17. M. Spite and F. Spite, Astron. Astrophys. 115, 357 (1982).
18. D. Kunth and W. Sargent, Astrophys. J. 273, 81 (1983); J. Lequeux etal., Astron. Astrophys. 80, 155 (1979).
19. S. Hawking and R. Tayler, Nature 209, 1278 (1966); J. Barrow, Mon. Not. R. Astron. Soc. 175, 359 (1976).
20. R. Matzner and T. Rothman, Phys. Rev. Lett. 48, 1565 (1982).
21. Y. David and H. Reeves, in Physical Cosmology, see ref. 6.
22. S. Faber and J. Gallagher, Ann. Rev. Astron. Astrophys. 17, 135 (1979).
23. M.Davis and P. J. E. Peebles, Ann. Rev. Astron. Astrophys. 21, 109 (1983).
24. K. Freese and D. N. Schramm, see ref. 1.
25. V. F. Shvartsman, JETP Lett. 9, 184 (1969).
26. G. Steigman, D. N. Schramm, and J. Gunn, Phys. Rev. Lett. 43, 202 (1977); J. Yang, D. N. Schramm, G. Steigman, and R. T. Rood, Astrophys. J. 227, 697 (1979); refs. 13, 15.
27. G. Arnison etal., Phys. Lett. 126B, 398 (1983).
28. D. N. Schramm and G. Steigman, Phys. Lett. 141B, 337 (1984).
29. S. Wolfram, Phys. Lett. 82B, 65 (1979); G. Steigman, Ann. Rev. Nucl. Part. Sci. 29, 313 (1979).
30. B. Lee and S. Weinberg, Phys. Rev. Lett. 39, 169 (1977).
31. E. W. Kolb, Ph.D. thesis (Univ. of Texas, 1978).
32. M. S. Turner, in Proceedings of the 1981 Int'l. Conf. on ν Phys. and Astrophys., eds. R. Cence, E. Ma, and A. Roberts, 1, 95 (1981).
33. G. Steigman, Ann. Rev. Astron. Astrophys. 14, 339 (1976).
34. F. Stecker, Ann. NY Acad. Sci. 375, 69 (1981).
35. T. Gaisser, in Birth of the Universe, eds. J. Audouze, J.Tran Thanh Van (Editions Frontiers: Gif-sur-Yvette, 1982).
36. A. Sakharov, JETP Lett. 5, 24 (1967).
37. M. Yoshimura, Phys. Rev. Lett. 41, 281; (E) 42, 746 (1978).
38. D. Toussaint, S. Treiman, F. Wilczek, and A. Zee, Phys. Rev. D19, 1036 (1979).
39. S. Dimopoulos and L. Susskind, Phys. Rev. D18, 4500 (1978).
40. A. Ignatiev, N. Krasnikov, V. Kuzmin, and A. Tavkhelidze, Phys. Lett. 76B, 486 (1978).
41. J. Ellis, M. Gaillard, and D. V. Nanopoulos, Phys. Lett. 80B, 360; (E) 82B, 464 (1979).
42. S. Weinberg, Phys. Rev. Lett. 42, 850 (1979).
43. E. Kolb and S. Wolfram, Nucl. Phys. 172B, 224; Phys. Lett. 91B, 217 (1980); J. Harvey, E. Kolb, D. Reiss, and S. Wolfram, Nucl. Phys. 201B, 16 (1982).
44. J. N. Fry, K. A. Olive, and M. S. Turner, Phys. Rev. D22, 2953; 2977; Phys. Rev. Lett. 45, 2074 (1980).
45. D. V.Nanopoulos and S. Weinberg, Phys. Rev. D20, 2484 (1979).
46. S. Barr, G. Segrè and H. Weldon, Phys. Rev. D20, 2494 (1979).
47. G. Segre and M. S. Turner, Phys. Lett. 99B, 339 (1981).
48. J.Ellis, M. Gaillard, D. Nanopoulos, S. Rudaz, Phys. Lett. 99B, 101 (1981).

49. E. W. Kolb and M. S. Turner, Ann. Rev. Nucl. Part. Sci. 33, 645 (1983).
50. P. A. M. Dirac, Proc. Roy. Soc. A133, 60 (1931).
51. G. 't Hooft, Nucl. Phys. B79, 276 (1974).
52. A. Polyakov, JETP Lett. 20, 194 (1974).
53. T. Kibble, J. Phys. A9, 1387 (1976).
54. J. Preskill, Phys. Rev. Lett. 43, 1365 (1979).
55. Ya. B. Zel'dovich and M. Yu. Khlopov, Phys. Lett. 79B, 239 (1978).
56. M. Einhorn, D. Stein, and D. Toussaint, Phys. Rev. D21, 3295 (1980).
57. A. Guth and E. Weinberg, Nucl. Phys. B212, 321 (1983).
58. J. Harvey and E. Kolb, Phys. Rev. D24, 2090 (1981).
59. T. Goldman, E. Kolb, and D. Toussaint, Phys. Rev. D23, 867 (1981).
60. D. Dicus, D. Page, and V. Teplitz, Phys. Rev. D26, 1306 (1982).
61. J. Fry, Astrophys. J. 246, L93 (1981).
62. J. Fry and G. Fuller, Astrophys. J. 286, 397 (1984).
63. F. Bais and S. Rudaz, Nucl. Phys. B170, 149 (1980).
64. A. Linde, Phys. Lett. 96B, 293 (1980).
65. G. Lazarides and Q. Shafi, Phys. Lett. 94B, 149 (1980).
66. P. Langacker and S.-Y. Pi, Phys. Rev. Lett. 45, 1 (1980).
67. E. Weinberg, Phys. Lett. 126B, 441 (1983).
68. M. Turner, Phys. Lett. 115B, 95 (1982).
69. G. Lazarides, Q. Shafi, and W. Trower, Phys. Rev. Lett. 49, 1756 (1982).
70. J. Preskill, in The Very Early Universe, eds. G. W. Gibbons, S. Hawking, and S. Siklos (Cambridge Univ. Press: Cambridge, 1983).
71. A. Goldhaber and A. Guth, in preparation (1983-84).
72. W. Collins and M. Turner, Phys. Rev. D29, 2158 (1984).
73. M. Turner, E. Parker, T.Bogdan, Phys. Rev. D26, 1296 (1982).
74. J. P. Vallee, Astrophys. Lett. 23, 85 (1983).
75. K. Freese, J. Frieman, and M. S. Turner, U. of Chicago preprint (1984).
76. M. S. Turner, Nature 302, 804 (1983).
77. M. S. Turner, Nature 306, 161 (1983).
78. J. Bahcall, IAS preprint (1984).
79. E. N. Parker, Astrophys. J. 160, 383 (1970).
80. G. Lazarides, Q. Shafi, and T. Walsh, Phys. Lett. 100B, 21 (1981).
81. Y. Rephaeli and M. Turner, Phys. Lett. 121B, 115 (1983).
82. J. Arons and R. Blandford, Phys. Rev. Lett. 50, 544 (1983).
83. E. Salpeter, S. Shapiro, and I. Wasserman, Phys. Rev. Lett. 49, 1114 (1982).
84. V. A. Rubakov, JETP Lett. 33, 644 (1981); Nucl. Phys. B203, 311 (1982).
85. C. Callan, Phys. Rev. D25, 2141 (1982); D26, 2058 (1982).
86. E. Kolb and M. S. Turner, Astrophys. J. 286, 702 (1984).
87. S. Dawson and A. Schellekens, Phys. Rev. D27, 2119 (1983).
88. E. Weinberg, D. London, and J. Rosner, Nucl. Phys. B236, 90 (1984).
89. J. Harvey, M. Ruderman, and J. Shaham, in preparation (1983-85).
90. K. Freese, M. S. Turner, and D. N. Schramm, Phys. Rev. Lett. 51 1625 (1983).
91. E. Kolb, S. Colgate, and J. Harvey, Phys. Rev. Lett. 49, 1373 (1982).
92. S. Dimopoulos, J. Preskill, and F. Wilczek, Phys. Lett. 119B, 320 (1982).
93. K. Freese, Astrophys. J. 286, 216 (1984).
94. B. Cabrera, Phys. Rev. Lett. 48, 1378 (1982).
95. B. Cabrera, M. Taber, R. Gardner, and J. Bourg, Phys. Rev. Lett. 51, 1933 (1983).
96. P. B. Price, S. Guo, S. Ahlen, and R. Fleischer, Phys. Rev. Lett. 52, 1265 (1984).
97. Magnetic Monopoles, eds. R. Carrigan and W. Trower (Plenum: NY, 1983).

98. Proceedings of Monopole '83, eds. J. Stone (Plenum: NY, 1984).
99. S. Dimopoulos, S. Glashow, E. Purcell, and F. Wilczek, Nature 298, 824 (1982).
100. K. Freese and M. Turner, Phys. Lett. 123B, 293 (1983).
101. A. Guth, Phys. Rev. D23, 347 (1981).
102. A. Linde, Phys. Lett. 108B, 389 (1982).
103. A. Albrecht and P. Steinhardt, Phys. Rev. Lett. 48, 1220 (1982).
104. R. Sachs and A. Wolfe, Astrophys. J. 147, 73 (1967).
105. P. J. E. Peebles, The Large-Scale Structure of the Universe (Princeton Univ. Press: Princeton, 1980).
106. R. H. Dicke and P. J. E. Peebles, in General Relativity: An Einstein Centenary Survey eds. S. Hawking and W. Israel (Cambridge Univ. Press: Cambridge, 1979).
107. S. Hawking, I. Moss, and J. Stewart, Phys. Rev. D26, 2681 (1982).
108. A. Albrecht, P. Steinhardt, M. Turner, and F. Wilczek, Phys. Rev. Lett. 48, 1437 (1982).
109. L. Abbott, E. Farhi, and M. Wise, Phys. Lett. 117B, 29 (1982).
110. A. Dolgov and A. Linde, Phys. Lett. 116B, 329 (1982).
111. M. S. Turner, Phys. Rev. D28, 1243 (1983).
112. J. Uson and D. Wilkinson, Astrophys. J. 277, L1 (1984).
113. D. Wilkinson, in Inner Space/Outer Space, eds. E. Kolb, M. Turner, D. Lindley, K. Olive, D. Seckel (U. of Chicago Press: Chicago, 1984).
114. G. Lazarides and Q. Shafi, in The Very Early Universe (see ref. 70).
115. J. Bardeen, Phys. Rev. D22, 1882 (1980).
116. Ya. B. Zel'dovich, Mon. Not. R. Astron. Soc. 160, 1p (1972); E. R. Harrison, Phys. Rev. D1, 2726 (1970).
117. S. Hawking, Phys. Lett. 115B, 295 (1982).
118. A. Starobinskii, Phys. Lett. 117B, 175 (1982).
119. A. Guth and S.-Y. Pi, Phys. Rev. Lett. 49, 1110 (1982).
120. J. Bardeen, P. Steinhardt, and M. Turner, Phys. Rev. D28, 679 (1983).
121. Ya. B. Zel'dovich, Mon. Not. R. Astron. Soc. 192, 663 (1980); A. Vilenkin, Phys. Rev. Lett. 46, 1169, 1496 (E) (1981).
122. A. Linde, Phys. Lett. 116B, 335 (1982).
123. A. Vilenkin and L. Ford, Phys. Rev. D26, 1231 (1982).
124. A. Albrecht etal., Nucl. Phys. B229, 528 (1983).
125. J. Ellis, D. Nanopoulos, K. Olive, and K. Tamvakis, Nucl. Phys. B221, 524 (1983); Phys. Lett. 120B, 331 (1983); D. Nanopoulos etal., Phys. Lett. 127B, 30; 123B, 41 (1983); G. Gelmini, etal, Phys. Lett. 131B, 161 (1983).
126. B. Ovrut and P. Steinhardt, Phys. Lett. 133B, 161 (1983); R. Holman, P. Ramond, and G. Ross, Phys. Lett. 137B, 343 (1984).
127. Q. Shafi and A. Vilenkin, Phys. Rev. Lett. 52, 691 (1984); S.-Y. Pi, Phys. Rev. Lett. 52, 1725 (1984).
128. P. Steinhardt and M. Turner, Phys. Rev. D29, 2162 (1984).
129. W. Boucher and G. Gibbons in The Very Early Universe, ref. 70; S. Hawking and I. Moss, Phys. Lett. 110B, 35 (1982).
130. J. Frieman and M. Turner, Phys. Rev. D30, 265 (1984); R. Brandenberger and R. Kahn, Phys. Rev. D29, 2172 (1984).
131. G. Steigman and M. Turner, Phys. Lett. 128B, 295 (1983).
132. J. Hartle, in The Very Early Universe, ref. 70.
133. P. J. E. Peebles, Astrophys. J. 243 L119 (1981); M. L. Wilson and J. Silk, Astrophys. J. 243, 14 (1981); 244, L37 (1981).
134. N. Vittorio and J. Silk, Astrophys. J. 285, L39 (1984); Phys. Rev. Lett. 54, 2269 (1985).
135. J. R. Bond and G. Efstathiou, Astrophys. J. 285, L44 (1984).
136. L. Abbott and M. Wise, Astrophys. J. 282, L47 (1984).
137. P. J. E. Peebles, Large Scale Structure of the Universe (Princeton Univ. Press, Princeton, 1980).
138. M. Davis and P. J. E. Peebles, Astrophys. J. 267, 465 (1983); also, see ref. 137.

139. N. Bahcall and R. Soneira, Astrophys. J. 270, 20 (1983); M. Hauser and P. J. E. Peebles, Astrophys. J. 185, 757 (1973).
140. N. Kaiser, Astrophys. J. 284, L9 (1984).
141. A. Szalay and D. N. Schramm, Nature 314, 718 (1985).
142. S. Gregory and L. Thompson, Astrophys. J. 222, 784 (1978); S. Gregory, L. Thompson, and W. Tifft, Astrophys. J. 243, 411 (1981); M. Tarenghi, W. Tifft, G. Chincarini, H. Rood, and L. Thompson, Astrophys. J. 234, 793 (1979); 235, 724 (1980).
143. R. P. Kirshner, A. Oemler, P. L. Schechter, and S. A. Shectman, Astrophys. J. 248, L57 (1981).
144. See, e.g., E. T. Vishniac, in Proceedings of Inner Space/Outer Space, eds. E. Kolb etal. (Univ. of Chicago Press, Chicago, 1985).
145. S. Faber and J. Gallagher, Ann. Rev. Astron. Astrophys. 17, 135 (1979).
146. V. Rubin etal., Astrophys. J. 225, L107 (1978); 261, 439 (1980); A. Bosma, Astron. J. 86, 1721; 86, 1825 (1981).
147. J. Bahcall, Astrophys. J. 287, 926 (1984).
148. Proceedings of IAU Symposium 117, eds, J. Kormendy and E. Turner (1985).
149. See pp. 280-284 of ref. 137.
150. M. Davis and P. J. E. Peebles, Astrophys. J. 267, 465 (1983).
151. B. Carr, J. Bond, and W. Arnett, Astrophys. J. 277, 445 (1984); C. J. Hogan, Mon. Not. r. Astron. Soc. 188, 781 (1979); A. Doroshkevich, Ya B. Zel'dovich, and I. Novikov, Sov. Astron. 11, 231 (1967).
152. J. M. Bardeen, Phys. Rev. D22, 1882 (1980).
153. W. Press and E. T. Vishniac, Astrophys. J. 239, 1 (1980).
154. J. Silk, Astrophys. J. 151, 459 (1968); also, see ref. 5, chapter 15.
155. B. J. Carr, Astrophys. J. 201, 1 (1975); 206, 8 (1976).
156. S. W. Hawking, Nature 248, 30 (1974).
157. E. R. Harrison, Phys. Rev. D1, 2726 (1970); Ya B. Zel'dovich, Mon. Not. r. Astron. Soc. 160, 1p (1972).
158. For a review of the pancake scenario, see, Doroshkevich etal., Ann. NY Acad. Sci. 375, 32 (1980); J. Bond and A. Szalay, Ann. NY Acad. Sci. 422, 82 (1984).
159. S. D. M. White, in Proceedings of Inner Space/Outer Space, eds, E. Kolb etal. (Univ. of Chicago Press, Chicago, 1985) and references therein.
160. C. Frenk, S. White, and M. Davis, Astrophys. J. 271, 417 (1983); J. Centrella and A. Melott, Nature 305, 196 (1983).
161. P. R. Shapiro, C. Struck-Marcell, and A. Melott, Astrophys. J. 275, 413 (1983).
162. J. R. Bond, J. Centrella, A. S. Szalay and J. R. Wilson, Mon. Not. r. Astron. Soc. 210, 515 (1984).
163. M. Davis, G. Efstathiou, C. Frenk, and S. White, Astrophys. J. in press (1985); also see ref. 159.
164. A. Melott etal., Phys. Rev. Lett. 51, 935 (1983).
165. G. Blumenthal, S. Faber, J. Primack, and M. Rees, Nature 311, 517 (1984).
166. M. S. Turner, G. Steigman, and L. Krauss, Phys. Rev. Lett. 52, 2090 (1984); M. S. Turner, Phys. Rev. D31, 1212 (1985).
167. P. J. E. Peebles, Astrophys. J. 284, 439 (1984).
168. D. Dicus, E. Kolb, and V. Teplitz, Astrophys. J. 221, 327 (1978); G. Gelmini, D. Schramm, and J. Valle, Phys. Lett. 146B, 311 (1984); M. Fukugita and T. Yanagida, Phys. Lett. 144B, 386 (1984); K. Olive, D. Seckel, and E. T. Vishniac, Astrophys. J., in press (1985).
169. A. Vilenkin, Phys. Rev. Lett. 53, 1016 (1984); but also see M. S. Turner, Phys. Rev. Lett. 54, 252 (1984).

170. N. Kaiser, in Proceedings of Inner Space/Outer Space, eds. E. Kolb etal. (Univ. of Chicago Press, Chicago, 1985).
171. A. Vilenkin, in Proceedings of Inner Space/Outer Space, eds. E. Kolb etal. (Univ. of Chicago Press, Chicago, 1985); Phys. Rep. in press (1985); Ya B. Zel'dovich, Mon. Not. r. Astron. Soc. 192, 663 (1980); A. Vilenkin, Phys. Rev.. Lett. 46, 1169, 1496(E) (1981); A. Vilenkin and Q. Shafi, Phys. Rev. Lett. 51, 1716 (1983); N. Turok and D. Schramm, Nature 312, 598 (1984); N. Turok, Phys. Lett. 126B (1983); Nucl. Phys. B242, 520 (1984); T. Vachaspati and A. Vilenkin, Phys. Rev. D30, 2036 (1984); A. Albrecht and N. Turok, in press (1985); R. Scherrer and J. Frieman, Phys. Rev. D, in press (1985).
172. J. P. Ostriker and L. Cowie, Astrophys. J. 243, 427 (1980); C. Hogan, Mon. Not. r. Astron. Soc. 202, 1101 (1983); S. Ikeuchi, Pub. Astr. Soc. Japan 33, 211 (1981).
173. P. Sikivie, Phys. Rev. Lett. 51, 1415 (1983).
174. B. J. Carr, Nucl. Phys. B252, 81 (1985) and references therein.

NEUTRINO MASSES AND MIXINGS: AN EXPERIMENTAL REVIEW

Michael Shaevitz

Columbia University
Nevis Labs
Irvington, NY 10533

INTRODUCTION

Over the past two decades, the neutrino has been used as a unique probe of the weak interaction. Since the postulation of the neutrino by Pauli, the question of whether the neutrino is massless or not has been asked and many experimental investigations of weak decay or scattering processes have sought to answer this question. In these lectures, we will review some of these experiments and give the status of our present knowledge.

In the standard electro-weak model, the neutrino is a weakly interacting, spin-1/2 particle with zero mass and definite lepton number. The neutrinos are in weak iso-spin doublets with their corresponding charged lepton partners and there are three generations of leptons and quarks:

$$\binom{e^-}{\nu_e} \quad \binom{\mu^-}{\nu_\mu} \quad \binom{\tau^-}{\nu_\tau}$$

$$\binom{d}{u} \quad \binom{s}{c} \quad \binom{b}{t}$$

Each neutrino type, lepton flavor, is unique with no mixing between flavors; lepton number is, therefore, conserved. In the minimal model, only V-A couplings contribute to the charged-current weak interaction and only left-handed neutrinos and right-handed anti-neutrinos participate in charged-current processes. The weak interaction is also assumed to be equal in strength for all flavors giving equal coupling for the three neutrinos ($e/\mu/\tau$ universality). The final assumption of the standard model is that the neutrino is a Dirac particle with the neutrino and anti-neutrino being distinct particles, i.e. $\nu_i \neq \bar{\nu}_i$.

Predictions of Grand Unified Theories

Beyond this standard model, some of the current grand unified theories based on larger symmetry groups do allow for finite neutrino masses and mixings between flavors. These theories allow transitions between quarks and leptons which can lead to lepton number non-conservation. At present none of these theories predict unambiguously the known mass spectrum of charged lepton and quarks or the number of generations, and so neutrino mass predictions and insights based on these models are somewhat speculative.

To illustrate the typical predictions, several examples are given below. A Majorana mass term can be constructed from dimensional arguments and added to the Lagrangian.[1)]

$$\mathcal{L}_{\text{Majorana}} = \frac{\langle\phi_o\rangle^2 \lambda}{M_{GUT}} \overline{\nu_L{}^c} \nu_R$$

where $\langle\phi_o\rangle$ = Higgs vacuum expectation value $\approx$ 300 GeV, M_{GUT} = grand unified theory mass scale $\approx 10^{15}$ GeV and λ is the coupling constant for the basic process. This term leads to lepton number violation and the equality of neutrino and antineutrino species, $\nu \equiv \bar{\nu}$. The predicted mass is small with $m_\nu = \lambda \times 0.1$ eV.

If both Majorana and Dirac terms are present, then the predictions depend on the symmetry group.[2)]

$$\mathcal{L}_{\text{Majorana}} = m_L(\bar{\nu}_R{}^c \nu_L) + m_R(\overline{\nu_L{}^c} \nu_R)$$

$$\mathcal{L}_{\text{Dirac}} = m_D(\bar{\nu}_L\nu_R + \bar{\nu}_R\nu_L)$$

For SU(5), baryon number minus lepton number is conserved which implies that m_L and m_R are zero. Further, no right-handed current exists in this theory so ν_R does not naturally exist leading to $m_D = 0$. In SO(10), diagonalizing the mass matrix, including both the Dirac and Majorana mass terms, leads to two Majorana neutrinos;[3] a heavy right-handed neutrino, $m_{\nu_R} \approx m_R$ and a light left-handed neutrino with $m_{\nu_L} \approx m_D^2/m_R$. If we associate m_R with the grand unified scale, M_{GUT}, and $m_D \approx m_{quark}$, this formalism leads to left-handed neutrino masses of order 10^{-6} eV. Some modifications by Witten[4] give $m_R \approx (\alpha/\pi)^2\ M_{GUT}$, which implies that $m_{\nu_L} \approx 1$ eV.

If neutrinos do have mass, then it is likely that the weak (or CP) eigenstates are not equal to the mass eigenstates. In general, then, we have a unitary transformation relating the two sets of eigenstates.

$$\begin{pmatrix} \nu_e \\ \nu_\mu \\ \nu_\tau \\ \nu_x \\ \vdots \end{pmatrix} = \begin{pmatrix} U_{e1} & U_{e2} & U_{e3} & U_{e4} \\ U_{\mu 1} & U_{\mu 2} & U_{\mu 3} & \cdots \\ U_{\tau 1} & U_{\tau 2} & U_{\tau 3} & \cdots \\ U_{x1} & \cdots & & \end{pmatrix} \begin{pmatrix} \nu_1 \\ \nu_2 \\ \nu_3 \\ \nu_4 \\ \vdots \end{pmatrix}$$

or

$$\nu_f = \sum_{i=1}^{N} U_{fi}\nu_i \quad .$$

Some GUTs models predict neutrino mixing to be similar to quark mixing, i.e. $|U_{e2}|^2 = \sin^2\theta_c = 0.05$. Other models relate the U_{ei}'s to the ratio of lepton masses, i.e.,

$$|U_{e2}|^2 \approx m_e/m_\mu \text{ or } (m_e/m_\mu)^{1/2} \quad .$$

<u>Neutrino Masses and Cosmology</u>

The ideal interrelation of particle physics and cosmology would be for particle physics to provide information on the number of neutrino flavors, their masses and lifetimes, and for cosmologists to then use this information to construct models of the universe. At present, since our knowledge of these parameters is limited, cosmological models have been used to put restrictions on the above quantities.

From models of big-bang nucleosynthesis, the number of neutrino flavors has been bounded, $n_\nu \leq 4$.[5] Mass constraints have been obtained from the expansion rate of the universe; these limits rely on calculations of the ν background and the measured expansion rate. This analysis[6] implies that the sum for all stable neutrino masses is less than 100 eV, $\Sigma m_{\nu_i} < 100$ eV. If neutrinos are assumed to be responsible for the dark matter halo of galaxies and clusters, lower mass limits can be obtained. Tremaine and Gunn[7] have shown that all the dark matter could be neutrinos if one neutrino had $m_\nu \geq 20$ eV or if $\Sigma m_{\nu_i}^4 \geq (20\text{ eV})^4$. Studies of how these neutrinos are trapped in galaxies have been made but the conclusions are very model dependent. A more complete discussion of the constraints imposed by cosmology on the number and masses of neutrinos can be found in the lectures of M. Turner in these proceedings.

Experimental Consequences

Finite mass neutrinos with mixing led to the following experimental consequences that will be addressed in this review:

1) changes in the decay spectra of leptons and mesons caused by finite masses for the dominantly coupled neutrinos;

2) secondary peaks or kinks in the decay spectra caused by finite masses of the subdominantly coupled neutrinos;

3) unstable heavy neutrinos (neutral heavy leptons) and neutrino decays;

4) neutrino oscillations caused by the mixing of different mass eigenstates;

5) neutrinoless double β-decay mediated by a Majorana neutrino with finite mass or right-handed couplings.

Lepton Number Conservation and e/μ/τ Universality

At present, there are three known types of neutrinos, ν_e, ν_μ, and ν_τ. (The existence of the ν_τ is inferred from the decay properties of the τ lepton.) Cosmological arguments limit the number of flavor to ≤ 4.[5] Whether or not a fourth generation exists will most likely be answered by measurements of the Z^o width or possibly direct observation in higher energy e^+e^- collisions. Limits on lepton number non-conservation can be made by searching for wrong-flavor or wrong-sign production in neutrino scattering experiments or lepton flavor changing decays. The present limits are:

Reaction	Upper Limit	Reference
$\nu_\mu + N \rightarrow e^- + X$	3×10^{-3}	8
$\nu_\mu + N \rightarrow \tau^- + X$	6×10^{-3}	9
$\nu_e + N \rightarrow \tau^- + X$	0.3	8
$\nu_\mu + N \rightarrow \mu^+ + X$	1.6×10^{-4}	10
$\mu^+ \rightarrow e^+ + \gamma$	1.7×10^{-10}	11

The most precise test of e/μ universality comes from measurement of the ratio:[12)]

$$R = \frac{\Gamma(\pi \rightarrow e\nu)}{\Gamma(\pi \rightarrow \mu\nu)} = \left(\frac{f_\pi^{\,e}}{f_\pi^{\,\mu}}\right)^2 \underbrace{(1.233 \times 10^{-4})}_{\text{Theory}}$$

Universality predicts that the pion decay constants are equal, $f_\pi^{\,e} = f_\pi^{\,\mu}$. The best current measurement is $R_{exp} = (1.218 \pm 0.014) \times 10^{-4}$,[13)] which implies that $f_\pi^{\,e}/f_\pi^{\,\mu} = 0.994 \pm 0.006$.

ELECTRON NEUTRINO MASS MEASUREMENTS

Tritium β-Decay Experiments

The most accurate studies of the electron neutrino mass have been done using tritium β-decay:

$$^3H \rightarrow {}^3He^+ + e^- + \bar{\nu}_e$$

In these studies, the electron spectrum is measured near the end-point energy; a finite neutrino mass is inferred from any deviations from the expected spectrum. The present experiments grew out of the pioneering work of Bergkvist[14)] who performed the first measurement with good resolution and high statistics. The results of these older experiments are given below:

Experiment	Source	Resolution (FWHM)	Result	Measured End-point
Bergkvist[14)]	^3H in Al_2O_3	50 eV	$m_{\bar{\nu}} \leq 60$ eV	18610 ± 16 eV
Lubimov et al[15)]	^3H in valine	45 eV	$13\text{eV} < m_{\bar{\nu}} < 46\text{eV}$	18575 ± 13 eV
Simpson[16)]	^3H in silicon detector	200-300eV	$m_{\bar{\nu}} < 65\text{eV}$	18567 ± 5 eV

There are several experimental limitations to the accuracy of these measurements. First, the energy resolution of the apparatus tends to distort the spectral shape near the end point. The resolution function must be accurately known and unfolded from the observed data in order to make precise measurement. Second, the atomic excitation spectrum of the final state $^3He^+$ must also be known. These states can be modified by the molecular environment of the source and calculations of the transition probabilities are difficult for complex molecules. For atomic tritium decay, the probabilities are 70% to the ground state of $^3He^+$ and 25% to the first excited state. The first excited state is 41 eV above the ground state implying that transitions to this first excited state will have an end-point 41 eV lower than those to the ground state. Finally, the background rate in the final electron detector limits how close one can probe near the end point. The β-decay spectrum falls rapidly with electron energy and limits the measurements to the region where the signal is bigger than the background.

New ITEP-83 Result

The ITEP Group has substantially improved their apparatus and conducted a new set of measurements (ITEP-83).[17] The main change was the introduction of an electrostatic spectrometer. This allows the β-spectrum to be measured by changing the electric field between the source and the magnetic field while keeping the magnetic field constant. This method has several advantages over the old system. For a given scan, the detector always sees the same energy electron, thus keeping the detection efficiency constant. Background is reduced by accelerating the electrons in the electrostatic spectrometer and holding the magnetic field at a value above the end point. Other improvements were made to give better focusing at the source and higher counting rates with better resolution. The quantitative comparison of the ITEP-83 and ITEP-80 systems is: the resolution has been improved from 45 eV to 20 eF FWHM; the background has been reduced by a factor of 12; the region scanned near the end point has been increased from 700 eV to 1750 eV.

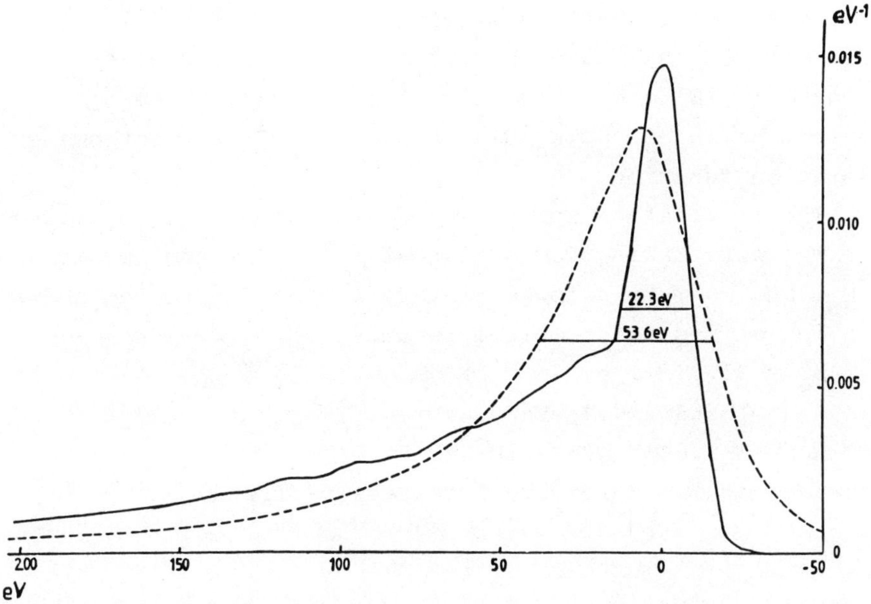

Fig. 1. The overall resolution function for the working source corresponding to the energy E=18430 eV. The solid curve is for the ITEP-83 apparatus; the dotted curve is for the ITEP-80.

The resolution function for the new electrostatic-magnetic spectrometer was measured using the M_1 line of a ^{169}Yb calibration source. The intrinsic width of the M_1 line was measured in a special experiment and found to be 14.7 eV. The final resolution function used in the analysis included ionization losses. These losses were measured by studying the ^{169}Yb spectrum measured after traversing thin layers of valine. This resolution function, along with the ITEP-80 one, is shown in Fig. 1.

Figure 2 shows the end-point spectrum for a series of thirty-five measurements. Each measurement has been fit to a theoretical β spectrum with corrections as follows:

$$S(KE)_e = A\int F(E')P_e E_e \sum_{\ell} W_{\ell}\Delta E_{\ell}(\Delta E_{\ell}^2 - m_{\bar{\nu}}^2)^{1/2}$$

$$\cdot\ \psi(E')\varphi(E')R(KE_e,\ E')dE' \quad ,$$

where $\Delta E_\ell = (E_o - E_\ell - E')$; and A ≡ normalization constant; F ≡ Fermi function; $m_{\bar{\nu}}$ ≡ neutrino mass; E_o ≡ end point energy of β spectrum at $m_{\bar{\nu}} = 0$; E_ℓ, W_ℓ ≡ energy and weight of the final state of the valine molecule; ψ(E) ≡ correction for defocusing of the magnetic system due to electrostatic potential; φ(E) ≡ correction of the standard shape of β spectrum having a $1+\alpha(E_o-E)^2$ form; R(E', E) ≡ spectrometer resolution function.

The final state spectrum and transition probabilities (E_ℓ, W_ℓ) were taken from Kaplan et al.[18] Three parameters, α, E_o, and $m_{\bar{\nu}}$ were allowed to vary in the fit to the above equation giving the combined results:

$$m_\nu = 33.0 \pm 1.1 \text{ eV}$$
$$E_o = 18583.2 \pm 0.3 \text{ eV}$$
$$\alpha = (-1.84 \pm 0.42) \times 10^{-9} \text{eV}^{-2}$$

with a χ^2 of 10798 for 10355 degrees of freedom.

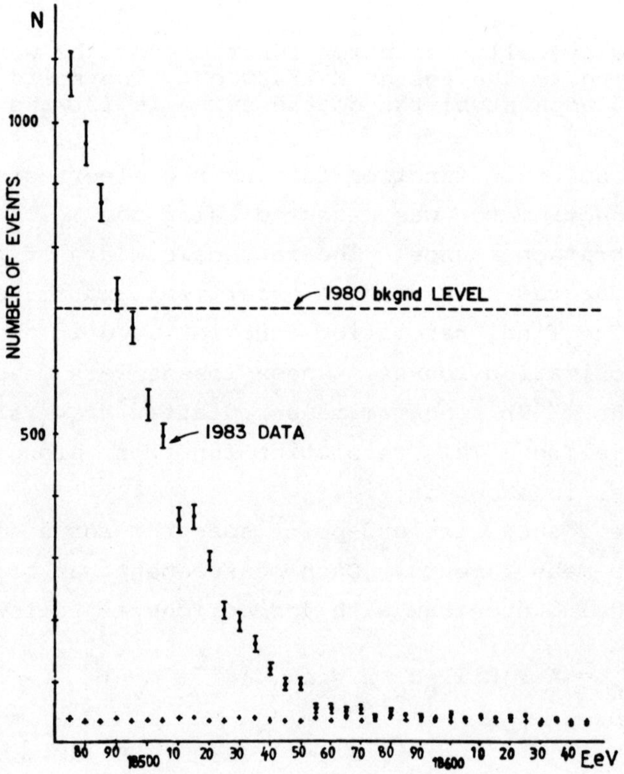

Fig. 2. The end point of the β spectrum.

Other final state spectra were also tried. Table I gives the results of fits using valine molecular states,[18] tritium molecular states (T^2), tritium atomic states and finally nuclear T^+ decay with only one final state.

Table I

	Valine Molecular	Tritium Molecular	Tritium Atomic	Tritium Nuclear	
$m_{\bar{\nu}}$	33 ± 1.1	35.8 ± 1	27.3 ± 1.2	$6\,^{+4}_{-12}$	(eV)
E_o	$18583.2\pm.3$	$18585\pm.3$	$18580.3\pm.3$	$18567.3\pm.3$	(eV)

As one can see, the values of $m_{\bar{\nu}}$ and E_o depend on the assumed spectrum of final states. With the low background and better resolution of the new data, detailed fits can be made within 50 eV of the end point. These fits have been made for the four models of Table I; the results give a 95% C.L. lower limit of $E_o \geq 18575$ eV for the end point energy. The experimenters, thus, conclude that the nuclear (one-state) model is eliminated by this limit, and further, that the data show strong indications of a neutrino mass with $m_{\bar{\nu}} > 20$ eV.

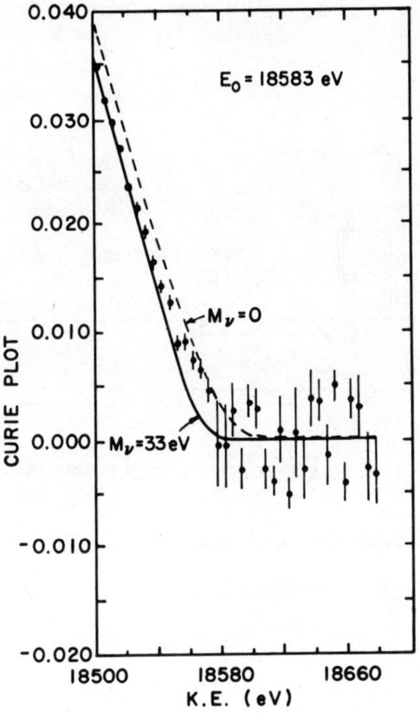

Fig. 3. The edge of the spectrum (Curie plot). The solid line: M_ν=33 eV; E_o=18583 eV; α=-1.84 x 10^{-9}eV^{-2}. The broken line: M_ν=0; E_o=18583 eV; α=-1.84 x 10^{-9} eV.

The new ITEP-83 data are much better than older β decay results. The statistics, background and resolution have been substantially improved and allow a much more precise probe of non-zero neutrino masses. There is still some question of the one-mass fits to the data. Figure 3 shows a curie plot near the end point of the combined data with the 33 eV and 0 eV fits superimposed. Neither curve fits the data very well; the χ^2 for the 33 eV fit is 522 for 295 degrees of freedom. The source of this discrepancy is at present unknown. One possibility is that the valine final state spectrum is not completely correct. It has also been pointed out[19)] that the ITEP result for the end point is not in good agreement with mass spectrometer measurements of the ^3H-^3He mass difference.[20)] The ITEP result of 18583 ± 0.2 eV is higher than the value predicted from mass spectrometer, (18573±7 eV - 18 eV for ground state differences) = 18555±7 eV by almost 30 eV. If the end point is constrained to be the mass spectrometer value, the resulting fitted neutrino mass would be substantially smaller.

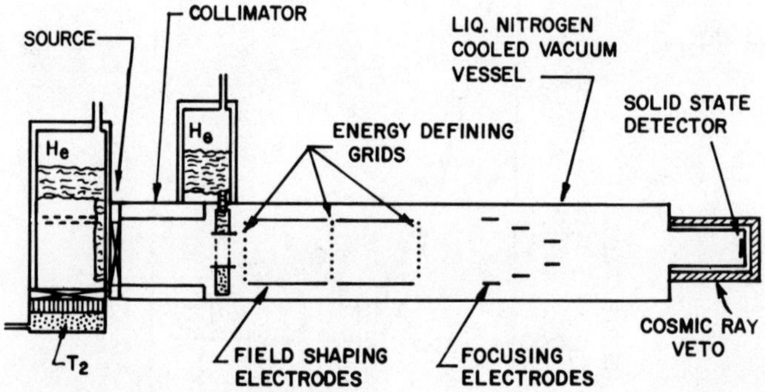

SOURCE : PURE FROZEN TRITIUM (T_2) (10-50 Ci)

RESOLUTION : σ_E = 1.3 eV

SENSITIVITY : $m_{\bar{\nu}_e} \geq 4$ eV

Fig. 4. The Rockefeller-Fermilab-Lawrence Livermore Lab Tritium β-Decay Experiment.

Future β-Decay Experiments

Several new experiments will be coming on-line over the next few years. A list of these experiments is given in Table II. The new experiments have better resolution (down to 2 eV), high count rates with low backgrounds (allowing the study of the region within 5 eV of the end point) and many of the experiments use molecular or atomic tritium (simplifying the final state spectra calculations). The best new experiment, on paper, is the Rockefeller-Fermilab-Lawrence Livermore Lab experiment shown in Fig. 4. This experiment uses an electrostatic spectrometer, a pure frozen tritium source of 10-50 Ci activation, and has a resolution of σ=1.3 eV for the electron energy. All of the experiments listed in Table II have good sensitivity in the region of the ITEP-83 result and their forthcoming measurements should be able to address the effect seen by the ITEP group.

Table II: Future β-Decay Experiments

Experiment	Source	Resolution (rms)	Sensitivity
Fackler et al Rock-FNAL-LLL	Solid Molecular ^3H	1-2 eV	$m_\nu > 4$ eV
Boyd Ohio State	"	10 eV	$m_\nu >$ 10eV
Bowles et al LAMPF	Atomic ^3H	40 eV	$m_\nu >$ 10eV
Clark IBM	Solid ^3H	5 eV	-
Heller et al UC Berkeley	^3H in Semi-conductor	100 eV	$m_\nu >$ 30eV
Graham et al Chalk River	-	10 eV	$m_\nu > \sim$ 20eV
Bergkvist	^3H in Valine	~25 eV	$m_\nu >$ 19eV
Kundig	^3H Implanted	5 eV	$m_\nu >$ 10eV
INS Japan	-	13 eV	$m_\nu >$ 25eV

Electron Neutrino Mass Measurements Using Electron Capture

Electron capture processes are also affected by a finite mass for the electron neutrino. In principle, these studies have some advantages over the tritium β-decay experiments: 1) there is no energy loss or scattering in the source; 2) final state spectral effects are small; 3) these experiments can probe masses down to 0.2-0.4 eV. In addition, the electron capture measurements offer an independent method with different systematic uncertainties from the tritium experiments. The low Q-value for electron capture makes ^{163}Ho particularly attractive for observing effects of a non-zero neutrino mass. This Q-value has been measured by several groups[21,22,23]; the results are consistent and give a value of $Q = \sim 2.4 \pm 1$ keV.

Two possible types of studies can be used to isolate the finite neutrino mass effects. In the first type, the relative rate of captures from the various atomic states is measured. Bennett et al[23] have estimated that a precision of 0.12% for the relative capture rates in ^{163}Ho could be necessary to measure neutrino masses greater than 35 eV; this group is currently attempting these measurements.

The second type of measurement was proposed by DeRujula.[24] In this method, the internal bremsstrahlung (IB) spectrum following electron capture (EC) is measured. This spectrum near the end point is sensitive to the mass of the electron neutrino.

$$Z \xrightarrow[\text{capture}]{\text{electron}} (Z-1) + \nu_e + \gamma$$

Again, the best candidate isotope is ^{163}Ho. A figure of merit, G,[24] can be constructed as the fraction of decays with IBEC within 30 eV of the end point. Using the above Q-value, $G=1.8 \times 10^{-11}$ for ^{163}Ho as compared[22] with a similarly defined G of 10^{-8} for tritium β-decay. This low value of G for ^{163}Ho may be enhanced because the Q-value is very close to the $p \rightarrow s$ x-ray transition energy. At present, the measurements of neutrino masses using EC are much less sensitive than the ^3H decay experiments but several groups are continuing their experimental programs.

Recently, Raghavan[25)] discovered an ultra-low energy K-capture branch in the decay of ^{158}Tb with a Q-value of 156±17 eV. This decay may, in the future, offer the best possibility for observing finite mass effects in an electron capture process.

MUON NEUTRINO MASS MEASUREMENTS

The most accurate measurements of the muon neutrino mass come from pion decay at rest or in-flight and from $K_{\pi 3}$ decays. The best limits at present are from two experiments. The first by Anderhub et al[26)] uses pion decay in-flight with an average pion momentum of 350 MeV/c. The apparatus is a combined spectrometer decay channel in which the pion momentum, muon momentum, and muon decay angle are simultaneously measured. The result is almost independent of the pion and muon masses, and the primary source of systematic error, the relative pion-muon momentum difference,

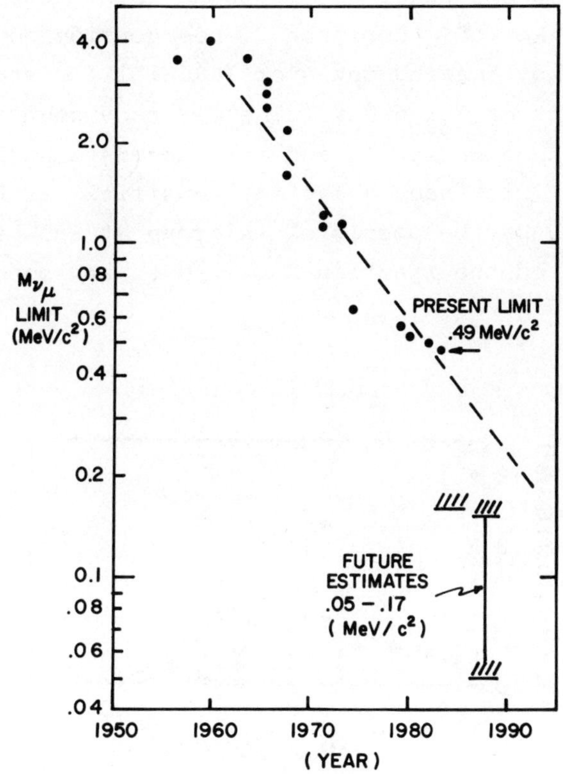

Fig. 5. Muon neutrino mass limits vs. time.

has been minimized. The final result based on 1000 decays is:

$$\langle m_{\nu_\mu}^2 \rangle = 0.14 \pm 0.20 \, (\text{MeV}/c^2)^2$$

$$m_{\nu_\mu} < 0.50 \ \text{MeV}/c^2 \quad 90\% \ \text{C.L.}$$

The second new measurement comes from Abela et al[27] and is a pion decay at rest study at SIN. The experiment has made a precision measurement of the decay muon momentum, $P_\mu = 29.7873 \pm 0.0008$ MeV/c. Combining this with the pion mass measurement of Lu et al[19] gives a neutrino mass limit of:

$$m_{\nu_\mu} < 0.49 \ \text{MeV}/c^2 \quad 90\% \ \text{C.L.}$$

Several authors have discussed improvements of the above measurements. The pion decay in-flight measurement at SIN is mainly limited by statistics and chamber non-linearities. A future experiment[29] with higher statistics and improved spectrometer calibrations could reach masses as low as 160 keV/c^2. Robinson[30] has considered an optical readout Cerenkov imaging detector (ORCHID) for studying π decay in flight or $K_{\ell 3}$ decay. The Cerenkov counter gives better background rejection and more accurate angle measurements. With such a system, sensitivities between 50 and 100 keV may be possible. A group at Snowmass[31] has investigated the reaction $K^+ \to \pi^0 \mu^+ \nu_\mu$ and conclude that

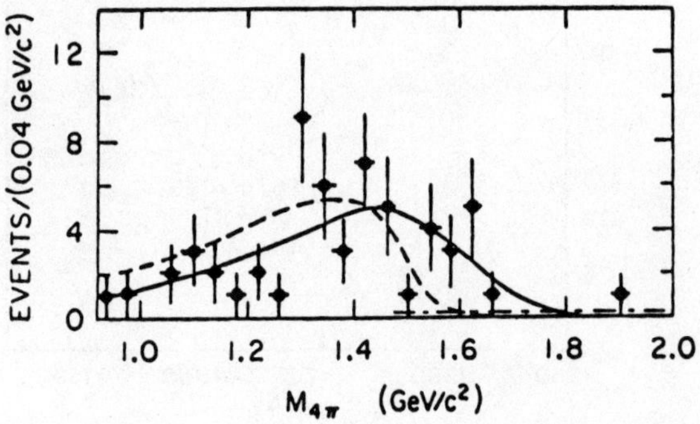

Fig. 6. Invariant mass of the $3\pi^\pm-\pi^0$ system together with the curves expected for $m_{\nu_\tau} = 0$ (solid) and $m_{\nu_\tau} = 250$ MeV/c^2 (dotted).

limits of a few hundred keV/c^2 may be feasible. Figure 5 shows the history of m_{ν_μ} measurements along with these future estimates.

TAU NEUTRINO MASS MEASUREMENTS

Up until a year ago, the best limit on the ν_τ mass came from two studies at PEP[32)] of $\tau \rightarrow \nu_\tau e \nu_e$ decays giving $m_{\nu_\tau} < 250$ MeV/c^2 at 95% C.L. These measurements have been superceded by a new study[33)] of the low Q-value decay, $\tau^+ \rightarrow \pi^+\pi^+\pi^-\pi^o\nu_\tau$, by the Mark II collaboration at PEP. A limit on the ν_τ mass is found by observing the four pion invariant mass distribution near the kinematic limit. Figure 6 shows the $m_{4\pi}$ distribution with curves for $m_{\nu_\tau} = 0$ (solid) and $m_{\nu_\tau} = 250$ MeV/c^2 (dashed). Including systematic uncertainties, the experiment sets an upper limit of

$$m_{\nu_\tau} < 164 \text{ MeV/}c^2 \qquad 95\% \text{ C.L.}$$

NEUTRINO OSCILLATIONS

Neutrino oscillations provide the only possible tool for studying muon and tau neutrino masses in the eV and below mass range. In addition, these studies also can be used to observe the effects of a massive fourth generation neutrino or wrong-handed neutrino. Neutrino oscillations can take place if two conditions are satisfied.

1) The weak eigenstates are quantum mechanical mixtures of mass eigenstates:

$$\nu_\alpha = \Sigma U_{\alpha j}\nu_j \quad \text{(as described in the previous section)}$$

with $\alpha = e,\mu,\tau,\ldots$ and $j = 1,2,3\ \ldots$

This condition implies that lepton number is not strictly conserved.

2) At least two eigenstates, ν_i, have different masses.

$$\Delta m_{ij}^2 = m_{\nu_i}^2 - m_{\nu_j}^2 \neq 0$$

For the case of two component mixing, the initial weak eigenstate is given by

$$|\nu_\alpha\rangle = \cos\theta|\nu_1\rangle + \sin\theta|\nu_2\rangle \ ,$$

where $|\nu_1\rangle$ and $|\nu_2\rangle$ are the mass eigenstates and $\sin\theta = U_{\alpha 2} = U_{\beta 1}$ describes the two component mixing. At a later time, the state evolves to:

$$|\nu_\alpha(t)\rangle = \cos\theta\, e^{-i\phi_1}|\nu_1\rangle + \sin\theta\, e^{-i\phi_2}|\nu_2\rangle$$

where the time-dependent phases are of the form

$$\phi_1(t) = E_1 t - p_1 L \approx L(E_1 - p_1) \approx L\,\frac{m_1^2}{2E}$$

From these equations, one can then derive the following probability distributions for the transitions of a weak eigenstate $|\nu_\alpha\rangle$:

$$\text{i) } \mathrm{Prob}(\nu_\alpha \to \nu_\alpha) = 1 - \sin^2(2\theta)\sin^2\left(\frac{1.27\,\Delta m^2 L}{E}\right)$$

and

$$\text{ii) } \mathrm{Prob}(\nu_\alpha \to \nu_\beta) = \sin^2(2\theta)\sin^2\left(\frac{1.27\,\Delta m^2 L}{E}\right)$$

where L is the distance from the ν source to the detector in km, E is the energy of the ν in GeV, and Δm^2 is in eV^2. An extension to more than one active oscillation channel can easily be made but the number of parameters to be experimentally determined increases. For this reason, most experimental papers present results only for the one channel assumption.

For three generations, the mixing matrix is similar to the Kobayashi-Maskawa[34)] matrix for quark mixing and has seven parameters: three neutrino masses, three mixing angles and one phase. As in the case of quark mixing, a non-zero value of the phase implies CP violation. For neutrino oscillations, CP violations could be observed if, for example, $\mathrm{Prob}(\nu_\alpha \to \nu_\beta) \neq \mathrm{Prob}(\bar{\nu}_\alpha \to \bar{\nu}_\beta)$. It should be noted, though, that CPT invariance implies that:

$$\mathrm{Prob}(\nu_\alpha \to \nu_\beta) = \mathrm{Prob}(\bar{\nu}_\beta \to \bar{\nu}_\alpha)$$

and

$$\mathrm{Prob}(\nu_\alpha \to \nu_\alpha) = \mathrm{Prob}(\bar{\nu}_\alpha \to \bar{\nu}_\alpha)$$

For n-generations, the inclusive probability is given by:

$$\mathrm{Prob}(\nu_\alpha \to \nu_\alpha) = 1 - \sum_{i<j} 4|U_{\alpha i}|^2|U_{\alpha j}|^2 \sin^2\left(\frac{1.27\,\Delta m_{ij}^2}{E}\,L\right),$$

which simplifies at large Δm^2 (where the $\sin^2$ factor averages to 1/2) to:

$$\mathrm{Prob}(\nu_\alpha \to \nu_\alpha) = 1 - \sum_{i<j} 2|U_{\alpha i}|^2|U_{\alpha j}|^2 \geq 1/n \quad .$$

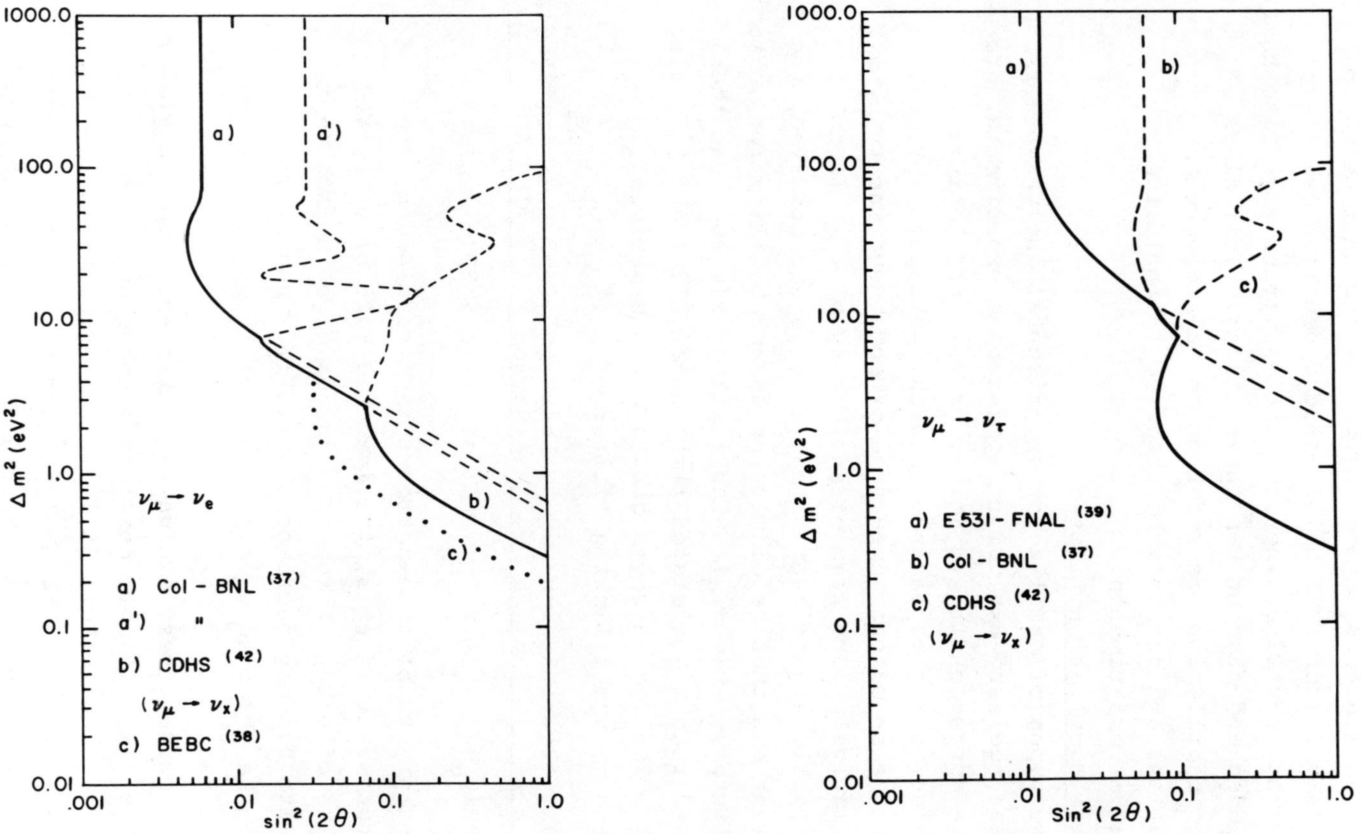

Fig. 7. Exclusive oscillation limits for a) $\nu_\mu \to \nu_e$ and b) $\nu_\mu \to \nu_\tau$.

The oscillations between different types (or flavors) of neutrinos are called flavor oscillations. If there are Majorana mass terms in the Lagrangian, then it is possible to have oscillations between neutrinos and anti-neutrinos of the same flavor. If a left-handed neutrino oscillates into an anti-neutrino, then the anti-neutrino will also be left-handed and have no V-A interactions. This class of neutrino oscillations is referred to as second class oscillations and can only be observed in inclusive oscillation experiments.

Exclusive Oscillation Limits

Two types of oscillation experiments, inclusive (or disappearance) and exclusive (appearance) experiments, are possible corresponding to Eq. i) or ii), respectively. In the exclusive (or appearance) search, an experiment must isolate the anomalous appearance of some neutrino type, ν_β, in a relatively pure beam of another type, ν_α. If no statistically significant signal of ν_β's are observed, the experiment can then set limits on the probability for ν_α to change into ν_β, $P(\nu_\alpha \to \nu_\beta)$. The limit of $P(\nu_\alpha \to \nu_\beta)$ can then be turned into a correlated limit on $\sin^2(2\theta)$ and Δm^2 using Eq. ii). The region covered by a given experiment is bracketed by the following limits:

Large $\sin^2(2\theta)\approx 1$ Region	Large Δm^2 ($L \gg L_{osc} = \frac{2.5E_\nu}{\Delta m^2}$) Region
$\Delta m^2 > [P(\nu_\alpha \to \nu_\beta)]^{\frac{1}{2}}/1.27(\frac{L}{E_\nu})$	$\sin^2(2\theta) > 2P(\nu_\alpha \to \nu_\beta)$

The sensitivity of an experiment is determined by the background rate of ν_β's, the measured rate of ν_α's, and the value of L/E_ν. Large L/E_ν values give the smallest Δm^2 limits but may decrease the $\sin^2(2\theta)$ sensitivity because the observed ν_α event rate falls with increasing L or decreasing E_ν.

The current best exclusive limits are given in Fig. 7 and Table III. (A more complete discussion of exclusive limits is given in Ref. 35.) All experiments assume two component mixing in the analysis and the limits are at the 90% C.L. For several of the channels, there exist more sensitive inclusive limits. (Exclusive limits on $\nu \to \bar{\nu}$, i.e. $\nu_\mu^{L.H.} \to \bar{\nu}_e^{L.H.}$, are not presented since they depend in detail on the interactions of wrong-handed neutrinos.)

Table III. Present Exclusive Limits for Neutrino Oscillations.

	Δm^2 at $\sin^2(2\theta)=1$	Large Δm^2 $\sin^2(2\theta)$ Limit	Experiment
$\nu_\mu \to \nu_e$	0.6 eV^2	6×10^{-3}	Columbia-BNL[37)]
	0.16 eV^2	0.03	BEBC-PS[38)]
$\bar{\nu}_\mu \to \bar{\nu}_e$	1.7 eV^2	8×10^{-3}	ITEP-FNAL-Michigan[40)]
	0.49 eV^2	0.028	LAMPF-UCI-Maryland[41)]
$\nu_\mu \to \nu_\tau$	3 eV^2	0.013	E531-FNAL[39)]
$\bar{\nu}_\mu \to \bar{\nu}_\tau$	2.2 eV^2	0.044	ITEP-FNAL-Michigan[40)]
	7.4 eV^2	0.088	Hawaii-LBL-FNAL [36)]
$\nu_e \to \nu_\tau$	8.0 eV^2	0.6	Columbia-BNL[37)]
$\bar{\nu}_e \to \bar{\nu}_\tau$	None	None	None

Inclusive Oscillation Limits

An inclusive (or disappearance) oscillation study is made by measuring the change in the number of a given type of neutrino, ν_α, with distance or energy. These experiments are sensitive to all the oscillation channels probed by exclusive experiments and, in addition, to oscillation in which the final state neutrino does not interact with the usual strength. Examples of the latter are: 1) a ν_x associated with a new heavy lepton with mass too heavy to be produced in present neutrino experiments; and 2) wrong-handed neutrinos (i.e. $\bar{\nu}_\mu$ left-handed) with suppressed weak interaction couplings.

The small Δm^2 sensitivity of an inclusive experiment is dependent on the statistical and systematic errors present for the comparison to Eq. i). Measurements of the ν interaction rate at two distances minimizes the dependence of the measurement on knowledge of the energy spectrum and absolute flux at the source. Further, if the data at the two distances are taken simultaneously by using two detectors, any systematic time dependence errors are also eliminated. For an oscillation measurement using data at two distances, there is a high Δm^2 limit to the sensitivity of the experiment. This limit corresponds to the point at which both detectors are many oscillation lengths from the source. At this point, the finite neutrino source size and detector energy resolution reduce the oscillatory behavior of Eq. i). For good sensitivity, the following inequalities should be satisfied:

$$L_{source} < L_{osc} = \frac{2.5\ E_\nu}{\Delta m^2}$$

and

$$\text{Energy Resolution} = \frac{\Delta E_\nu}{E_\nu} < \frac{0.2}{\begin{array}{c}\text{\# of oscillation lengths } (L_{osc}) \\ \text{between source and detector}\end{array}}$$

The high Δm^2 limit is absent if the neutrino intensity as a function of E_ν at the source is known. This knowledge is typically model dependent and therefore introduces some additional systematic uncertainties.

The effect of an oscillation is maximal when $1.27\ \Delta m^2$ L/E is near $\pi/2$; therefore, E/L serves as a figure of merit for the measurement of Δm^2. The Fermilab neutrino beam ($E/L=50$) is sensitive to Δm^2 in the 10-1000 eV^2 range, the CERN PS and BNL AGS ($E/L=3$) to 1 to 100 eV^2, and reactors ($E/L=1$) to 0.1-5 eV^2.

Inclusive Oscillation Experiments

The possibility of measuring neutrino oscillation gained interest after the report of a positive result by the Savannah River reactor group.[43] In that experiment, both the charged-current and neutral-current reactions were measured. The ratio of the two reactions differed from that expected and the authors claimed evidence for neutrino instability ($\sin^2(2\theta) > 0.5$ for the neutrino oscillation hypothesis), although only at the two to three standard deviation level. Future studies at the Grenoble, Goesgen and Bugey reactors failed to confirm this effect.

The current situation is that, except for the Bugey reactor experiment, all other experiments observe no positive signal and exclude regions in the $\Delta m^2/\sin^2(2\theta)$ parameter plane. Table IV gives a summary of the current best limits again assuming only two component mixing. Most of these experiments will be described more in the following sections.

Table IV: Present Inclusive Limits

	Lower Limit Δm^2 for $\sin^2(2\theta)=1$	Best Limit $\sin^2(2\theta)$	Experiment
$\nu_\mu \to \nu_x$	10^{-3} eV^2	$.6/\Delta m^2$ large	IMB[45] (cosmic ray ν's)
	.3 eV^2	$.06/\Delta m^2=3$eV^2	CDHS-PS[42]
	15 eV^2	$.02/\Delta m^2=120$eV^2	CCFR (NBB)[43]
$\bar{\nu}_\mu \to \bar{\nu}_x$	15 eV^2	$.02/\Delta m^2=200$eV^2	CCFR (NBB)[44]
$\nu_e \to \nu_x$	10 eV^2	$.34/\Delta m^2$ large	BEBC (WBB)[46]
$\bar{\nu}_e \to \bar{\nu}_x$	.015 eV^2	$.17/\Delta m^2$ large	Goesgen (reactor)[47] (spectrum analysis)
	.12 eV^2	$.2/\Delta m^2=1$eV^2	Bugey (reactor)[48] (two detector analysis)
	[positive signal (3.2σ) $\Delta m^2=.2$eV^2	$\sin^2 2\theta=.25$]	

(The BEBC limit given in Table IV for $\nu_e \to \nu_x$ oscillations is modified from the published limit of $\sin^2(2\theta) < 0.07$ at high masses given in Ref. 46. In this measurement, the ν_e flux from K_{e3} decay is estimated from the observed number of muon neutrino events from $K_{\mu 2}$ decay. The limit is based on a measured number of 110 ± 13 ν_e events as compared to a predicted number of 91 ± 10. A proper statistical analysis of this result should include the constraint that the underlying rate of observed events must be less than or equal to the true predicted rate. Including this constraint in the analysis produces a more reasonable $\sin^2(2\theta)$ limit of 0.34 at 90% C.L.)

The CCFR $\nu/\bar{\nu}$ Oscillation Experiment[44] at Fermilab

The CCFR (Chicago-Columbia-Fermilab-Rochester) group has made a search for inclusive oscillations of muon neutrinos in the mass range, $30 < \Delta m^2 < 900$ eV^2. The search (Fermilab E701) is based on the analysis of 150K charged current events in each of two detectors running simultaneously in a dichromatic neutrino beam. The neutrino beam results

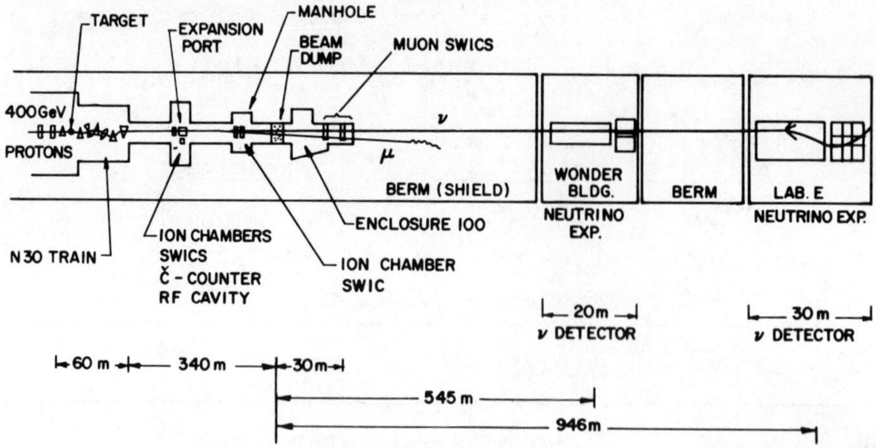

Fig. 8. Narrow band beam configuration for the CCFR Expt.

from the decay of monoenergetic ($\sigma(E_{\pi,k})/E_{\pi,k}$=11% rms) pions and kaons in a 352 m evacuated decay pipe. Data for several secondary energies were recorded corresponding to neutrino energies between 30 and 230 GeV.

The two neutrino detectors are located at 715 m and 1115 m from the center of the decay pipe (Fig. 8). Each detector consists of a target calorimeter instrumented with spark chambers and scintillation counters, followed by a solid iron muon spectrometer. Analysis requirements are imposed to equalize the acceptance of the two detectors and to ensure that data come from simultaneous good running in both detectors. The fiducial volumes are scaled so that both detectors subtend the same solid angle. All events are translated in software to both detectors and are required to pass cuts in both places. These requirements ensure that all accepted events would trigger and be fully reconstructed in both detectors. After these cuts, the sample consists of 36,100 events in the downstream detector and 35,500 events in the upstream detector.

The oscillation measurement is performed by comparing the number of events in the two detectors as a function of neutrino energy. The neutrino energy for this comparison is obtained from the dichromatic beam's correlation of neutrino energy and angle (interaction radius). The measured energy and radius of interaction of each event is used to determine whether the event is due to a neutrino from pion or kaon decay. The neutrino events are then separated into energy bins by using the pion/kaon category and the event radius.

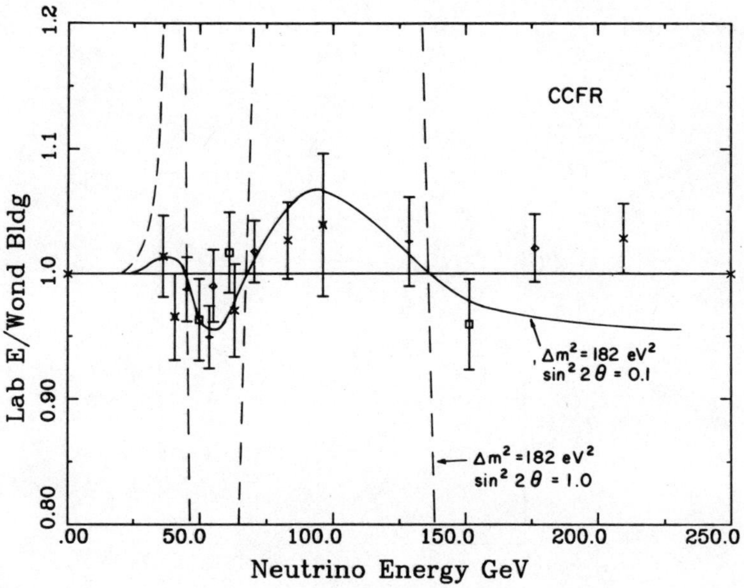

Fig. 9. Ratio of events in two detectors from the CCFR experiment. A curve (solid) for a typical Δm^2 value and 10% mixing is shown. The curve for maximal mixing (dashed) at the same Δm^2 is largely off-scale.

The ratio of the number of events in the two detectors is corrected for systematic differences (i.e. detector live-time, reconstruction and target mass) and the effects of the finite length decay pipe and secondary beam angular divergence are calculated by Monte Carlo. The event ratios after all corrections are applied are given in Fig. 9. A comparison of the corrected data with the hypothesis of no oscillations yields a χ^2 of 12.9 for 15 degrees of freedom, and is therefore consistent with no oscillations. The hypothesis of $\Delta m^2 = 182\ eV^2$ and $\sin^2(2\theta) = .1$ gives a χ^2 of 13.0/15 D.F.

After incorporating systematic uncertainties in the beam divergence and relative normalization, the 90% C.L. limits are calculated by comparing the data to oscillation hypotheses with various Δm^2 and $\sin^2 2\theta$ parameters. Two limits are presented in Fig. 10 corresponding to the likelihood ratio test (solid curve) and the Pearson's χ^2 test (dotted curve).[49] The likelihood ratio test compares the data to all two-component hypotheses; the Pearson's test compares the data to all possible mixing hypotheses. The experiment, thus, excludes oscillations of muon neutrinos in the region $15 < \Delta m^2 < 1000\ eV^2$ for $\sin^2(2\theta) > 0.02$-0.2.

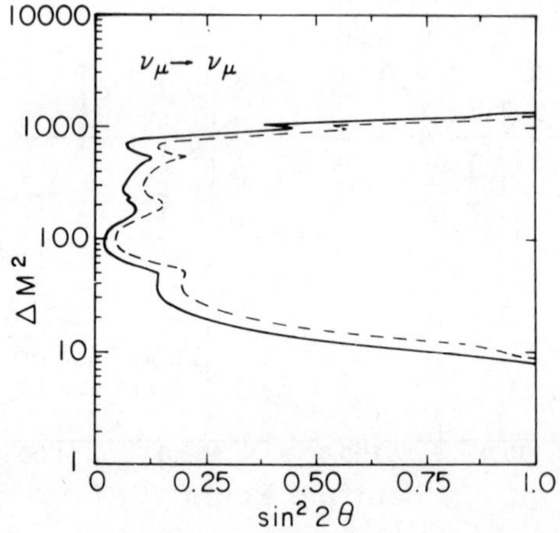

Fig. 10. 90% C.L. limits on possible $\nu_\mu \rightarrow \nu_x$ oscillations as a function of Δm^2 and $\sin^2(2\theta)$, calculated using the likelihood ratio test (solid line). The region to the right of the curve is excluded. These limits are calculated including systematic errors. The limits obtained from the Pearson χ^2 test are also shown (dotted line).

The CCFR group has also made the first experimental investigation of $\bar{\nu}_\mu \rightarrow \bar{\nu}_x$ oscillations. Anti-neutrino data were only taken for one secondary energy giving three anti-neutrino energy points for the oscillation analysis between the energies of 40 and 150 GeV. An analysis similar to the neutrino data yielded a χ^2 of 4.7/3 D.F. for the no oscillation hypothesis and the limits presented in Fig. 11.

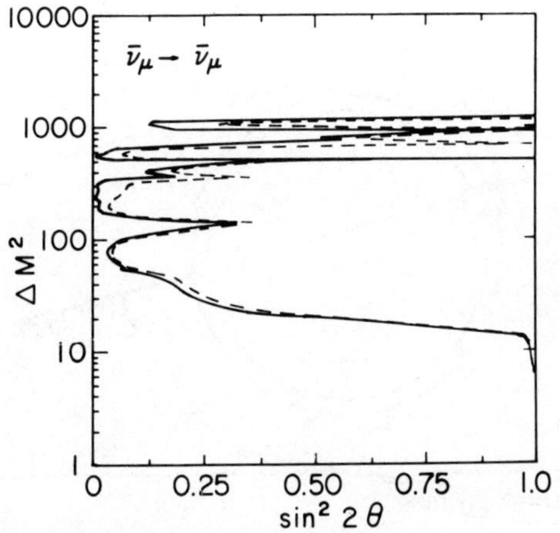

Fig. 11. 90% C.L. limits on possible $\bar{\nu}_\mu \rightarrow \bar{\nu}_x$ oscillations (curves are as in Fig. 10).

The CDHS and CHARM ν-Oscillation Experiments at the CERN PS

Another inclusive oscillation result[42] has recently been presented by the CDHS group. A low energy, bare target beam produced by 19.6 GeV incident protons was used as a source for this measurement ($\langle E_\nu \rangle$=3 GeV). This beam has no magnetic elements after the secondary production target and gives a neutrino flux that approximately falls with the inverse of the distance squared:

$$\text{Flux } \nu\text{'s} \propto 1/L^2 \quad .$$

The neutrino source is a 52m long decay tunnel and two detectors are placed at 130m and 880m from the source (Fig. 12). The far detector is the CDHS WA1 detector consisting of Fe scintillator calorimeter modules with 2.5, 5 or 15 cm sampling. The near detector is made of a smaller number of identical modules. The detector is run in a mode different from the high energy running; no magnetic field is present and no drift chamber information is used. The experiment triggers on a muon with minimal projected range, and thus, selects out charged current events. Cosmic ray and beam correlated muons are cut or subtracted from the data leaving a sample of 29K events in the near detector

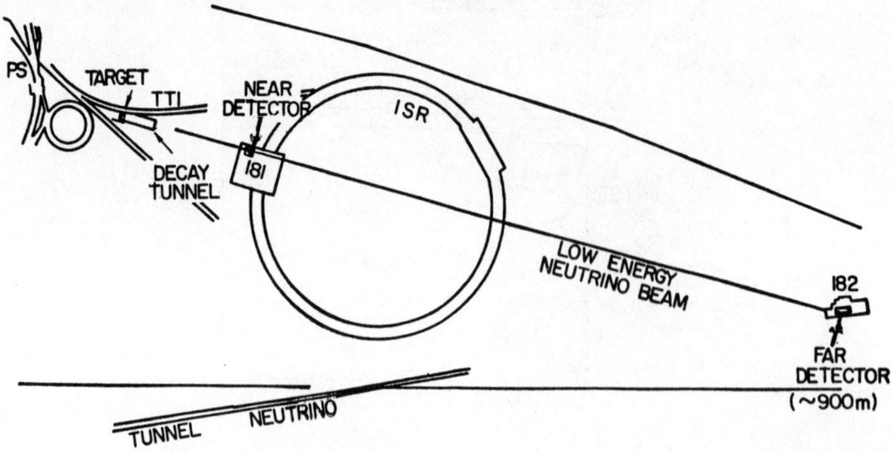

Fig. 12. Layout for the CDHS ν-oscillation experiment at the CERN PS.

and 3.6K events in the far within a restricted fiducial volume. The ratio of events in the two detectors is then calculated in bins of projected range for the outgoing muon. (The projected muon range is used because the total neutrino energy of each event is not measured; the strong correlation between the muon and neutrino energies implies that this procedure is almost equivalent to neutrino energy bins.) A Monte Carlo calculation is used to correct for the non-$1/L^2$ behavior of the neutrino flux (approximately 5-10% in the ratio). The data are displayed in Fig. 13 with statistical errors only, along with a representative curve of neutrino oscillations. No significant deviations from the expected ratio of one for no oscillations is observed; the combined Monte Carlo corrected ratio for the global data set is 1.026 ± 0.022. Systematic uncertainties have been estimated to be 2.5% and when included produce the 90% C.L. limits shown in Figs. 7 and 14. (For this experiment, no separation of neutrino and anti-neutrino data is made but the beam is predominantly neutrinos.) This result significantly improves the existing $\nu_\mu \rightarrow \nu_\tau$ oscillation limit for $\Delta m^2 < 4$ eV^2 (see Fig. 7a).

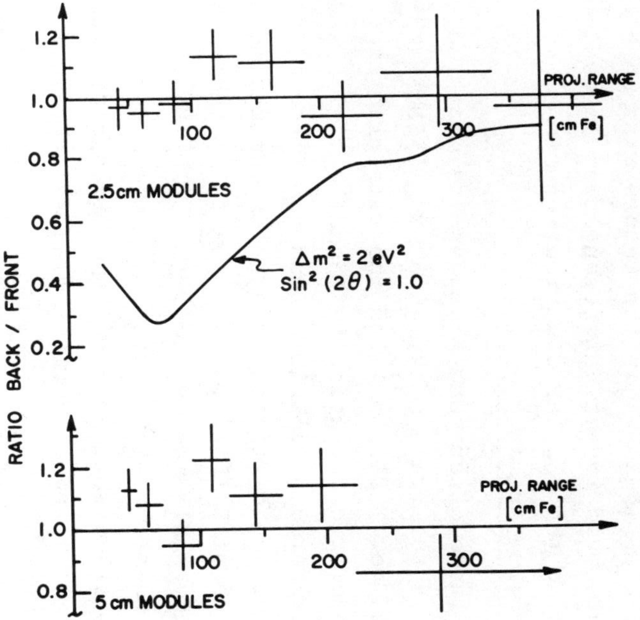

Fig. 13. Ratio of events in back to front detector vs. projected range for the CDHS ν oscillation experiment.

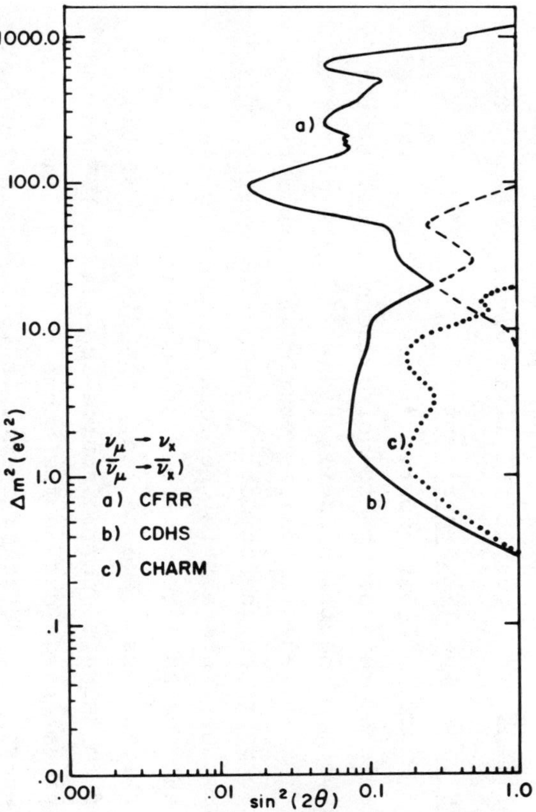

Fig. 14. Inclusive oscillation limits for $\nu_\mu \to \nu_x$. (The experiments also include a small admixture of $\bar{\nu}_\mu \to \bar{\nu}_x$.)

The CHARM collaboration has also investigated neutrino oscillations[50)] in the same bare target beam as CDHS. Again, two detectors at 123m and 903m are used to detect charged current events. The number of events (2043 in the near detector and 270 in the far) is somewhat less than the CDHS experiment and leads to the limit given in Fig. 14.

Cosmic Ray ν's - The IMB Experiment

The IMB proton decay experiment has used the measured versus predicted rate of upward-going muons to set limits on inclusive muon neutrino oscillations.[45)] The source of neutrinos (or anti-neutrinos) is cosmic ray interactions in the atmosphere on the opposite side of the earth. The neutrinos, then, interact in the rock surrounding the IMB detector and the subsequent muons produced in the charged current interactions are detected. The distance of the detector from the atmospheric source is $\sim 1.3 \times 10^4$ km and with a minimum energy cut on the muons of 2 GeV, the average neutrino energy for detected events is 15 GeV. In 276 days of running, 71 upward-going events ($-1.0 < \cos\theta < -0.1$) were observed. A calculation using cosmic ray flux data predicts 55 ± 11 events. Combining these two numbers, the experiment excludes inclusive $\nu_\mu \to \nu_x$ oscillations for $\Delta m^2 > 10^{-3}$eV^2 and $\sin^2(2\theta) > .6$-1.0. In the future with more events, the experiment could investigate even lower Δm^2 values ($> 10^{-4}$eV^2) by measuring the angular distribution of muons.

The Solar ν Experiment

The large discrepancy between the measured[51] solar-neutrino capture rate of 2.1 ± 0.3 SNU (1 SNU=10^{-36} ν captures per second per ^{37}Cl atom) and the rate predicted from the standard solar calculation by Bahcall et al,[52)] 7.9 ± 1.5, has been interpreted as possible evidence for ν oscillations. In terms of ν oscillation, the solar ν experiment has a very small value of E_ν/L=10 MeV/10^{11}m = 10^{-10} and, therefore, probes very small mass values, $\Delta m^2 > 10^{-11}$ eV^2.

For neutrino oscillations to make a significant reduction in the observed solar-ν rate, the mixings must be large, $\sin^2(2\theta) \approx 1.0$ and either the mass must be small $\Delta m^2 \approx$

10^{-10} eV^2 or there must be more than two component mixing. For the latter case, if $\Delta m^2 > 10^{-10}$ eV^2,

$$\text{Prob}(\nu_e \rightarrow \nu_e) > \frac{1}{\#\ \text{components}}$$

A probability of $\nu_e \rightarrow \nu_e$ less than ~ 30% implies greater than three generation mixing. The uncertainties in the flux calculation also make the oscillation interpretation difficult. Bahcall et al[47] concluded that the "...^{37}Cl experiment is not suitable for studying neutrino oscillations because of uncertainties in the prediction of the (relatively small) B neutrino flux." The proposed ^{71}Ga and ^{115}In experiments are sensitive to neutrinos from the p-p solar reaction. The calculation of this flux is much more reliable and these experiments should be sensitive to neutrino oscillations with $\Delta m^2 > 10^{-12}$ eV^2.

Beam Dump ν Experiments

Various experiments[53] have measured the rate of prompt electron and muon neutrino production from high energy proton interactions in a high density target/beam dump. Non-prompt sources (i.e. $\pi/k \rightarrow \mu\nu_\mu$) are removed by calculation and/or changing the density of target dump. Since the source of prompt neutrino/anti-neutrinos is charmed particle decay, there should be equal numbers of prompt electron and muon neutrinos. An investigation of inclusive ν_e oscillations can, thus, be made by comparing the number of observed electron and muon neutrinos some distance from the dump. The current situation is summarized in Table V. For neutrino oscillations to make measurable changes in the ν_e flux, $\Delta m^2 > 100$ eV^2 and $\sin^2(2\theta) > 0.3$ for these experiments. The wide range in the 1982 CERN results and the difficulties in subtracting the non-prompt sources make it very difficult to make any strong conclusions about possible neutrino oscillations.

Nuclear Reactor Oscillation Experiments

Nuclear reactors provide a high flux of $\bar{\nu}_e$ at low energy (1-5 MeV) that can be used to search for neutrino oscillations. In these experiments, the flux of $\bar{\nu}_e$ is either measured at several distances from the reactor

Table V: Beam Dump Measurements of ν_e/ν_μ

Experiment	Distance from Dump	Ratio $\left(\frac{\nu_e \text{ events}}{\nu_\mu \text{ events}}\right)$
CERN Exp's ('77-'79) (CDHS/CHARM/BEBC)	~ 860m	$0.57 \pm \sim 0.20$
BEBC ('82)	400m	$1.35^{+.65}_{-.35}$
CDHS ('82)	470m	$0.83 \pm 0.13 \pm 0.12$
CHARM ('82)	490m	$0.59 \pm 0.11 \pm 0.08$
FNAL (E613)('82) (FFMOW)	56m	$1.09 \pm 0.09 \pm 0.10$

core or compared to a calculated spectrum. The experiments measure the $\bar{\nu}_e$ flux by detecting the reaction:

$$\bar{\nu}_e + p \rightarrow e^+ + n\ ,$$

using liquid scintillation counter target cells to detect the e^+ and ^3H multiwire chambers for the n. A time correlated e^+, n event constitutes a valid signature. Pulse shape discrimination in the scintillation counters is used to eliminate correlated neutron background events.[47)]

Two methods have been used by these experiments to search for neutrino oscillations. For the first method (two position method), event rates at two positions are compared versus neutrino energy. This would be best done by using two detectors run simultaneously but the measurements so far use one detector moved to two different locations. Since the data at the two distances are taken at different times, there could be systematic changes in the reactor flux. The two position method, though, has the advantage of being independent of the knowledge of the reactor $\bar{\nu}_e$ flux spectrum.

The second method (spectrum method) involves comparing the observed number of events with a calculated number from a reactor $\bar{\nu}_e$ flux model. The calculated spectrum is determined by taking the measured e^+ spectrum obtained from the BILL spectrometer [54] for ^{235}U and ^{239}Pu, the composition of the reactor core, and calculations[55] for ^{238}U and ^{241}Pu and unfolding the neutrino spectrum. The measured ^{235}U and ^{239}Pu components contribute 89% of the $\bar{\nu}_e$ flux with an error of ±3% and the ^{238}U and ^{241}Pu give 11% with an error of ±10-20%. The total error on the calculated spectrum is 5.3%. There are also uncertainties in the absolute normalization from: 1) the $\bar{\nu}_e p$ cross section, 2) the number of free protons in the target, 3) neutron detection efficiency, and 4) reactor core composition and time variation; these uncertainties contribute an error of 5-10% for the present measurements.

Results have recently been presented from two high statistics studies. The first is by a Caltech/SIN/TU/Munich group[47] using the Goesgen power reactor (2806 MW). Data have been recorded with one detector placed either 37.9m or 45.9m from the reactor core. The ratio of events as a function of positron energy is shown in Fig. 15. The χ^2 for no oscillations is acceptable χ^2=.48/D.F. and using the "two position method", the data set the correlated limits shown as curve a) of Fig. 16. The experiment has also set more restrictive limits by the "spectrum method". For this measurement, the overall normalization and spectrum uncertainty

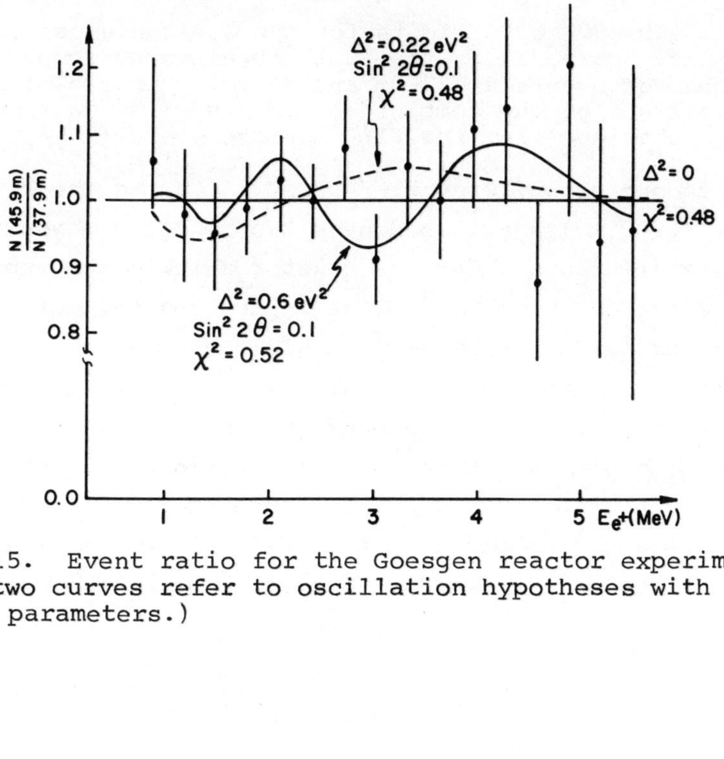

Fig. 15. Event ratio for the Goesgen reactor experiment. (The two curves refer to oscillation hypotheses with the given parameters.)

is estimated to be 7.4% and leads to the limits shown as curve b) in Fig. 16.

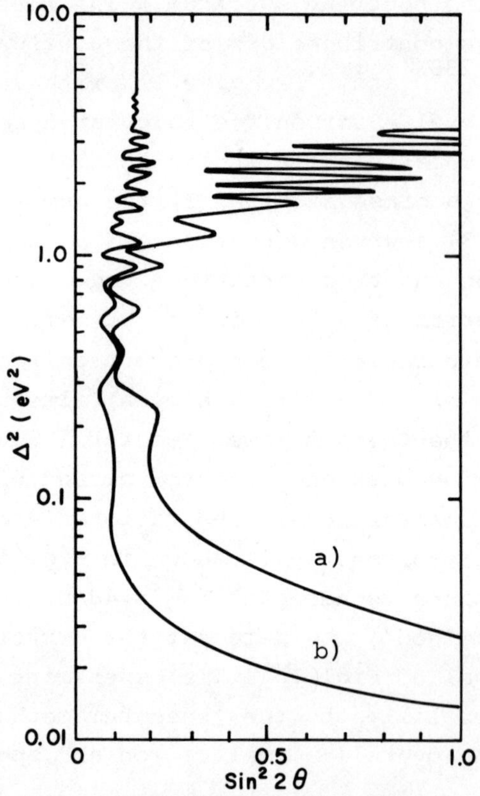

Fig. 16. The 90% C.L. limits for the Goesgen ν oscillation experiment. Curve a) is the limit obtained by comparing the number of events at 37.9m and 45.9m. Curve b) is from the comparison of the combined 37.9m and 45.9m data to the spectrum obtained from the BILL spectrometer.[57)]

A second group, Grenoble/Annecy,[48)] using the Bugey reactor (2800 MW) has also done a two-distance measurement. Several small rooms under the reactor allowed the experimenters to place the detector very close to the reactor core, either at 13.6m or 18.3m. From solid angle coverage, this gives the Bugey experiment about eight times more $\bar{\nu}_e$ flux at the detector as compared to the Goesgen experiment. The Bugey experiment quotes an absolute normalization uncertainty of 8% but also claims that the large uncertainties in the calculated spectrum preclude any oscillation limits from the "spectrum method" (Fig. 17). The ratio of observed events

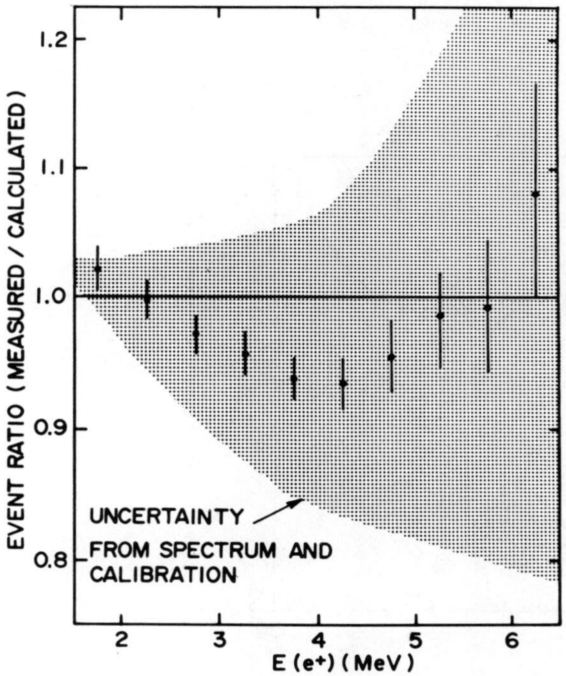

Fig. 17. Ratio R_1 for position 1 between measured and calculated event yields as a function of the positron energy. The shaded area shows the domain where the points move when different spectrum calculations are used for the U^{238} and Pu^{241} isotopes and when a systematic shift up to 100 keV is applied to the absolute energy calibration.

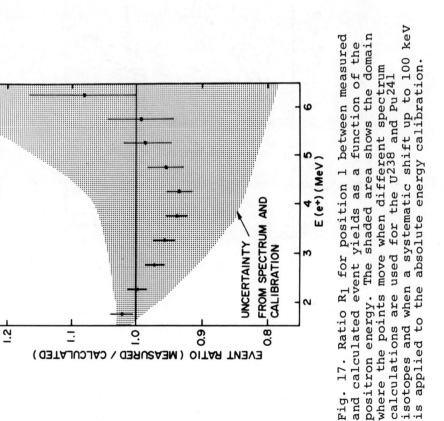

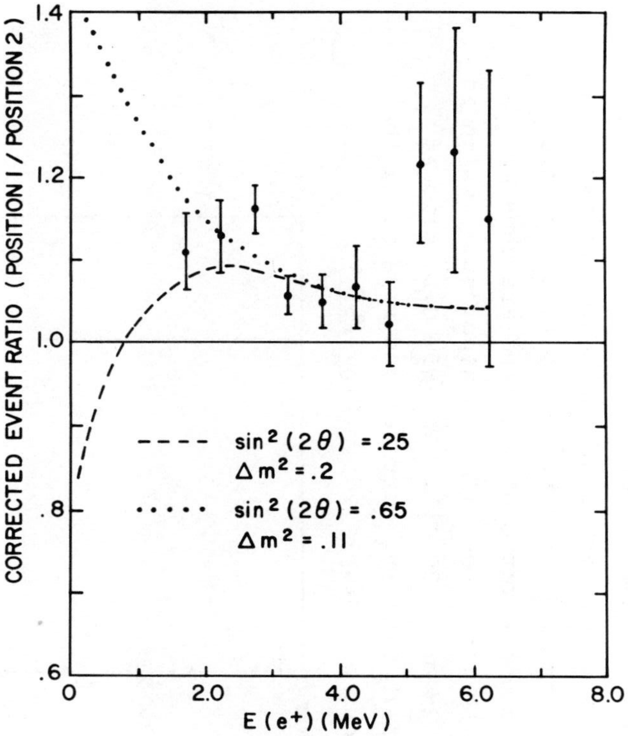

Fig. 18. Ratio of events in position 1 and position 2, as a function of positron energy.

at the two positions, corrected for solid angle differences and fuel burn-up, is shown in Fig. 18. Using the "two position method", these data can be compared to various neutrino oscillation hypotheses. The data points lie systematically above a ratio of one, and are thus inconsistent with the no-oscillation hypothesis, which is ruled out by 3.2σ or 99.8% C.L. The 95% C.L. allowed region is shown in Fig. 19, curve c) and several of the allowed solutions are displayed in Fig. 18.

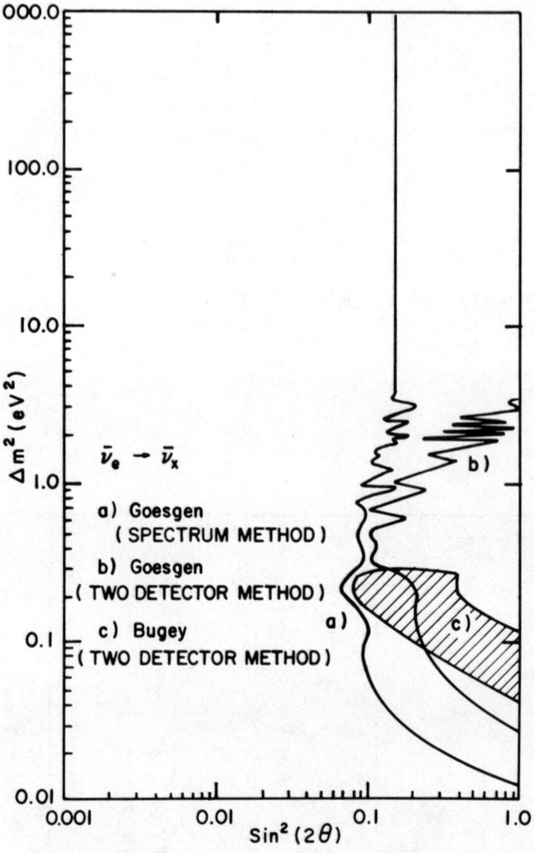

Fig. 19. Inclusive oscillation limits for $\bar{\nu}_e \rightarrow \bar{\nu}_x$.

Comparing the "two position method" limits of the Goesgen and Bugey groups shows that there is a region near $\Delta m^2 \approx 0.2$ and $\sin^2 2\theta \approx 0.2$ which is compatible with both experiments. This region, though, is ruled out by the "spectrum method" limit of the Goesgen experiment. There could be uncertainties in both experiments that are larger than estimated. The Goesgen limit relies heavily on the calculated spectrum and the Bugey result is sensitive to time variations in the reactor flux and differences in background for the two positions. Both groups are continuing their measurements and additional information should be available in one to two years. The Goesgen group is making new measurements at 65m and the Bugey group plans to repeat the 13.6 and 18.3m experiments possibly with two detectors run simultaneously to eliminate reactor power time variations. In addition, a group (U Cal at Irvine) at the Savannah River reactor is now taking data at 17.5 and 24m.

Future Oscillation Experiments

Several new experiments will be taking data over the next two years. These experiments are typically sensitive to a region of Δm^2 and $\sin^2(2\theta)$ approximately an order of magnitude lower than the present limits. Tables VI and VII list the proposed sensitivities of these experiments.

The most ambitious program is the BNL long baseline experiment, Columbia/Illinois/Johns Hopkins (E776). Two detectors at 300m and 1000m will run simultaneously in either a low energy dichromatic or wideband horn beam ($\langle E_\nu \rangle \approx$ 1-2 GeV). If the improvements of the AGS at BNL continue to increase the accelerator intensity, this program will be able to significantly improve the sensitive region for $\nu_\mu \to \nu_e$, $\nu_\mu \to \nu_x$, and $\nu_e \to \nu_x$ oscillations.

Table VI: Upcoming Exclusive Oscillation Experiments

		Large Δm^2 $\sin^2(2\theta)$ Limit	$\sin^2(2\theta)=1$ Δm^2 Limit
BEBC ($\nu_\mu \to \nu_e$)	Low Energy Horn $<E_\nu> \approx 1$ GeV	5×10^{-3}	0.04 eV^2
BNL E776 ($\nu_\mu \to \nu_e$)	NBB $<E_\nu> \approx 1.5$ GeV	5×10^{-4}	0.03 eV^2
BNL E775 ($\nu_\mu \to \nu_e$)	NBB $<E_\nu> \approx 1.5$ GeV	1×10^{-3}	0.2 eV^2
LAMPF E645 ($\bar{\nu}_\mu \to \bar{\nu}_e$)	Beam Stop Beam $<E_\nu> \approx$ 0-53 MeV	8×10^{-4}	0.05 eV^2
	Old Limits	6×10^{-3}	0.06 eV^2

Table VII: Upcoming Inclusive Experiments

Accelerator Experiments	Source	Sensitivity Δm^2 Range	Best $\sin^2(2\theta)$
BNL E776 ($\nu_\mu \to \nu_x$)	NBB $<E_\nu> = 1.5$ GeV	.13-100 eV^2	$.02/\Delta m^2 = 2$ eV^2
BNL E775 ($\nu_\mu \to \nu_x$)	WBB $<E_\nu> = 2$ GeV	5-100 eV^2	$.2/\Delta m^2 = 25$ eV^2
	NBB <1.5 GeV>	12-100 eV^2	$.03/\Delta m^2 = 1$ eV^2
LAMPF E645 ($\nu_e \to \nu_x$)	Beam Stop Beam	1-25 eV^2	$.07/\Delta m^2 = 10 eV^2$
BNL E776 ($\nu_e \to \nu_x$)	WBB $<E_\nu> = 2$ GeV	.5-30 eV^2	$.04/\Delta m^2 = 7$ eV^2

SEARCHES FOR SUBDOMINANTLY COUPLED ν's

In 1980, Shrock[56)] and others pointed out that correlated tests for neutrino masses and mixings could be carried out by searching for secondary peaks in the lepton momentum spectrum from pion and kaon decays. If the weak and mass eigenstates of the neutrino are different, then the muon and electron neutrino are mixtures of mass eigenstates:

$$\nu_e = \sum_i U_{ei}\nu_i$$

$$\nu_\mu = \sum_i U_{\mu i}\nu_i$$

where ν_i are the various mass eigenstates and $U_{\ell i}$ is the unitary mixing matrix.

In a two body decay, each mass eigenstate would show up as a peak in the lepton recoil spectrum with intensity governed as $|U_{\ell i}|^2$. For example, in $\pi \to \mu\nu_i$ decay, a spectrum such as schematically represented below could be observed.

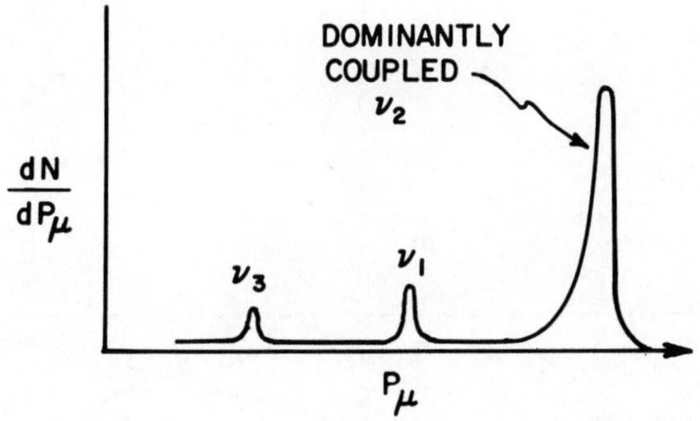

In three body decays, mixtures of finite mass eigenstates will cause cusps (or kinks) in the decay lepton spectrum because each eigenstate will have a different end point energy. The limits for heavy neutrinos coupled to electron and muon neutrinos are given in Figs. 20 and 21, respectively.

Another type of correlated mass mixing search can be carried out by looking for the decays of heavy neutrinos in various neutrino beams. Heavy neutrinos could be present in these beams due to mixing of the mass eigenstates and these heavy neutrinos could then decay into lepton pairs, i.e. $e^+e^-\nu_e$ and, thus, be detected. Gronau[60)] has used this

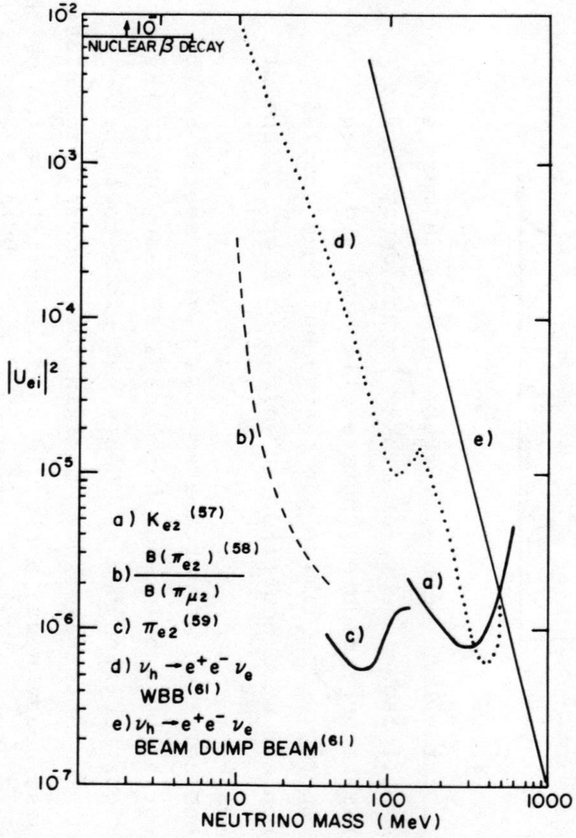

Fig. 20. Limits on the masses of heavy neutrino coupled to electron neutrinos.

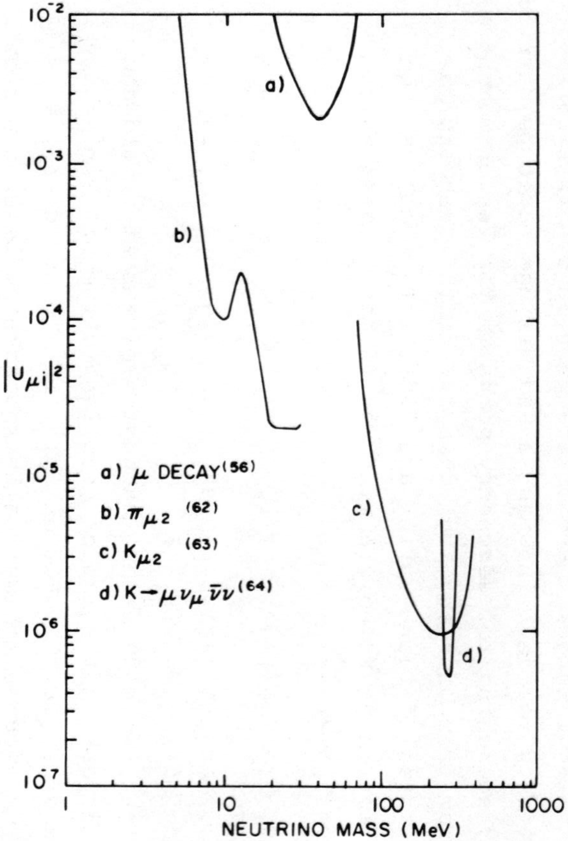

Fig. 21. Limits on the masses of heavy neutrinos coupled to muon neutrinos.

idea to analyze previous data and obtained the following limits:

$$310 \le m_h \le 370 \text{ MeV} \quad |U_{\mu h}|^2 \le 10^{-6}$$

$$160 \le m_h \le 480 \text{ MeV} \quad |U_{eh}|^2 \le 10^{-6} - 10^{-5}$$

The CHARM collaboration[61] has also used this technique to search for heavy neutrinos produced in π/K decay that decay into a pair of electrons, $\nu_h \to e^+e^-\nu_e$. The fine grain CHARM calorimeter was used to isolate candidates for the above decay mode in the CERN wideband neutrino beam. The number of events attributed to heavy neutrino decay was compatible with zero giving the limits shown in Fig. 20. The CHARM group has also set correlated limits on m_{ν_τ} and $|U_{\tau e}|^2$ with a new detector optimized for detecting $\nu_\tau \to e^+e^-\nu_e$. Data were taken with this detector during the beam dump running at CERN. From these data, the group sets very tight limits on $|U_{\tau e}|^2 < 10^{-6}$-10^{-10} for m_{ν_τ} between 50-250 MeV. These ν_τ limits rely critically on the assumptions made in calculating the ν_τ flux from F production and decay in the beam dump. Using this same data set, the group has also set limits on the coupling and mass of a heavy neutrino produced in D-meson decays, $D \to e\nu_h$. These limits are most sensitive at high mass and reach a level for $|U_{he}|^2$ of 10^{-6}-10^{-7} for masses between 0.5-2 GeV.

NEUTRINOLESS DOUBLE β DECAY

Double β decay offers another window to possible neutrino mass effects. Isotopes can undergo double β decay if the decays to adjacent nuclei are energetically forbidden. In the standard model, these decays are second-order weak and involve the emission of two electrons and two neutrinos. It is also possible that double β decays can occur in which no neutrino is emitted, neutrinoless double β decay.

$$[A,Z] \to [A,Z+2] + e^- + e^- \quad .$$

This neutrinoless decay violates lepton number conservation and requires that a virtual neutrino be emitted and reabsorbed at the two decay vertices. For this to happen, the neutrino must be a Majorana particle, $\bar{\nu} \equiv \nu$. A further constraint is made because of the (V-A) structure of the weak current; either the neutrino must have mass or right-handed couplings (RHC) must exist. Thus, limits on the rates for neutrinoless double β decay can be turned into correlated limits on the neutrino mass and RHC for Majorana neutrinos. These limits do rely on theoretical calculations of the nuclear matrix elements. Several calculations[65)] have been done giving similar results to about a factor of three.

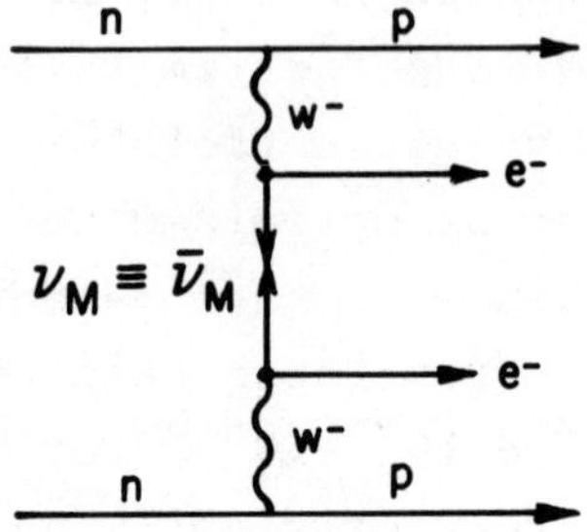

Experimental limits for the rate of neutrinoless double β decay have been set by a number of groups. Geochemical measurements[66)] of the excess of daughter nuclei in ore $\sim 10^9$ years old are consistent with calculations of 2ν double β decay and can be used to set limits on the rate of neutrinoless decay. Laboratory measurements, though, have better control of systematic errors and several studies have been made. Backgrounds from cosmic rays and natural radioactivity can be minimized by performing the experiments in tunnels or mines and using active local shielding. The amount of background from other decays inside the experiment is directly related to the energy resolution of apparatus and how well the experiment can isolate the monoenergetic, two electron signal.

The current best limits for neutrinoless β decay are given in Table VIII. The limits use the matrix element calculations of Haxton et al,[65)] and η is the RHC coupling parameter given by $(M_{W_{LEFT}}/M_{W_{RIGHT}})^2$.

Table VIII: Laboratory Limits of Neutrinoless Double Beta Decay.

Source	Half Life Limit for 0ν Decay	m_ν Limit ($\eta = 0$)	RHC Limit ($m_\nu = 0$)	Ref.
^{48}Ca	$> 10^{21.3}$ years	< 30 eV	-	67)
^{82}Se	$> 10^{21.5}$ "	< 12 eV	$\eta < 1 \times 10^{-5}$	68)
^{76}Ge	$> 10^{22.5}$ "	< 8 eV	$\eta < 1.3 \times 10^{-5}$	69)

The ^{76}Ge experiment[69)] uses two $\sim$ 120 cm^3 Ge(Li) crystal source detectors with energy resolutions of 0.2%. The experiment has run for about $\sim$ 13K hours in Mont Blanc tunnel and these measurements are part of an ongoing program to increase the detector size and running time. Several other experiments using Ge(Li) detectors are also underway; the size of the detectors range from 400 to 1500 cm^3. In addition, studies of $^{136}X \rightarrow {}^{136}Ba + 2e^- + (2\nu)$ are planned by four groups using either high pressure xenon gas or liquid xenon chambers.

Kayser[1)] has pointed out that one possible theoretical uncertainty in setting limits on neutrino masses from these double β decay searches is the unknown CP eigenvalues of the various mass eigenstates. The effective mass that comes into the 0ν double β decay calculation is:

$$\langle m_\nu \rangle_{\text{Effective}} = \sum_{i=1}^{n} \lambda^i_{CP} m_i |U_{ei}|^2$$

where λ^i_{CP} is the CP eigenvalue of the ith mass eigenstate. Two mass eigenstates that couple to electrons could have opposite λ_{CP} and give an effective mass close to zero even though the mass eigenstates have large masses. For these reasons, the 0ν double β decay limits cannot restrict the eigenstate masses directly but only the combination giving the above effective mass. With this caveat, though, the neutrinoless double β decay limits place severe restrictions on the masses and interactions of Majorana neutrinos, and suggest that $m_\nu < 3$-10 eV. If other types of measurements (decay and oscillation experiments) do detect neutrino masses greater than 3-10 eV, then a possible explanation is that the neutrino is a Dirac rather than a Majorana particle.

CONCLUSIONS

Over the past five years, experimental investigations have been made of all the phenomena listed in the introduction as relating to finite mass neutrinos. Almost all studies obtained negative results. A large region of possible masses and mixings have been excluded but a still larger region remains to be explored. The small mass region (tens of eV) for muon and tau neutrinos is not accessible to decay experiments and, therefore, can only be probed by neutrino oscillation experiments. Future experiments, now underway or being planned, should be sensitive to masses and mixings about an order of magnitude below the present limits.

Two experiments have reported positive evidence for finite mass neutrinos, the ITEP-83 tritium β decay experiment and Bugey reactor neutrino oscillation experiment. The ITEP group claim indications of a non-zero mass with $m_{\bar{\nu}_e} > 20$ eV. With regard to this result, there are still questions concerning whether valine molecular or resolution smearing effects have been accounted for properly in the end point spectrum. Several new experiments with sensitivities in the ITEP mass range are presently running or under construction. Some of these will use either atomic or molecular tritium sources which have well understood final state spectra. There is also some controversy concerning the Bugey result and whether it conflicts with the Goesgen limits. Both groups will continue to run in the future with improved control of systematic uncertainties. Taking the Bugey result at face value, $\Delta m^2 = 0.2$ eV and $\sin^2(2\theta) = 0.25$, and combining it with the ITEP-83 value of $m_{\nu_e} = 30$ eV implies two neutrino mass eigenstates with $m_1 = 30$ eV and $m_2 = 30.003$ eV. These masses are only consistent with the neutrinoless double β decay limit of $m_{\nu_e} < 8$ eV if the ν_e is a Dirac particle or if there is some cancellation in the mixture of CP eigenstates. (See the previous section.)

At present, there is no strong theoretical argument why neutrinos should be massless; in fact, it is natural to assume that they are massive with mixing between generations of weak eigenstates. Setting the scale of these masses and mixings is another problem with little theoretical guidance. For these reasons, experimental measurements should continue probing ever further into the unexplored regions.

Acknowledgments

I would like to thank my colleagues in the CCFR collaboration and especially Carl Haber for help with the preparation of and collection of data for these lectures. Also, I thank A. Therrien, J. Therrien and A. Vermeulen for their work preparing this manuscript. This work has been supported in part by the National Science Foundation.

References

1) B. Kayser, Fourth Moriond Workshop, La Plagne, France (January 1984).
2) R.E. Shrock, Quarks-84 Conference, Tbilisi, USSR (1984).
3) M. Gell-Mann, P. Ramond, R. Slansky, in Supergravity (North-Holland, 1979), p. 315.
4) E. Witten, Phys. Lett. 91B, 81 (1980).
5) K.A. Olive, D.N. Schramm, G. Steigman, Nucl. Phys. B180, 497 (1981).
6) R. Cowsik, J. McClelland, Phys. Rev. Lett. 29, 669 (1972).
7) S. Tremaine, J.E. Gunn, Phys. Rev. Lett. 42, 407 (1979).
8) N.J. Baker et al, Phys. Rev. Lett. 47, 1576 (1981).
9) N. Ushida et al, Phys. Rev. Lett. 47, 1694 (1981).
10) M. Holder et al, Phys. Lett. 74B, 277 (1978).
11) W. Kinnison et al, Phys. Rev. D25, 2846 (1982).
12) T. Kinoshita, Phys. Rev. Lett. 2, 477 (1959).
13) D.A. Bryman et al, Phys. Rev. Lett. 50, 7 (1983).
14) K. Bergkvist, Nucl. Phys. B39, 317 (1972).
15) V.A. Lubimov et al, Phys. Lett. 94B, 266 (1980).
16) J.J. Simpson, Phys. Rev. D23, 649 (1981).
17) S. Boris et al, Proceedings of HEP 83, Rutherford Appleton Lab (1983).
18) I.G. Kaplan, V.N. Smutny, G.V. Smelov, Phys. Lett. 112B, 417 (1982).
19) J.J. Simpson, P. Vogel, Low Energy Tests of Conservation Laws, AIP Conference Proceedings # (AIP, New York 1984).
20) L.G. Smith, E. Koets, A.H. Wapstra, Phys. Lett. 102B, 114 (1981).
21) J.U. Andersen et al, Phys. Lett. 113B, 72 (1982).
22) S. Yasumi et al, Proc. HEP 83, Rutherford Appleton Lab (1983).

23) C.L. Bennett et al, Phys. Lett. 107B, 19 (1981).
24) A. DeRujula, Nucl. Phys. 188B, 414 (1981).
25) R.S. Raghaven, Phys. Rev. Lett. 51, 975 (1983).
26) H.B. Anderhub et al, Phys. Lett. 114B, 76 (1982).
27) R. Abela et al, SIN Newsletter 15, p. 26.
28) D.C. Lu et al, Phys. Rev. Lett. 45, 1066 (1980).
29) P. Seiler, Neutrino Mass and Gauge Structure of Weak Interactions, Telemark, Wisconsin (1982), p. 41.
30) B. Robinson, Ibid., p. 25.
31) C. Hoffman, V. Sandberg, Proc. of 1982 DPF Summer Study, Snowmass, Colorado, p. 552.
32) W. Bacino et al, Phys. Rev. Lett. 42, 749 (1979).
C.A. Blocker et al, Phys. Lett. 109B, 119 (1982).
33) C. Matteuzzi et al, Phys. Rev. Lett. 52, 1869 (1984).
34) M. Kobayashi, K. Maskawa, Prog. Th. Phys. 49, 652 (1973).
35) C. Baltay, Neutrino 81, Maui, Hawaii (1981);
A. Bodek, ICOBAN 84, Park City, Utah (1984);
M. Shaevitz, Proc. of the Lepton-Photon Symposium, Cornell Univ., Ithaca, NY (1983).
36) G.N. Taylor et al, Phys. Rev. D28, 2705 (1984)
37) N.J. Baker et al, Phys. Rev. Lett. 47, 1576 (1981).
38) BEBC Collaboration (Athens-Padova-Pisa-Wisconsin). See J. Wotschack review talk, Neutrino 84, Dortmund, Germany.
39) N. Ushida et al, Phys. Rev. Lett. 47, 1694 (1981).
40) A.E. Asratyan et al, Phys. Lett. 105B, 301 (1981).
41) H. Chen et al, Irvine-LosAlamos-Maryland experiment E225 at LAMPF. See J. Wotschack review talk at Neutrino 84, Dortmund, Germany.
42) F. Dydak et al, Phys. Lett. 134B, 34 (1984).
43) F. Reines et al, Phys. Rev. Lett. 45, 1307 (1980).
44) I.E. Stockdale et al, Phys. Rev. Lett. 52, 1384 (1984);
C. Haber et al, Fourth Moriond Workshop, LaPlagne, France (1984).
45) J. Learned, Neutrino 84, Dortmund, Germany.
46) O. Erriquez et al, Phys. Lett. 102B, 73 (1981).
47) K. Gabathuler et al, Phys. Lett. 138B, 449 (1984).
H. Kwon et al, Phys. Rev. D24, 1097 (1981);
F. Boehm et al, Phys. Lett. 97B, 310 (1980).
J.L. Vuillemier et al, Phys. Lett. 114B, 298 (1982).
48) J.F. Cavaignac et al, LAPP-EXP-84-03 (1984), submitted to Phys. Lett. B.
49) A.G. Frodessen et al, Probability and Statistics in Particle Physics, Universitets-Forlaget, Bergen (1979), p. 415. W.T. Eadie et al, Statistical Methods in Experimental Physics, North Holland, Amsterdam (1971), p. 230.
50) F. Bergsma et al, Phys. Lett. 142B, 103 (1984).
51) R. Davis et al, Proc. of the Telemark Neutrino Mass Miniconference, Telemark, Wisconsin (1980).
52) J.N. Bahcall et al, Phys. Rev. Lett. 45, 945 (1980).
J.N. Bahcall, Neutrino 81, Maui, Hawaii (1981).
53) K. Winter, Proc. of the Lepton-Photon Symposium, Cornell Univ., Ithaca, NY (1983).
54) K. Schreckenbach et al, Phys. Lett. 99B, 251 (1981);
W. Mampe et al, NIM 154, 127 (1978).

55) B.R. Davis et al, Phys. Rev. C19, 2259 (1979);
F.T. Avignone et al, Phys. Rev. C22, 594 (1980).
56) R. Shrock, Phys. Lett. 96B, 159 (1980); Phys. Rev. D24, 1232, 1275 (1981); Proc. 1982 DPF Summer Study, Snowmass, Colorado, pp. 264-273.
57) T. Yamazaki, Neutrino 84, Dortmund, Germany (1984).
58) D.A. Bryman et al, Phys. Rev. Lett. 50, 1546 (1983).
59) J. Deutsch et al, Lovain-Lausanne-Zurich experiment at SIN. See T. Yamazaki, Review Talk at Neutrino 84, Reference 57.
60) M. Gronau, SLAC-PUB-2967, August 1982.
61) F. Bergsma et al, Phys. Lett. 128B, 361 (1983).
K. Winter, Proc. of Lepton-Photon Symposium, Cornell Univ. (1983).
62) R. Abela et al, Phys. Lett. 105B, 263 (1981).
63) R.S. Hayano et al, Phys. Rev. Lett. 49, 1305 (1982).
64) C.Y. Pang et al, Phys. Rev. D8, 1989 (1973).
65) S.P. Rosen, Neutrino 81, Vol. II, p. 76, Maui, Hawaii.
W.C. Haxton et al, Phys. Rev. Lett. 47, 153 (1981); Phys. Rev. D25, 2360 (1982).
M. Doi et al, Phys. Lett. 103B, 219 (1981).
66) E. Hennecke et al, Phys. Rev. C11, 1378 (1975).
T. Kirsten et al, Phys. Rev. Lett. 50, 474 (1982).
67) R. Barden et al, Nucl. Phys. A158, 337 (1970).
68) B.T. Cleveland et al, Phys. Rev. Lett. 35, 737 (1975).
69) E. Belloti, Neutrino 84, Dortmund, Germany.
E. Belloti et al, Phys. Lett. 121B, 72 (1983).

CALORIMETRY IN HIGH-ENERGY PHYSICS

Christian W. Fabjan

CERN

CH-1211 Geneva 23

1. INTRODUCTION

Much of our present knowledge about elementary particles has been established through a continuing refinement of techniques for measuring the trajectories of individual charged particles. Only in recent years has a different class of detectors—calorimeters—been widely employed, but these have already greatly influenced the scope of experiments.

Conceptually, a calorimeter is a block of matter which intercepts the primary particle, and is of sufficient thickness to cause it to interact and deposit all its energy inside the detector volume in the subsequent cascade or 'shower' of increasingly lower-energy particles. Eventually, most of the incident energy is dissipated and appears in the form of heat. Some (usually a very small) fraction of the deposited energy is detectable in the form of a more practical signal (e.g. scintillation light, Cherenkov light, or ionization charge), which is proportional to the initial energy.

The first large-scale detectors of this type were used in cosmic-ray studies [1]. Interest in calorimeters grew during the late 1960's and early 1970's in view of the new accelerators at CERN [the Intersecting Storage Rings (ISR) and the Super Proton Synchrotron (SPS)] and at the Fermi National Accelerator Laboratory (FNAL), with their greatly changed experimental directions and requirements [2]. One consequence of the new fixed-target accelerators was the advent of intense, high-energy neutrino beams with the need for very massive detectors to study their interactions. This detector development was paralleled by the rapid growth of analog signal-processing techniques: during the last decade the typical number of analog signal-channels of nuclear spectroscopy quality has increased from about 10 to 10^4 in high-energy physics experiments.

Calorimeters offer many attractive capabilities, supplementing or replacing information obtained with magnetic spectrometers:

1) They are sensitive to charged and neutral particles.
2) The 'energy degradation' through the development of the particle cascade is a statistical process, and the average number $\langle N \rangle$ of secondary particles is proportional to the energy of the incident particle. In principle, the uncertainty in the energy measurement is governed by statistical fluctuations of N, and hence the relative energy resolution σ/E improves as $1/\sqrt{\langle N \rangle} \sim E^{-1/2}$.
3) The length of the detector scales logarithmically with particle energy E, whereas for magnetic spectrometers the size scales with momentum p as $p^{1/2}$, for a given relative momentum resolution $\Delta p/p$.
4) With segmented detectors, information on the shower development allows precise measurements of the position and angle of the incident particle.
5) Their different response to electrons, muons, and hadrons can be exploited for particle identification.
6) Their fast time response allows operation at high particle rates, and the pattern of energy deposition can be used for rapid on-line event selection.

In these notes we comment first on the principal features of detectors designed to measure the energy of photons and electrons, the 'electromagnetic shower detectors' (ESD). The underlying physics has been understood for many years, and such detectors were the main components in many experiments—some of which were credited with important discoveries. Recent developments have been emphasizing precision measurements of energy and position in large arrays.

In the subsequent section the physics of 'hadronic calorimeters' is reviewed. Progress during the last decade contributed to an understanding of the physics of this technique and to a steadily growing range of applications.

The final section concentrates on the technical issues of information processing from calorimeters. We can only select representative examples from the numerous and ingenious methods devised to extract the energy information. We end with a discussion on the state-of-the-art Monte Carlo simulation of electromagnetic and hadronic showers.

These notes follow an earlier review [3], emphasize recent developments, update the bibliography, but do not supersede other excellent introductions to this field [4, 5].

2. ELECTROMAGNETIC SHOWER DETECTORS

2.1 Energy Loss Mechanism

The contributions of the various energy loss mechanisms as a function of particle energy are given in Fig. 1 for electrons and positrons and in Fig. 2 for photons [6]. Above approximately 1 GeV, the principal processes—bremsstrahlung for electrons and positrons, pair production for photons—become energy independent. It is through a succession of these energy loss mechanisms that the electromagnetic cascade (EMC) is propagated, until the energy of the charged secondaries has been degraded to the regime dominated by ionization loss. Within this description, the combined energy loss of the cascade particles in the detector equals the energy of the incident electron or photon. The measurable signal—excitation or ionization of the medium—can be considered as the sum of the signals from the track segments of the positrons and electrons. Naively one might therefore expect that this signal should be equivalent to that produced by muons traversing the detector and whose combined track length equals that of the track

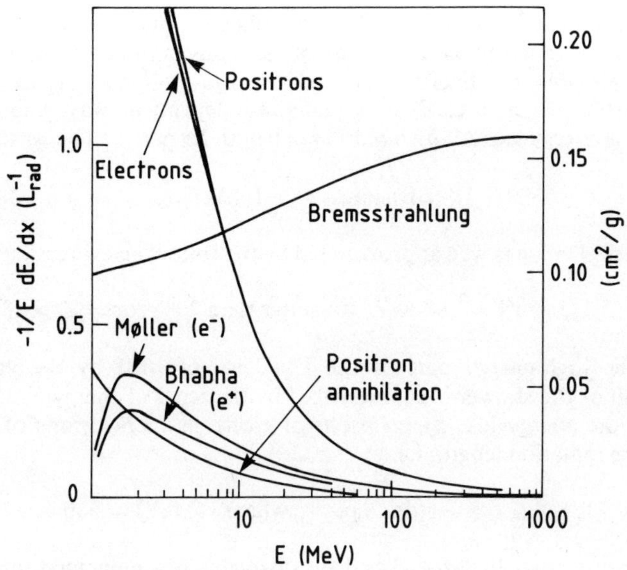

Fig. 1: Fractional energy loss per radiation length (left ordinate) and per g/cm^2 (right ordinate) in lead as a function of electron or positron energy. (Review of Particle Properties, April 1982 edition).

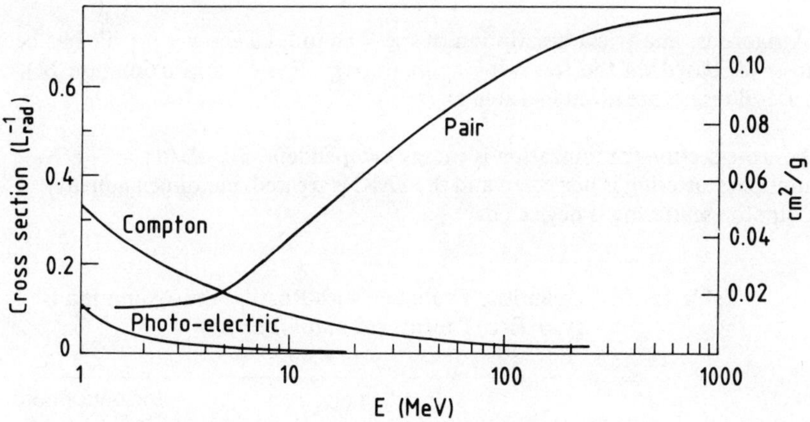

Fig. 2: Photon cross-section σ in lead as a function of photon energy. The intensity of photons can be expressed as $I = I_0 \exp(-\sigma x)$, where x is the path length in radiation lengths. (Review of Particle Properties, April 1980 edition).

segments of the EMC. This picture ignores finer points concerning low-probability photo-nuclear interactions, non-linear response of the ionizing medium as a function of ionization density or the detailed response to the very low energy e^+'s or e^-'s of the last generation of the shower. It emphasizes, however, the concept of 'track length T': a calorimeter is a useful device, because in the process of cascade formation the total track length T required for absorption is broken up into a 'tree' of many individual segments.

The electromagnetic cascading is fully described by quantum electrodynamics (QED) [7], and depends essentially on the density of electrons in the absorber medium. For this reason it is possible to describe the characteristic longitudinal dimensions of the high-energy EMC ($E > 1$ GeV) in a material-independent way, using the 'radiation length X_0'. The energy loss ΔE *by radiation* in length Δx can then be written

$$(\Delta E)_{radiation} = -E\,(\Delta x/X_0)$$

and the numerical value is well approximated by the following expression:

$$X_0\ [g/cm^2] \simeq 180\ A/Z^2 \text{ (to better than 20\% for } \simeq Z > 13).$$

Whilst the high-energy part of the EMC is governed by the value of X_0, the low-energy tail of the shower is characterized by the 'critical energy ϵ' of the medium. It is defined as the energy loss *by collisions* of electrons or positrons of energy ϵ in the medium in one radiation length, i.e.

$$(dE)_{collision} = -\epsilon(dx/X_0), \qquad \text{where } \epsilon\ (MeV) \simeq 550 \times Z^{-1}$$

(accurate to better than 10% for $Z > 13$). This value of ϵ coincides approximately with that value of the electron energy below which the ionization energy loss starts to dominate the energy loss by bremsstrahlung. The critical energy ϵ is seen to define the dividing line between shower multiplication and the subsequent dissipation of the shower energy through excitation and ionization.

A rigorous, analytical description of the longitudinal shower profile has been given by Rossi [8] based on the following assumptions ('Rossi's approximation B'), and the most useful results are given in Table 1:

i) the cross-section for ionization is energy independent, i.e. $dE/dx = -\epsilon/X_0$;
ii) multiple scattering is neglected and the EMC is treated one-dimensionally;
iii) Compton scattering is neglected.

Table 1: EMC Quantities Evaluated with Rossi's Approximation B
($y = E/\epsilon$; T measured in units of X_0)

	Incident electron	Incident photons
Peak of shower, t_{max}	$1.0 \times (\ln y - 1)$	$1.0 \times (\ln y - 0.5)$
Centre of gravity, t_{med}	$t_{max} + 1.4$	$t_{max} + 1.7$
Number e^+ and e^- at peak	$0.3\,y \times (\ln y - 0.37)^{-1/2}$	$0.3\,y \times (\ln y - 0.31)^{-1/2}$
Total track length T	y	y

The characteristic longitudinal EMC profile is shown in Fig. 3 for four very different materials and demonstrates the 'longitudinal scaling in radiation length'. A convenient analytical description of the profile has been given in the form [9]

$$dE/dt = E_0\, b^{\alpha+1}/\Gamma(\alpha+1) t^{\alpha} e^{-bt}\ ;\ t = x/X_0\ ,\ \alpha = bt_{max}, \text{ and } b \simeq 0.5\ .$$

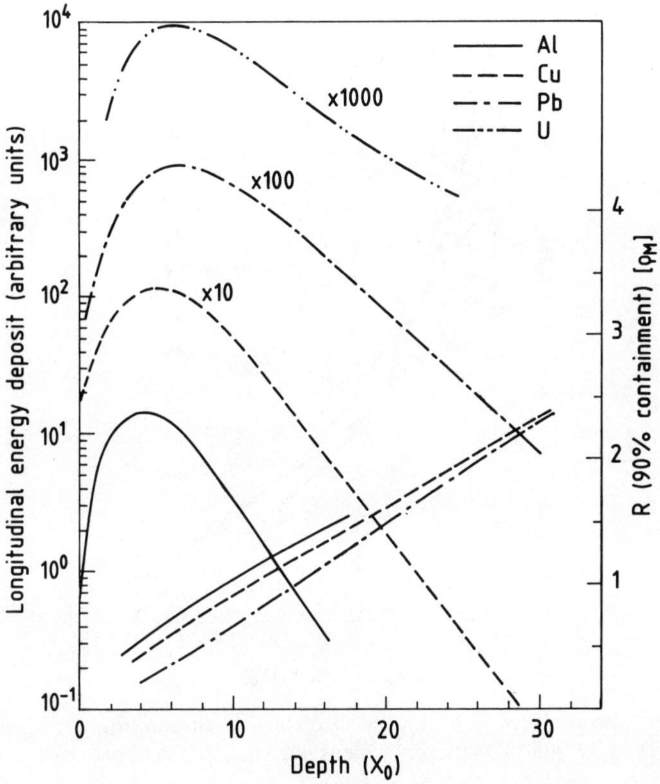

Fig. 3: Longitudinal shower development (left ordinate) of 6 GeV/c electrons in four very different materials, showing the scaling in units of radiation lengths X_0. On the right ordinate the shower radius for 90% containment of the shower is given as a function of the shower depth. In the later development of the cascade, the radial shower dimensions scale with the Molière radius $\varrho_M \sim 7A/Z$. [Al, Cu, and Pb, adapted from G. Bathow et al., Nucl. Phys. B20:592 (1970). Uranium data from G. Barbiellini et al., Ref. [127].

The transverse shower properties, which are not described within the framework of Rossi's 'approximation B' can also be easily understood qualitatively. In the early, most energetic part of the cascade the lateral spread is characterized by both the typical angle for bremsstrahlung emission, $\theta_{brems} \sim p_e/m_e$, and multiple scattering in the absorber. This latter process increasingly influences the lateral spread with decreasing energy of the shower particles and causes a gradual widening of the shower. For the purpose of total energy measurement, the EMC occupies a cylinder of radius R

$$R \approx 2\varrho_M ; \quad \varrho_M = 21X_0/\epsilon \simeq 7A/Z \ [\mathrm{g\,cm^{-2}}],$$

ϱ_M being the 'Molière Radius', which describes the average lateral deflection of electrons of energy ϵ after traversing one radiation length. In Fig. 4, the transverse shower profile as a function of depth clearly exhibits the rather pronounced central and energetic core surrounded by a low-energy 'halo'.

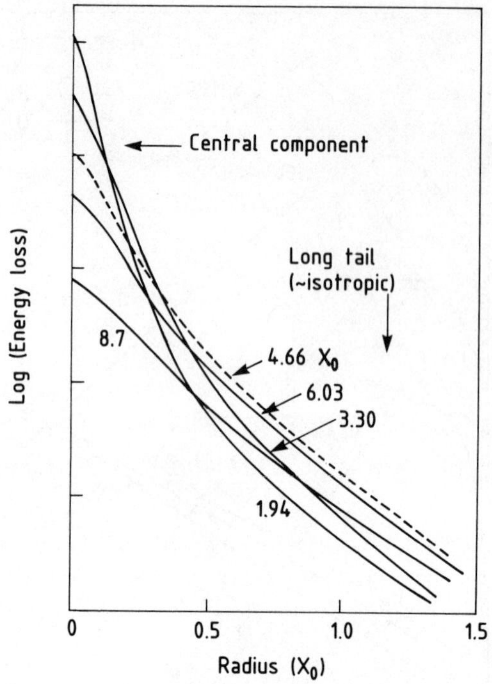

Fig. 4: Radial shower profile of 1 GeV electrons in aluminium; a pronounced central core, surrounded by 'halo', gradually widens with increasing depths of the shower [17].

2.2 Limits on Energy Resolution of EMCs

In the discussion in the previous section we have represented the shower by a total track length T, which could be expressed as $T(X_0) = E_{particle}(MeV)/\epsilon(MeV)$. The 'detectable' track length T_d, i.e. the equivalent track length which corresponds to the measured signal in a particular detector, will in general be shorter, $T_d \leq T$, as practical devices are only sensitive to the cascade particles above a certain threshold energy η. The fractional reduction in visible track length as a function of η/ϵ [4] is indicated in Fig. 5. The dotted lines are the result of an analytic calculation [8] for $E \gg \eta$, and $F(\xi)$ is given by $F(\xi) = [1+\xi \ln(\xi/1.53)] \exp \xi$ ($\xi = 2.29\ \eta/\epsilon$). The points are obtained by Monte Carlo calculations [9–11].

The average detectable track length $\langle T_d \rangle$ is given by $\langle T_d \rangle (X_0) = F(\xi) \cdot E/\epsilon$ and calorimetric energy measurements are possible because $\langle T_d \rangle \propto E$ for any reasonable value of ϵ. The *resolution* of the energy measurement is determined by the *fluctuations* in the shower propagation. The intrinsic component of the resolution is caused by the fluctuations in T_d. This represents the lower bound on the energy resolution and may be qualitatively estimated in the following way: the maximum number of track segments $N_{max} = E/\eta$ hence $\sigma(E)/E \geq \sigma(N_{max})/N_{max}$. In a lead-glass shower counter for which $\eta \sim 0.7$ MeV, one estimates for a 1 GeV shower, $N_{max} = 1000/0.7 = 1.5 \times 10^3$, implying an energy resolution at the level of one to two percent, somewhat higher than the level computed by detailed Monte Carlo calculations [9].

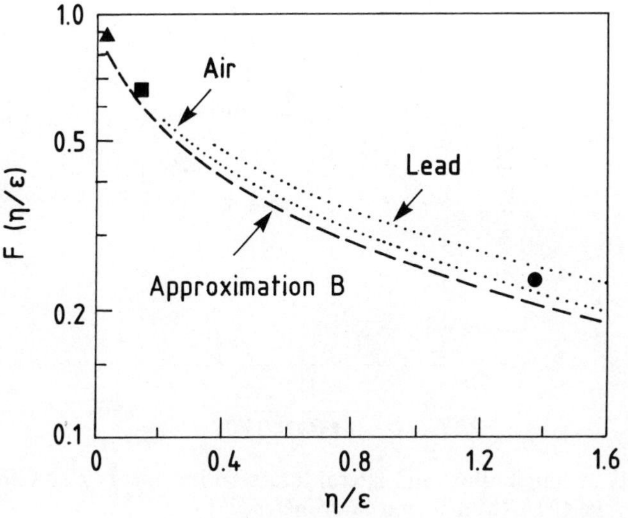

Fig. 5: Fraction F of the total track length which is seen on the average in a fully contained electromagnetic shower. The dotted lines represent analytical calculations and the points represent Monte Carlo results. ▲: lead glass; ■ and ●: lead sampling devices (see Ref. [4]).

In practical detectors, however, usually a number of additional components must be considered, which may conspire to affect the resolution. One important instrumental contribution to the energy resolution comes from incomplete containment of the showers ('energy leakage'), as can be seen from Fig. 6. Available information on average longitudinal containment (both experimental and calculational) may be parametrized as

$$L(98\%)_{av} \simeq t_{max} + 4\,\lambda_{att},$$

where L(98%) gives the length for 98% longitudinal containment. The quantity λ_{att} characterizes the slow, exponential decay of the shower after the shower maximum (see Fig. 3) expressed as $\exp(-t/\lambda_{att})$. The values of λ_{att} are found to be rather energy independent, but material dependent and close in value to the mean free path of photons that have minimum attenuation in a given material. Experimental values cluster around $\lambda_{att}\ [X_0] \simeq 3.4 \pm 0.5X_0$. This estimate is in reasonable agreement with other parametrizations [12], e.g. $L(98\%) \simeq 2.5\ t_{max}$ for E in the 10 to 1000 GeV range. The effect of longitudinal leakage on the energy resolution is consistent with the parametrization

$$\sigma(E)/E \simeq [\sigma(E)/E]_{f=0} \times [1 + 2\sqrt{E(GeV)} \times f]$$

for values f of the fractional energy loss through leakage, $f < 0.2$ and $E < 100$ GeV. One notes that longitudinal leakage is more critical than transverse leakage due to the fact that fluctuations about the average longitudinal loss are much larger than for transverse leakage.

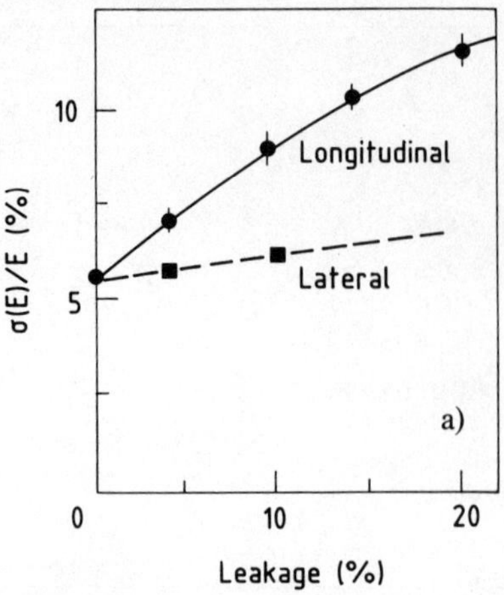

Fig. 6a: Effects of longitudinal and lateral losses on the energy resolution as measured for electrons in the CHARM neutrino calorimeter [23].

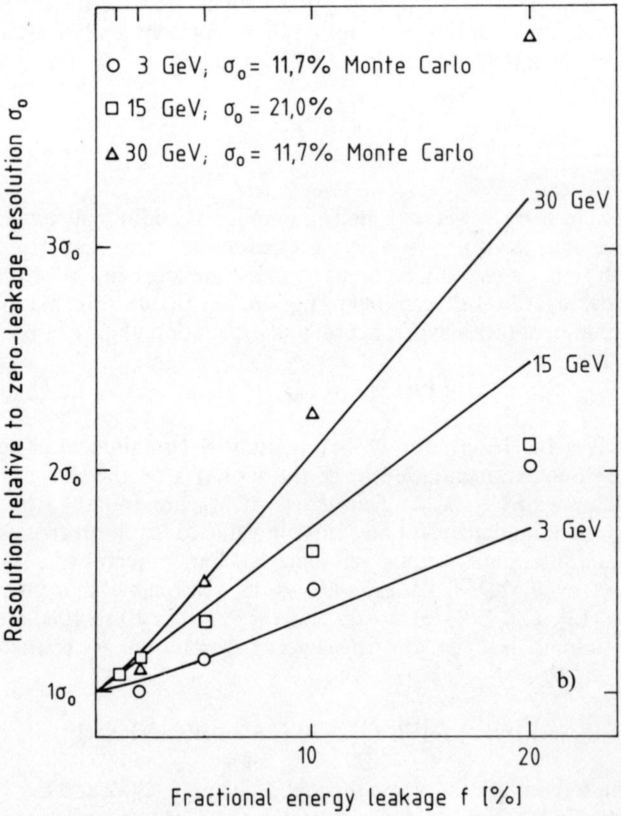

Fig. 6b: Deterioration of the zero-leakage resolution σ_0 as a function of fractional energy leakage f for three different electron energies [21, 23].

Homogeneous detectors have always played a very important role as e.m. calorimeters. Historically, it was NaI that was used as one of the first calorimetric detectors, still unsurpassed in energy resolution. For energies $E \simeq 1$ GeV, one obtains a value close to the intrinsic limit; at higher energies, the full statistical gain $\sim 1/\sqrt{E}$ is not obtained, even for very carefully tuned instruments [13] for which the resolution is quoted as $\sigma(E)/E \simeq 0.009 \times E^{-1/4}$ (GeV). Such a behaviour is characteristic of contributions other than those governed by statistical processes, such as non-uniformity in the signal collection, energy leakage, etc. A second, very frequently used homogeneous detector is lead glass, i.e. glass loaded with 50–60% PbO. The EMC is sampled by the production of Cherenkov light emitted from the relativistic electron–positron pairs. These detectors are therefore characterized by a relatively low light yield—typically 1000 photoelectrons per GeV are measured—and a relatively large cut-off energy η. These two effects combined give a resolution of

$$\sigma_{tot} = (\sigma_\eta^2 + \sigma_{ph}^2)^{1/2} \simeq (0.020^2 + 0.032^2)^{1/2} = 3.8\% \text{ at } 1 \text{ GeV},$$

in agreement with the best values reported, of $\gtrsim 4\%$. Recent developments of new scintillating crystals have stimulated interest in such homogeneous detectors, which promise exceedingly good performance in the 1 to 100 GeV regime (see the discussion in Section 5).

2.3 Energy Resolution in 'Sampling' Calorimeters

'Sampling' calorimeters are devices in which the functions of energy degradation and energy measurement are separated in alternating layers of different substances. This allows a considerably greater freedom in the optimization of detectors for certain specific applications. The choice of a 'passive' absorber—typically plates made of Fe, Cu, or Pb, each ranging in thickness from a fraction of X_0 to a few X_0—makes it possible to build rather compact devices, and it permits optimization for a specific experimental requirement such as electron/pion discrimination or position measurement. Independently of the choice of the absorber, the readout method may be selected for best uniformity of signal collection, high spatial subdivision, rate capability or other criteria. The disadvantage is that only a fraction of the total energy of the EMC is 'sampled' in the active planes, resulting in additional *'sampling' fluctuations* of the energy determination.

These general comments apply to both electromagnetic and hadron calorimeters. The following discussion of sampling fluctuations is specifically valid for the measurement with e.m. calorimeters, for which sampling fluctuations have been rather carefully studied. Today we know that they depend on the characteristics of both the passive and the active medium (in particular, thickness and density) and that several effects contribute to the 'total' sampling fluctuation.

The *'intrinsic sampling' fluctuations* express the statistical fluctuations in the number of e^+e^- pairs traversing the active signal planes and can be estimated in the spirit of approximation B. The number N_x of crossings is ($\eta = 0$)

$$N_x = T \text{ (total track length)}/d \text{ (distance between active plates)},$$

where $T = E/\epsilon$ and hence $N_x = E/\epsilon d = E/\Delta E$, ΔE being the energy loss per unit cell.

The contribution to the energy resolution is

$$\sigma(E)/E_{sampling} = \sigma(N_x)/N_x = 1/\sqrt{N_x} = 3.2\% \, [\Delta E\,(\mathrm{MeV})/E\,(\mathrm{GeV})]^{1/2}.$$

This expression has to be regarded as a lower bound on the sampling fluctuations for the following reasons:

- tracks originate from pair-produced particles and therefore the number of independent gap crossings would be only $N_x/2$ for totally correlated production;
- approximation B ignores multiple scattering, which increases the effective distance d to $d = d/\langle\cos\theta\rangle$, where the characteristic multiple scattering angle θ is given by $\langle\cos\theta\rangle \sim \cos(21\,\mathrm{MeV}/\epsilon\pi)$ [4];
- for $\eta \neq 0$, $T_d = F(\xi)T$.

Considering these effects, the contribution of sampling fluctuations to the energy resolution is evaluated as

$$[\sigma(E)/E]_{sampling} \gtrsim 3.2\% \, \{\Delta E\,(\mathrm{MeV})/[F(\xi) \times \cos(21/\epsilon\pi)\, E\,(\mathrm{GeV})]\}^{1/2}.$$

This expression does not include possible additional effects due to *'Landau' fluctuations* of the energy deposit in the active signal planes, which can be estimated to contribute

$$[\sigma(E)/E]_{Landau} \simeq 3/[\sqrt{N_x} \times \ln(1.3 \times 10^4\,\delta)],$$

where δ (MeV) gives the energy loss per active detector plane. Such additional fluctuations are small for energy losses δ of a few MeV (e.g. a few millimetres of scintillator), but may become comparable to the 'intrinsic' sampling fluctuations for very thin detectors, e.g. gaseous detectors with δ in the keV range. In addition to these 'Landau' fluctuations there is a further source of errors which also depends on the density of the active medium, *'path-length' fluctuation:* low-energy electrons may be multiply scattered into the plane of the active detector and then travel distances much larger than, for example, the gap thickness in gaseous detector planes, depositing considerably more energy compared to that deposited under perpendicular traversal. This effect is quantitatively less significant in dense active layers, because the range of the low energy electrons is comparable to the thickness of these layers. Moreover, increased multiple scattering in dense detector planes will also tend to reduce this effect relative to light absorbers. From Fig. 7 it can be seen that path-length fluctuations may contribute as much as Landau fluctuations to the resolution [14] in detectors with gaseous readouts.

Concluding this section, we compare in Table 2 the measured performance of some characteristic sampling devices, and compare it with the estimated contributions using the formulae given here. The energy resolution is seen to be rather well described by these estimates, provided that instrumental effects (such as calibration errors, photon statistics, leakage, etc.) do not dominate.

It is interesting to note that the path length and the Landau fluctuations are not negligible even in the case of dense active readout gaps, if these are very thin (e.g. measurements with the W/Si calorimeter).

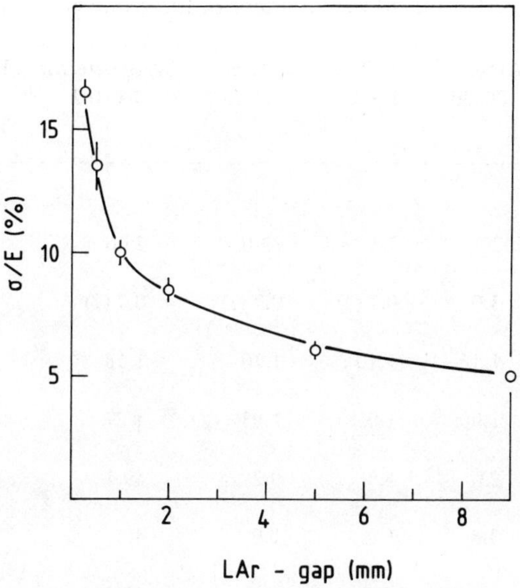

Fig.7a: Energy resolution versus thickness of the active liquid-argon layer for 1 GeV electrons in an iron/argon sampling calorimeter.

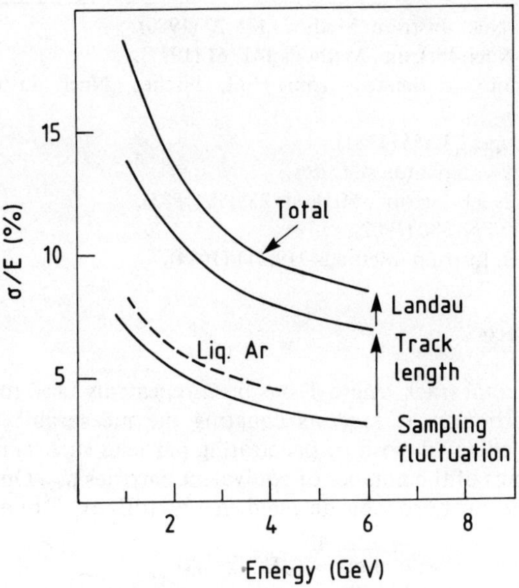

Fig. 7b: Contributions of sampling, path length, and Landau fluctuations to the energy resolution of a lead/MWPC sampling calorimeter. The latter contribute approximately equally (~ 12% at E = 1 GeV), and combined quadratically with the sampling fluctuations (~ 7%) they account for the overall resolution of ~ 18%/√E [14].

Table 2: Measured and Estimated Performance of Electromagnetic Sampling Calorimeters

Device passive/active (mm)	Al/scint. 89/30	Fe/LAr 1.5/2.0	Cu/scint. 5/2.5	W/Si detector 7.0/0.2	$Pb/Ar/CO_2$ at NTP 2.0/10.0	U/scint. 1.6/2.5
Energy resolution measured at 1 GeV(%)	20	7.5	13.0	25.0	$\leqslant$ 20.0	11.0
η (MeV)	3.0	0.7 (?)	0.7 (?)	0.7 (?)	$\leqslant$ 0.6 (?)	0.7 (?)
$F(\xi)^{-1/2}$	1.16	1.10	1.10	1.18	1.18	1.20
$\langle\cos\theta\rangle^{-1/2}$	1.00	1.03	1.03	1.27	1.36	1.51
σ^{sample}	23	4.8	9.2	19.1	8.2	10.6
σ_{Landau}	3.8	1.0	1.0	4.5	8.70	1
$\sigma_{\text{path length}}$		5.7	6	17.5	13.0	6 (?)
$\sigma_{\text{estimated}}$	23	7.5	10.0	25.9	17.7	12.2
Note	a	b, c	c, d, e	c, f	c, g	c, e, h

a) A.N. Diddens et al., Nucl. Instrum. Methods 178:27 (1980).
b) C.W. Fabjan et al., Nucl. Instrum. Methods 141: 61 (1977).
c) Path-length fluctuations estimated from H.G. Fischer, Nucl. Instrum. Methods 156:81 (1978).
d) O. Botner, Phys. Scripta 23:555 (1981).
e) Difference consistent with photon statistics.
f) G. Barbiellini et al., Nucl. Instrum. Methods 235:55 (1983).
g) J.A. Appel, Fermilab FN-380 (1982).
h) R. Carosi et al., Nucl. Instrum. Methods 219:311 (1984).

2.4 'Transition' Effects

The concept of total track length T has been repeatedly used to estimate properties of the EMC. In particular, it suggests equating the measurable signal in a specific calorimeter with the energy deposit of penetrating particles such as muons of equivalent path length, or in terms of the number of equivalent particles n_{ep}. One equivalent particle (1 ep) is defined as the most probable detected energy, $(dE/dx)^{mp}$ of a penetrating muon

$$1\ \text{ep} = (dE/dx)^{mp}_{\text{visible},\ \mu}\,.$$

The expected number n_{ep} for electrons with kinetic energy E^{e}_{kin} would be

$$n^{el}_{ep}(\text{expected}) = E^{el}_{kin}\,/1\ \text{ep}\,.$$

Experimentally, yet, one always observes:

$$n_{ep}^{el}(\text{visible}) < n_{ep}^{el}(\text{expected}) \quad \text{or} \quad `e/\mu < 1'.$$

A summary of some representative measurements is given in Table 3.

The calibration of an EMC with muons is one way of establishing an absolute energy scale. If carried out in a reproducible and consistent way, it would allow us to compare, on an absolute energy scale, the electron response of different calorimeters—a crucial ingredient also in the understanding of hadronic calorimeters (see Section 3). As an example, the energy scale for muons quoted in Table 3 is based on the most probable

Table 3: Average Calorimeter Response for Pions and Muons*)
Relative to Electrons [38]

Type of particle (energy)	Sampling calorimeter structure scintillator / liquid argon		
	with Fe (Cu)	with Pb	with ^{238}U
Electrons (10 GeV/c) scintillator	1	1	1
Electrons (10 GeV/c) liquid argon	1	1	1
Pions (10 GeV/c) scintillator	0.63 ± 0.03[a,b)]	0.68 ± 0.04[b)]	0.93 ± 0.03[b,d)]
Pions (10 GeV/c) liquid argon	0.7[c)]	not yet measured	1.0 ± 0.05[c)]
Muons (~ 10 GeV/c) scintillator	1.15[b)]	1.26[b)]	1.29[b,d)]
Muons (~ 10 GeV/c) liquid argon	1.1	1.4[e,f)]	1.65[g)]

*) See text for definition of muon response.
NB: Errors of typically 10% have to be assumed for figures quoted without error.
a) A. Beer et al., Nucl. Instrum. Methods 224:360 (1984).
b) O. Botner, Phys. Scripta 23:555 (1981).
c) C.W. Fabjan et al., Nucl. Instrum. Methods 141:61 (1977).
d) T. Akesson et al., Properties of a fine-sampling uranium-copper scintillator hadron calorimeter, submitted to Nucl. Instrum. Methods (1985).
e) J. Cobb et al., Nucl. Instrum. Methods 158:93 (1979).
A. Lankford, CERN–EP Internal Report 78–3 (1978).
C. Kourkoumelis, CERN Report 77–06 (1977).
f) P. Steffen (NA31 Collaboration, CERN), private communication.
g) C.W. Fabjan and W. Willis, unpublished note on measurements reported in c).

energy loss evaluated for the total thickness of the device, applying the energy loss formula [15] for the appropriate muon momentum, including the non-negligible relativistic rise. The energy loss in the active medium was estimated to follow the ratio of the respective mass of the passive and active materials. Table 3 shows that the discrepancies from the 'naïve' expectations are substantial, with some indication that the response depends on the sampling thickness and the atomic number of the active and passive materials.

These discrepancies have been repeatedly attributed to 'transition effects' at the boundary between the different layers [16–18], often characterized by very different critical energies and hence different collision losses per radiation length. One expects that at a boundary from high Z to low Z (e.g. Fe to scintillator), the increased collision losses in the low-Z substance will reduce the electron flux, in agreement with measurements [17, 18] and some recent Monte Carlo calculations [19]. Apart from local disturbances of the EMC, multiple scattering tends to increase the effective path length in the high-Z absorber relative to the low-Z readout and this mechanism may also suppress the electron response relative to muons, for increasing Z. Furthermore, a considerable fraction of the energy is deposited by the last generation of the cascade, consisting of low-energy particles, and saturation in the response of readout substances (which occur in scintillator or liquid argon) will further suppress the measured response relative to muons.

It may be concluded that for a more refined understanding of sampling detectors it will be important to calibrate carefully the electron response on an absolute energy scale with a reproducible standard, as provided e.g. by muons.

2.5 Spatial Resolution

In subsection 2.1 we described in general the physical processes contributing to the shower propagation and its characteristic dimension. Typical angles for bremsstrahlung emission and multiple scattering depend on the energy of the shower particles and hence alter the transverse shower profile as a function of longitudinal depth inside the shower. Before the shower maximum, typically more than 90% of the energy is contained in a cylinder of radius $r \simeq 0.5X_0$, whereas the radius for 90% containment of the total energy is $r \simeq 2\varrho_M$. For the localization of the impact point of a photon it is therefore advantageous to probe the shower in the early part before the shower maximum. In principle, given sufficiently fine-grained instrumental resolution, the localization σ_x of the centre of gravity of the transverse distribution is determined by signal/noise considerations and, therefore, should improve with increasing particle energy E as $\sigma_x \sim E^{-1/2}$, which is confirmed experimentally [20], reaching sub-millimetre accuracy for 100 GeV showers [21]. If position resolution is the principal criterion, one may achieve very high spatial subdivision using multiwire proportional chamber (MWPC) techniques [22] allowing localization at the millimetre level.

Somewhat different criteria apply if both good position and energy resolution are required, e.g. for the determination of the invariant mass of particles such as π^0's, $J/\psi \to e^+e^-$, etc. In this case the centres of gravity of the complete showers have to be determined—frequently with the constraint of minimizing the sharing of energy between the neighbouring showers—; even then, excellent spatial resolutions of the order of 1 mm have been reported [20], e.g. in an array of lead-glass blocks of 35 × 35 mm cross-section.

Given simultaneous information on the transverse and longitudinal shower development, the *direction* of a shower and hence the angle of incidence of the particle may be reconstructed. As an example, for the CHARM neutrino calorimeter [23] an angular resolution of $\sigma(\theta_e)$ (mrad) $= 20/E^{1/2} + 560/E$ (E in GeV) was measured; for a FNAL neutrino calorimeter [24] the following result is quoted:

$$\sigma(\theta_e)\,(\text{mrad}) = 3.5 + 53/E\,(\text{GeV}).$$

3. HADRONIC SHOWER DETECTORS

3.1 General properties

Conceptually, the energy measurement of hadronic showers is analogous to that of EMCs, but the much greater variety and complexity of the hadronic processes propagating the hadronic cascade (HC) complicate the detailed understanding. No simple analytical description of hadronic showers exists, but the elementary processes are well studied.

Typical of hadronic interactions is multiple particle production with transverse momentum $\langle p_T \rangle \simeq 0.35$ GeV/c, for which about half of the available energy is consumed (the inelasticity $K \simeq 0.5$). The remainder of the energy is carried by fast, forward-going (leading) particles. The secondaries are mostly pions and nucleons, and their multiplicity is only weakly energy-dependent. The characteristic stages in the HC development are summarized in Table 4. Two specific features have been identified as the principal physics limitations to the energy resolution of hadronic calorimeters:

i) A considerable part of the secondaries are π^0's, which will propagate electromagnetically without any further nuclear interactions; the average fraction converted into $f_{\pi^0} \simeq 0.1 \ln E$ (GeV), for energies E in the range of a few to several hundred GeV. The size of the π^0 component is largely determined by the production in the first interaction, and event-by-event fluctuations about the average value are, therefore, important.
ii) A sizeable amount of the available energy is converted into excitation or break-up of the nuclei, of which only a fraction will result in detectable ('visible') energy.

The two processes, intimately correlated, may lead, for a given entering hadron, to a very different shower composition, which has a very different detectable response. Together they impose the intrinsic limitation on the performance of hadronic calorimeters.

Table 4 gives some indications of the relative importance of these competing processes. Considerable insight has been gained from very detailed Monte Carlo calculations, which in their most ambitious form aim to simulate the full nuclear and particle physics aspects of the hadronic cascade based on the measured cross-sections of the elementary processes (see also Section 6) [25]. Examples showing the energy dependence of the principal effects are given in Fig. 8. It should be noted that these various processes contribute in varying degrees to the visible energy of the HC, and that a considerable fraction—such as nuclear binding energy, muons, and neutrinos—will be lost in the form of 'invisible' or undetectable energy.

Table 4: Characteristic Properties of the Hadronic Cascade

Reaction	Properties	Influence on energy resolution	Characteristic time (s)	Characteristic length (g/cm^2)
Hadron production	Multiplicity $\simeq A^{0.1} \ln s$ Inelasticity $\simeq 1/2$	π^0/π^+ ratio Binding energy loss.	10^{-22}	Abs. length $\lambda \simeq 35A^{1/3}$
Nuclear de-excitation	Evaporation energy $\simeq$ 10% Binding energy $\simeq$ 10% Fast neutrons $\simeq$ 40% Fast protons $\simeq$ 40%	Binding energy loss. Poor or different response to n, charged particles, and γ's.	10^{-18}–10^{-13}	Fast neutrons $\lambda_n \simeq 100$ Fast protons $\lambda_p \simeq 20$
Pion and muon decays	Fractional energy of μ's and ν's $\simeq$ 5%	Loss of ν's	10^{-8}–10^{-6}	$\gg \lambda$
Decay of c, b particles produced in multi-TeV cascades	Fractional energy of μ's and ν's at percent level	Loss of ν's. Tails in resolution function.	10^{-12}–10^{-10}	$\ll \lambda$

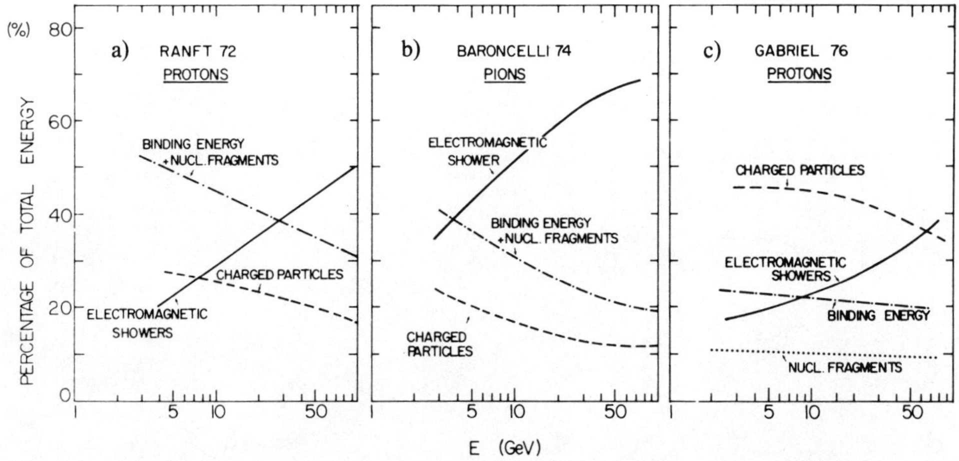

Fig. 8: Relative contributions of the most important processes to the energy dissipated by hadronic showers, as evaluated from three representative Monte Carlo calculations [25].

Average longitudinal and transverse distributions (Fig. 9) are useful estimates of the characteristic dimensions for near-complete shower containment. The average longitudinal distribution exhibits 'scaling in units of absorption length λ'. The transverse distributions depend—as in the case of EMCs—on the longitudinal depth: the core of the shower is rather narrow (FWHM from 0.1 to 0.5λ), increasing with shower depth. The highly energetic, very collimated core is surrounded by lower-energy particles, which extend a considerable distance away from the shower axis, such that for 95% containment a cylinder of radius R ~ 1λ is required.

Experimental data are consistent with the following parametrization:

a) the shower maximum, measured from the face of the calorimeter, is given by

$$t_{max}\,(\lambda) \sim 0.2 \ln E\,(\mathrm{GeV}) + 0.7;$$

it occurs at a smaller depth in high-Z materials due to the smaller ratio of X_0/λ.

b) The longitudinal dimension required for almost full containment is approximated by

$$L_{0.95}\,(\lambda) \simeq t_{max} + 2.5\,\lambda_{att},$$

again measured from the face of the calorimeter. The quantity λ_{att} describes the exponential decay of the shower beyond t_{max} and increases with energy approximately as $\lambda_{att} \simeq \lambda[E\,(\mathrm{GeV})]^{0.13}$, with an indication of a weaker energy dependence for high-Z absorbers. The expression for $L_{0.95}$ describes available data in the energy range of a few GeV to a few hundred GeV to within 10%.

c) The transverse radius R of the 95%-containment cylinder is very approximately $R_{0.95} \lesssim 1\lambda$; it does not scale with λ and is smaller in high-Z substances.

d) A useful parametrization of the longitudinal shower development is

$$dE/ds = K[wt^a\,e^{-bt} + (1-w)\,\ell^c\,e^{-d\ell}],$$

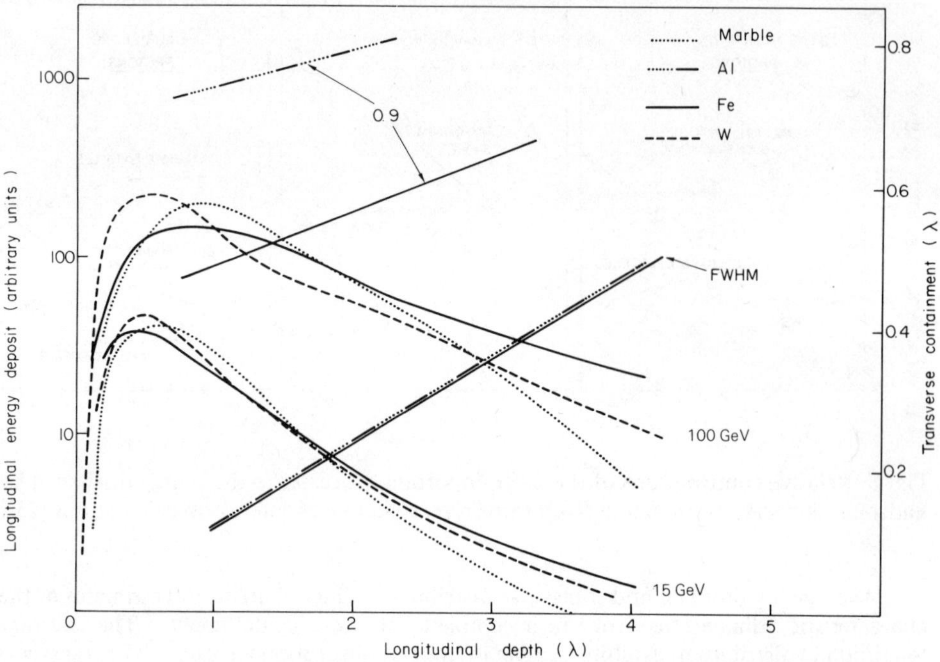

Fig. 9: Longitudinal shower development (left ordinate) induced by hadrons in different materials, showing approximate scaling in absorption length λ. The shower distributions are measured from the vertex of the shower and are therefore more peaked than those measured with respect to the face of the calorimeter. For the transverse distributions as a function of shower depth, scaling in λ is found for the narrow core (FWHM) of the showers. The radius of the cylinder for 90% lateral containment is much larger and does not scale in λ. [10 GeV/c π's: B. Friend et al., Nucl. Instrum. Methods 136:505 (1976)]. Note that marble and aluminium have almost identical absorption and radiation lengths [Marble: M. Jonker et al., Nucl. Instrum. Methods 200:183 (1982); Fe: M. Holder et al., Nucl. Instrum. Methods 151:69 (1978); W: D.L. Cheshire et al., Nucl. Instrum. Methods 141:219 (1977)].

where t is the depth, starting from the shower origin, in radiation lengths, and ℓ is the same depth in units of absorption lengths. The parameters a,b,c,d are fits to the data and are given a logarithmic energy dependence. Crude shower fluctuations may be simulated by i) randomly varying the depth of the shower origin; ii) smearing the incident particle energy to simulate the calorimeter energy resolution; iii) randomly varying the length of the shower by scaling the values of t and l [26].

Although the total depth needed for near-complete absorption increases only logarithmically with energy, it does require, for example, about 8λ to contain, on the average, more than 95% of a 350 GeV pion.

3.2 Intrinsic Energy Resolution

In the previous subsection we indicated that the fluctuations in the HC development, producing a range of different particles—from π^0's to slow neutrons, muons, and neutrinos—with vastly different detection characteristics, are the principal limitations to the energy resolution. These fluctuations have been found to be large—of the order of 50% at 1 GeV—in strong contrast with the measurement of e.m. calorimeters, where the *intrinsic* fluctuations of the visible track length are less than 1% at 1 GeV. This understanding of hadronic cascades emerged from studies in which the various possible contributions could be individually identified and measured [27]. The dominant influence of the nuclear processes manifests itself also in the shape of the response function (Fig. 10) and is corroborated by detailed Monte Carlo estimates. Available experimental evidence indicates that the intrinsic hadronic energy resolution is

$$\sigma(E)/E|_{\text{intrinsic}} \simeq 0.45/\sqrt{E}\ (\text{GeV})\ .$$

This relation describes devices made from materials covering almost the complete periodic table from aluminium [23] to lead [28]. Only in hydrogen are these nuclear effects absent, but they are already sizeable in hydrogen-rich absorbers (e.g. scintillators): one measurement in an homogeneous liquid-scintillator calorimeter is reported [29] for which the quoted energy resolution is also consistent with the above-quoted value. The sole known exception from this rule is given by uranium-238, for reasons that are explained in the next subsection.

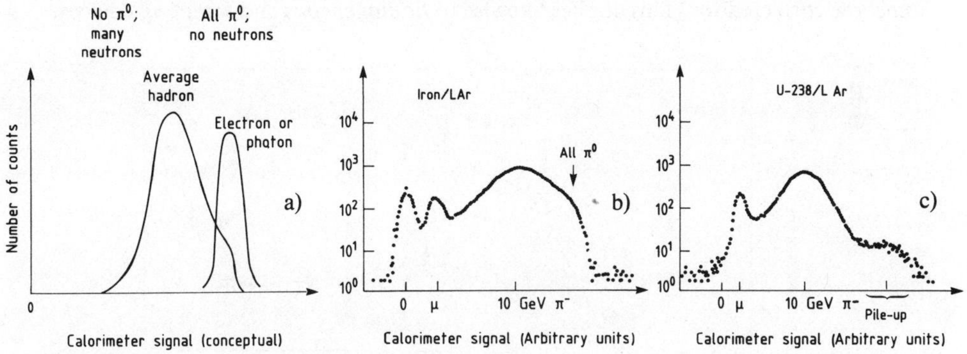

Fig. 10: Calorimeter response for 10 GeV/c pions.
a) Conceptually, the fluctuations are dominated by the nature of the first inelastic interaction. On the average, a certain number of π^0's, charged pions, nuclear fragments, and slow protons and neutrons will be produced. In one extreme case, no π^0's will be produced—only charged pions, neutrons, etc. (low-energy side of response function). At the other extreme the reaction products are mostly π^0's, and the energy deposit will be very similar to that of electrons or photons of equivalent energy.
b) The measured response function is shown for a calorimeter using iron as absorber. The logarithmic ordinate exposes a break in the resolution function corresponding to the case of mostly electromagnetic propagation.
c) The response for a U-238 sampling calorimeter is shown. The nuclear losses are effectively compensated, leading to a response that is nearly equal for charged and neutral pions. The essentially Gaussian response function is also an indication of this uniform response. (The muon peak sets the energy scale corresponding to 'one equivalent particle').

The level of these nuclear effects and, more generally, the level of 'invisible' energy is sensitively measured by comparing the response of a calorimeter for electrons and hadrons at the same 'available' energy, which is the kinetic energy of electrons and nucleons, the total energy for mesons, and the total energy plus the rest mass for antinucleons. A summary of some representative data is shown in Fig. 11. Two features deserve a comment: all calorimeters except those made from uranium show a visible energy of approximately 70% relative to electrons, which slowly increases owing to the rise in the electromagnetic component at higher energies. On the other hand, with energies decreasing below $E \sim 1.5$ GeV, the nature of the hadronic cascade changes: to a larger measure, the energy is degraded by ionization alone, with the hadron response approaching that of muons and being above that of electrons (see subsection 2.4). In this low-energy limit all calorimeters, including those using uranium as a degrader, are expected to give similar responses. This interpretation is confirmed by the fact that in this low-energy regime the relative resolution improves, $\sigma/E < 0.45/\sqrt{E}$ [30–32], as well as by quantitative Monte Carlo estimates [32].

In summary, the response of hadrons relative to electrons is a sensitive probe of the level of nuclear effects. Typical values are $e/h \simeq 1.4$ for most materials. Strongly correlated with this average suppression are fluctuations in the response; these are due to large fluctuations of the electromagnetic component. The intrinsic resolution of hadron calorimeters is, for these reasons, limited to $\sigma(E)/E_{\text{intrinsic}} \simeq 0.45/\sqrt{E}$ (GeV), unless event-to-event fluctuations in the electromagnetic component of hadron cascades are somehow corrected for. This applies likewise to *homogeneous* and *sampling* devices.

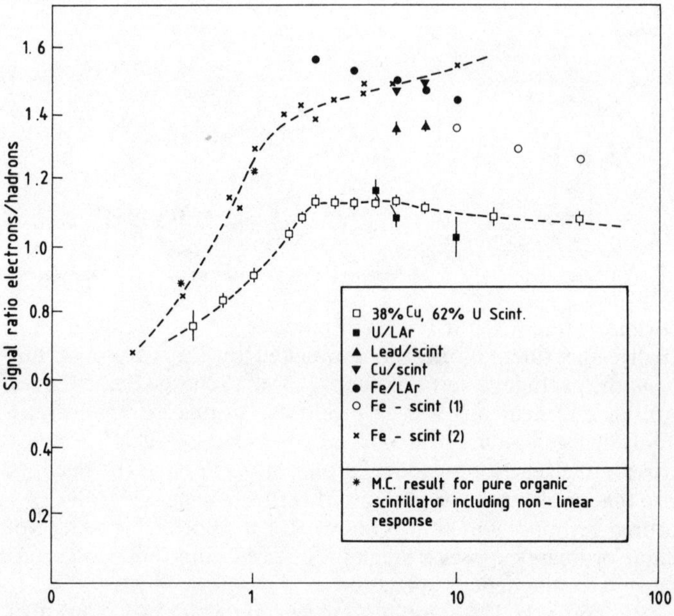

Fig. 11: The ratio of electromagnetic to hadronic energy response as a function of energy for different calorimeter systems: 38% Cu, 62% U/scint. [32]; U/LAr [37]; Cu/scint. [88]; Fe/LAr [37]; Fe/scint. (1) [35]; Fe/scint. (2) [31].

3.3 Compensating Fluctuations

We have emphasized that the relative response between electrons and pions is a sensitive measure of the level of nuclear interactions. An improvement in the energy resolution would be expected if the response of the electromagnetic cascade were identical compared with the purely hadronic one, i.e. if devices with an e/h ratio equal to one were available. Alternatively, given sufficiently detailed information on the individual hadronically induced shower, one would be able to assess the relative components and apply suitable corrections to improve the energy resolution. Both approaches have been explored and are described here.

Several suggestions have been made for monitoring the level of the electromagnetic component event-by-event. One suggestion was to use, as an indicator, the Cherenkov light from relativistic particles dominantly produced by e^+e^- pairs [33], but Monte Carlo estimates [33] for practical devices suggest that it is difficult to obtain a very useful correlation and to improve the resolution significantly. Another suggestion was to monitor the level of the nuclear component by associating heavily ionizing particles with the 'late' component of the hadronic shower [34]. If a calorimeter is instrumented so as to provide detailed longitudinal information, then some useful compensation on a shower-by-shower basis is possible for the electromagnetic/hadronic fluctuations. The most successful attempt was made by the CERN–Dortmund–Heidelberg–Saclay (CDHS) collaboration using the longitudinal information from their relatively fine-grained neutrino calorimeter [35, 36]. This was done by a weighting algorithm applied to the individual longitudinal measurements relative to the total energy measured. In Fig. 12 the unweighted and weighted results for the energy resolution are presented. Firstly, it can be seen that the raw results show a marked deviation from the expected $E^{-1/2}$ dependence.

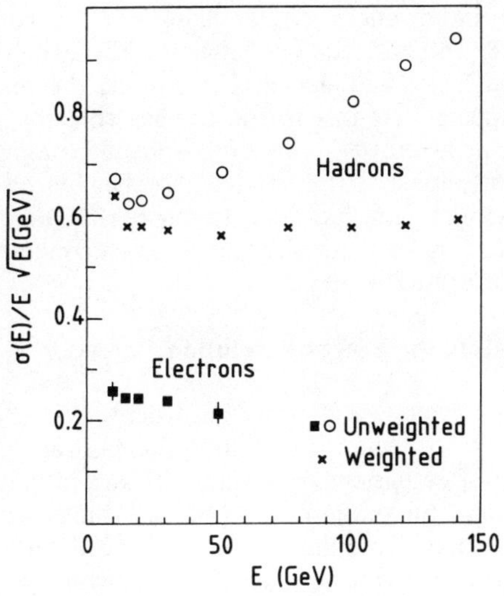

Fig. 12: Energy resolution measured in the CDHS neutrino calorimeter (2.5 cm Fe/scint.) versus the energy of the incident particle. Note that the uncorrected energy resolution for hadrons does not improve as $1/\sqrt{E}$. With a weighting procedure to reduce the large fluctuation due to the e.m. component the resolution is improved, and is consistent with the $1/\sqrt{E}$ scaling up to the highest energies measured [35].

This cannot be ascribed to instrumental effects given the reported result for the electron resolution, which is well described by an $E^{-1/2}$ law. Secondly, the weighting algorithm improves the resolution, particularly at the highest energies, to a level that would be expected from extrapolating the low-energy resolution according to an $E^{-1/2}$ law. In subsection 3.5 we comment further about this deviation from the naïve $1/\sqrt{E}$ behaviour of hadronic calorimeters at high energies.

The more direct cure for these fluctuations would be to equalize the response for electrons and hadrons. In principle, equalizations of these differences, which are at the 30% to 40% level, may be accomplished in two ways: either by decreasing the electron response—typically 20% to 40% lower relative to a minimum-ionizing particle calibration (subsection 2.2); or by boosting the hadronic signal. This latter aspect is being exploited by using uranium-238 as the energy degrader [37]. In that material (and probably also to a lesser extent in thorium) some of the normally invisible energy expended in the nuclear break-up leads to neutron-induced fission, which in turn produces detectable energy in the calorimeter. It can be estimated that on the average 40 fissions are induced per GeV of energy deposited, which altogether liberate about 10 GeV of fission energy. Only a very small fraction (300 to 400 MeV) needs to be detected in order to compensate the nuclear deficit; this could be done either by the few-MeV γ-component or through the fission neutrons liberated in the fission process. Which component and what fraction of the fission contributions are measured depends on the nature of the active sampler. One may achieve essentially complete compensation not only on average but also event-by-event, because the intrinsic resolution is measured to be [32, 37]

$$\sigma(E)^{\text{uranium}}_{\text{intrinsic}} \simeq 0.22/\sqrt{E}\ (\text{GeV}) .$$

The fundamental importance of equalizing the hadronic and electromagnetic response should again be emphasized. The latter sensitively depends on the details of the low-energy part of the EMC and hence critically on the material and the sampling frequency. It would appear that this is one further contribution to the tuning of the e/h ratio. Hence for hadron showers, the level of visible compensation is expected to be affected not only by the choice of the passive absorber but also by the response of the active readout to densely ionizing particles (from the HC) and to the electromagnetic component. We do not yet have a complete set of measurements, but Table 3 attempts to organize the available information [38].

3.4 Instrumental Effects to the Energy Resolution

Most hadronic calorimeters are 'sampling' detectors, using preferentially rather dense passive absorbers to reduce the linear dimensions of the instrument. As a consequence, sampling fluctuations of statistical origin analogous to the case of electromagnetic sampling fluctuations (section 2.3) may contribute to the energy resolution, although, for the sampling of the HC we do not have a similar detailed description. Available measurements (see Fig. 13 with the quoted references) are consistent with a parametrization of the form

$$\sigma(E)/E/\text{hadron-sampling} \simeq 0.09\ [\Delta E\ (\text{MeV})/E\ (\text{GeV})]^{1/2} .$$

The quantity ΔE expresses the energy loss per unit sampling cell for minimum ionizing particles. Hadronic sampling fluctuations are approximately twice as large as the

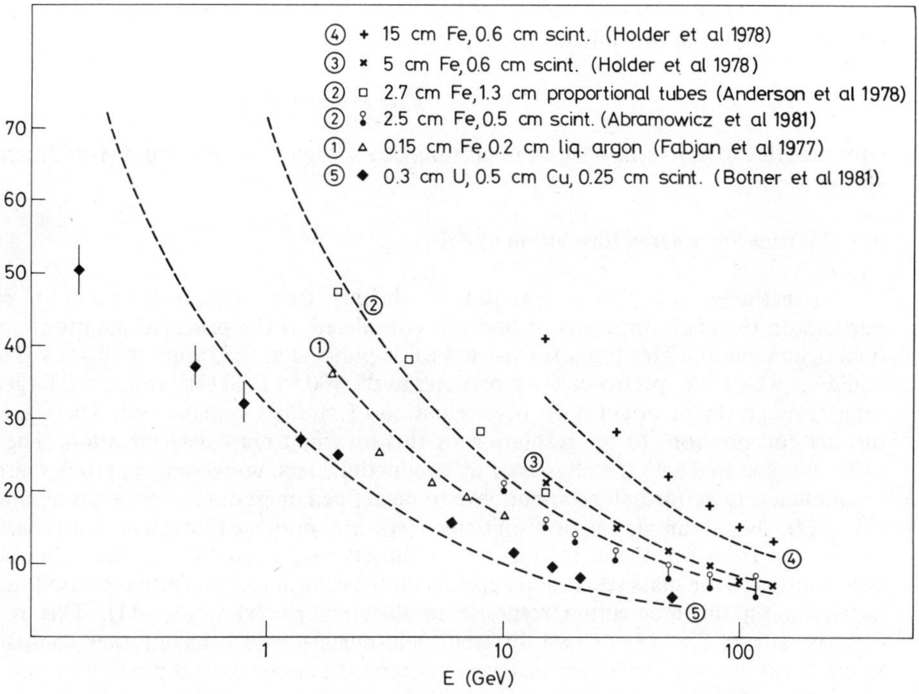

Fig. 13: Energy resolution for hadrons measured with iron and uranium sampling calorimeters. Curves 1–4 are calculated with the values for intrinsic and sampling fluctuations as given in Table 5. For the data on curve 2 [35], the open circles are the raw data; the solid circles are the results of the off-line analysis, using the longitudinal shower information to correct for fluctuations in the electromagnetic/hadronic energy ratio. For curve 5 the intrinsic fluctuation is assumed to be $0.2/\sqrt{E}$, and does not take account of the 35% (in units of λ) admixture of Cu. Below 1 GeV the resolution improves over the expected value and indicates the influence of mechanisms such as ranging and reduced nuclear effects [32, 88]. The data labelled 3 and 4 are by M. Holder et al., Nucl. Instrum. Methods 151:69 (1978), the open squares refer to R.L. Anderson et al., IEEE Trans. Nucl. Sci. NS–25:1–340 (1978).

electromagnetic sampling fluctuations for the same detector; unlike the e.m. case however, where sampling is the predominant contribution to the resolution, sampling in hadronic detectors can be made small relative to the large intrinsic component, and energy resolution need not be sacrificed in hadronic sampling calorimeters.

Energy leakage due to partial shower containment will not only degrade the energy resolution, but will also give rise to very asymmetric resolution functions with low-energy tails. Calorimetric experiments, which emphasize measurements such as neutrino detection based on missing energy or hadronic high-p_T jet production have therefore particularly stringent requirements to achieve very close to 100% containment. Again, as already noted for the measurement of e.m. calorimeters, longitudinal fluctuations are larger than transverse fluctuations and hence longitudinal leakage is more critical to the performance. For values of fractional leakage $f \leq 0.3$ the degradation of the energy resolution follows approximately the expression

$$\sigma(E)/E \simeq [\sigma(E)/E]_{f=0} \times (1+4f) ,$$

with the effect being somewhat more pronounced at higher energies for a given fractional energy leakage.

3.5 Calorimetric Energy Resolution of Jets

Increasingly, the physics emphasis is shifting from the measurement of single particles to the analysis of jets of hadrons considered as the principal manifestation of quarks and gluons. This trend is expected to be pursued at the future multi-TeV hadron colliders, where the spectroscopy of particles in the 100 to 1000 GeV range will largely be done through the invariant mass determination of multijet systems [39]. There are two distinct contributions to the resolution of this invariant mass determination. The first effect is associated with the physics of jet production. Jets, unlike single particles, are not unambiguously defined objects, but have to be defined operationally by a 'jet algorithm' (Fig. 14). For example, hadron-initiated jets are produced together with particles originating from peripheral interactions; multijets may partially coalesce. The second contribution to the mass resolution depends on the calorimeter performance itself, and in particular on the momentum response to different particles (Fig. 11). This is seen conceptually in Fig. 15 for two different calorimeters, which have rather comparable nominal resolutions but a very different response to electrons and pions. For very low

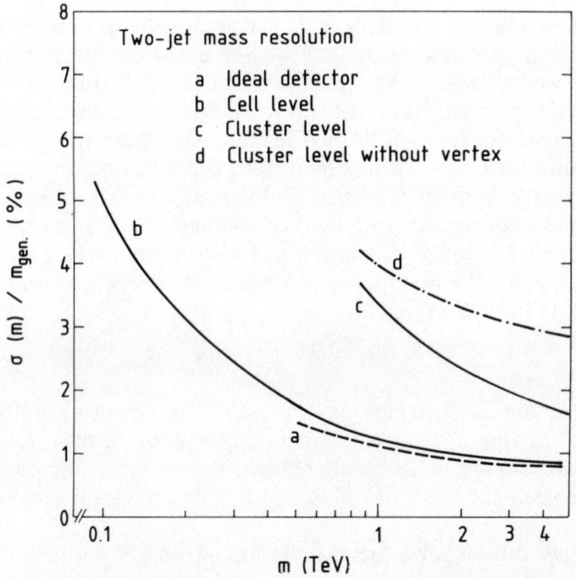

Fig. 14: Two-jet mass resolution as a function of the two-jet mass and for various assumptions on the detector performance. The ideal detector measures the jet mass at the individual particle level; at the cell level the energy information from each cell ($\Delta\phi \times \Delta\eta = 5° \times 0.05$) is considered. At the 'cluster level' certain pattern recognition criteria are introduced. Precise knowledge of the event vertex is important [39].

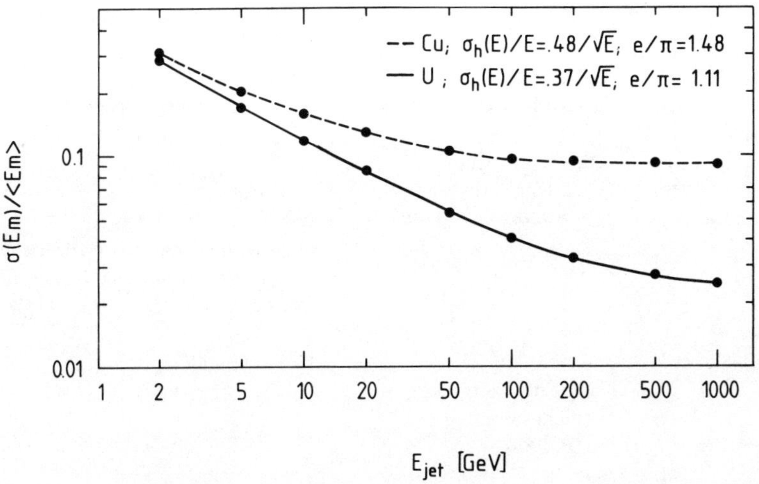

Fig. 15: Jet resolution for an 'infinitely' thick, 4π calorimeter, assuming a Feyman–Field-like fragmentation function. The advantage of a (nearly) compensated calorimeter is particularly evident at very large jet energies. For calorimeters with e/h very different from one, the resolution ceases to improve as $E^{-1/2}$ in the high-energy limit [39].

energies of the jets ($E_{jet} \leq 5$ GeV), the jet resolution is dominated by the very non-linear response to low-momentum particles, and is similar for both calorimeters. At very high energies, the performance is dominated by the relative electron/hadron response. In particular, for the e/h = 1.48 calorimeter it is estimated that the energy resolution levels at approximately 10%. Qualitatively, such a strong influence is expected, because a large fraction of the jet energy is carried by only a few high-momentum particles; fluctuations in the charged hadron/π^0 ratio of these leading particles at sufficiently high energies will dominate over the simple statistical $E^{-1/2}$ improvement.

A similar argument should also be valid for single, very energetic hadrons, which after the first inelastic interaction in a calorimeter will be similar to a jet of particles of comparable energy. Therefore, hadron calorimeters with e/h $\neq$ 1 are expected to show an intrinsic energy resolution $\sigma(E)/E \sim c \times E^{-\alpha}$, with $\alpha < 1/2$ for large energies ($E > 50$ GeV). This behaviour is consistent with the careful analysis (Refs. [35, 36], and Fig. 12) discussed in the previous subsection. Table 5 summarizes the contributions to the energy resolution for both electromagnetic and hadronic calorimeters.

3.6 Spatial Resolution for Hadronic Showers

This discussion follows closely the related comments on e.m. calorimeters (subsection 2.5). Hadron showers are found to consist of a narrow core surrounded by a 'halo' of particles extending to several times the dimensions of the core. Consequently, somewhat different criteria apply to the measurement of the position and to considerations of shower separation. Measurements on the spatial resolution of the impact point [23, 32] may be parametrized approximately in the form

$$\sigma(\text{vertex})\ (\text{cm}) \simeq \langle\lambda\rangle/[4\sqrt{E}\ (\text{GeV})]\,.$$

Table 5: Principal Contributions to Energy Resolution in Electromagnetic and Hadronic Calorimeters

Mechanisms (add in quadrature)	Electromagnetic showers	Hadronic showers
Intrinsic shower fluctuations	Track-length fluctuations: $\sigma/E \gtrsim 0.005/\sqrt{E}$ (GeV).	Fluctuations in the energy loss: $\sigma/E \simeq 0.45/\sqrt{E}$ (GeV). Scaling weaker than $1/\sqrt{E}$ for high energies. With compensation for nuclear effects: $\sigma/E \simeq 0.22/\sqrt{E}$ (GeV).
Sampling fluctuations	$\sigma/E \simeq 0.04\sqrt{\Delta E/E}$. Nature of readout may augment sampling fluctuations.	$\sigma/E \simeq 0.09\sqrt{\Delta E/E}$
Instrumental effects	Noise and pedestal width: $\sigma/E \sim 1/E$ – determine minimum detectable signal; – limit low-energy performance. Calibration errors and non-uniformities: $\sigma/E \sim$ constant and therefore limits high-energy performance.	
Incomplete containment of shower	$\sigma/E \sim E^{-\alpha}$, $\alpha < 1/2$ (see subsec. 2.2, resp. 3.4). For leakage fraction $\geq$ few %: non-linear response and non-Gaussian 'tail'.	

a) $\Delta E \simeq$ energy loss of a minimum ionizing particle in one sampling layer, measured in MeV; E = total energy, measured in GeV.

In compact calorimeters, where the average interaction length may be as low as $\langle\lambda\rangle \leq 20$ cm, spatial resolutions in the range of a few centimetres at 1 GeV are achievable. The influence of the transverse segmentation has also been studied [40] and the following dependence can be derived:

$$\sigma(\text{vertex}) \simeq \sigma_0\,(\text{vertex}) \exp(2d),$$

where the segmentation d is expressed in units of absorption length and σ_0 refers to the intrinsic vertex resolution in the absence of instrumental effects due to finite segmentation. This expression suggests that the improvement becomes rather modest if the lateral segmentation is increased beyond $d(\lambda) \leq 0.1$, even disregarding other aspects such as photon statistics or noise.

Finally, the angular resolution of hadron showers has been carefully studied for several calorimeters used to investigate neutrino scattering. The limitations again stem from fluctuations in the $\pi^{\pm}/\pi^0$ composition of the HC, because of their (usually) very different spatial shower developments. These effects were purposely minimized in the case of the CERN–Hamburg–Amsterdam–Rome–Moscow (CHARM) neutrino calorimeter with the choice of marble as the passive absorber material in which EMCs and HCs have approximately the same dimensions [$3X_0$ (cm) $\sim \lambda$ (cm)]. An angular resolution of

$$\sigma(\theta)_{\text{hadron}}\ (\text{mrad}) \simeq 160/\sqrt{E}\ (\text{GeV}) + 560/E\ (\text{GeV})$$

is reported [23], and a similar result was obtained with a detector constructed for the same purpose at FNAL [24].

4. PARTICLE IDENTIFICATION

With hadronic calorimeters it is possible to identify a class of particles which are not always easily identified by other methods, and which may be particularly interesting for very topical physics studies, as summarized in Table 6.

In the following, we discuss in some detail the identification of electrons, muons, and neutrinos.

Table 6: Particle Identification with Calorimeters

Particle produced	Calorimeter technique	Comment
Electron, e	Charged particle initiating the electromagnetic shower	Background from charge exchange $\pi^{\pm}N \rightarrow \pi^0 + X$ in calorimeter; π discrimination of ~ 10–1000 possible
Photon, γ	Neutral particle initiating the electromagnetic shower	Background from photons from meson decays
$\pi^0, \eta, \ldots \rightarrow \gamma\gamma$ $\varrho, \phi, J/\psi, \Upsilon, \ldots \rightarrow e^+e^-$	Invariant mass obtained from measurement of energy and angle	Classical application for electromagnetic calorimeters;
Protons, deuterons, tritons, ... and their antiparticles	Comparison of visible energy E_{vis} in calorimeter with momentum of particle	$E_{vis}^{b(\bar{b})} = (\vec{p}_b^{\,2} + m_b^2)^{1/2} - (+)\, m_b$ Protons (antiprotons) identified up to 4 (5) GeV/c; deuterons (antideuterons) correspondingly higher
(Anti)neutrino	Visible energy E_{vis} in calorimeter compared with missing momentum	Important tool for $e^+e^- \rightarrow \nu(\bar{\nu}) + X$ and at CERN Collider (FNAL $p\bar{p}$ collider, $pp(p\bar{p}) \rightarrow \nu(\bar{\nu}) + X$
Muon	Particle interacting only electromagnetically (range). E_{vis} compared to $\vec{p}$.	Background from non-interacting pions
Neutron or $K_L^0(\bar{n}, \bar{K}_L^0)$	Neutral particle initiating hadronic shower	Some discrimination perhaps possible based on detailed (longitudinal) shower information

4.1 Discrimination between Electrons (Photons) and Hadrons

The discrimination is based on the difference in the shower profiles, accentuated in materials with very different radiation and absorption lengths. One finds approximately

$$\lambda\ (\text{g/cm}^2)/X_0\ (\text{g/cm}^2) \sim 35A^{1/3}\, Z^2/180A \sim 0.12Z^{4/3},$$

which explains that heavy materials (lead, tungsten, or uranium) are best suited for electron–hadron discrimination.

The principal physics limitation is imposed by the charge exchange reaction $\pi^- p \rightarrow \pi^0 n$ (or $\pi^+ n \rightarrow \pi^0 p$), which may, under unfavourable circumstances, simulate an electromagnetic shower, closely matching the energy of the incident pion. For pion energies in the few-GeV range, the cross-section for this process is at the one percent level of the total inelastic cross-section, and decreases logarithmically with energy [41]. Typical values for pion discrimination are of the order of 1 in 10^2 in the 1 to 10 GeV region and of 1 in 10^3 or more for particles energies beyond 100 GeV [42]. Considerably better performance (close to 10^3 pion rejection for few-GeV particles) is reported for instruments with very fine longitudinal subdivision [43], which helps to recognize the hadronic origin of a charge-exchange-dominated cascade. Only relatively small further improvements (a factor of 3 to 5) are obtained if transverse shower profile information is available; this is because of the very high degree of correlation between transverse and longitudinal profile [44, 45]. The quoted values apply to electron–hadron discrimination based on shower shape analysis only. If, in addition, energy information can be used, for example knowledge of the electron and pion momenta from magnetic spectroscopy, a further improvement in the rejection of typically one order of magnitude is obtained.

4.2 Muon Identification

Several calorimetric methods exist for discriminating between muons and hadrons or electrons, all based on the very large differences of energy deposit.

i) Calorimeters with fine longitudinal subdivision: such calorimeters, typically many tens of absorption lengths long, have been used predominantly in experiments on incident neutrinos. Energetic muons are very clearly recognized as isolated, minimum-ionizing tracks, frequently ranging far beyond the tracks from hadronic showers.

ii) Muon penetration through active or passive absorbers: the absorbers or calorimeters are deep enough to contain the hadrons adequately and to reduce the 'punch through' probability P of pions ($P \sim e^{-d/\langle\lambda\rangle}$). The observed path length d is measured in units of 'detectable' absorption lengths $\langle\lambda\rangle$, which is found to agree closely with tabulated values [46]. The detailed rejection power against hadrons depends critically on the experimental precautions taken and may be improved by

a) reducing the background from pion and K decay before the calorimeter; the active 'beam dump' experiments have refined this method [47];

b) measuring the muon momentum after the calorimeter in, for example, magnetized iron [48] or a precision magnetic spectrometer [49]; momentum matching of muon candidates before and after the absorber may further improve the rejection [50];

c) correlating the direction of the particle before and behind the absorber. The applicability of the method is limited by multiple scattering of the muons in the absorber (Fig. 16), and accidental overlap with nearby tracks before the absorber.

Very good muon identification will be of increasing importance for experimentation at the storage rings under construction [the FNAL 2 TeV $p\bar{p}$ collider, the CERN Large Electron–Positron Storage Ring (LEP)], or *a fortiori* at those being discussed [the Superconducting Super Collider (SSC) in the USA, the Large Hadron Collider (LHC) in the LEP tunnel]. The very high particle density will make the identification of electrons inside jets extremely difficult, leaving possibly only the muon as a charged lepton signature. In addition, accurate momentum measurement of the muon will be inevitable

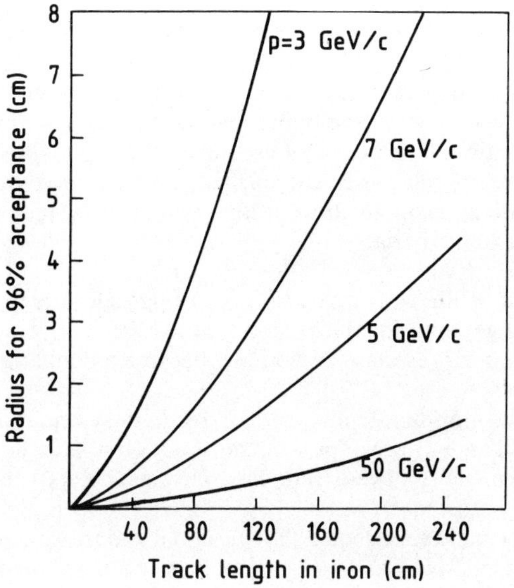

Fig. 16: Radius of 96% acceptance circle for multiply scattered muons as a function of muon track length in iron and of muon momentum. [H. Burmeister et al., CERN/TCL/Int. 74–7 (1974)].

for those experiments which will increasingly exploit very good total energy measurement, for which, of course, the muon momentum has to be included [50, 51].

4.3 Neutrino Identification

The recent discovery of W production based on 'missing momentum' analysis [52] has reminded us of the power of such information. Two related methods can be used:

i) total energy measurement can be accomplished provided 4π calorimetric coverage in the c.m. system is available for all particles (charged, neutrals, muons). This can be practically achieved at e^+e^- storage rings (although 4π hadron calorimetry is not the forte of the forthcoming LEP experiments) or in a fixed-target environment [50]. Neutrino production is implied whenever the measured energy is lower than the total available energy and incompatible with the resolution function of the detector. Total energy measurement does not work well at a hadron collider, such as the CERN $p\bar{p}$ Collider, because a significant fraction of the total energy is always produced at very small angles relative to the incident beams, making a calorimetric measurement impractical. Fortunately, help is provided by

ii) a missing transverse momentum measurement. In this method, clearly related to method (i), the production of a neutrino is signalled by $\Sigma p_{T,i} \neq 0$ to a degree which is incompatible with the detector resolution. Very good missing-momentum resolution at the level of $\sigma(p_{T,miss})/p_{total} \gtrsim 0.3/\sqrt{E}$ has been estimated [53]. This is to be contrasted with the actual performance of a much cruder device for which a $\sigma(p_{T,miss}) \sim 0.7\sqrt{p_T}$ (GeV) is quoted, which is still adequate for a range of striking experimental results [52].

5. SIGNAL READOUT TECHNIQUES FOR CALORIMETERS

During the last 10 years, considerable effort has been devoted to calorimeter instrumentation with a view to developing readout techniques which optimally match an experimental application. The principal goal is to develop methods which will minimize the instrumental effects, relative to the intrinsic performance, caused by the physics of the detectors. Modern beam facilities with their steadily increasing particle energies impose ever more taxing criteria:

- the response must be linear as a function of the particle energy, frequently over a very large dynamic range; only for the exceptional case of energy measurement on isolated particles is a non-linear response acceptable, because it could be remedied (in principle) by calibration;
- the 'noise' or the non-uniformity of the readout system (photoelectron statistics, equivalent noise charge of preamplifiers) must not dominate the energy resolution;
- the readout system must have a rate capability adapted to the observed interaction rate;
- provision must be made for adequate longitudinal and transverse segmentation;
- the absolute and relative energy response must be monitored and maintained with sufficient accuracy;
- other operational characteristics, such as sensitivity to magnetic fields, radiation and temperature, have to be considered.

We have already emphasized the fundamental distinction between homogeneously active or sampling devices. The former have been used in many practical applications for the measurement of electromagnetic showers, whilst the latter represent the only really practical form of hadronic calorimeters. The instrumentation used can be categorized as 'light-collecting' devices measuring scintillator or Cherenkov light or as 'charge-collecting' methods, operating in the ion chamber, proportional, streamer, or saturated Geiger modes. In the following discussion, only recent developments or novel applications are emphasized, as witnessed by the choice of a very restricted number of references from the vast amount of literature.

5.1 Homogeneous Calorimeters

Some of the active readout materials have a density that is high enough for them to be used as homogeneously sensitive calorimeters. The properties of most frequently used materials are summarized in Table 7 [54–79].

Among recent noteworthy developments we find:
- a programme to use BGO crystals (amongst the presently known, optically transparent materials, the one with the shortest radiation length) for large 4π photon calorimeters with excellent space and energy resolution [49, 61];
- materials with considerably improved radiation resistance [64, 65, 74, 75];
- the use of BaF_2 crystals, coupled to very fast UV-sensitive light detectors for very high rate applications [73, 74].

5.2 Readout systems for sampling calorimeters

A great and very diversified number of readout systems have been developed, reflecting the desire to tailor the systems performance to a physics application.

Table 7: Properties and Performances of Homogeneous e.m. Shower Detectors

Detector type	NaI(Tl)	CsI(Tl)	BaF_2	$Bi_4Ge_3O_{12}$	Scintillating glass	Lead glass 55% PbO + 45% SiO_2	$Tl(HCO_2)$-liquid 'Helicon'	Liquid argon
Radiation length (cm)	2.59	1.86	2.1	1.12	~ 4	2.36	~ 1.9	14
Density (g/cm^3)	3.7	4.51	4.9	7.13	~ 3.5	4.08	~ 4.3	1.4
Detection mechanism	Scintillation	Scintillation	Scintillation (20% around 210 nm, 80% around 310 nm)	Scintillation	Scintillation	Cherenkov light	Cherenkov light	Ionization charge
Energy resolution (E in GeV)	$\sim 0.015\,E^{-1/2} < 1$ $< 0.015\,E^{-1/4} > 1$	Comparable to NaI(Tl)	Comparable to NaI(Tl)	Comparable to NaI(Tl)	$\sim 0.002\,E^{-1/2}$	$\sim 0.04\,E^{-1/2}$	Comparable to lead glass	$\gtrsim 0.02\,E^{-1/2}$
Principal limitation to $\sigma(E)$	Shower fluctuations optically non-uniform	Similar to NaI(Tl)	Light collection non-uniformities	Similar to NaI(Tl)	Photon statistics	Photon statistics	Photon statistics	Effect of shower fluctuation on electron collection
Signal[a)] (photo-el/GeV)	$\sim 10^7$	$\sim 5 \times 10^6$	$\sim 10^6$	$\sim 10^6$	Few $\times 10^3$	10^3	$\lesssim 10^3$ (?)	$\lesssim 2 \times 10^6$
Characteristic time (ns)	250	900	0.6 ; 300	350	~ 70	~ 20	~ 20	$\gtrsim 100$
Rad. damage at appr. dose[b)] (Gy)	$\leqslant 10$	$\leqslant 10$	$\sim 10^5$	~ 10	$\sim 10^4$	$\sim 10^2$	$\gtrsim 10^4$	Not measured; expected to be very large
Mechanical stability	Hygroscopic, fragile	Very good	Good	Good	Very good	Very good	Toxic liquid	Cryogenic liquid
References	[54, 55, 57]	[72]	[73, 74]	[61–63]	[64, 65]	[66–69]	[75]	[76–79]

a) Values are approximate, and depend on spectral matching between light source and photon detector.
b) Values are guidelines only and very substantially depending on experiment and measuring conditions.

5.2.1 Light-collecting sampling calorimeters. The renaissance of such calorimeters started with the introduction of cheap 'plastic scintillators' and elegant light-readout techniques using 'wavelength shifters' (WLSs) to replace the technique of scintillator plates individually coupled to a lightguide [80] (Fig. 17a). The principle is indicated in Fig.17b [81–89]. Scintillation light crosses an air gap and enters the WLS, where it is absorbed and subsequently re-emitted at longer wavelengths; a fraction of this 'wavelength shifted' light is then internally reflected to the light detector. This scheme avoids complicated and costly optical contacts between the scintillators and the light collectors, and minimizes dead spaces. A variety of scintillators have been developed for the large calorimeter facilities. They are based on polymethyl methacrylate (PMMA) [90] or polystyrene [30] as the matrix for the primary scintillating agent. The light yield is close to that of more conventional organic scintillators (usually based on a polyvinyl toluene solvent) if certain aromatic compounds, e.g. up to about 20% naphthalene, are added. These new scintillators are more easily mass-produced, hence cheaper, and have superior mechanical properties. Some of the limitations of the WLS method may be removed after further development: better spectral matching between the scintillator emission and the WLS absorption, and also between the WLS emission and the photocathode sensitivity, will increase the number of detected photons, which is marginal in present systems. Related developments might result in the use of thinner yet more uniform WLSs; increased granularity might be achieved with WLSs having spatially

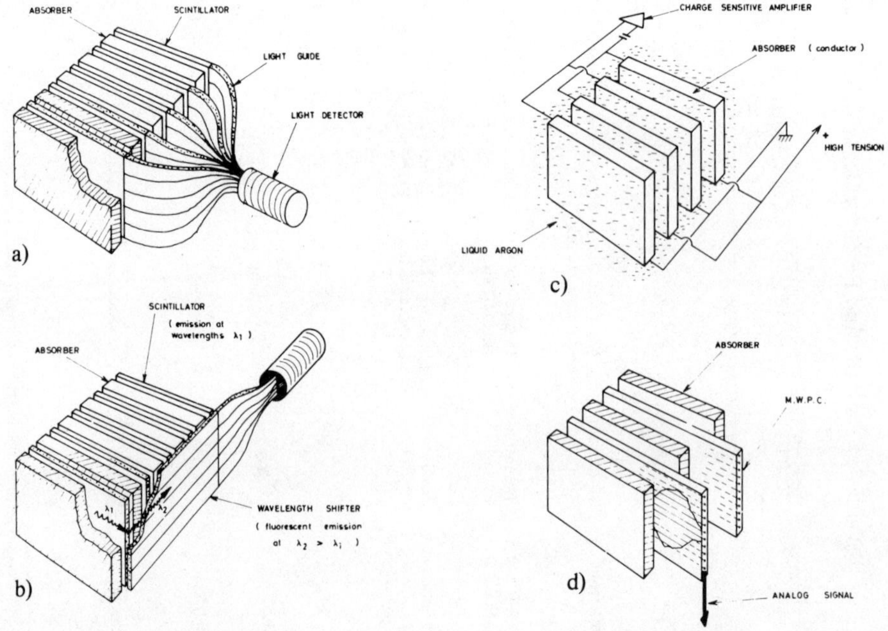

Fig. 17: Schematic representation for frequently used calorimeter readout techniques: a) Plates of scintillator optically coupled individually to a photomultiplier. b) Plates of scintillator read out by photon absorption and conversion in a wavelength shifter plate. c) Charge produced in an electron-transporting medium (e.g. liquefied or high-pressure argon) collected at electrodes, which may also function as the passive absorber plates. d) Charge produced in a proportional gas and amplified internally on suitable readout wires (proportional or saturated gas amplification).

different spectral sensitivities [87], or with very thin foils of WLS [91]. Potentially the most promising developments concern scintillators. They are still rather inefficient (only a few percent of the energy loss is converted into visible photons), and reduced saturation of the response to densely ionizing nuclear fragments should improve the energy resolution of hadron calorimeters (see subsection 3.3). The scintillator properties are important for the energy resolution of calorimeters, and need to be carefully investigated and specified when comparing various seemingly equivalent calorimeters.

Interest in high granularity and very compact readouts has recently led to 'double wavelength-shifter' applications [92–95] (Fig. 18). Although the second shifting reduces the number of photons by a further factor of ~ 5, the compression of the light into a very small cross-section light-pipe is attractive for several reasons, such as small insensitive zones and the possibility to use light detectors with small active areas. Such schemes favour the light registration with vacuum photodiodes [56, 70, 96] or silicon photodiodes [59, 93]. These devices—lacking an internal charge amplification mechanism—are operated with low-noise charge-sensitive preamplifiers, and are therefore more stable compared with photomultipliers; furthermore, they are insensitive to the commonly used levels of magnetic fields (vacuum diodes with some restrictions). These light detectors are therefore particularly attractive for the photon calorimeters of storage ring detectors, which most frequently are operated inside magnetic spectrometer fields [49, 56, 59, 60, 70].

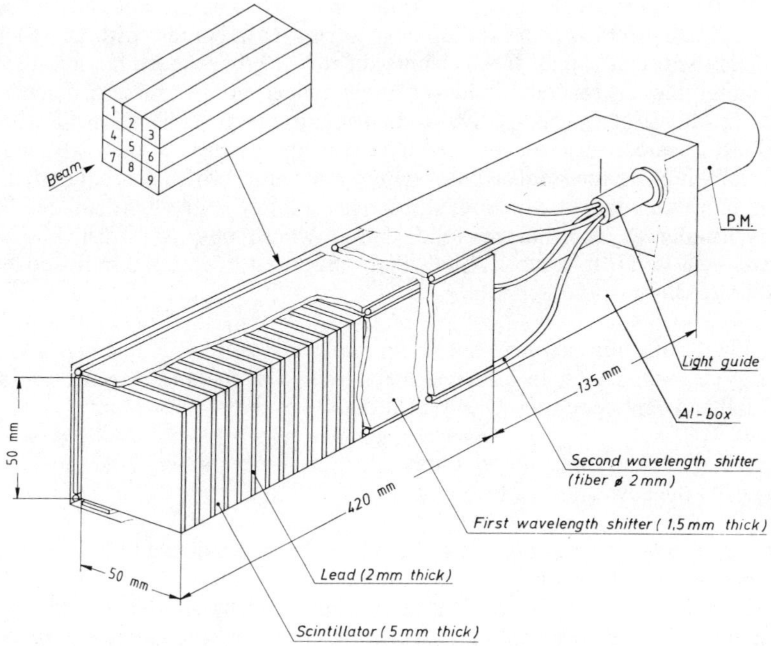

Fig. 18: Schematic view of a 'tower' of a sampling calorimeter array, using double-wavelength shifting techniques for the light measurement. The thin plates of the first WLS cover all four sides of the tower, whereas the second WLS, in the form of a fibre, registers the light emerging from the first shifter and guides it to the photomultiplier [92].

Another line of study concerns the innovative use of scintillators. Very long, narrow, Teflon tubes filled with liquid scintillator have been used for a large photon calorimeter [97]. The Teflon tubes define the spatial granularity of the active element and guide the light through total internal reflection to photomultipliers. A logical refinement consists in using small scintillating fibres [98] embedded in a metal matrix. With this technique a photon detector was constructed with the very short average radiation length of $X_0 = 14.5$ mm and an energy resolution of $\sigma_\gamma/E = 0.11/\sqrt{E}$—the modest man's BGO [99].

In recent years considerable effort was devoted to minimizing two disadvantages of the scintillator readouts; namely, the inherent non-uniformity in the light collection, and the difficulty of energy calibration.

The principal source of non-uniform light collection is not primarily the attenuation of light propagating in thin scintillator sheets, but is usually due to the light collection geometry. The non-uniform response, measured for example by scanning the active surface with a monochromatic electron beam, is at a level of $\pm 5\%$ in finely tuned instruments [32], and one representative example is shown in Fig. 19. Such a level of non-uniform response will of course influence the energy resolution for electrons with $p \geq 10$ GeV/c, if no correction for the impact point is applied. Hadron showers are much less affected, if the geometric extension of the non-uniformities is comparable to or smaller than the shower size. Such problems may be considerably aggravated if the usually sufficiently high transmission of the scintillator is affected, e.g. by radiation damage [32], surface cracking, or for other reasons. Plastic scintillators will show radiation damage after exposure to less than 100 Gy, if in contact with air [100], whilst toluene-based scintillator may sustain approximately 10 times more radiation [101, 102] before its usefulness is severely limited. Closely connected with these problems of light collection is the strategy of relative and absolute energy calibration. For precision applications, it is necessary to expose each calorimeter cell at least once to some kind of particle beam in order to establish an absolute calibration, which subsequently has to be transferred and maintained with some kind of absolutely stable light source. This light source is usually an external reference lamp, whose output is distributed to the calorimeter cells [103]. In some cases, the light produced by internal radioactive sources [32] has served this purpose.

The limitations outlined here become a major concern for applications where very high energy deposits could, in principle, be measured at the one percent accuracy level [50, 95, 104] and correspondingly benefit the quality of the physics data.

Increasingly, therefore, the experimental teams are evaluating alternative solutions, as discussed in the next subsection.

5.2.2. Charge collection readout. The ionization charge produced by the passage of the charged particles of the shower may be collected from solids, liquids, or gases. Solids [105] and liquids can only be used in an ionization chamber mode with no internal amplification. The best known and, to date, the only practical example is based on the use of liquid argon [106]. In specific cases, liquid xenon may be used [107–110]. The use of room-temperature liquids has also been repeatedly advocated [95, 111], but with increasing operating temperature the tolerable level of impurities decreases strongly. If gas is used as the active sampling medium, internal amplification to various degrees is usually exploited: proportional chambers or tubes provide a signal proportional to the

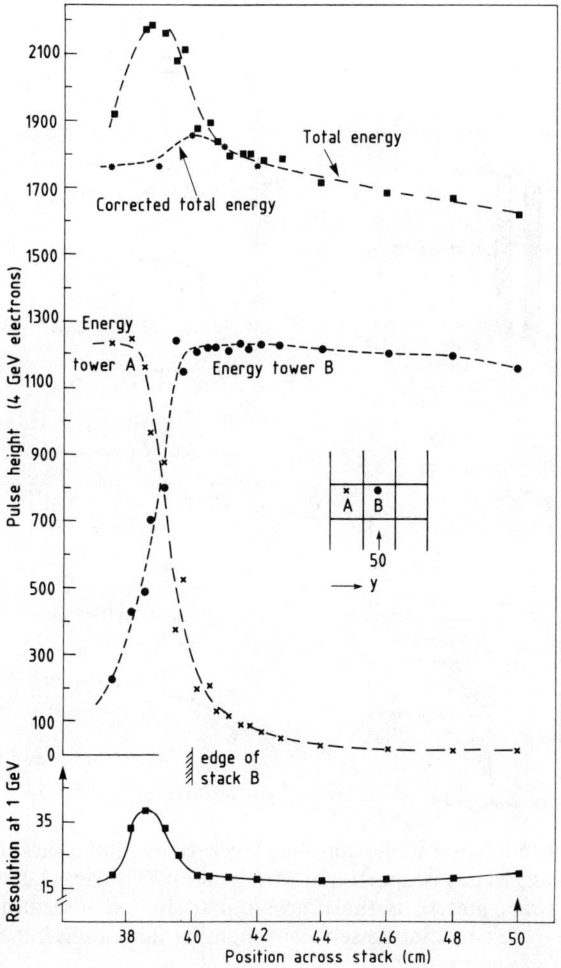

Fig. 19. The worst-case optical non-uniformity obtained by scanning with a 4 GeV/c electron beam under normal incidence across the gap between neighbouring stacks. The approximately 15% non-uniformity can be further reduced with a correction algorithm using the two signals in each tower, A and B (short-dashed curve) [32].

energy loss. At higher gas gain, with devices operating in a controlled streamer or Geiger mode, the measured signal is related to the number of shower particles which traverse the active medium ('digital readout') [112]. The conceptual arrangements are shown in Fig. 20.

The principal advantages common to all these charge collection methods are seen in the ease of segmentation of the readout and the capability to operate in magnetic fields. Some features specific to the various types are:

a) operation in the ionization mode, i.e. liquid-argon calorimeters, provides the best control of systematic effects [44, 113–116];

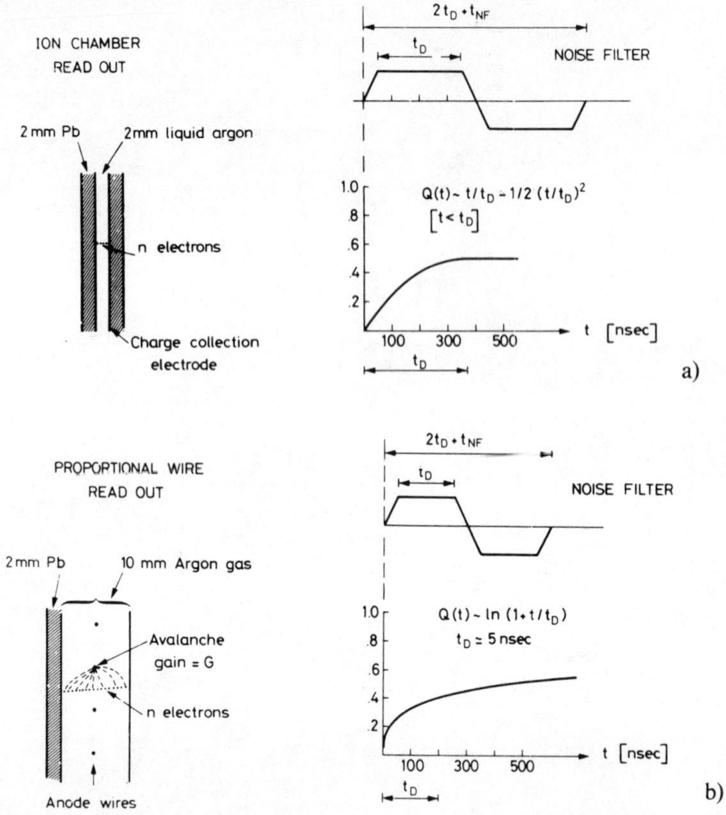

Fig. 20: Charge collection in a single sampling layer for a) a liquid-argon calorimeter with ion chamber readout; b) gas proportional wire readout. The signal charge Q(t) is shown as a function of time, and t_D is the time required for all ionization electrons to be collected. For each case a bipolar noise-filter weighting function is indicated (see text).

b) gas proportional devices offer a wide variety of relatively inexpensive construction methods [117];
c) digital operation, in the Geiger or streamer mode, allows for very simple and cheap signal-processing electronics [112, 118, 119].

The ion chamber technique is the preferred solution whenever optimum performance is at a premium because it excels in the following points, all of which have been realized in practice in devices using liquid argon as the active medium:

- uniformity of response at the fractional percent level;
- ease in fine-grained segmentation with a minimum of insensitive area;
- excellent long-term operating stability (radiation damage absent in liquid argon; response of active medium controllable).

Experimentation at today's fixed-target or storage ring facilities often justifies the use of such high-performance calorimeters, and as a consequence the liquid-argon ion chamber technique has been adopted for several large facilities, both in existence [44, 113–116] or planned [51, 120, 121]. One noteworthy exception is made by the teams preparing the four LEP detectors—they have opted for scintillator or proportional chamber readout solutions.

For many years, several liquid substances have been known, in which electrons may be drifted over large (several centimetres) distances, given adequate electric fields and levels of purity [122]. Generally, it may be said that operation at cryogenic temperatures, $T < 100$ K, eases the purification problem. Liquid-argon calorimeters, for example, may be operated with levels of ~ 1 ppm O_2, but the requirements are already considerably more severe for liquid xenon (~ 10 ppb O_2 tolerable) and develop into a major engineering difficulty for room-temperature liquids [95, 123]. In practical applications one has to weigh the complexity of a cryogenic detector (difficult access to components inside, extra space for the cryostat, increased mechanical engineering problems) against the difficulties arising from impurity control (large multistage purification plant, use of ultra-high vacuum techniques and components in the construction) [123]. It was thought that some of the lighter room-temperature liquids [such as tetramethylsilane (TMS)] might be intrinsically more advantageous for hadron calorimeter readouts, because one expected relatively small saturation for densely ionizing particles (e.g. recoil protons) [124]. Recent measurements [95, 125] however, have indicated that in practical electric drift fields such an advantage may not exist.

The very low cost of sufficiently pure liquid argon and the relative ease of maintaining it in operating conditions suggest its use in large-mass detectors for neutrino-scattering or proton-decay experiments [77–79, 126]. As with all rare-event detectors, the possibility of very large drifts of the ionization [time projection chamber (TPC)] have been studied with the aim of reducing the number of electronic channels to an 'acceptable' level [76, 77].

A recent extension of the ion chamber techniques to the use of solid dielectrics has met with considerable success [127], the signal being measured with Si surface barrier detectors in sampling calorimeter detectors. This scheme combines the advantages of a room-temperature ion chamber readout, with the attractive feature that these Si detectors are usually extremely thin, less than 500 μm, permitting the realization of detectors with $X_0 \leq 4$ mm, and hence offering the ultimate localization of showers.

For ion chambers and proportional wire readout, the measured signal typically amounts to a few picocoulombs of charge per GeV of shower energy. Since sampling calorimeters are inherently devices having a large capacitance, the optimum charge measurement requires careful consideration of the relationship between signal, noise, resolving time, and detector size. A detailed noise analysis gives [128]

$$ENC_{opt} = k \times 10^6 (C_D/t_{NF})^{1/2},$$

the 'equivalent noise charge' ENC being the input signal level which gives the same output as the electronic noise. The parameter k is proportional to the r.m.s. thermal noise of the input field-effect transistor; for practical detector arrangements a value of $k \simeq 5$ is realized. [C_D is the detector capacitance (in μF), and t_{NF} is the noise filter-time (in ns)]. This relation determines the fundamental lower limit to the noise, which is achieved with optimal capacitance matching between the detector and amplifier. The noise figure grows with increasing detector capacitance, but can be reduced at the expense of augmenting the resolving time.

Charge collection in gases, usually followed by some degree of internal amplification, forms the basis of another important category of calorimeter readouts [117]. The method lends itself naturally to highly segmented construction, of particular value for the topological analysis of the energy deposit (e.g. γ/hadron discrimination, muon identification). The technique has profited from the diversified developments in gaseous position detectors [129] over the last fifteen years: the versatility of arranging readout anode wires combined with the ease of gain control have produced a great variety of different solutions tailored to the specific requirements of an experiment.

With these types of detectors, spatial segmentation and localization can be easily implemented. This may be achieved in a projective geometry using strips, or in a 'tower' arrangement, e.g. by measuring the signal charge induced on a pattern of cathode pads [130, 131]. The tower arrangement is mandatory for reducing ambiguities and confusion in multiparticle events.

Another technique, currently being pursued, aims at achieving a very high degree of spatial segmentation, and uses a TPC method, the so-called 'drift-collection' calorimeter [132]. The ionization produced by the charged particles of a cascade is drifted over very long distances and collected on a relatively small number of proportional wire planes, equipped with a two-dimensional readout, while the third shower coordinate is determined by a drift-time measurement. The conceptual configuration is shown in Fig. 21. An example of the pictorial quality of shower reconstruction expected with this technique is given in Fig. 22.

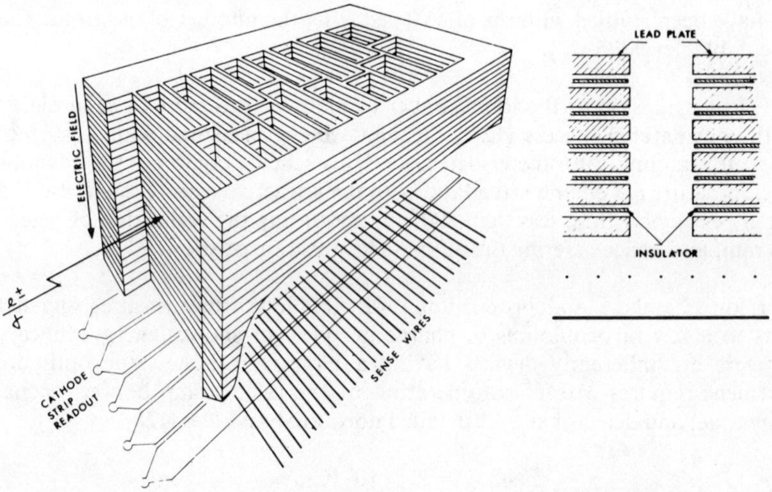

Fig. 21: Geometrical arrangement for the high-density drift calorimeter. Cavities between absorber plates allow the drifting of ionization electrons over long distances on to MWPC-type detectors. Very high spatial granularity can be achieved at the cost of mechanical complexity and rate capability. A very uniform magnetic guidance field parallel to the drift direction is usually required [132].

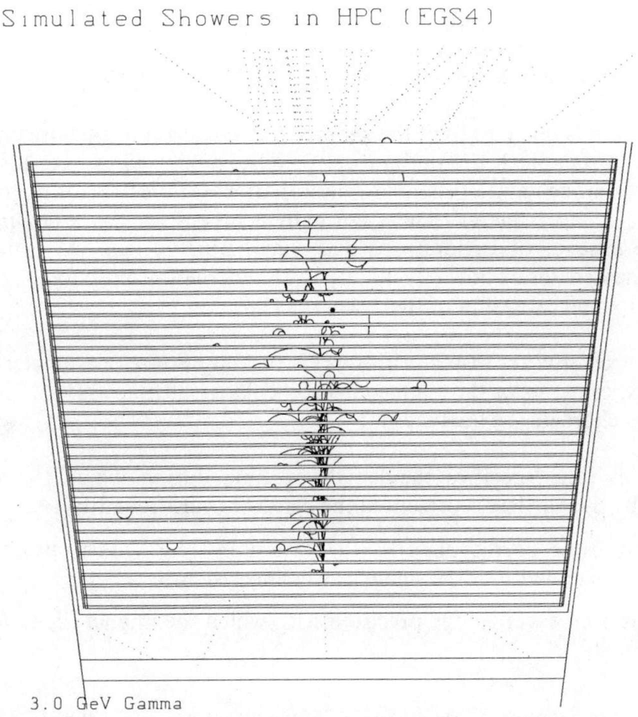

Fig. 22: Simulation of an electromagnetic shower in the calorimeter of the LEP DELPHI collaboration using EGS IV. The bubble-chamber-like pictorial quality of the information allows the individual shower particles to be distinguished. (Courtesy of H. Burmeister, CERN.)

Gas sampling calorimetry in a digital mode (using Geiger, streamer, or flash-tube techniques) is a means of simplifying the signal processing circuitry, and offers an expedient method for achieving a high degree of segmentation. Flash chambers consist of an array of tubular cells filled with a mixture of about 96% neon + 4% helium; a pulsed high-voltage is applied across each cell after an external event trigger. In the presence of ionization charge, a signal-producing plasma discharge propagates over the full length of the cell. In one such array for a large neutrino experiment at Fermilab, 608 flash-chamber planes with a total of some 400,000 cells are sandwiched between absorber layers of sand and steel shot [24]. The pattern of struck cells in each plane is read out by sensing induced signals on a pair of magnetostrictive delay lines. Wire readout planes may also be operated in the 'streamer mode', where the charge gain is controlled to cover only a segment around the particle impact point [133, 134]. Common to these saturated gas-gain readout schemes are the following properties:

- the energy resolution is in principle better than in a proportional gain system, because Landau fluctuations are reduced or suppressed;
- the mechanical tolerances of the readout system are less stringent than for proportional systems;

- these readouts are inherently non-linear. One charged particle causes an insensitive region along the struck wire, which prevents other nearby tracks from being registered. Typically, non-linearities become measurable above ~ 10 GeV.

Usually, care is taken to limit the geometrical extension of the discharge region, e.g. with various mechanical discontinuities (beads, nylon wires, etc.). A calorimeter operated in this mode gave an energy resolution of $\sigma \simeq 14\%/\sqrt{E}$ for electron energies up to 5 GeV [135]. This is better than is normally achieved for gas sampling calorimeters, reflecting the absence of Landau and path-length fluctuations. At higher energies the calorimeter showed saturation effects due to the increasing probability of multiple hits over the geometrical extension of the discharge region.

The *rate capability* of calorimeters is an important parameter for high-rate fixed-target experiments or the planned hadron colliders (Table 8) [136]. Several different time constants characterize the readout:

1) 'occupation' time specifies the length of time during which the physical signal produced by the particle is present in the detector (pulse length);
2) 'integration' time corresponds to the externally (e.g. electronically) chosen time defining the bandwidth of the signal processing system;
3) 'time resolution' specifies the precision with which the impact time of a particle may be determined.

For scintillator-based methods the signal duration is typically about 20–50 ns but, with special care, signals of about 10 ns length have been achieved [94].

For liquid-argon devices, the charge-collection time is ~ 200 ns/mm gap; it may be reduced by a factor of 2 by adding ~ 1% of methane. Sometimes it is acceptable not to integrate over the full signal length, entailing a reduction in the signal-to-noise ratio. For liquid-argon devices, the noise is proportional to $(\tau_{\text{integ}})^{-1/2}$ and the signal is almost proportional to $(\tau_{\text{integ}})^{1/2}$, if $\tau_{\text{integ}} < \tau_{\text{tot. coll.}}$. A very short integration time is sometimes chosen ('clipping' of a signal) to enable very fast trigger decisions; only if the event information is of interest is the signal processed with a longer integration time, in order to obtain an optimum signal-to-noise ratio.

The time resolution achievable with calorimeters may help to associate the calorimeter information with different events, separated by a time interval much shorter than the integration time. For scintillator and liquid-argon calorimeters, time resolutions of 2–3 ns have been measured for few-GeV energy deposits [44, 137].

The ultimate rate limitation is, however, determined by the physics to be studied with a calorimeter. In collisions involving hadrons in the initial state, reactions which occur with very different cross-sections may be characterized by very different event topologies. A typical example is the production of several high-p_T jets in a pp collision, the cross-section of which is very small in comparison with the total inelastic cross-section. In such a case, several events may be recorded within the occupation time of the detector without serious effects (Fig. 23). This figure suggests that calorimeters may still be very useful even if, during the occupation time of an interesting event, several other events produce energy deposits and are recorded in the detector [32, 39, 138].

Table 8: Temporal Response of Readout Systems

Calorimeter system	Occupation time (ns)	Pulse width (integration) (ns)	Timing resolution σ (ns)	Radiation resistance
Metal/scintillator with WLS readout	50	50	2.5 for few-GeV deposit; better at higher energy, if not limited by PM	Depends on scintillator, dose rate, environment; $> 10^3$ Gy appear achievable [102].
^{238}U/scintillator with fast WLS readout	~ 100	100	2.5	As above
Metal/fast scintillator with fast WLS readout	$\leqslant 20$	$\leqslant 20$	< 2 (?)	
Metal/proportional or saturated gas-gain readout	50–100	100–200 bipolar shaping	$\leqslant 10$	Adequate for chamber; lifetime of on-detector electronics may be a limitation; readout elements need to be shielded from U radioactivity.
^{238}U/proportional or saturated gas-gain readout	$\geqslant 100$	$\geqslant 200$ bipolar shaping	$\leqslant 10$	
Metal/LAr ion chamber	~ 200 per 1 mm gap	$2\lambda = 400$ bipolar shaping	~ 2 for few-GeV deposit	Lifetime of on-detector electronics may be a limitation
Metal/LAr–CH_4 ion chamber	~ 100 per 1mm gap	$2\lambda \leq 200$	$\lesssim 2$ for few-GeV deposit	NB: shorter pulse width (2λ) possible the expense of signal/noise $\simeq \lambda^{\alpha}$, at $\alpha \sim 1$.

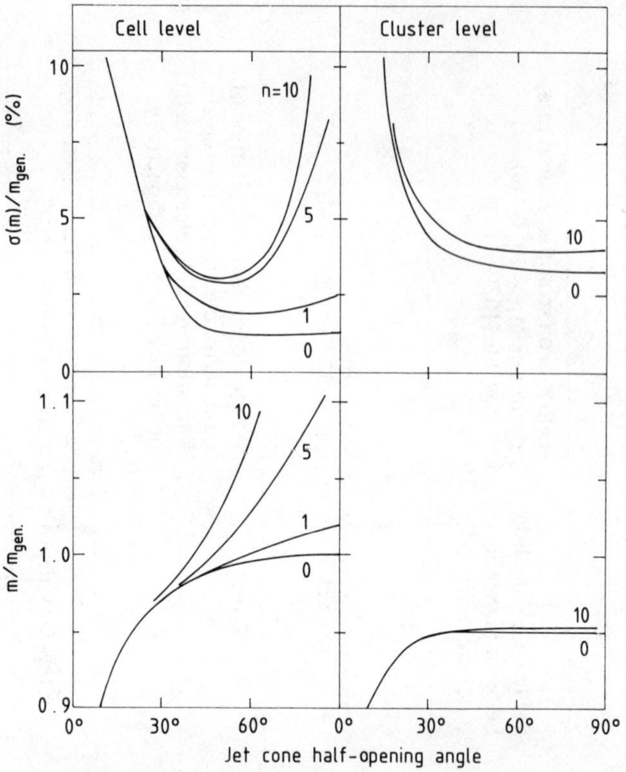

Fig. 23: Two-jet mass resolution (upper part) and reconstructed mass value (lower part) for a 1 TeV system at the cell and at the cluster level as a function of the opening angle between the jets, and the number n of additional accidental minimum-bias events [39] (see Fig. 14 for definition of cell and cluster level).

5.2.3. Bolometric Readout. Repeatedly, particles have been detected through the temperature rise in a calorimeter, caused by the absorption of the particles. Such experiments contributed decisively to our understanding of radioactivity and to the concept of the neutrino [139]. Already 50 years ago it was recognized that owing to the large reduction in heat capacity of many materials at cryogenic temperatures, very sensitive instruments were permitted [140]. More recently, interest has refocused on cryogenic calorimetry, operated in the temperature range of 1 mK to ~ 1 K. Such detectors [141–143] are considered as possible particle detectors offering an energy resolution of ≤ 10 eV and having a time response in the 100 μsec to 1 msec range. Possible experiments include the search for double-beta decay or the measurement of the end point of tritium beta decay. More ambitiously, multiton silicon detectors operated at mK temperatures are considered as a solar neutrino observatory to measure the spectrum of solar neutrinos and to probe for neutrino mass differences at the 10^{-6} eV level. Cryogenic calorimetry promises far more precise energy measurements than any of the other 'standard' techniques and allows us to contemplate some of the most challenging and fundamental experiments in particle physics.

6. SYSTEM ASPECTS OF CALORIMETERS

6.1 General Scaling Laws of the Calorimeter Dimensions

A number of parameters which may be external to the calorimeter design, e.g. the dimensions of a charged-particle spectrometer, may ultimately determine the size of a calorimeter.

If, however, global optimization of an experiment is attempted, one should aim for the most compact calorimeter layout that may lead to the achievement of the physics goals of the detector.

The distance D of a calorimeter from the interaction vertex, and hence its necessary size, is determined by the achievable useful segmentation and the characteristic angular dimension θ required to be resolved in the measurement. The useful segmentation d is determined by the shower dimensions, approximately $d \sim 2\varrho_M$ for electromagnetic detectors and $d \sim \lambda$ for hadronic ones. The characteristic angular dimension θ may be the minimum angular separation of the photons from π^0 decay, or the typical angle between the energetic particles in a hadron jet.

For correctly designed detector systems, the calorimeter dimensions are determined by the angular topology and size of the showers to be measured. The minimum detector distance is then $D \geq d/\mathrm{tg}\,\theta$, and the required calorimeter volume V is found to be

$$V \propto D^2 \times L \text{ (depth of calorimeter required for total absorption)}$$

whence

$$V \propto \varrho_M^2 \times X_0 \text{ for e.m. detectors,}$$

and

$$V \propto \lambda^2 \times \lambda \text{ for hadronic detectors.}$$

The third-power dependence of the calorimeter volume on shower dimensions implies that it may be economically advantageous to select very compact calorimeter designs, even if the price per unit volume is very high. Table 9 explains why sometimes only the most expensive materials (per cm^3) can be afforded.

6.2. Calorimeter Systems for Physics Applications

6.2.1 Neutrino physics and nucleon stability. Detectors for these studies share some common features: the event rate is low and is proportional to the total instrumented mass; the physics requires a fine-grained readout system, which permits detailed three-dimensional pattern reconstruction. Bubble chambers have therefore been used extensively but are limited by the long analysis time per event. The present generation of neutrino detectors makes extensive use of wire-chamber techniques to approach the intrinsic spatial and angular resolution of calorimeters. New 'visual' electronic techniques are being developed for proton decay experiments, for which very massive detectors with a high density of signal channels are required.

A neutrino detector, even when exposed to present-day intense neutrino beams, must have a very large mass (typically hundreds of tons), and its volume must be

Table 9: Characteristic Dimensions and Price Comparisons for e.m. Calorimeters

Material / Quantity	NaI	BGO	U/Si sampling calorimeter
X_0 (mm)	26	11	4
ϱ_M (mm)	44	23	11
Reference volume (cm^3) for 95% containment of ~ 5 GeV electrons	1600	180	15
Approx. price/Ref. vol. (arbitrary units)	1	1	0.1

uniformly sensitive to the signature that an interaction has occurred and to the characteristics of the reaction products. These requirements explain the modular construction typical of modern electronic neutrino detectors. The general form of the interaction is $\nu_{(\mu,e)}$ + nucleon = $\ell_{(\mu,e)}$ + X, where ℓ is a charged lepton (charged-current interaction) or a neutrino (neutral-current interaction), and X represents the hadronic system. The signature for a neutrino interaction is the sudden appearance, in the detector, of a large amount of energy in a few absorber layers. If the scattered lepton is a muon, it leaves in each layer the characteristic signal of a single minimum-ionizing particle. In some detectors the absorber layers are magnetized iron, which makes possible a determination of the muon momentum from the curvature of its trajectory. For a detailed study of neutral-current interactions, a very fine grained subdivision of the calorimeter system is required for measuring the energy and direction of the hadronic system X and for reconstructing the 'missing' transverse momentum of the final-state neutrino.

Clearly the scope and sensitivity of neutrino experiments would be much improved if even the most massive detectors could resolve final state particles with the reliability and precision typical of a bubble chamber. With this goal in mind, some schemes are currently being investigated that use drift chamber methods with large volumes of liquid argon [79], which provides a visual quality characteristic of homogeneously sensitive detectors. In another case [144] a cylindrical detector, 3.5 m in diameter and 35 m long, is foreseen, containing about 100 tons of argon gas at 150 atm pressure, with ionization electrons collected on planes of anode wires. The idea has also been advanced [145] of using compressed mixtures of more common gases (air or freon), large liquid-argon TPCs [78], or room-temperature liquid hydrocarbons [146] to detect the ions that migrate away from charged-particle tracks (positive and/or negative ions produced by electron attachment).

Detectors to search for the decay of nucleons are also characterized by a very large instrumented mass, varying from a hundred to several thousand tons. Current theoretical estimates place the proton lifetime at around $\tau_p \simeq 3 \times 10^{29\pm1.5}$ y compared with present experimental limits [147] of $\tau_p \geq 10^{32}$, reached by detectors of at least 1000 tons mass, or approximately $> 10^{33}$ nucleons. These detectors have to be instrumented to search sensitively for some of the expected decay modes such as $p \to e^+ + h^0$, where h^0 is a neutral meson (π^0, η, ϱ^0, ω^0). The signature of such a decay is clear, provided a detector is sufficiently subdivided to recognize the back-to-back decay into a lepton and a hadron with the relatively low energy deposit of about 1 GeV. The sensitivity is limited by the flux of muons and neutrinos originating from atmospheric showers. Only muons can be shielded by placing the experiments deep underground in mine shafts or in road tunnels beneath high mountains. The ν_μ-induced rate simulating nucleon decay is estimated at $\sim 10^{-2}$ events per ton per year if energy deposition alone is measured. If complete event reconstruction is possible, an experimental limit of $\tau \geq 10^{33}$ y may be reached [147].

The most massive calorimetric detectors to date have been conceived to explore the very high energy cosmic-ray spectrum. The detector volumes needed are so large that only the sea water [148] and air [149] are available in sufficiently large quantities. The interaction provoked by cosmic-ray particles is detected through the Cherenkov radiation emitted in the ensuing particle cascade in the case of the deep underwater array. In the case of the atmospheric detector it is the light from the excited N_2 molecules which is detected and measured with great ingenuity. The 'air calorimeter' represents the largest ($V \simeq 10^3$ km^3) and most massive ($W \sim 10^9$ tons) calorimeter conceived to date; already it has produced evidence about the cosmic-ray flux at $E > 10^{20}$ eV.

6.2.2. Calorimeter facilities for storage rings. At hadron machines the studies focus on reactions that are characterized by a large transverse energy (E_T) flow, as a signal for an inelastic interaction between the nucleon constituents. The signature appears in many different characteristic event structures and may therefore be efficiently selected with hadron calorimeters : examples are single high-p_T particles, 'jets' of particles, or events exhibiting large E_T, irrespective of their detailed structure. Topical applications include invariant mass studies of multijet events, often in conjunction with electrons, muons, or neutrinos (missing E_T). The power of this approach has been demonstrated by the results obtained in recent years at the CERN $p\bar{p}$ Collider [52], which in turn have led the UA1 and UA2 collaborations to proceed with major upgradings or replacements of their calorimeter detectors [95, 150]. This central role of calorimetry in exploratory hadronic physics programmes [39, 151, 152] is also recognized in the planning for the second detector facility for the FNAL Tevatron. The group is planning a 400 ton uranium/liquid-argon hadron calorimeter, which is expected to be the most advanced calorimeter facility in use during the coming years [51].

At electron–positron colliders, electromagnetic detectors are frequently used to measure the dominant fraction of neutral particles, the π^0's. They are also the ideal tool for detecting electrons, which may signal decays of particles with one or more heavy (c, b, ...) quarks. Unique investigations of cc and bb quark spectroscopy were accomplished with high-resolution NaI shower detectors [153].

For the physics programmes at LEP [154–156] and at the SLAC Linear Collider [120, 157] extensive use of hadron calorimetry will be made. At e^+e^- machines the event topology—production of particles at relatively large angles, with a total energy equal to the centre-of-mass (c.m.) collision energy, favours the experimental technique of total energy measurement. For future e^+e^- physics this method will be important because

- the fraction of neutron and K_L^0 production, measurable only with hadronic calorimeters, increases with energy [158];

- a large and most interesting fraction of events will contain neutrinos in the final state; missing energy and momentum analysis provides the sole handle for such reactions;

- a considerable fraction of events will show good momentum balance but large missing energy—these may be two-photon events or, above the Z^0 pole, events on the radiative tail; total energy will provide the cleanest signature;

- hadronic calorimeters will be the most powerful tool for measuring the reaction $e^+e^- \rightarrow W^+W^-$, either in channels where each W decays hadronically (a total of four jets) or through leptonic decay channels;

- most importantly, a measurement of energy topology is a powerful way of unravelling very rare and unexpected physics phenomena [159].

The technically most difficult requirements for the calorimetry will be imposed by the physics programme at HERA (DESY, Hamburg) [160]. Owing to the very asymmetric energy of the beams (800 GeV protons on 30 GeV electrons), the jets of particles fragmenting from the scattered quarks will have to be measured with the greatest possible energy and *angular* precision in a geometry similar to that of a fixed-target experiment. This implies that the detectors will have a very asymmetrical arrangement, a very large dynamic range, and very high granularity. Innovative developments [93, 127] signal the HERA groups' anticipation of this challenge.

At e^+e^- machines the performance with respect to the energy and space resolution of a well-designed calorimeter is matched to the physics programme foreseen at LEP and at the SLC. Consider, as an example, a 100 GeV multijet event of which the total energy can be measured with an accuracy of $\sigma \simeq$ 3–5 GeV and the total momentum balance checked at a level of $\sigma \simeq$ 3 GeV/c. These are intrinsic performance figures, disregarding possible instrumental effects (see Section 3). At hadron colliders, however, a further serious difficulty arises from the convolution of the energy response function with the steeply falling p_T distribution of hadronically produced secondaries [161–164]. As a consequence, the measured energy deposit E' in the detector will originate predominantly from incident particles with energy $E < E'$; count rates and trigger rates are higher than the true physics rates. The result can be devastating for detectors with poor energy resolution or a non-Gaussian response function ('tails' in energy resolution), introducing large errors in the deconvolution. The problem is compounded if the calorimeter response is different for charged and neutral pions (see Section 3): without adequate precautions, these detectors would preferentially select π^0's, making the use of calorimeters marginal for general trigger applications.

The trigger capability is a unique and perhaps the most important requirement of hadron calorimeters employed at hadron machines. For satisfactory operation, one needs uniform response irrespective of event topology and particle composition; good energy resolution at the trigger level to minimize effects of the response function; and adequate granularity for the selection of specific event topologies. For high selectivity, rather complex analogue computations are required, as may be seen from the examples in Table 10 [165]. A tabular summary on calorimeter facilities may be found in Ref. [3].

Table 10: Triggering with Hadron Calorimeters

Experiment	Trigger
Single-particle inclusive distribution; correlations	Localized energy deposit in spatial coincidence with matching track; several thresholds used concurrently.
Jet studies	Extended (~ 1 sr) energy deposit; several thresholds and multiplicities.
Inclusive leptons, multileptons	Electromagnetic deposit in spatial coincidence with matching track; several thresholds and multiplicities.
Heavy flavour jets; correlations	Various combinations of above triggers.

6.3 Monte Carlo Simulations

The development of calorimeters from crudely instrumented hadron absorbers to finely tuned precision instruments owes much to the development of a number of simulation codes. The relatively simple physics governing the electromagnetic showers has facilitated their Monte Carlo simulation. Today, one program has emerged as the world-wide standard for simulating e.m. calorimeters [166]. It has progressed through several improvement stages up to the currently used version, EGS IV, which allows one to follow the shower history, tracking electron pairs down to zero kinetic energy and photons down to ~ 100 keV. It has successfully passed many very detailed tests, including the perhaps ultimate one, that of simulating absolutely the response of electrons relative to muons [167].

The physics and consequently its simulation are considerably more complex for hadronic showers. Over the last decade several programs have been developed, the aim of which is to simulate fairly accurately the detailed particle production of a hadronic cascade [25, 168, 169]. As an example, the flow chart of one of the most detailed simulations is shown in Fig. 24. The attentive reader will no doubt realize that even the most faithful physics simulation of the hadronic process will not guarantee unconditional success. Already the uncertainties associated with the sampling medium (relative response to minimum and heavily ionizing particles) and the complexities of the nuclear interactions are too large to make an *ab initio* calculation possible at present. These programs therefore do require careful tuning against many different measurements, before they become a reliable guide for designing new facilities. The results of a sample calculation are shown in Fig. 25 and give an impression of the power of this code. The status of these shower calculations has recently been extensively discussed [170].

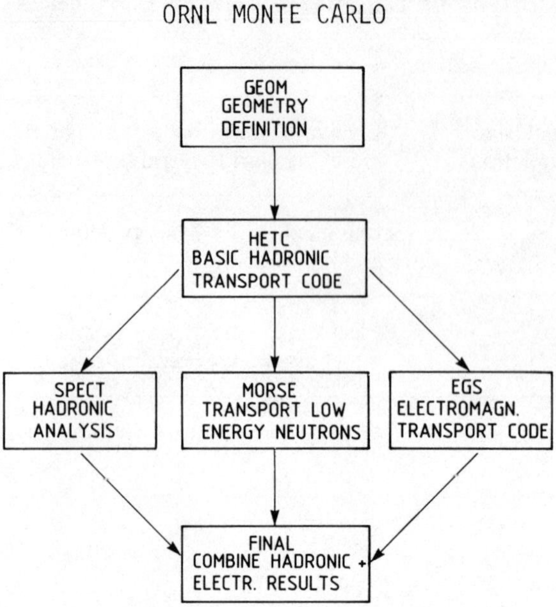

Fig. 24: Flow chart of the hadronic cascade Monte Carlo [T.A. Gabriel et al., Nucl. Instrum. Methods 195:461 (1982)].

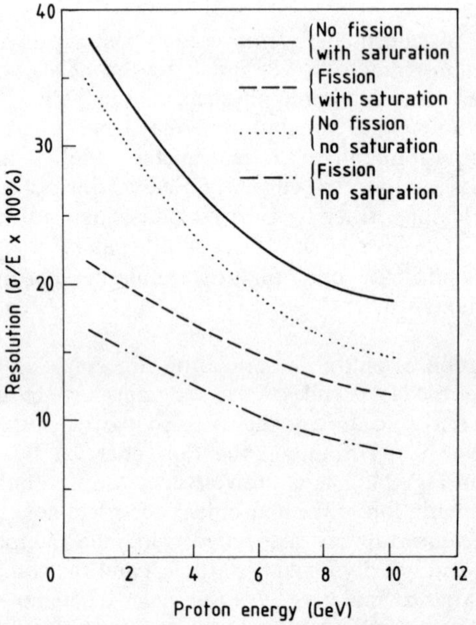

Fig. 25: Results of a 'Gedanken-experiment' using HETC to study the importance of the fission contribution and its manifestation is different readouts, with and without saturation [T. Gabriel et al., Nucl. Instrum. Methods. 164:609 (1979)].

7. OUTLOOK

The use of calorimetric methods in high-energy physics began with rather specialized applications, which capitalized on some unique features not attainable with other techniques: electromagnetic shower detectors for electron and photon measurements; neutrino detectors; and muon identifiers.

The evolution towards a more general detection technique—similar in scope to magnetic momentum analysis—had to wait for extensive instrumental developments driven by strong physics motivations. The application of this technique to the hadron storage rings (which were inaugurated in the early 1970's) was therefore delayed, but it has shaped the detectors for the CERN $p\bar{p}$ Collider and has proved to be of major importance for the detector facilities currently in preparation. At the same time, physics studies have evolved along lines where measurements based on 'classic' magnetic momentum analysis are supplemented or replaced by analyses which are based on precise global measurements of event structure, frequently requiring extraordinary trigger selectivity, and which are much more suitable for calorimetric detection. During the next decade, experiments will rely increasingly on these more global studies, with properties averaged over groups of particles and with the distinction between individual particles blurred, unless they carry some very specific information. This role of calorimetry in both present and planned physics programmes is summarized in Table 11; the impact of instrumental advances and technology is outlined in Table 12.

Table 11: Future Role of Absorptive Spectroscopy

Source of particles	Physics emphasis	Calorimeter properties	Technical implications
pp ($p\bar{p}$) collider	Rare processes: high-p_T lepton, photon production; manifestations of heavy quarks $W^{\pm}$, Z^0, ...	~ 4π coverage with e.m. and hadronic detection; high trigger selectivity	Approach intrinsic resolution in multicell device; control of inhomogeneities, stability
e^+e^- collider	Complex high-multiplicity final states (multi-jets, electrons in jet, neutrinos)	Precision measurements of total visible energy and momentum	Very high granularity; particle identification
ep colliders (secondary beams, $p \geq 1$ TeV/c)	High-multiplicity final states; strong emphasis on global features	Calorimeter becomes primary spectrometer element	High granularity, high rate operation
Penetrating cosmic radiation; proton decay	Detailed final-state analysis of events with extremely low rate	Potentially largest detector systems ($\geq$ 10,000 tons) with fine-grain readout	Ultra-low-cost instrumentation

Table 12: Interdependence of Detector Physics and Technologies for Calorimeters

Principles	Electronics	Mode of operation
Improved understanding of limitation to energy resolution in hadron calorimeters	Gain stability at ~ 1% gain monitoring at 0.1% of ~ 10^5 analogue channels	
Calorimeters replace magnetic spectrometers at high energies	Operational systems of ~ 10^5 light- or charge-measuring channels	Very high spatial resolution for electromagnetic and hadronic showers
Particle identification through energy deposit pattern (μ, e, γ, π^0, n, $\bar{p}$, K_L^0, ν)	Cheap, fast ADC for high-level fast trigger decisions	Helps or allows pattern recognition of very complex events (jets, ...)

Despite recent conceptual and technical progress, a number of questions deserve further attention:

1) What is the precise energy dependence of hadronic energy resolution?
2) What improvement in energy, position, and angular resolution could be obtained with complete information on the individual shower distributions? With such information, could we minimize the effect of increased longitudinal leakage on the energy resolution?
3) What contributes to the measured energy resolution $\sigma(E) \simeq 0.2 \times E^{-1/2}$ of a fission-compensated hadron calorimeter?
4) Is there a cure for the low-energy non-linearities?
5) Can we understand hadronic sampling fluctuations to the same degree as we understand electromagnetic ones?
6) Can particle identification and separation be improved if more detailed shower information is available?
7) Are there any advantages in mixing different absorber materials, or in changing the sampling step inside a calorimeter?
8) Can we tailor the signal response of the readout to improve the calorimeter response?

The very diverse applications of calorimetric techniques will ensure continued study of these and the many technical questions connected with the signal processing of calorimeter information. There can never be a unique solution, but there should always be a search for the most suitable method. We hope that the information provided in this review will be useful for attaining this goal.

ACKNOWLEDGEMENTS

I wish to express my appreciation to Professor T. Ferbel who, as director of the 'Advanced Institute', has developed a School, which by its size, its students, its style, and its 'ambiance' appears optimally conducive to the pursuance of a variety of very topical themes.

R. Wigmans helped me considerably with a critical reading of these notes. C. Comby, L. Karen-Alun, M. Mazerand, M.-S. Vascotto, and K. Wakley efficiently and cheerfully converted my manuscript into printable form.

REFERENCES

[1] V.S. Murzin, Progress in elementary particle and cosmic-ray physics, J.G. Wilson and I.A. Wouthuysen, eds., North Holland Publ. Co., Amsterdam (1967), Vol. 9, p. 247.

[2] M. Atac, ed., Proc. Calorimeter Workshop, Batavia, 1975, FNAL, Batavia, Ill. (1975).

[3] C.W. Fabjan and T. Ludlam, Ann. Rev. Nucl. Part. Sci. 32:335 (1982).

[4] U. Amaldi, Phys. Scripta 23:409 (1981).

[5] S. Iwata, Nagoya University report DPNU-3-79 (1979).

[6] H. Messel and D.F. Crawford, Electron-photon shower distribution: Function tables for lead, copper and air absorbers, Pergamon Press, London (1970).

[7] Y.S. Tsai, Rev. Mod. Phys. 46:815 (1974).

[8] B. Rossi, High-energy particles, Prentice Hall, New York (1964).

[9] E. Longo and I. Sestili, Nucl. Instrum. Methods 128:283 (1975).

[10] H.H. Nagel, Z. Phys. 186:319 (1965).

[11] D. F. Crawford and H. Messel, Phys. Rev. 128:2352 (1962).

[12] Yu.D. Prokoshkin, Proc. Second ICFA Workshop on Possibilities and Limitations of Accelerators and Detectors, Les Diablerets, 1979, U. Amaldi, ed., CERN, Geneva (1980), p. 405.

[13] E.B. Hughes et al., IEEE Trans. Nucl. Sci. NS-19: 126 (1972).

[14] H.G. Fischer, Nucl. Instrum. Methods 156:81 (1978).

[15] R.W. Sternheimer et al., Phys. Rev. B3:3681 (1971).

[16] K. Pinkau, Phys. Rev. B139:1548 (1965).

[17] T. Yuda, Nucl. Instrum. Methods 73:301 (1969).

[18] C.J. Crannel, Phys. Rev. 182:1435 (1969).

[19] T. Kondo et al., A simulation of electromagnetic showers in iron-lead and uranium-liquid argon calorimeters using the EGS, and its implication for e/h ratios in hadron calorimetry, contributed paper to the Summer Study on the Design and Utilization of the Superconducting Super Collider, Snowmass, Colo. (1984).

[20] G.A. Akopdjanov et al., Nucl. Instrum. Methods 146:441 (1977).
S.R. Amendolia et al., Pisa 80-4.
R. Rameika et al., Measurement of electromagnetic shower position and size with a saturated avalanche tube hodoscope and a fine grained scintillator hodoscope, to be published in Nucl. Instrum. Methods.

[21] T. Kondo and K. Niwa, Electromagnetic shower size and containment at high energies, contributed paper to the Summer Study on the Design and Utilization of the Superconducting Super Collider, Snowmass, Colo. (1984).

[22] E. Gabathuler et al., Nucl. Instrum. Methods 157:47 (1978).
[23] A.N. Diddens et al., Nucl. Instrum. Methods 178:27 (1980).
[24] D. Bogert et al., IEEE Trans. Nucl. NS-29:336 (1982).
[25] J. Ranft, Particle Accelerators 3:129 (1972);
A. Baroncelli, Nucl. Instrum. Methods 118:445 (1974);
T.A. Gabriel et al., Nucl. Instrum. Methods 134:271 (1976).
[26] R. Bock et al., Nucl. Instrum. Methods 186:533 (1981).
[27] C.W. Fabjan and W.J. Willis, Proc. Calorimeter Workshop, Batavia, 1975, M. Atac ed., FNAL, Batavia, Ill. (1975), p. 1.
C.W. Fabjan et al., Phys. Lett. 60B:105 (1975).
[28] O. Botner, Phys. Scripta 23:555 (1981).
[29] A. Benvenuti et al., Nucl. Instrum. Methods 125:447 (1975).
[30] M.J. Corden et al., Phys. Scripta 25:5 (1982).
[31] A. Beer et al., Nucl. Instrum. Methods 224:360 (1984).
[32] T. Akesson et al., Properties of a fine sampling uranium-copper scintillator hadron calorimeter, submitted to Nucl. Instrum. Methods (1985).
[33] T. Gabriel and W. Selove, private communication.
[34] U. Amaldi and G. Matthiae, private communications.
[35] H. Abramowicz et al., Nucl. Instrum. Methods 180:429 (1981). See also, for earlier work, J.P. Rishan, SLAC 216 (1979).
[36] Results similar to those given in [35] were recently obtained by the WA78 Collaboration at the CERN SPS (P. Pistilli, private communication).
[37] C.W. Fabjan et al., Nucl. Instrum. Methods 141:61 (1977).
[38] W.J. Willis, Invited talk given at the Discussion Meeting on HERA Experiments, Genoa (1984).
[39] T. Akesson et al., Proc. ECFA-CERN Workshop on a Large Hadron Collider in the LEP Tunnel, Lausanne and Geneva, M. Jacob, ed., CERN 84-10 (1984).
[40] F. Binon et al., Nucl. Instrum. Methods 188:507 (1981).
[41] A.V. Barns et al., Phys. Rev. Lett. 37:76 (1970).
See also T. Ferbel, Understanding the fundamental constituents of matter, A. Zichichi, ed., Plenum Press, New York, NY (1978).
[42] J.A. Appel et al., Nucl. Instrum. Methods 127:495 (1975).
D. Hitlin et al., Nucl. Instrum. Methods 137:225 (1976).
R. Engelmann et al., Nucl. Instrum. Methods 216:45 (1983)
U. Micke et al., Nucl. Instrum. Methods 221: 495 (1984).
[43] M. Basile et al., A limited-streamer tube electron detector with high rejection power against pions, to be published in Nucl. Instrum. Methods (1985).
[44] J. Cobb et al., Nucl. Instrum. Methods 158:93 (1979).
[45] J. Ledermann et al., Nucl. Instrum. Methods 129:65 (1975).
[46] L. Baum et al., Proc. Calorimeter Workshop, Batavia, 1975, M. Atac, ed., FNAL, Batavia, Ill. (1975), p. 295.
A. Grant, Nucl. Instrum. Methods 131:167 (1975).
M. Holder et al., Nucl. Instrum. Methods 151:69 (1978).
[47] A. Bodek et al., Phys. Lett. 113B:77 (1982).
[48] K. Eggert et al., Nucl. Instrum. Methods 176 (1980) 217.
[49] Technical Proposal of the L3 Collaboration, CERN/LEPC/83-5 (1983).
[50] H. Gordon et al. (HELIOS Collaboration), Lepton production, CERN/SPSC 83-51 (1983).
[51] Design Report: An experiment at D0 to study antiproton-proton collisions at 2 TeV, December 1983.

[52] G. Arnison et al., Phys. Lett. 139B:115 (1984).
P. Bagnaia et al., Z. Phys. C. 24:1 (1984).
[53] W.J. Willis and K. Winter, *in* Physics with very high energy e^+e^- colliding beams, CERN 76-18 (1976).
[54] B.L. Beron et al., Proc. 5th Int. Conf. on Instrumentation for High-Energy Physics, Frascati, 1973. Laboratori Nazionali del CNEN, Frascati (1973), p. 362.
[55] Y. Chan et al., IEEE Trans. Nucl. Sci. NS-25:333 (1978).
[56] R. Batley et al., Performance of NaI array with photodiode readout at the CERN ISR, to be submitted to Nucl. Instrum. Methods
[57] M. Miyajima et al., Number of photo-electrons from photomultiplier cathode coupled with NaI (Tl) scintillator, KEK (Japan) 83-36 (1983).
[58] G.J. Bobbink et al., Nucl. Instrum. Methods 227:470 (1985).
[59] G. Blanar et al., Nucl. Instrum. Methods 203:213 (1982).
E. Lorenz, Nucl. Instrum. Methods 225:500 (1984).
[60] J. Ahme et al., Nucl. Instrum. Methods 221:543 (1984).
[61] J.A. Bakken et al., Nucl. Instrum. Methods 228:294 (1985).
[62] C. Laviron and P. Lecoq, Radiation damage of bismuth germanate crystals, CERN-EF/84-5 (1984).
[63] Ch. Bieler et al., Nucl. Instrum. Methods 234:435 (1985).
[64] M. Kobayashi et al., Proc. Int. Symp. on Nuclear Radiation Detectors, INS Tokyo, (1981). Inst. for Nuclear Study, Tokyo (1981), p. 465.
[65] D.E. Wagoner et al., A measurement of the energy resolution and related properties of an SCG1-C scintillation glass shower counter array for 1-25 GeV positrons, to be published in Nucl. Instrum. Methods (1985).
[66] B. Powell et al., Nucl. Instrum. Methods 198:217 (1982).
[67] W. Bartel et al., Phys. Lett. 88B:171 (1979).
[68] P.D. Grannis et al., Nucl. Instrum. Methods 188:239 (1981).
[69] K. Ogawa et al., A test of dense lead glass counters, to be published in Nucl. Instrum. Methods.
[70] R.M. Brown et al., An electromagnetic calorimeter for use in a strong magnetic field at LEP based on CEREN 25 lead glass and vacuum photo-triodes, presented at the IEEE Meeting on Nuclear Science, Orlando, Fla., 1984.
[71] C.A. Heusch, The use of Cherenkov techniques for total absorption measurements, preprint CERN-EP/84-98 (1984): invited talk given at the Seminar on Cherenkov Detectors and their Application in Science and Technology, Moscow, 1984.
[72] H. Grassmann, Untersuchung der Energieauflösung eines CsI(Tl) Testkalorimeters für Elektronen zwischen 1 GeV und 20 GeV, Universität Erlangen (1984).
H. Grassmann et al., Nucl. Instrum. Methods 228:323 (1985).
[73] M. Laval et al., Nucl. Instrum. Methods 208:169 (1983).
[74] D.F. Anderson et al., Nucl. Instrum. Methods 228:33 (1985)
[75] A. Kusumegi et al., Nucl. Instrum. Methods 185:83 (1981).
[76] H.H. Chen et al., Nucl. Instrum. Methods 150:579 (1984).
[77] E. Gatti et al., Considerations for the design of a time projection liquid argon ionization chamber, BNL 23988 (1978).
[78] K. L. Giboni, Nucl. Instrum. Methods 225:579 (1984).
[79] C. Cerri et al., Nucl. Instrum. Methods 227:227 (1984).
[80] J. Engler et al., Phys. Lett. 29B: 321 (1969).
[81] W.A. Shurcliff, J. Opt. Soc. Am. 41:209 (1951).

[82] R.C. Garwin, Rev. Sci. Instrum. 31:1010 (1960).
[83] G. Keil, Nucl. Instrum. Methods 89:111 (1970).
[84] W.B. Atwood et al., SLAC-TN-76-7 (1976).
[85] A. Barish et al., IEEE Trans. Nucl. Sci. NS-25:532 (1978).
[86] W. Selove et al., Nucl. Instrum. Methods 161:233 (1979).
[87] V. Eckardt et al., Nucl. Instrum. Methods 155:353 (1978).
[88] O. Botner et al., IEEE Trans. Nucl. Sci. NS-28:510 (1981).
[89] W. Hofmann et al., Nucl. Instrum. Methods 195:475 (1982).
[90] W. Kienzle, Scintillator development at CERN, CERN-NP Int. Report 75-12 (1975).
[91] W. Viehmann and R.L. Frost, Nucl. Instrum. Methods 167:405 (1979).
[92] J. Fent et al., Nucl. Instrum. Methods 225:509 (1984).
[93] H. Spitzer, Contribution to the Discussion Meeting on HERA Experiments, Genoa (1984);
J. Ahme et al., Novel readout schemes for scintillator sandwich shower counters, to be published.
[94] H.A. Gordon et al., Phys. Scripta 23:564 (1981).
[95] UA1 Collaboration, Technical report on the design of a new combined electromagnetic/hadronic calorimeter for UA1, CERN/SPSC/84-72 (1984).
[96] W. Kononnenko et al., Nucl. Instrum. Methods 214:237 (1983).
[97] L. Bachman et al., Nucl. Instrum. Methods 206:85 (1983).
[98] J. Borenstein et al., Phys. Scripta 23:549 (1981).
[99] H. Blumenfeld et al., Nucl. Instrum. Methods 225:518 (1984).
H. Burmeister et al., Nucl. Instrum. Methods 225:530 (1984).
[100] H. Schönbacher and W. Witzeling, Nucl. Instrum. Methods 165:517 (1979).
Y. Sirois and R. Wigmans, Radiation damage in plastic scintillators, submitted to Nucl. Instrum. Methods.
[101] G. Marini et al., Radiation damage of organic scintillation materials, CERN 'Yellow' Report, in preparation (1985).
[102] Usually, 'accelerated' tests are performed with levels of irradiation 10 to 10^6 times higher than those encountered in an experiment. Because radiation damage is frequently a function of both dose rate and integral dose, such tests are likely to indicate a higher dose tolerance than in the actual lower dose-rate experimental environment. R. Wigmans, private communication and Ref. [100].
[103] R.J. Madaras et al., Nucl. Instrum. Methods 160:263 (1979).
A.E. Baumbaugh et al., Nucl. Instrum. Methods 197:297 (1982).
A.M. Breakstone et al., Nucl. Instrum. Methods 211:73 (1982).
[104] Design report for the Fermilab Collider Detector Facility (CDF), FNAL (1981).
[105] V. Brisson et al., Phys. Scripta 23:688 (1981).
[106] W.J. Willis and V. Radeka, Nucl. Instrum. Methods 120:221 (1974).
[107] L.W. Alvarez, LRL Physics Note 672 (1968) unpublished.
[108] S.E. Derenzo et al., Nucl. Instrum. Methods 122:319 (1974).
[109] K. Masuda et al., Nucl. Instrum. Methods 188:629 (1981).
[110] T. Doke et al., Nucl. Instrum. Methods 134:353 (1976).
T. Doke, Portugal Phys. 12,1:9 (1981).
[111] G.R. Gruhn, private communication (1973).
[112] M. Conversi, Nature 241:160 (1973);
M. Conversi and L. Frederici, Nucl. Instrum. Methods 151:193 (1978).
[113] G.S. Abrams et al., IEEE Trans. Nucl. Sci. NS-27:59 (1980).
[114] V. Kadansky et al., Phys. Scripta 23:680 (1981).

[115] H.J. Behrend et al., Phys. Scripta 23:610 (1981).
[116] C. Nelson et al., Nucl. Instrum. Methods 216:381 (1983).
[117] J.A. Appel, Summary Session of the Gas Sampling Calorimeter Workshop, Fermilab FN-380 (1982);
J. Engler, Nucl. Instrum. Methods 217:9 (1983).
[118] M. Jonker et al., Nucl. Instrum. Methods 215:361 (1983).
[119] G. Battistoni et al., Nucl. Instrum. Methods 202:459 (1982).
[120] SLD Design Report SLAC-273 (1984).
[121] Presentations at the Discussion Meeting on HERA Experiments, Genoa, 1984.
[122] W.F. Schmidt and A.O. Allen, J. Chem. Phys. 52:4788 (1970).
[123] J. Engler and H. Keim, Nucl. Instrum. Methods 223:47 (1984).
[124] L. Onsager, Phys. Rev. 54:554 (1938).
[125] R.C. Munoz et al., Ionization of tetramethylsilane by alpha particles, Brookhaven Nat. Lab. C-2911 (1984), submitted to Chemical Physics Letters.
[126] G.G. Harigel, Nucl. Instrum. Methods 225: 641 (1984).
[127] P.G. Rancoita and A. Seidman, Nucl. Instrum. Methods 226:369 (1984).
G. Barbiellini et al., Nucl. Instrum. Methods 235:55 (1985).
[128] E. Gatti and V. Radeka, IEEE Trans. Nucl. Sci. NS-25:676 (1978).
[129] G. Charpak and F. Sauli, Ann. Rev. Nucl. Part. Sci. 34:285 (1984).
[130] H. Videau, Nucl. Instrum. Methods 225:481 (1984).
[131] G. Battistoni et al., Nucl. Instrum. Methods 176:297 (1980).
[132] H.G. Fischer and O. Ullaland, IEEE Trans. Nucl. Sci. NS-27:38 (1980);
M. Berggren et al., Nucl. Instrum. Methods 225:477 (1984).
[133] E. Iarocci, Nucl. Instrum. Methods 217:30 (1983).
[134] P. Campana, Nucl. Instrum. Methods 225:505 (1984).
[135] H. Aihara et al., Nucl. Instrum. Methods 217:259 (1983).
[136] B. Pope, Proc. DPF Workshop, Berkeley (1983), LBL-15973, p. 49.
[137] O. Botner and C.W. Fabjan, Measurements with the AFS calorimeter, unpublished note (1982).
[138] R. Diebold and R. Wagner, Physics at 10^{34} cm^{-2} s^{-1}, ANL-HEP-CP-84-87 and Proc. of the 1984 Summer Study of the Design and Utilization of the Superconducting Super Collider, Snowmass, Colo. (1984).
[139] P. Curie and A. Laborde, C.R. Acad. Sci. 136:673 (1903).
[140] S. Simon, Nature 135:763 (1935).
[141] T.O. Niinikoski and F. Udo, Cryogenic detection of neutrinos, CERN/NP Internal Report 74-6 (1974).
E. Fiorini and T. Niinikoski, Nucl. Instrum. Methods 224:83 (1984).
[142] B. Cabrera et al., Bolometric detection of neutrinos, Harvard preprint HUTP-84/A077 (1984).
[143] N. Coron et al., A composite bolometer as a charged-particle spectrometer, preprint CERN-EP/85-15 (1985).
[144] A.V. Vishnevskii et al., Moscow preprint ITEP-53 (1979).
[145] R. Bouclier et al., CERN-EP Internal Report 80-07 (1980).
[146] W.J. Willis, private communication.
[147] D.H. Perkins, Ann. Rev. Nucl. Part. Sci. 34:1 (1984).
[148] See, for example, Proc. 1980 International DUMAND Symposium, ed. V.J. Stenger. Honolulu, Hawaii (1981).
[149] R. Cady et al., Proc. 1982 DPF Summer Study on Elementary Particle Physics and Future Facilities, eds. R. Donaldson, R. Gustavson and F. Paige, p. 630.
R.M. Baltrusaitas et al., Phys. Rev. Letters 52:380 (1984).

[150] UA2 Collaboration, Proposal to improve the performance of the UA2 detector, CERN/SPSC 84–30 (1984).
[151] W.J. Willis, BNL 17522:207 (1972).
[152] See for example, Proc. DPF Workshop, Berkeley (1983), LBL–15973.
[153] E.D. Bloom and C.W. Peck, Ann. Rev. Nucl. Part. Sci. 33:143 (1983).
[154] Physics with very high energy e^+e^- colliding beams, CERN 76–18 (1976).
[155] E. Picasso, General Meeting on LEP, Villars-sur-Ollon, 1981, ed. M. Bourquin. ECFA 81/54, CERN, Geneva (1981), p. 32.
[156] The technical proposals for the four LEP experiments have the following LEP Committee numbers:
ALEPH, CERN/LEPC 83–2(1983);
DELPHI, CERN/LEPC/83–3 (1983);
OPAL CERN/LEPC/83–4 (1983);
L3, CERN/LEPC/83–5 (1983).
[157] W.K.H. Panofsky, Proc. Int. Symp. on Lepton and Photon Interactions at High Energies, Bonn, 1981, ed. W. Pfeil, Phys. Inst., Bonn (1981), p. 957.
[158] S.L.Wu, Phys. Reports 107:61 (1984).
[159] C. Rubbia, Physics results of the UA1 Collaboration at the CERN proton–antiproton Collider preprint CERN-EP/84–135 (1984): invited talk given at the Int. Conf. on Neutrino Physics and Astrophysics, Nordkirchen near Dortmund,1984.
[160] Experimentation at HERA, Proceedings of a Workshop jointly organized by DESY, ECFA, and NIKHEF, Amsterdam (1983).
[161] W. Selove, CERN/NP Internal Report 72–25 (1972).
[162] S. Almehed et al., CERN/ISRC/76–36 (1976).
[163] M.A. Dris, Nucl. Instrum. Methods 161:311 (1979).
[164] M. Block, UA1 Collaboration (CERN), unpublished note UA1-6, (1977).
[165] L. Rosselet, Proc. Topical Conf. on the Applications of Microprocessors to High-Energy Physics Experiments, Geneva, 1981, CERN 81–07 (1981), p. 316.
[166] R.L. Ford and W.P. Nelson, Stanford preprint SLAC–210 EGS, Version IV.
[167] T. Kondo et al., Talk given at the 1984 Summer Study on the Design and Utilization of the Superconducting Super-Collider, Snowmass, Colo., 1984.
DELPHI Progress Report, CERN/LEPC 84–16 (1984).
[168] A. Grant, Nucl. Instrum. Methods 131:167 (1975).
[169] H. Fesefeldt, Proc. Workshop on Shower Simulation for LEP Experiments, eds. A. Grant et al. CERN report in preparation.
[170] Proc. Workshop on Shower Simulation for LEP Experiments, eds. A. Grant et al. CERN report in preparation.

TOPICS IN THE PHYSICS OF PARTICLE ACCELERATORS

Andrew M. Sessler

Lawrence Berkeley Laboratory
University of California
Berkeley, CA 94720

I. INTRODUCTION

High energy physics, perhaps more than any other branch of science, is driven by technology. It is not the development of theory, or consideration of what measurements to make, which are the driving elements in our science. Rather it is the development of new technology which is the pacing item.

Thus it is the development of new techniques, new computers, and new materials which allows one to develop new detectors and new particle-handling devices. It is the latter, the accelerators, which are at the heart of the science.

Without particle accelerators there would be, essentially, no high energy physics. In fact, the advances in high energy physics can be directly tied to the advances in particle accelerators. Looking terribly briefly, and restricting one's self to recent history, the Bevatron made possible the discovery of the anti-proton and many of the resonances, on the AGS was found the μ-neutrino, the J-particle and time reversal non-invariance, on Spear was found the ψ-particle, and, within the last year the Z_0 and $W^{\pm}$ were seen on the CERN SPS p-$\bar{p}$ collider. Of course one could, and should, go on in much more detail with this survey, but I think there is no need. It is clear that as better acceleration techniques were developed more and more powerful machines were built which, as a result, allowed high energy physics to advance.

What are these techniques? They are very sophisticated and ever-developing. The science is very extensive and many individuals devote their whole lives to accelerator physics. As high energy experimental physicists your professional lives will be dominated by the performance of "the machine"; i.e. the accelerator. Primarily you will be frustrated by the fact that it doesn't perform better. Why not?

In these lectures, six in all, you should receive some appreciation of accelerator physics. We cannot, nor do we attempt, to make you into accelerator physicists, but we do hope to give you some insight into the machines with which you will be involved in the years to come. Perhaps, we can even turn your frustration with the inadequacy of these machines into marvel at the performance of the accelerators. At the least, we hope to convince you that the accelerators are central, not peripheral, to our science and that the physics of such machines is both fascinating and sophisticated.

The plan is the following: First I will give two lectures on basic accelerator physics; then you will hear two lectures on the state of the art, present limitations, the specific parameters of LEP, HERA, TEV2 and SLC, and some extrapolation to the next generation of machines such as the Large Hadron Collider (LHC), Superconducting Super Collider (SSC), and Large Linear Colliders; finally, I will give two lectures on new acceleration methods.

On basic accelerator physics (which material is encompassed by this article) I must, clearly, select some topics. Notice that everyone of the machines mentioned in the last paragraph is a colliding-beam device. The day of fixed-target machines, for high energy physics, seems over! My choice of topics, and emphasis, will be made with this trend very much in mind. In addition, I shall not go into anything in sufficient detail to allow you to go out and design an accelerator, but I do plan to present the basic physics and, thus, hopefully, give you some appreciation of the limits and performance capabilities of colliders. Lastly, I shall cover some topics where discrete particle effects are dominant, such as in stochastic cooling.

The plan of topics is to first cover single particle dynamics. A complete understanding of single particles is possible, and essential, to the design of particle-handling devices. We can, conveniently, break this up into transverse motion and longitudinal motion. Then I shall cover some topics in collective effects. These effects can be treated by perturbation theory (on the single particle motion), but one must note that it is the collective phenomena which produce the limits on performance and, hence, that it is a proper understanding of this perturbation theory which becomes the essence of accelerator physics. Lastly, I shall cover some topics where discrete particle effects are dominant, such as in stochastic cooling.

There is much material on particle accelerators for this is, after all, an art which is half-a-century old. The student might do well to firstly, consult the five general references listed here.[1,2,3,4,5] These books then give references to original papers and other books. The interested person will want to study the proceedings of the International Conferences on High Energy Accelerators (there have been 12 of them going back to 1956) and the many proceedings of the National Accelerator Conferences. (Published as special volumes, by the IEEE Trans. on Nuclear Science.)

Finally, by way of introduction, I shall not be elegant in my treatment. Rather than presenting Hamiltonians and formalism, I shall give the simplest approach which is adequate. Sometimes, this means using a physical argument, or simply stating, that one can ignore one thing or another. The doubtful reader, or the reader wishing a better treatment, is invited to read the literature where he probably will find what he desires.

PART A: SINGLE PARTICLE DYNAMICS

II. TRANSVERSE MOTION: LINEAR ANALYSIS

In analyzing the transverse motion it is convenient to break this up into linear and non-linear effects. The linear approximation, i.e. linear in the amplitudes of oscillation about a reference orbit, is an exceedingly good approximation and serves to give one a great deal of insight into particle motion in an accelerator. For this reason, the linear theory has been highly developed and is, by now, quite sophisticated. Furthermore, some very comprehensive computer programs have been developed and are now used, throughout the world, to quickly perform linear design of devices.

2.1 Equations of Motion

Everyone knows that in a homogeneous magnetic field B, a charged particle moves in a circle of radius ρ where

$$\rho = \frac{p}{eB} , \tag{2.1}$$

where p is the momentum of the particles. The angular frequency of the particle, its cyclotron frequency, is

$$\omega_c = \frac{eB}{m\gamma} , \tag{2.2}$$

where $\gamma = (1 - \beta^2)^{-1/2}$ is the relativistic factor and β is its velocity in units of the velocity of light.

A convenient set of units is

$$\rho(m) = \frac{p(GeV/c)}{(0.3)B(T)} . \tag{2.3}$$

The first circular accelerator, the cyclotron, was based upon the observation that for a non-relativistic particle ω_c is a constant and hence that fixed frequency radio frequency could be employed to accelerate these particles.

A modern accelerator, again as everyone knows, does not, at all subject the particles to a constant field. In a general magnetic field there exists a "closed orbit," or periodic solution of the equations of motion. This orbit is usually planar and transverse motion in this plane is described by the displacement x. Vertical motion (i.e. perpendicular to the median, equilibrium, plane) is described by the displacement y.

From the Lorentz force or, more elegantly, from the Hamiltonian for a particle in a static (but spacially varying) magnetic field one obtains (keeping only linear terms):

$$\frac{d^2x}{ds^2} + \left[\frac{1}{\rho^2(s)} - k(s)\right] x = \frac{1}{\rho(s)} \frac{\Delta p}{p} ,$$

$$\frac{d^2y}{ds^2} + k(s)y = 0 \quad . \tag{2.4}$$

The arc length along the equilibrium orbit is s, $\rho(s)$ is the radius of curvature of this orbit, and k(s) which describes the focusing property of the magnetic field is given by

$$k(s) = -\frac{1}{B\rho} \frac{\partial B_y}{\partial x} \quad . \tag{2,5}$$

From Eq. (2.1) one sees that $B\rho$, the "magnetic rigidity" is just the momentum of a particle and, of course, a constant in a static magnetic field. The "momentum error" (i.e. from that of the "particle" which defines the equilibrium orbit) is just Δp.

2.2 Matrix Formulation

Taking $\Delta p = 0$, at first, we note that both of Eqs. (2.4) are of the form

$$\frac{d^2z}{ds^2} + k(s)\, z = 0 \quad , \tag{2.6}$$

with suitable definition of k(s) and with z being either x or y. From now we work with Eq. (2.6). The function k(s) is periodic with period, C, the circumference of the machine. Perhaps k(s) is periodic in a length $L \leq C$ corresponding to super periods.

We can write the solution of this second order equation as

$$(Z(s), Z'(s)) = \begin{pmatrix} M_{11}(s,s_0) & M_{12}(s,s_0) \\ M_{21}(s,s_0) & M_{22}(s,s_0) \end{pmatrix} \begin{pmatrix} Z(s_0) \\ Z'(s_0) \end{pmatrix} ,$$

$$\underset{\sim}{Z}(s) = M(s,s_0)\, \underset{\sim}{Z}(s_0) \tag{2.7}$$

where we have, for convenience used a matrix notation. All of the properties of the machine lattice are in the matrix M which is independent of particular orbits. The determinant of M is proportional to the Wronskian of the two linearly independent solutions obtained by starting with (1) unit amplitude and zero slope and (2) zero amplitude and unit slope. Thus the determinant of M is a constant and equal to unity.

It is clear that if the lattice is made up of sections, then the matrix M for transport through the full lattice is just obtained by successively multiplying the matrices for each section.

If the focusing function k(s) is constant, and piece-wise constants cover just about all cases one meets in practice, then the matrix M is simply

$$M(S,S_0) = \begin{pmatrix} \cos\phi & \frac{1}{\sqrt{k}}\sin\phi \\ -\sqrt{k}\sin\phi & \cos\phi \end{pmatrix}, \tag{2.8}$$

where $\phi = \sqrt{k}\,(s - s_0)$. This form is convenient if $k > 0$. If $k < 0$ then M simply transforms to:

$$M(s,s_0) = \begin{pmatrix} \cosh\theta & \frac{1}{\sqrt{-k}}\sinh\theta \\ \sqrt{-k}\sinh\theta & \cosh\theta \end{pmatrix}, \tag{2.9}$$

where $\theta = \sqrt{-k}\,(s - s_0)$. Note that both of these specific forms for M satisfy the requirement det M = 1.

2.3 The α, β, γ Formalism

Any 2 x 2 matrix can be written in the form

$$M = \begin{pmatrix} \cos\mu + \alpha\sin\mu & \beta\sin\mu \\ -\gamma\sin\mu & \cos\mu - \alpha\sin\mu \end{pmatrix}. \tag{2.10}$$

Since our M has unit determinant

$$\gamma = \frac{1+\alpha^2}{\beta} \quad . \tag{2.11}$$

Notice that M^k can be written in the compact form:

$$M^k = \begin{pmatrix} \cos k\mu + \alpha\sin k\mu & \beta\sin k\mu \\ -\gamma\sin k\mu & \cos k\mu - \alpha\sin k\mu \end{pmatrix}. \tag{2.12}$$

Thus, for example, if M describes the motion once around the accelerator then M^k describes k-circuits of the machine. It is evident, from Eq. (2.12), that if μ is real then all the elements of M^k are bounded and hence that the motion is stable. A necessary and sufficient condition for stability is simply

$$|\ \text{Trace } M\ | \leq 2 \quad . \tag{2.13}$$

It is evident that we can find differential equations that are satisfied by α, β, γ. Why, you ask, should the single equation (Eq. (2.6)) be replaced with equations for α, β, γ? The thought is simple: The original equation was for a particle orbit and hence the initial conditions of special orbits come in. If we get equations for α, β, γ we can get free of particular orbits.

Perhaps the easiest way to proceed is to invoke Floquet's theorem, which states that two independent solutions of Eq. (2.6), with k(s) satisfying a periodic condition with L the period, are

$$Z_1(s) = P_1(s)\, e^{+i\frac{\mu s}{L}} ,$$

$$Z_2(s) = P_2(s)\, e^{-i\frac{\mu s}{L}} , \qquad (2.14)$$

where P_1 and P_2 are periodic functions having the same period L. Now one can show that (a problem for the reader!):

$$\frac{d\beta}{ds} = -2\alpha ,$$

$$\frac{d\alpha}{ds} = k\beta - \gamma , \qquad (2.15)$$

and, of course, Eq. (2.11).

One can now write the solution, Eq. (2.14) in terms of these parameters and obtains (another problem for the reader!)

$$Z_{\binom{1}{2}}(s) = C\beta(s)^{1/2}\, e^{\pm i\phi(s)} , \qquad (2.16)$$

where C is a constant and the phase advance $\phi(s)$ is given by

$$\phi(s) = \int \frac{ds}{\beta(s)} . \qquad (2.17)$$

The form of Eq. (2.16) is often used in describing accelerators. Notice that all the machine focusing parameters are in the function $\beta(s)$. The amplitude is proportional to $\beta^{1/2}$ ("Low β" makes small beams which is good for a colliding beam point) and the local wavelength is $2\pi\beta$. (At a "low-β" point, a focus, the particles have a lot of transverse momenta, go at large angles, which is good for a crossing because then the kick from the other beam is relatively less effective.)

From Eq. (2.16) and Eq. (2.14) we see that if we integrate Eq. (2.17) over one machine circumference then

$$2\pi Q \equiv \mu\Big|_{\text{one circumference}} = \int_0^C \frac{ds}{\beta(s)} , \qquad (2.18)$$

where the "Q-value" is simply the number of betatron oscillations per turn. Often Q is called the "tune."

In terms of α, β, γ we can form the quantity

$$\varepsilon = \pi\, \frac{Z^2 + (\alpha Z + \beta Z')^2}{\beta} , \qquad (2.19)$$

which is an invariant, first introduced by Courant & Snyder. They, with Livingston, invented the very concept of "strong focusing"; i.e. allowing k to be a function of S rather than a constant. Demonstration that the emittance, ε, is a constant of the motion is left as still another exercise for the reader.

Typically ε will be limited by some aperture stop; i.e. the accelerator has an acceptance which is less than the emittance of the beam (ion source). Matching of acceptance (or admittance) and emittance is a big subject. Clearly, it is relevant to many aspects of beam handling.

Since Eq. (2.19) is simply an ellipse for any values of α, β, γ the focusing properties of a lattice can be completely described by the motion of this ellipse. The relation between the invariant ellipse and the parameters α, β, γ is shown in Fig. 1. Since particles are usually completely distributed in phase the ellipse curve is uniformly occupied. Thus we only need to study how the ellipse rotates, stretches, and distorts (but keeps the same area!) as we move through the lattice.

2.4 The Dispersion Function, Momentum Compaction, & Chromaticity

In the last section we took a particle with $\Delta p = 0$; i.e. having the same energy as that of the reference particle. We now want to consider the effect of a non-zero Δp.

The dispersion, η (s), is defined as the periodic solution of the first equation of Eq. (2.4). It isn't difficult to show that

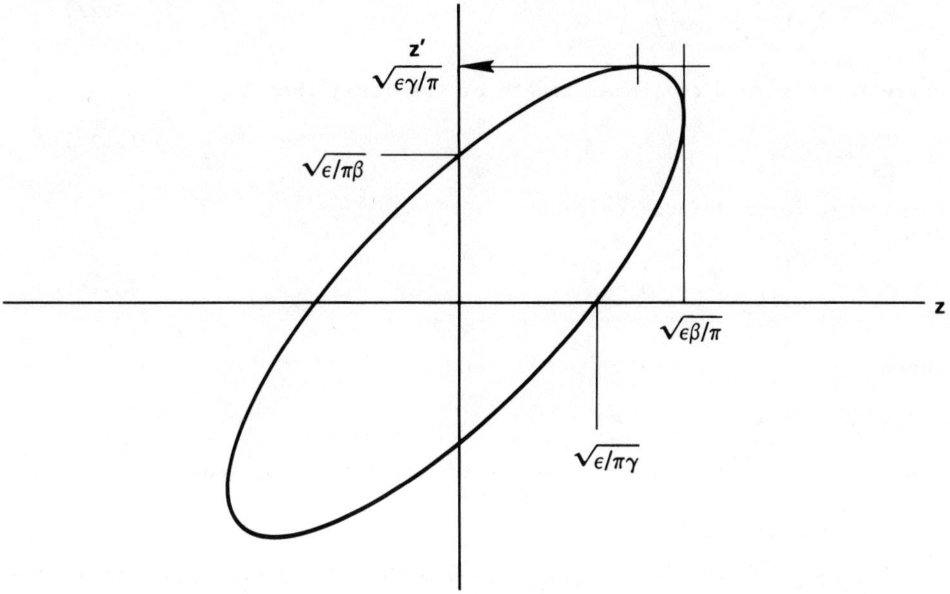

Fig. 1. The invariant ellipse, of area ε, and some of its special dimensions in terms of α, β, γ.

$$\eta(s) = \frac{\beta^{1/2}(s)}{2 \sin \pi Q} \int_0^C \frac{\beta^{1/2}(x)}{\rho(x)} \cos [\mu(x)-\mu(s)-\pi Q] \, dx \quad , \tag{2.20}$$

The average value, over a circumference, of the dispersion is called the momentum compaction times the radius, $R\alpha$. Since $\bar{\eta} \approx R/Q^2$, $\alpha \approx 1/Q^2$. One can see that the name is very physical for α is a measure of how compacted in transverse space are particles of different momenta.

The chromaticity, ξ, is defined by

$$\xi = \frac{\Delta Q/Q}{\Delta p/p} \; ; \tag{2.21}$$

i.e. by the variation in tune with momenta. We will see that ξ comes into an analysis of the coupling between energy oscillations and transverse oscillations (The Head-Tail Effect).

2.5 Thin Lens Approximation

We can often employ thin lenses, rather than the thick lenses of Eqs. (2.8) and (2.9). This is a good approximation frequently, and even if it isn't adequate for detailed numerical predictions it can be used to give insight.

In the thin lens approximation we simply let the arc length, $s-s_0$, in Eqs. (2.8) and (2.9) go to zero. Thus we get, for a focusing lens

$$M_+ = \begin{pmatrix} 1 & 0 \\ -1/f_+ & 1 \end{pmatrix} , \tag{2.22}$$

where f_+ is simply the focal length of the lens; that is

$$1/f_+ = k\ell \quad . \tag{2.23}$$

Similarly, for a defocusing lens

$$M_- = \begin{pmatrix} 1 & 0 \\ 1/f_- & 1 \end{pmatrix} , \tag{2.24}$$

where

$$1/f_- = |K|\ell, \tag{2.25}$$

and $(-f_-)$ is simply the focal length of the defocusing lens.

2.6 The FODO Lattice

A very good lattice for high energy machines, it has been used in the FNAL main ring and PEP and the SPS, is the FODO lattice which is shown in Fig. 2.

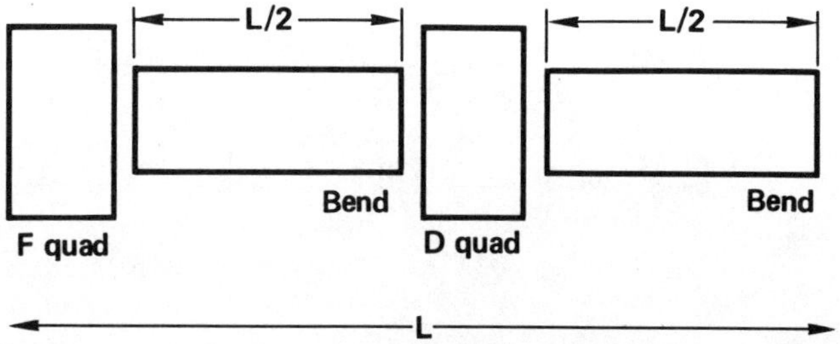

Fig. 2. The FODO Lattice

We can make a thin lens approximation to this lattice and determine all of its properties. Thus, starting at a bend magnet we have:

$$M = \begin{pmatrix} 1 & 0 \\ -1/f & 1 \end{pmatrix} \begin{pmatrix} 1 & L/2 \\ 0 & 1 \end{pmatrix} \begin{pmatrix} 1 & 0 \\ 1/f & 1 \end{pmatrix} \begin{pmatrix} 1 & L/2 \\ 0 & 1 \end{pmatrix} , \qquad (2.26)$$

where we have introduced (the obvious) matrices for the drift through a bend magnet which doesn't focus but simply bends.

The reader might want to carry this through and determine α, β, γ as functions of position, as well as to determine conditions on f and L to give focusing (in both planes!). One finds that

$$\mu = 2 \sin^{-1} \left(\frac{L}{4f} \right) . \qquad (2.27)$$

In addition, the dispersion $\eta(s)$ and then the momentum compaction and chromaticity can be determined.

2.7 Straight Sections and Low-β Insertions

Accelerators need free spaces for injection, rf cavities, extraction, etc. They also need low-β intersection regions. We have developed a formalism which covers these cases, but we haven't explicitly discussed the subject.

Generally, one proceeds by making "insertions"; i.e. sections which don't affect particle motion in the rest of the lattice. Thus one can first optimize the lattice and then design inserts as needed.

A possible long straight section is shown in Fig. 3. This insert has a phase advance of $\pi/2$ and in thin lens approximation

$$s_1 = 1/\gamma ,$$

$$s_2 = \alpha^2 \, 1/\gamma ,$$

$$\frac{1}{f} = \frac{\gamma}{\alpha} , \qquad (2.28)$$

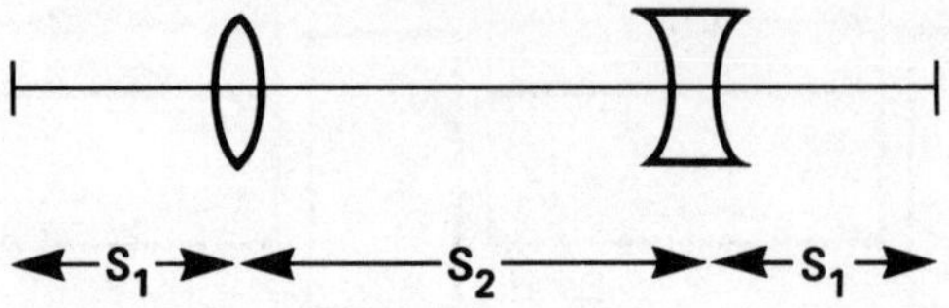

Fig. 3. A Collins Straight section

where α, γ are the parameters of the normal lattice at the "break point" where the straight section is inserted.

Notice that the length of the straight section is (roughly) β. Matching can be achieved in both transverse planes if $\alpha_x = -\alpha_y$, $\gamma_x = \gamma_y$ at the break point. If the dispersion and its derivative (η, η') are both zero at the break point then the insert will not change η (or α).

The design of lattices is an art to which people devote their whole lives. A number of computer programs (Synch, Magic, AGS, Transport) exist to aid in this process for it is exceedingly complicated since many choices are at hand and better choices cost less, accomplish more, or do both simultaneously.

The design of low-β sections is dominated by the variation of β in a free-space region. (We consider a free region for detection of the reaction products; having nearby magnets would help!). From Eq. (2.15) we obtain in a region where $k = 0$

$$\beta \frac{d^2\beta}{ds^2} - \frac{1}{2}\left(\frac{d\beta}{ds}\right)^2 - 2 = 0 \quad . \tag{2.29}$$

This has solution

$$\beta(s) = \beta(0) + \frac{s^2}{\beta(0)} \quad , \tag{2.30}$$

where $\beta(0)$ is the value of β at the crossing point. The very first quadrupoles must turn β around and, because β is large there (probably larger than anyplace else in the lattice), the beam is most sensitive to imperfections in the field or displacements of that quadrupole. This sensitivity; i.e. tolerances in the construction of the first quadrupoles, is what puts the limit on how low $\beta(0)$ can be made.

2.8 Machine Imperfections and Coupling Resonances

Up to this point we have been thinking of a perfect machine. No such thing exists and there is, naturally, a highly developed field of random errors, misalignments, etc.

Suppose there is an error in field, ΔB, which extends over a length L. It is easy to show that just at the error position the periodic orbit is displaced from the periodic orbit when $\Delta B = 0$ by

$$u = \frac{1}{2} \beta \frac{(\Delta B)}{B} \frac{L}{\rho} \cot \pi Q \quad . \tag{2.31}$$

Thus integral values of Q are bad because the new equilibrium orbit is outside the machine. In designing a machine one must stay away from integral values of Q.

Suppose there is a gradient error; i.e. a thin lens of focal length f is introduced into the lattice. Then, the new tune value is given by

$$\cos \mu = \cos \mu_0 + \frac{\beta_0}{2f} \sin \mu_0 \quad . \tag{2.32}$$

(Remember $Q = \mu/2\pi$.) If we let

$$\mu = \mu_0 + \Delta\mu \quad , \tag{2.33}$$

then for $\Delta\mu << \mu_0$ we have, provided $\sin \mu_0 \neq 0$,

$$\Delta\mu = - \frac{\beta_0}{2f} \tag{2.34}$$

$$\text{or } \Delta Q = - \frac{\beta_0}{4\pi f} \quad . \tag{2.35}$$

Thus there is a tune shift provided Q_0 is not an integer or a half-integer. If Q_0 is half-integral or integral that is bad for the motion becomes unstable. One must design machines to avoid these values. How wide is the stop band? Clearly

$$\delta Q = \frac{\beta_0}{2\pi f} , \tag{2.36}$$

and this small band must be avoided.

In an ideal machine the two transverse planes are independent. Field errors introduce coupling and the Eqs. (2.4) become:

$$\begin{aligned} \frac{d^2x}{ds^2} + k_x x &= \frac{y}{B\rho} \frac{\partial B_y}{\partial y} - \frac{B_s}{B\rho} \frac{dy}{ds} , \\ \frac{d^2y}{ds^2} + k_y y &= - \frac{x}{B\rho} \frac{\partial B_x}{\partial x} + \frac{B_s}{B\rho} \frac{dx}{ds} , \end{aligned} \tag{2.37}$$

where B_s is the field along an orbit. (For example a solenoid.)

The analysis of coupling shows that this can be large when

$$\begin{aligned} Q_x + Q_y &= \text{integer} \quad , \\ Q_x - Q_y &= \text{integer} \quad . \end{aligned} \tag{2.38}$$

These values, also, must be avoided; especially the "sum resonance" (the + sign) where both x and y motion becomes unstable and grows exponentially.

Non-linearities add to this subject, but we can't go into that in this article.

III. TRANSVERSE MOTION: NONLINEARITIES

Is the solar system stable? Clearly it is for short times ($\approx 10^{10}$ years), but what about for long times? It consists of 10 major bodies, and thousands of minor bodies. What is the eventual configuration? Will, for example, one of the larger bodies fall into the sun and the other planets escape to infinity (while overall energy is conserved)?

The planets move near equilibrium orbits which are, approximately, ellipses. The small deviations from these orbits can be shown to be linearly stable. What, however, will the effect be of the non-linearities?

Thus one can see that the situation is very similar to that in a particle accelerator which has been designed, according to the theory of Section II, so as to be linearly stable. In this section we address the effect of non-linearities on the linearly stable motion of particles. The first part is devoted to a general discussion and the second to a particular, but terribly important, aspect of non-linear phenomena.

3.1 The KAM Theorem

Starting in the 1960s there has been a revolution in classical mechanics. Whole conferences are now devoted to stochasticity, solitons, strange attractors, and the routes to turbulence. Much of this work is built upon that of Kolmogorov, while the mathematicians Arnold and Moser, who rigorously proved the (correct) intuitive approach of Kolmogorov, where both, even at the time, very aware -- even motivated by -- the importance of this work to accelerators.

The KAM Theorem, which takes hundreds of pages to prove, and many pages to state precisely, is (roughly) the following: Consider a system of N degrees of freedom governed by the Hamiltonian

$$H = \left(1/2\right) \sum_{k=1}^{N} \left(P_k^2 + \omega_k^2 Q_k^2 \right) + \lambda \, [V_3 + V_4 + \ldots] \quad , \qquad (3.1)$$

where ω_k are real frequencies, V_3 and V_4 are cubic and quartic polynomials in P_k and Q_k, and λ is a measure of the nonlinearities. Suppose that:

1) $\sum_k n_k \omega_k \neq 0$; for any integers n_k such that $\sum_k |n_k| \leq 4$

(That is; that low-order linear resonances are avoided.),

2) λ is sufficiently small (but not infinitesimal, so the nonlinearities are small),

3) V_3 is non-zero. (So there is some cubic non-linearity.)

Then, the theorem states that except for a set of small measure, the trajectories are quasi-periodic orbits lying on a smooth N-dimensional surface in the 2 N-dimensional phase space.

Thus it has been proved that some non-linear (i.e. "real") dynamical systems are not ergodic. In particular, a one-dimensional non-linear system has a non-zero stable region around a linear stable equilibrium point, with the phase plane looking like that in Fig. 4. For an N-dimensional system -- there may be instability due to "resonance streaming," "Arnold diffusion," or by "modulational diffusion." Generally, for small non-linearities these processes are negligibly small.

For large non-linearity the system is wildly unstable; i.e. trajectories which are initially close to each other separate at an exponential rate. The system is said to be stochastic.

The dividing line between these two situations is a much-studied subject. It is given (roughly) by the Chirikov condition which is just that the separation between stable regions (buckets) be equal to the extent of the stable region (height of the bucket). (Clearly one only takes the nearest resonance into account when computing the extent of the stable region.)

3.2 Beam-Beam Interaction

The interaction of one beam with another is, generally, a many-body problem, but we can consider (and still get most of the physics) the interaction of a single particle with a charge and current distribution ("weak-strong" beams). If we restrict ourselves to this case we are simply studying a non-linear dynamics problem and it has many of the features outlined above.

Consider, for simplicity, the head-on collision of one particle with a bunch of N_b particles. Suppose the bunch has length ℓ, width w, and height h. If w >> h we need only consider the vertical force which is linearly varying inside the bunch and then drops-off for larger vertical distance y. The electric field inside the beam is

$$E_y = \left(\frac{4\pi N_b e}{\ell wh}\right) y \quad . \tag{3.2}$$

This electric field will change the particles momentum by

$$\Delta p_y = (eE)\left(\frac{\ell}{2c}\right)(2) \quad , \tag{3.3}$$

where the time factor comes from the fact that the particle and the bunch are both moving at the velocity c, and the factor of 2 comes because the magnetic force is just equal to the electric force.

$$\Delta\left(\frac{dy}{ds}\right) \equiv \Delta y' = \frac{\Delta p_y}{p} = \frac{4\pi N_b e^2 \ell y}{\ell whcmc\gamma} \quad . \tag{3.4}$$

Fig. 4. The phase plane of one dimensional motion around a stable fixed point. Notice the regions of instability (x-points) alternating with regions of stability.

The change in tune, ΔQ, is just

$$\Delta Q = \left(\frac{\beta}{4\pi}\right)\left(\frac{\Delta y'}{y}\right) , \tag{3.5}$$

in terms of the β-function at the crossing point. Combining Eq. (3.5) and (3.4) we have

$$\Delta Q = \frac{N_b \, r_0 \beta}{w \, h \, \gamma} , \tag{3.6}$$

where we have introduced the classical particle radius $r_0 = e^2/mc^2$.

Now, if the kick were really linear then as long as ΔQ is not large enough to shift one to a machine resonance there is no problem. But experimentally it is observed that even with $\Delta Q = 0.02$ (a very small number!) there is wild blow-up of the beam. Thus the "linear tune shift," Eq. (3.6), is a measure of non-linear phenomena. Consequences of Eq. (3.6), and the development of "low-β" sections (clearly a good direction) will be covered in the lectures on specific machines.

Theoretical study of the beam-beam phenomena is a very large subject. (Whole conferences have been devoted to it.) Suffice it to say that many dimensional (3) particle simulations are in good accord with the observations and with the basic physics that is described in Sec. 3.1.

IV. LONGITUDINAL MOTION

The longitudinal motion of particles in an accelerator received special attention with the discovery of phase stability by McMillan and Veksler. Subsequently, the special problems associated with storage rings caused a re-examination and new formulation of the theoretical framework of the subject.

4.1 Basic Equations

A particle, of energy γmc^2, circulating in a particle accelerator will have an orbital frequency f. A particle of slightly different energy will have a slightly different frequency and the relation between these two quantities is the dispersion in revolution frequency, η, defined by

$$\frac{df}{dp} = - \frac{f}{p} \eta . \tag{4.1}$$

The dispersion η has two contributions: A particle of energy slightly larger than that of the reference particle, will slip out to an equilibrium orbit of greater length (and hence have its frequency reduced), but move at a greater speed (and hence have its frequency increased). These two effects fight each other with the first clearly winning at high energies where a particle hardly can increase its speed. A typical figure is as shown in Fig. 5. We can show that

$$\eta = \alpha - \frac{1}{\gamma^2} \equiv \frac{1}{\gamma_t^{\,2}} - \frac{1}{\gamma^2} , \tag{4.2}$$

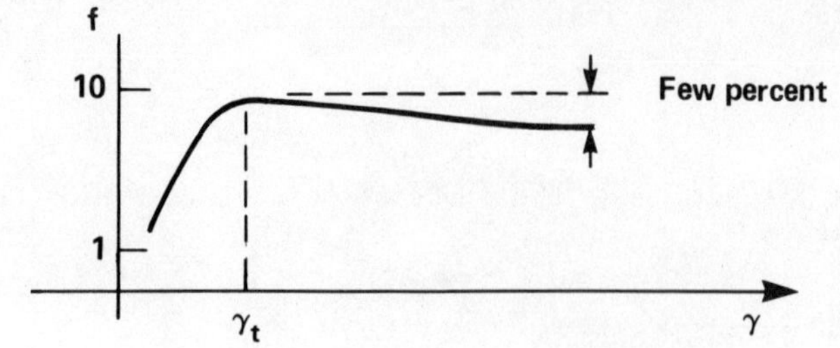

Fig. 5. Frequency vs energy in a typical accelerator.

where α is the momentum compaction factor and the transition energy, γ_t, is defined by $\gamma_t^2 = 1/\alpha$.

Consider an accelerator with only one accelerating cavity which is localized in azimuth and has an rf voltage impressed upon it so that the voltage across the cavity is

$$V = V_0 \sin 2\pi f_{rf} t \quad , \tag{4.3}$$

where the amplitude V_0 and the frequency f_{rf} could be (slowly) changing in time.

Let ϕ be the phase of a particle relative to the rf. Clearly, on passing the cavity the particle will have its energy augmented so that

$$\frac{dE}{dt} = eV_0 f \sin \phi \quad , \tag{4.4}$$

where f is the particle revolution frequency. The phase, ϕ, changes if f is not equal to f_{rf}. Thus

$$\frac{d\phi}{dt} = 2\pi (f_{rf} - hf) \quad , \tag{4.5}$$

where the integer h is the harmonic number of the rf.

These two equations, Eqs. (4.4) and (4.5), are the basic equations describing the energy oscillations, or longitudinal motion, of a particle. Combining them, by taking the derivative of Eq. (4.5):

$$\frac{d^2\phi}{dt^2} = 2\pi \left(\frac{df_{rf}}{dt} - h \frac{df}{dE} \frac{dE}{dt} \right) \quad . \tag{4.6}$$

Using Eq. (4.4):

$$\frac{d^2\phi}{dt^2} = 2\pi \frac{df_{rf}}{dt} - 2\pi h \frac{df}{dE} eV_0 f \sin \phi \quad . \tag{4.7}$$

If the rf frequency is not modulated, and using the definition of Eq. (4.1),

$$\frac{d^2\phi}{dt^2} - (2\pi)\frac{hf^2}{E}\eta e V_0 \sin\phi = 0 \quad . \tag{4.8}$$

Thus for ϕ near zero there are small stable oscillations at the synchrotron frequency:

$$\omega_s^{\,2} = (2\pi f)^2 \frac{\hbar e V_0(-\eta)}{2\pi E} \quad , \tag{4.9}$$

provided $(-\eta)$ is positive. At transition the sign of η changes and, hence, the phase of the rf must be quickly changed by 90° so as to keep the particle motion stable.

The motion is described by Eq. (4.8), which is an integrable system, so that one can easily study non-linear phenomena. These phenomena are very important in the design of systems which efficiently trap particles at injection, or in systems which properly "stack" particles; i.e. make the intense hadron beams one needs for collisions. A Hamiltonian formalism is most advantageous for these studies since the motion is rather complicated and the use of general theorems, such as Liouville's, proves very powerful.

4.2 Hamiltonian Formalism

Introduce the "energy" variable w, which is like an angular momentum,
by

$$w = \frac{E}{2\pi f} \quad . \tag{4.10}$$

Then the Hamiltonian

$$H(\phi,w) = \left(\frac{h\eta 2\pi f}{2pR}\right) w^2 + \frac{eV_0}{2\pi}\left[\cos\phi - \cos\phi_0 + (\phi-\phi_0)\sin\phi_0\right] \quad , \tag{4.11}$$

where ϕ_0 is the phase of the reference particle, gives the equations of motion

$$\frac{d\phi}{dt} = \frac{\partial H}{\partial w} = \frac{h\eta(2\pi f)}{pR} w \quad ,$$

$$\frac{dw}{dt} = -\frac{\partial H}{\partial \phi} = \frac{eV_0}{2\pi}\left[\sin\phi - \sin\phi_0\right] \quad , \tag{4.12}$$

which are just our basic equations in a slightly different notation.

Using this Hamiltonian we can, easily, study the size of the stable region of oscillation (the "bucket"), non-linear variation of the synchrotron frequency, criteria for adiabatic variation of V_0 and ϕ_0, etc.

V. ADIABATIC VARIATION

Accelerators are designed, after all, to accelerate particles. So far, our discussion has been restricted to parameters which are constant in time and we now must generalize our work. The point is, of course, that the synchrotron oscillation frequency, the circulation frequency, and the betatron frequencies are all larger than the frequencies with which parameters change. (This doesn't have to be so for rf modulation and synchrotron frequencies, but one tries to observe the inequality in practice.)

Thus acceleration, and injection and rf stacking, etc. are all adiabatic processes. Thus all of our previous analysis is valid; i.e. it describes properly the oscillations of particles. Any conditions we derived, and we really determined many conditions on the lattice elements and rf system, must all be observed.

What happens as we change parameters slowly; i.e. adiabaticly? If we formulate the theory in a Hamiltonian formalism then the answer is very simple for the theory of adiabatic invariants, or the change of diverse variables under change of parameters, is well worked out.

Take, for example, the longitudinal motion. The canonical variables are ϕ and w and hence

$$\int w \, d\phi = \text{constant} \quad . \tag{5.1}$$

Thus if we take particles from one energy E, to a second energy E_2, and in both cases they occupy all phases then the spread at energy E_2, ΔE_2, is related to the spread at energy E_1, ΔE_1 by

$$\Delta E_2 = \Delta E_1 \left(\frac{f_2}{f_1} \right) \quad . \tag{5.2}$$

A very significant question in building-up intense hadron beams for collision is how best to do this. The Hamiltonian formulation gives insight into this process which has, quite naturally, been studied very extensively.

For transverse motion it is easy to show that the emittance, ε, decreases as $\beta\gamma$. (Here β and γ are the relativistic factors not the orbit parameters α, β, γ.) Thus the transverse size of beams damps and beams get smaller as the energy is increased. Thus one has millimeter size beams at Fermilab as contrasted with the many centimeter size beams at the Bevatron. (Note that the minor radius only damps (for $\beta \approx 1$) as the square root of the energy.)

PART B: COLLECTIVE EFFECTS

VI. EQUILIBRIUM LIMITS

The performance of accelerators is, in general, determined by collective phenomena. (The beam-beam phenomena, considered in Sec. III, which is often a limit on performance but is, of course, a collective effect.)

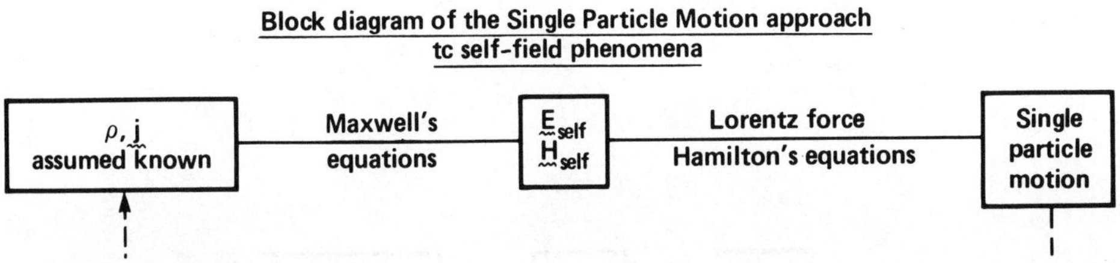

Fig. .6 Block diagram of the single-particle approach self-field phenomena.

Analysis proceeds by considering, firstly, equilibrium limits and then, secondly, possible instabilities of the equilibrium. This was, in fact, the historical path for it wasn't until the 1950s that it was realized that instabilities of relativistic particle beams were possible and, furthermore, imposed severe limits on accelerator performance.

Simple-mindedly, one can proceed as shown in Fig. 6. This approach can always be used, but most of the literature employs the approach shown in Fig. 7; i.e. using the collisionless Boltzmann equation or the Vlasov equation. We shall, in this section, follow the approach of Fig. 6; in the next section we shall follow the method Fig. 7.

Firstly, however, we must find the equilibrium configuration, and as we shall quickly see, there is a space charge limit. In fact, this was realized first, in 1940, by Kerst and Serber. Subsequently, of course, this calculation has been done for much more general geometry.

A very simple approach assumes that we have N particles in a cylindrical beam of minor radius, a, going in a circle of radius, R. For R >> a, we can ignore the curvature and taking a coordinate system as indicated in Fig. 8, we have;

$$\rho = \begin{cases} \dfrac{Ne}{(2\pi R)(\pi a^2)} & ; \quad r \leq a \\ 0 & ; \quad r > a \end{cases} , \tag{6.1}$$

$$\underset{\sim}{j} = \rho\, \beta\, c\, \hat{\underset{\sim}{j}} \, ,$$

where $(\rho, \underset{\sim}{j})$ are the beam charge density and current density.

From Maxwell's equations, the self electric and magnetic fields are:

$$\underset{\sim}{E}_{self} = 2\pi\rho(z\,\hat{\underset{\sim}{k}} + x\,\hat{\underset{\sim}{i}}) \, ,$$

$$\underset{\sim}{H}_{self} = 2\pi\rho\beta(z\,\hat{\underset{\sim}{i}} - x\,\hat{\underset{\sim}{k}}) \, , \tag{6.2}$$

for r less than a.

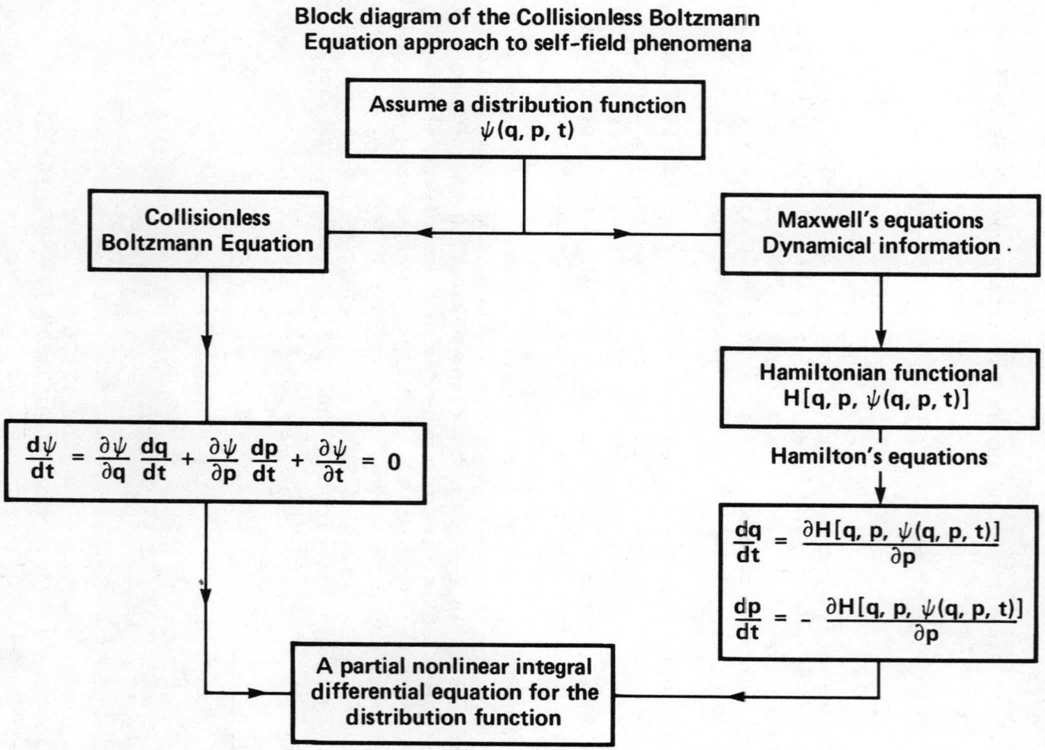

Fig. 7. Block diagram of the collisionless Boltzmann equation approach to self-field phenomena.

The Lorentz force and Hamilton's equations, note how we are just following the procedure of Fig. 6, imply

$$\gamma m \frac{d^2 z}{dt^2} = e\,(E_z - \beta\, H_x) \quad , \tag{6.3}$$

or using Eq. (5.2) and remembering that there is an external guide field:

$$\gamma m \frac{d^2 z}{dt^2} = 2\pi\, \rho e\,(1 - \beta^2) z - e\beta \frac{\partial B_0}{\partial z}\, z \quad . \tag{6.4}$$

Letting

$$Q_0{}^2 = \frac{R}{B_0} \frac{\partial B_0}{\partial z} \quad , \tag{6.5}$$

and introducing the classical electron (proton) radius

$$r_0 = \frac{e^2}{mc^2} \quad , \tag{6.6}$$

the solution is

$$z \approx e^{iQ\theta} \quad , \tag{6.7}$$

where we have now bent the coordinate system so that $z \to R\theta$. Clearly,

$$Q^2 - Q_0{}^2 = - \frac{2\pi R^2 e\rho}{\gamma^3 m\beta^2 c^2} \quad , \tag{6.8}$$

or, using Eq. (6.1),

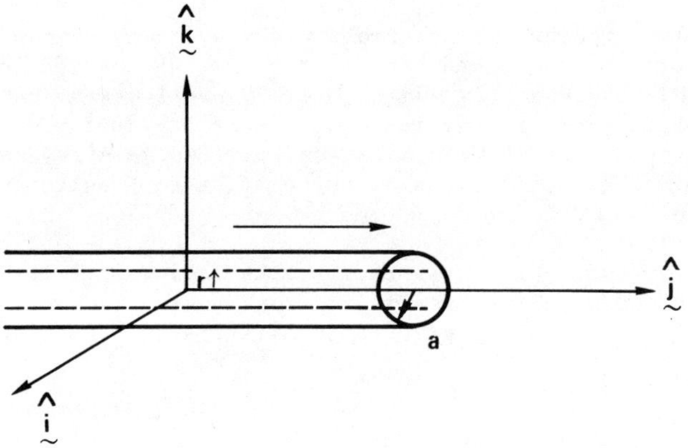

Fig. 8. Coordinate system for a simple derivation of the transverse space charge limit. The beam moves in direction j.

$$N = - \frac{\pi a^2 \gamma^3 \beta^2}{R^2 r_0} (\Delta Q^2) \quad . \tag{6.9}$$

Now we know from our work on machine resonances, Sec. 2.8, that there is a limit on ΔQ^2 in order not to displace the operating point, Q_0, to the nearest resonance; roughly that $\Delta Q \approx 1/4$. Thus we have found an equilibrium space charge limit. Our work needs to be, and has been, extended to elliptical beams and, also, curvature has been included. Most importantly, image effects have been included by Laslett, and these reduce the γ^3 to γ. (The reason is that the precise cancellation, seen in Eq. (6.4), is no longer true.) Of course this is a most significant modification at high energies although it is nevertheless generally true that the equilibrium space charge limit is not the limit in high energy accelerators -- other limits come in first.

VII. LONGITUDINAL INSTABILITY

The simplest collective effect is longitudinal in a beam which in equilibrium is uniform in azimuth. This is because, to good approximation, one has only one degree of freedom; namely the longitudinal coordinate or azimuthal angle. Much can be learned from studying this problem: all of the basic physics, really, and an approach which can be used in much more complicated situations. In fact, it was this instability which was first analyzed, and purely theoretically at that. Subsequently, the effect was found experimentally and still, to this day, imposes a severe limit on machine performance.

7.1 The Negative Mass Effect

In the first half of the nineteenth century the composition of Saturn's rings was a subject of considerable interest. Although one possibility was that a ring consists of many small rocks, this was considered unlikely for such a configuration is staticly unstable (since under mutual gravitational attraction the rocks will coalesce into one moon). Thus in 1856 the Adam's Prize was to be given to the most illuminating essay on subject of Saturn's rings.

The prize was won by a 25-year-old who was, however, not unknown. For 10 years previously he had his first paper read at the Royal Society of Edinborough. Subsequently, James Clerk Maxwell was to make many important contributions to physics, but his prize-winning essay was characterized by the great mathematician, Sir George Airy, as "one of the most remarkable applications of Mathematics to Physics that has ever been seen."

Maxwell's argument was, essentially, the following. Let M be the mass of Saturn and m be the mass of one rock moving with velocity v in a ring of radius R. Balancing gravitational attraction with centripetal force:

$$\frac{GMm}{R^2} = \frac{mv^2}{R} \quad . \tag{7.1}$$

The frequency, f, of the rock is:

$$f = \frac{v}{2\pi R} = \frac{1}{2\pi R} \sqrt{\frac{GM}{R}} \quad , \tag{7.2}$$

The energy, E, of the rock is:

$$E = (1/2)\, mv^2 - \frac{GMm}{R} = -\,(1/2)\, \frac{GMm}{R} \quad . \tag{7.3}$$

Combining these last two equations:

$$f = \frac{(-E)}{\pi GMm} \sqrt{\frac{2(-E)}{m}} \quad . \tag{7.4}$$

The variation of these last equations is shown in Fig. 9.

The many-body argument goes the following way. (Up to now the analysis, has been trivial. Now, suddenly, we are going to consider a many-body situation. The analysis jumps in complexity by orders of magnitude; yet it is all done in two sentences. Think deeply!) Consider a region of the ring where, for some reason (a fluctuation), the density is higher than the average. A rock in front of this increase will be pulled back and so its energy decreases and, hence, its frequency increases and so it moves ahead. (Clearly, an analagous argument can be made for a rock behind.) Hence a ring which is <u>staticly</u> unstable is <u>dynamically</u> stable.

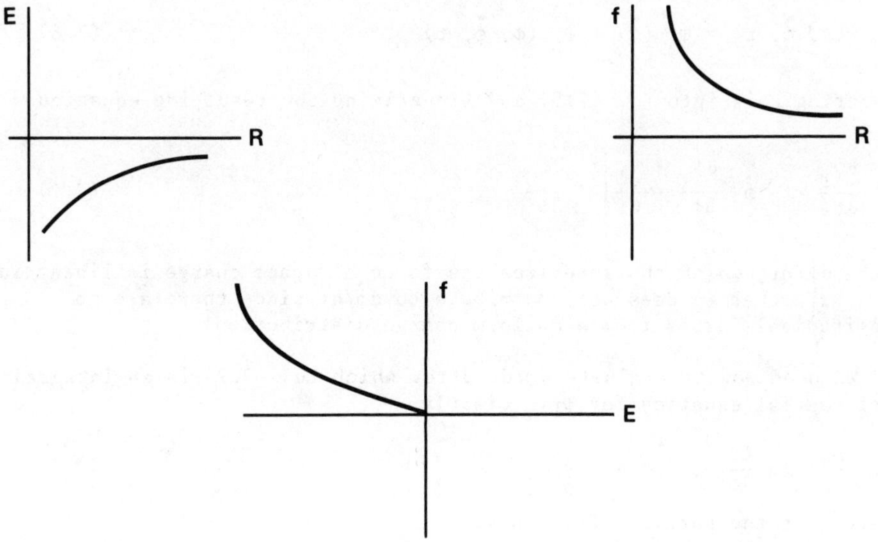

Fig. 9. Variations of f, E, and R for a satellite. It is well-known, in this space age, that as E decreases f increases; i.e. that satellites "go faster" (angularly) as they slow down.

Now it is clear that in a particle beam we have repulsion, not attraction, and thus a staticly stable situation is dynamically unstable. (Historically, this dynamical instability was discovered by accelerator physicists and it was only subsequently that they appreciated that Maxwell had been there 100 years earlier.) It is as if the particles have a "negative mass"; i.e. repulsive forces produce attraction.

We shall now give an analysis of the negative mass effect using the approach of Fig. 7. Thus we describe the particles by a distribution function $\psi\ (\phi,\ \dot{\phi},\ t)$, where ϕ is the azimuthal angle. The distribution function ψ satisfies to very good approximation the collisionless Boltzmann equation. (I realize we are not working in terms of canonical coordinates and momenta, but Liouville's theorem is valid in this ϕ-$\dot{\phi}$ space.) (By ignoring collisions we are neglecting effects which can, in fact, be treated separate. Some discussion of these effects is given in Sec. IX.

We thus have:

$$\frac{d\psi}{dt} \equiv \frac{\partial\psi}{\partial t} + \dot{\phi}\ \frac{\partial\psi}{\partial\phi} + \frac{d\dot{\phi}}{dt}\ \frac{\partial\psi}{\partial\dot{\phi}} \quad . \tag{7.5}$$

A stationary solution to this equation is given by the arbitrary function $\psi_0(\dot{\phi})$ since there are no forces which tend to change ϕ in time. In a "coasting beam" we can arbitrarily choose the distribution of particles over energy, which is, of course, the same statement.

Let us look for small variations from this solution by letting

$$\psi(\phi,\ \dot{\phi},\ t) = \psi_0\ (\dot{\phi}) + \psi_1\ (\phi,\ \dot{\phi},\ t) \quad . \tag{7.6}$$

Inserting this into Eq. (7.5) and linearizing the resulting equation we obtain

$$\frac{\partial\psi_1}{\partial t} = -\ \dot{\phi}\ \frac{\partial\psi_1}{\partial\phi} - \left.\frac{d\dot{\phi}}{dt}\right|_1 \frac{\partial\phi_0}{\partial\dot{\phi}} \quad , \tag{7.7}$$

where $d\dot{\phi}/dt$, which characterizes the force of space charge is linear in ψ_1. Note that ψ_0 does not contribute to $d\dot{\phi}/dt$ since there are no longitudinal forces from a uniform charge distribution.

We need now to evaluate $d\dot{\phi}/dt$ after which Eq. (7.7) is an integral-differential equation for ψ_1. Clearly,

$$\frac{d\phi}{dt} = 2\pi\ \frac{\partial f}{\partial t} \quad , \tag{7.8}$$

where f is the particle frequency.

We may write

$$\frac{\partial f}{\partial t} = \frac{df}{dE}\ \frac{\partial E}{\partial t} \quad , \tag{7.9}$$

and

$$\frac{\partial E}{\partial t} = 2\pi R\ f\ e\ E \quad , \tag{7.10}$$

where R is the radius of the orbit, and E is the electric field due to space charge and taken positive in the direction of particle motion. Thus, from Eq. (7.7), we obtain

$$\frac{\partial \psi_1}{\partial t} = - \dot{\phi}\ \frac{\partial \psi_1}{\partial \phi} - 4\pi^2 R \left(f\ \frac{df}{dE} \right)_e eE\ \frac{\partial \psi_0}{\partial \dot{\phi}} \quad , \tag{7.11}$$

where $(f\ df/dE)_e$ is evaluated for a typical particle in the equilibrium (stationary) distribution $\psi_0(\dot{\phi})$.

The electric field E may be evaluated very easily in the case that the wavelength of the perturbation is large compared to the gap G of the accelerator tank. In this case the electric field may be taken to depend only upon the gradient of the charge distribution at the azimuth in question. Thus

$$E \simeq - \frac{eg}{\gamma^2 R^2} \frac{\partial}{\partial \phi} \int \psi_1\ (\phi, \dot{\phi}, t)\ d\dot{\phi} \quad , \tag{7.12}$$

where $\underset{\sim}{g}$ is a geometrical factor which depends logarithmicly on the ratio of the gap G, to the radius $\underset{\sim}{a}$ of the coasting beam, and is given by

$$g = 1 + 2 \log_e \frac{2G}{\pi a} \quad . \tag{7.13}$$

Inserting this into Eq. (7.11) we obtain:

$$\frac{\partial \psi_1}{\partial t} = - \dot{\phi}\ \frac{\partial \psi_1}{\partial \phi} + \frac{4\pi^2 e^2 g}{\gamma^2 R} \left(f\ \frac{df}{dE} \right)_e \frac{\partial \psi_0}{\partial \dot{\phi}} \frac{\partial}{\partial \phi} \int \psi_1\ d\dot{\phi} \quad , \tag{7.14}$$

which is an linear integral-partial differential equation for ψ_1.

We now seek solutions to Eq. (7.14) of the form

$$\psi_1(\phi, \dot{\phi}, t) = \bar{\psi}_1\ (\dot{\phi})\ e^{i\ (n\phi\ -\ \omega t)} \quad , \tag{7.15}$$

where since Eq. (7.14) is linear with real coefficients we may use a complex solution, meaning always either the real or imaginary part. Clearly if ω is imaginary then the mode either grows or decays in time. Inserting Eq. (7.15) into Eq. (7.14) we obtain

$$\bar{\psi}_1(\dot{\phi})(n\dot{\phi}-\omega) = + \frac{4]^2 e^2 g}{\gamma^2 R} \left(f\ \frac{df}{dE} \right)_e \frac{\partial \psi_0}{\partial \dot{\phi}}\ n \int \bar{\psi}_1(\dot{\phi}) d\dot{\phi} \quad . \tag{7.16}$$

This integral equation for $\bar{\psi}_1(\dot{\phi})$ may be solved immediately since the dependence of $\bar{\psi}_1(\dot{\phi})$ on $\dot{\phi}$ is explicit. Let

$$\bar{\psi}_1(\dot{\phi}) = \frac{C\,\dfrac{\partial\psi_0(\dot{\phi})}{\partial\dot{\phi}}}{n\dot{\phi} - \omega} \quad , \tag{7.17}$$

where C is a constant, and insert this into Eq. (7.16) which yields a self-consistency requirement after cancelling C from both sides of the equation, namely:

$$1 = + \frac{4\pi^2 e^2 g}{\gamma^2 R} \left(f \frac{df}{dE}\right)_e n \int \frac{\dfrac{\partial\psi_0(\dot{\phi})}{\partial\dot{\phi}}\, d\dot{\phi}}{n\dot{\phi} - \omega} \quad . \tag{7.18}$$

This equation is a dispersion relation between the "wavelength" n and the "frequency" ω.

For a nearly monoenergetic coasting beam we may take

$$\psi_0(\dot{\phi}) = \begin{cases} \dfrac{N}{2\pi(2\Delta)} & \dot{\phi}_A - \Delta < \dot{\phi} < \dot{\phi}_A + \Delta \ , \\ 0 & \text{otherwise} \end{cases} \tag{7.19}$$

corresponding to a beam of N particles uniformly spread in "energy" over an interval of width 2Δ, about a mean value $\dot{\phi}_A$. In this case we may readily perform the integral in Eq. (7.18) since

$$\frac{\partial\psi_0(\dot{\phi})}{\partial\dot{\phi}} = \frac{N}{4\pi\Delta} \left\{ -\,\delta\,(\dot{\phi} - \dot{\phi}_A - \Delta) + \delta\,(\dot{\phi} - \dot{\phi}_A + \Delta) \right\} \quad . \tag{7.20}$$

Equation (7.18) becomes

$$1 = \frac{4\pi^2 e^2 g}{\gamma^2 R} \left(f \frac{df}{dt}\right)_e n \, \frac{N}{4\pi\Delta} \left\{ \frac{-1}{n(\dot{\phi}_A + \Delta) - \omega} + \frac{1}{n(\dot{\phi}_A - \Delta) - \omega} \right\} , \tag{7.21}$$

or simply, by solving for ω:

$$\omega = n \left\{ \dot{\phi}_A \pm \left[\frac{2\pi e^2 g N}{\gamma^2 R} \left(f \frac{df}{dE}\right)_e + \Delta^2 \right]^{1/2} \right\} \quad . \tag{7.22}$$

Thus the perturbation moves with the average speed of the beam, $\dot{\phi}_A$. If Δ is small then for real n, ω will be imaginary for $df/dE < 0$, i.e. for operation above the transition energy. It should be noted that increasing the spread in beam energy (increasing Δ) is always a stabilizing influence. Recalling that the energy spread in the beam ΔE is related to Δ by

$$\Delta E = \frac{2\Delta}{2\pi \left(\dfrac{df}{dE}\right)_e} \quad , \tag{7.23}$$

we may write Eq. (7.22) in the more convenient form:

$$\omega = n \left\{ \dot{\phi}_A \pm \left[\frac{2\pi e^2 gN}{\gamma^2 R} \left(f \frac{df}{dE} \right)_e + \pi^2 \left(\frac{df}{dE} \right)_e^2 (\Delta E)^2 \right]^{1/2} \right\} . \qquad (7.24)$$

Thus if the initial disturbance has n waves about the accelerator, the perturbation will grow as $e^{t/T}$ where the time to increase by a factor of e, T is given by:

$$T = \frac{1}{n} \left\{ + \frac{2\pi e^2 gN}{\gamma^2 R} \left(f \left| \frac{df}{dE} \right| \right)_e - \pi^2 \left(\frac{df}{dE} \right)_e^2 (\Delta E)^2 \right\}^{-1/2} , \qquad (7.25)$$

assuming that $\partial f/\partial E < 0$, and the bracket is positive. If the bracket is negative the motion, of course, is stable. Clearly this expression could be used to find a criterion for stability in any given accelerator.

7.2 Longitudinal Resistive Instability

The negative mass instability, as its very name implies, and consistent with Eq. (7.25) only occurs above the transition energy where $(\partial f/\partial E) < 0$. There, of course, the Landau damping, namely the energy spread within the beam, may be adequately large to prevent instability.

All of this is true on an ideal machine; that is a purely "reactive," in fact capacitive, environment where the electric field E is related to the spacial variation (azimuthally) of the charge density by Eq. (7.12). But this isn't the case in general: Resistive elements will produce a phase shift between E and the derivative of the charge density.

This space shift will produce an instability even below transition! The phase shift is usually quite small, but will have a dramatic effect; a fact that wasn't appreciated until a number of years after the work on the negative mass effect. A great deal of work has now been on this subject and, roughly, one can write that

$$\frac{Z_{\parallel}}{n} \leq \frac{m_0 c^2 |\eta| [\Delta(\beta\gamma)]^2}{e I_0 \gamma} , \qquad (7.26)$$

where I_0 is the beam current and η is the dispersion in revolution frequency; i.e.

$$\frac{\Delta f}{f} = - \eta \frac{\Delta p}{p} , \qquad (7.27)$$

and $Z_{\parallel}$ is the longitudinal coupling impedance of the ring. Note that this relation, which is the analog of the criterion of Eq. (7.25) is only dependent upon $|\eta|$ and must be observed below, as well as above, the transition energy.

7.3 The Coupling Impedance

It was an important step in the understanding of instabilities to realize that the analysis of Sec. 7.1 could be generalized so that one would obtain a relation such as Eq. (7.26) with a coupling impedance $Z_\parallel$. Thus the problem shifted to calculating and measuring $Z_\parallel$. In fact, in the construction of all modern machines careful track is kept of $Z_\parallel$ and of each element's (such as pickup electrodes, kickers, clearing electrodes, rf cavities, etc.) contribution to $Z_\parallel$.

For a smooth wall, such as we assumed in Sec. 7.1,

$$Z_\parallel = \frac{n Z_0 g}{2\beta\gamma^2} , \tag{7.28}$$

where Z_0 is the impedance of free space (377 Ω or $4\pi/c$ in Gaussian units).

For other structures, such as those mentioned above, $Z_\parallel$ has been calculated. Suffice it to say that if one is very careful then $Z_\parallel/n$ can be kept to 1 to 3 Ω.

VIII. TRANSVERSE INSTABILITY

A transverse instability of a beam; i.e. a coherent, collective, instability, is due to the reaction of the electromagnetic fields caused by an oscillating beam on the particle motion. In this regard it is just like the longitudinal instabilities considered in the last section. The situation is, however, more complicated now since both diverse amplitudes of transverse oscillation and diverse particle momenta will contribute to the Landau damping of the collective oscillation.

Clearly consideration of a uniform beam (a "rubber band") is simpler than consideration of a bunched beam. In the longitudinal case we didn't even treat in these lectures the case of bunched beams! In the transverse case, however, there are many different phenomena associated with bunches than are disclosed by the analysis of uniform beams. One of these phenomena, namely the head-tail effect, is discussed in Sec. 8.2.

8.1 Transverse Resistive Instability

Suppose an azimuthally uniform beam is displaced transversely, or kicked. It will start to oscillate, as we discussed in Sec. II, and will, as a result, excite electromagnetic fields. These fields should be calculated taking into account the surroundings of the beam.

The electromagnetic fields act back on the particle motion. If the fields are precisely in phase then they will simply change the frequency of the oscillation and provided this shift is not great enough to move one to a machine resonance there will be no bad effect. (The reader should contrast this with the longitudinal case where the negative mass made the situation very different indeed.)

If, however, there is an out-of-phase component due, perhaps, to the finite resistivity of the wall then the original oscillations can be reinforced; i.e. there can be an exponential growth of the oscillation amplitude; i.e. there can be an instability in which tiny oscillations (due to noise or most anything) grow until the beam hits the vacuum chamber walls (or is limited by non-linearities which is, also, usually, unacceptable since phase density has been greatly diluted).

The calculation of transverse instabilities is a complicated task, dozens of papers in the literature, with many interesting subtleties. Roughly, taking only energy spread (and not amplitude spread) into account, one obtains

$$|Z_\perp| < \frac{4mc^2}{e} \; | \, (n - Q)\,\eta + Q\,\xi \, | \; \frac{Q}{RI_0}\left(\frac{\Delta p}{mc}\right) , \tag{8.1}$$

where n is the mode number, Q is the tune, η is the dispersion in the revolution frequency, ξ is the chromaticity, Δp is the spread in momentum, and $Z_\perp$ is the perpendicular impedance. In accord with the physical discussion given above, instability only occurs if $Z_\perp$ has an imaginary component. However, the real part of $Z_\perp$ contributes to the frequency shift of the oscillation and must be "taken care of" by the Landau damping hence it is a good approximation to have the criteria involve the absolute value of $Z_\perp$ (similar to Eq. (7.26).

Note that Eq. (8.1) only depends linearly on Δp and, roughly, is independent of energy since Δp is an adiabatic invariant. Note that unless η and ξ are of the same sign then there will be trouble at some mode number.

The transverse impedance, $Z_\perp$, can be related to the longitudinal impedance $Z_{||}$ (Eq. 7.26 and Eq. 7.28). For a chamber of half width b, and for an oscillation of frequency ω (In Eq. 8.1) $\omega = n\,\omega_0$ where ω_0 is the revolution frequency:

$$Z_\perp = \frac{2c}{b^2}\;\frac{Z_{||}}{\omega} \quad . \tag{8.2}$$

For a cylindrical vacuum chamber, of radius b, containing a beam of radius a, one obtains:

$$Z_\perp = \frac{iRZ_0}{\beta^2\gamma^2}\left(\frac{1}{a^2} - \frac{1}{b^2}\right) + \frac{2cR}{\omega b^3}\,(1 - i)\,\rho/\delta \quad , \tag{8.3}$$

where ρ is the resistivity of the wall (Ωm) and δ is the skin depth. This expression is only valid if the skin depth is small compared to the thickness of the vacuum chamber.

8.2 The Head-Tail Effect

The head-tail instability is a coherent instability of the transverse motion of particles in a bunch. It is driven by a coupling between the

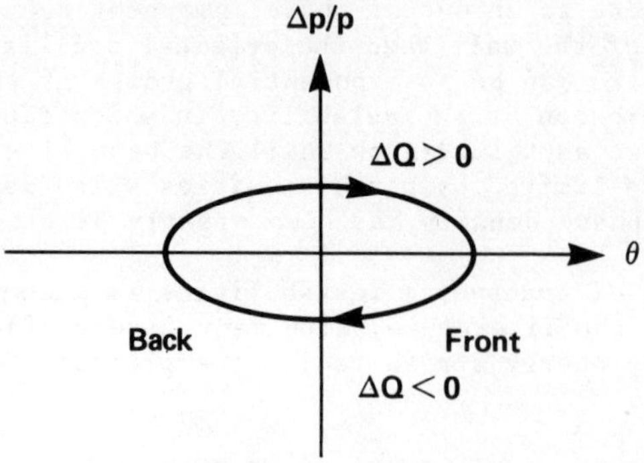

Fig. 10 A bunch showing the situation below transition ($\eta > 0$) and for positive chromaticity ($\xi > 0$).

frequency of transverse betatron oscillations and the momentum of particles; i.e. by the chromaticity. Because of this correlation, as particles move around a bunch, the phase of their betatron oscillations will change as shown in Fig. 10. One can see that the trailing particle always has the same phase shift with respect to the leading particle. This phase change is, clearly,

$$\Delta\phi = \left(\frac{\xi}{\eta}\right) Q\, \omega_0\, \tau \quad , \tag{8.4}$$

where τ is the bunch length (in units of time).

The leading particle will, through its electromagnetic wake, effect the trailing particle. Even in the absence of resistivity, since the chromaticity creates a phase shift, the resulting motion can be unstable. Thus since the effect is present even in ideal structures it can be quite large. This is in fact the case, and after a careful study of signs one can see that if $\xi < 0$ then there can be an instability, but if $\xi > 0$ there will be no trouble (above transition). Machines "naturally" have $\xi < 0$ but modern storage rings all are made with $\xi > 0$ just to avoid the head-tail effect.

PART C: DISCRETE PARTICLE EFFECTS

IX. STATISTICAL PHENOMENA

This article quite accurately mirrors the design of an actual accelerator. Firstly, one must be concerned with the motion of single particles, as described in Part A. Secondly, one examines collective phenomena, as in Part B; here the beam moves as a fluid but the single particle motion can lead to Landau damping of the collective mode. (Other means of damping, for example by feed-back, are not covered in this article.)

Finally, one examines discrete particle effects such as radiation by particles which can, of course, be coherent (as in Free Electron Lasers), but often is incoherent. Although the phenomena of incoherent radiation was known for a long time it was felt to be impossible to "direct each particle" which seemed to be necessary to beat Liouville's theorem (which was the basis for all the analysis of Part B). From this point of view (a very prevelant view up to 1972), stochastic cooling, the very concept and then its implementation, is quite remarkable. We shall describe the basic physics of cooling, in the last part of this section.

9.1 Radiation

An electron moving in a circular accelerator is, of course, accelerated and it will thus radiate. The amount of radiation is

$$P = \frac{2}{3} \frac{e^2 \beta^4 \gamma^4 c}{R^2} . \tag{9.1}$$

Hence the radiated energy per turn is

$$\delta E = \frac{4}{3} \frac{\pi e^2}{R} \beta^3 \gamma^4 , \tag{9.2}$$

or, in practical units (and taking $\beta = 1$):

$$(\delta E)(\mathrm{MeV}) = 8.85 \times 10^{-2} \frac{E^4(\mathrm{GeV})}{R\ (\mathrm{m})} . \tag{9.3}$$

The radiated power, which for electrons can be very significant, is

$$P(\mathrm{Watts}) = 10^6 (\delta E)(\mathrm{MeV})\ I\ (A), \tag{9.4}$$

where I is the beam current. This power loss must be made-up by the rf. In superconducting proton machines this power can be a significant source of heat at the cryogenic temperatures.

The frequency spectrum of the radiation is complicated; for low frequencies it varies as $\omega^{2/3}$. The radiation drops off exponentially for ω larger that a critical frequency, ω_c, where

$$\omega_c = \frac{3\gamma^3 c}{R} . \tag{9.5}$$

Note that this frequency varies as γ^3 times the circulation frequency. Thus the radiation can extend up to very high frequencies.

The high intensity of the radiation, Eq. (9.4) and the high frequency that the radiation extends to, Eq. (9.5), is the reason for the enthusiasm for synchrotron radiation sources. Special bending magnets, wigglers, are used to make the local radius of curvature as small as possible and hence to produce intense radiation of especially high frequency.

The emission of radiation has an effect on the radiating particle. It is only for electrons that this is a significant effect, but the effect is vital in determining the property of beams in an electron storage ring. The radiation reaction can cause either damping or undamping of the electrons' oscillations (transversely and in energy) about the equilibrium orbit. If we characterize this exponential damping rate by rate constants α_x, α_y, α_E then

$$\alpha_{\begin{pmatrix} x \\ y \\ E \end{pmatrix}} = \frac{3}{} \frac{mc^3}{e^2} \frac{\gamma^3}{R^2} J_{\begin{pmatrix} x \\ y \\ E \end{pmatrix}} , \tag{9.6}$$

where the damping partition numbers satisfy:

$$J_y = 1, \quad J_x + J_E = 3 \quad . \tag{9.7}$$

One can arrange by proper lattice design, as one must in a storage ring, to have damping in all three directions.

Thus, on the basis of the above analysis, an electron beam in a storage ring will just damp and damp so that its transverse size becomes smaller and smaller. This is approximately true, and beams become very small indeed, but they do not become arbitrarily small. Why not? Because quantum effects need to be taken into account; i.e. that electrons radiate discrete photons and that the hard photons, which are radiated statistically, kick the electron. In fact, the size of electron beams is determined by these quantum mechanical effects. The energy spread of the beam, which also damps to zero classically, is (in a uniform field).

$$\left(\frac{\sigma_E}{E}\right)^2 = \left(\frac{55}{32\sqrt{3}}\right) \left(\frac{\hbar}{mc}\right) \frac{\gamma^2}{J_E R} , \tag{9.8}$$

and one can see that the finiteness of σ_E is due to a quantum mechanical effect; i.e. to the non-zero nature of Phanck's constant $\hbar$. A similar formula can be given for the radial size of the beam. The vertical size is, clearly, determined by coupling to the horizontal motion.

Finally, the radiation reaction can lead to polarization of the electrons, but an exposition of this topic will not be given here.

9.2 Intra-Beam Scattering

Discrete particle effects must be invoked to understand intra-beam scattering. Generally, f course, we can ignore discrete particle phenomena, but in storage rings where particles are stored for many hours, or even days, attention must be given to even small effects.

The calculation of scattering must be done with careful attention to relativity and to small angle multiple scattering. For low-energy beams, say below 500 MeV, the scattering is important in determining equilibrium beam size (which comes into the luminosity of colliding beams or as the source size in synchrotron radiation sources).

One aspect of intra-beam scattering which is of historical interest (It was the effect which dominated the behavior of the early storage ring ADA.), but highlights a physical phenomena, is the Touschek Effect. In the beam frame of reference, we can conveniently speak of the particles' transverse and longitudinal temperatures. It is easy to see that these are quite different with the transverse temperatures being greater than the longitudinal temperature. Intra-beam scattering, by simple thermodynamic arguments, will tend to equalize these temperatures. As a consequence the longitudinal temperature will increase and, as a result, particles will be lost from the rf bucket. That is to say, intra-beam scattering can lead to a greatly reduced beam lifetime which is the Touschek Effect.

9.3 Stochastic Cooling

Stochastic cooling is the damping of transverse and energy oscillations by means of feedback. A pickup electrode detects (say), the transverse position of an electron and send this signal, after amplification, to a kicker downstream. The time delay is such that a particle is subject to its own signal, which is done by cutting across an arc of the accelerator. The idea was conceived by van der Meer.

Clearly if there is only one particle this will work. Equally clearly, by Liouville's theorem, if there are many particles so that the beam can be treated as a fluid, then there will be no damping. For a finite, but very large number of particles there is a residue of the single particle effect; i.e. some damping as was first realized by van der Meer.

Consider N particles in a ring where f = 1/T is the revolution frequency of particles. Suppose the electronics has a band width W. Then the pickup electrode effectively "sees" a number of particles.

$$n = \frac{N}{2WT} \quad . \tag{9.9}$$

Under the influence of the pickup and kicker this particle will have its transverse displacement, x_i, changed

$$x_i \to x_i - g \sum_{j=1}^{n} x_j \quad , \tag{9.10}$$

where g is the effective gain of the system. Consequently, the value of $(x_i)^2$ will change by:

$$\Delta(x_i^2) \equiv \left(x_i - g \sum_{j=1}^{n} x_j\right)^2 - x_i^2 \quad ,$$

$$\Delta x_i^2 = -2g\, x_i \left(\sum_{j=1}^{n} x_j\right) + g^2 \sum_{j=1}^{n} \sum_{k=1}^{n} x_j \cdot x_k \quad . \tag{9.11}$$

Initially there are no correlations between particles' positions and hence on averaging over all particles we have

$$< \Delta x_i^2 > = - 2g < x_i^2 > + ng^2 < x_i^2 > , \tag{9.12}$$

This maximizes at g = 1/n and the rate of damping of rms betatron amplitudes is

$$\frac{1}{\tau} = \frac{1}{4} \frac{1}{n} \frac{1}{T} = \frac{W}{2N} \tag{9.13}$$

where the T appears because the system works on any one particle once per turn and the factor of 4 reduction comes about from taking the rms (1/2) and the fact that phase space is two-dimensional and only x (not x′) is being damped (1/2).

The assumption of no correlations is not valid, in general, and is especially complicated if one has bunches. Much of the literature is devoted to analyzing this case, which will not be discussed, here, further.

Typically, $W \sim 1$ GHz and N varies from 10^7 to 10^{12}. Thus the cooling time varies from a few milliseconds to an hour. In the Fermilab cooling ring the method, which is applied to all 3 degrees of freedom, damps (for example) the momentum space phase density by a factor of 10^4. It is this large increase in density which has made $\overline{p} - p$ colliders possible.

In the above analysis, which is, of course, very simple, we have assumed no noise in the electronics. In real life there is noise and one might think that if the amplifier noise is greater than the stochastic beam signal then there will be no cooling because the feedback system will heat the beam faster than it cools it. Not so. All one needs to do is select a lower gain and there is cooling (of course, at a reduced rate). (An analogy with a refrigerator is, perhaps, more correct.)

If there is noise, then Eq. (9.10) goes into:

$$x_i \rightarrow x_i - g\left(\sum_{j=1}^{n} x_j + r\right) , \tag{9.14}$$

where r is the amplifier noise expressed as apparent average x-amplitude at the pickup. The analysis now proceeds exactly as before:

$$\Delta(x_i^2) = - 2g <x_i^2> + ng^2 <x_i^2> + ng^2 r^2 . \tag{9.15}$$

Assuming no correlations

$$\frac{1}{\tau} = - 2g + ng^2 \left(1 + \frac{r^2}{<x_i^2>}\right) , \tag{9.16}$$

and maximizing this one obtains

$$\frac{1}{\tau} = \frac{W}{2N\,(1 + \mathcal{N})} \quad , \tag{9.17}$$

where the factor

$$\mathcal{N} = \frac{\langle r^2 \rangle}{\langle x_i^2 \rangle} \quad ; \tag{9.18}$$

is simply noise power over signal power.

REFERENCES

1. H. Bruck, "Circular Particle Accelerators," PUF, Paris (1966). Translated by Los Alamos National Laboratory, LA-TR-72-10 Rev.
2. M. Sands, "The Physics of Electron Storage Rings, An Introduction," Stanford Linear Accelerator Center, SLAC-121 (1970).
3. "Theoretical Aspects of the Behavior of Beams in Accelerators and Storage Rings," CERN 77-13 (1977).
4. R. A. Carrigan, F. R. Huson, and M. Month, editors, "Physics of High Energy Particle Accelerators," American Institute of Physics Conference Proceedings No. 87 (1982).
5. M. Month, "Physics of High Energy Particle Accelerators," American Institute of Physics Conference Proceedings No. 105 (1983).

This work was supported by the U. S. Department of Energy under contract DE-AC03-76SF00098.

PROBLEMS

1. Employ the constancy of the Wronskian of solutions of

$$\frac{d^2 z}{ds^2} + k(s)\, z = 0 \quad ,$$

to show that the matrix in Eq. (2.7),

$$\underset{\sim}{z}(s) = M\,(s, s_0)\,\underset{\sim}{z}\,(s_0)$$

has unit determinant.

2. Show that emittance, ε, is a constant of the motion where

$$\varepsilon = \pi\,\frac{z^2 + (\alpha z + \beta z')^2}{\beta} \quad .$$

3. Derive the equations of motion for the parameters α, β, γ; namely:

$$\frac{d\beta}{ds} = -2\alpha \ ,$$

$$\frac{d\alpha}{ds} = k\beta - \alpha \ .$$

Show, also, that

$$z = C\,\beta^{1/2}\,e^{\pm i\phi(s)} \ ,$$

$$\phi(s) = \int \frac{ds}{\beta(s)} \ .$$

4. The sigma matrix formalism is often used in accelerator theory. Define a 2 x 2 matrix σ in terms of which the invariant ellipse parameters of Fig. 1 become:

$$\sqrt{\varepsilon\beta/\pi} \longrightarrow \sqrt{\sigma_{11}} \ ,$$

$$\sqrt{\varepsilon\gamma/\pi} \longrightarrow \sqrt{\sigma_{22}} \ ,$$

$$\sqrt{\varepsilon/\pi\beta} \longrightarrow \sqrt{\sigma_{22}\,(1 - r_{12}^2)} \ ,$$

$$\sqrt{\varepsilon/\pi\gamma} \longrightarrow \sqrt{\sigma_{11}\,(1 - r_{12}^2)} \ ,$$

where

$$r_{12} = \frac{\sigma_{12}}{\sqrt{\sigma_{11}\,\sigma_{22}}} \ ,$$

$$\varepsilon = \pi\left(\sigma_{11}\,\sigma_{22} - \sigma_{12}^2\right)^{1/2} \ .$$

Given the matrix M (s, s_0), which characterizes a machine lattice (Eq. (2.7)), show that σ (at the point s) is given in terms of σ_0 (at the point s_0) by

$$\sigma = M\,\sigma_0\,M^T \ .$$

5. Use the Hamiltonian formalism for longitudinal motion, Eq. (4.11), to evaluate the extent of the stable region in energy space; i.e. the "height of a bucket." Derive the small amplitude synchrotron frequency and, also, the synchrotron frequency as a function of synchrotron oscillation amplitude.

6. Take into account "images"; i.e. boundary conditions, using the geometry that is simplest for you, and show that the space charge limit of Eq. (5.9) has γ^3 replaced with only a γ-dependence at large γ. You will need to consider the appropriate boundary conditions for a bunched beam. (Assume the skin depth, at the bunch repetition frequency, is less than the vacuum chamber wall thickness.)

7. Work out the geometrical factor, g, in Eq. (7.12) for a cylindrical beam between conducting slabs (Eq. (7.13)) and in a cylindrical pipe where you will find

$$g = 1 + 2 \log_e (b/a)$$

8. Derive expressions for the transverse and longitudinal temperatures ($T_\perp$ and $T_\parallel$) of a beam in terms of the betatron frequency, amplitude of oscillations, etc. Make a numerical evaluation of these temperatures for some high energy machine.

NEW ACCELERATION METHODS

Andrew M. Sessler

Lawrence Berkeley Laboratory
University of California
Berkeley, CA 94720

I. INTRODUCTION

But a glance at the Livingston chart, Fig. 1, of accelerator particle energy as a function of time shows that the energy has steadily, exponentially, increased. Equally significant is the fact that this increase is the envelope of diverse technologies. If one is to stay on, or even near, the Livingston curve in future years then new acceleration techniques need to be developed.

What are the new acceleration methods? In these two lectures I would like to sketch some of these new ideas. I am well aware that they will probably not result in high energy accelerators within this or the next decade, but conversely, it is likely that these ideas will form the basis for the accelerators of the next century.

Anyway, the ideas are stimulating and suffice to show that accelerator physicists are not just "engineers," but genuine scientists deserving to be welcomed into the company of high energy physicists! I believe that outsiders will find this field surprisingly fertile and, certainly fun. To put it more personally, I very much enjoy working in this field and lecturing on it.

There are a number of review articles which should be consulted for references to the original literature.[1,2] In addition there are three books on the subject.[3,4,5] Given this material, I feel free to not completely reference the material in the remainder of this article; consultation of the review articles and books will be adequate as an introduction to the literature for references abound (hundreds are given).

At last, by way of introduction, I should like to quote from the end of Ref. 2 for I think the remarks made there are most germane. Remember that the talk was addressed to accelerator physicists:

"Finally, it is often said, I think by physicists who are not well-informed, that accelerator builders have used up their capital and now are bereft of ideas, and as a result, high energy physics will

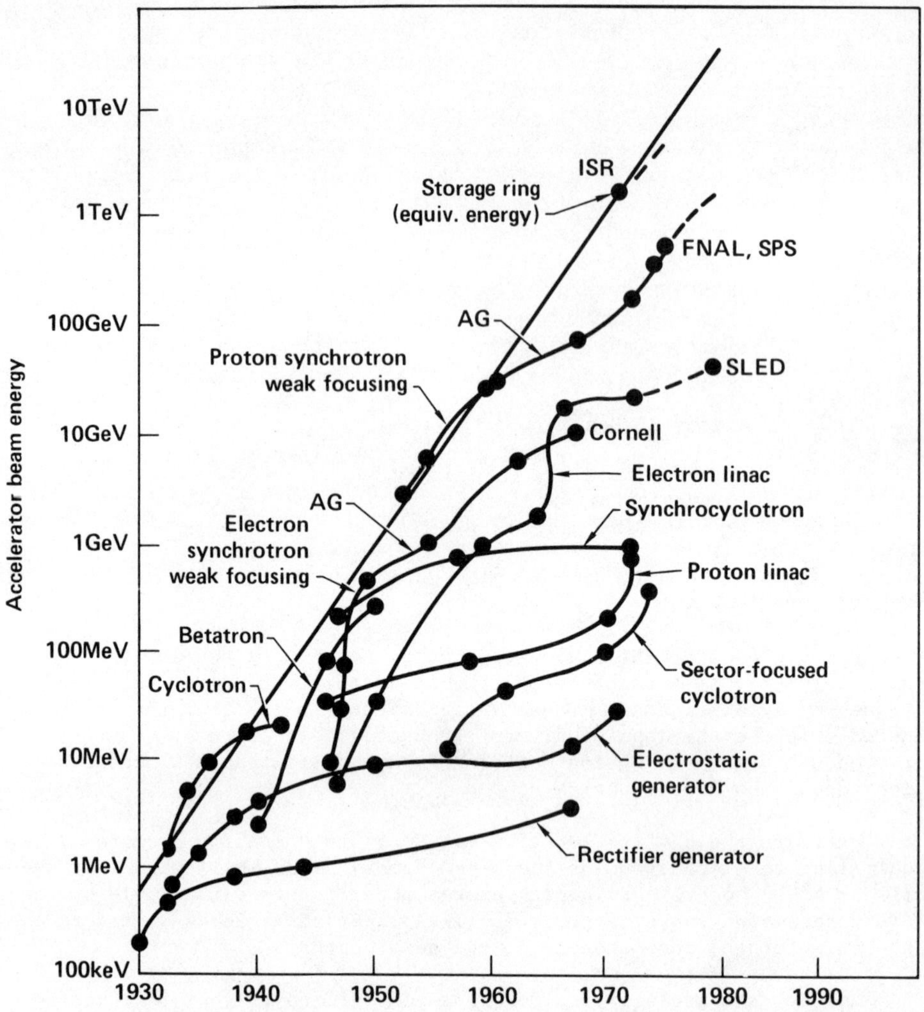

Fig. 1. The "Livingston chart" which shows energy of particle accelerators as a function of time.

eventually -- rather soon, in fact -- come to a halt. After all, one can't build too many machines greater than 27 km, and soon one will run out of space or money (almost surely money before space). This argument seems terribly wrong to me, and worse than that possibly destructive, for it will have a serious effect if it causes, as it well might, young people to elect to go into fields other than high energy physics. The proper response, I believe, is to point -- in considerable detail -- to some of the new concepts which show by example that we are far from being out of new ideas. Some of these concepts shall, in my view, be, or lead to, the "stocks in trade" of the next century, and thus they will allow high energy physics to be as exciting then as it is now. It is our job to make it all happen."

PART A: COLLECTIVE ACCELERATORS

II. OVERVIEW

Collective accelerators; i.e. accelerators which operate by accelerating the particles one really wants to accelerate by having them interact with other particles, have been considered for a very long time. The first thoughts were in the early 1950s and then the subject was given a considerable impetus by the work of Veksler, Budker and Fainberg in the mid 1950s. In the early 1960s both workers in the US and in the USSR discovered "naturally occurring" collective acceleration (described in Section III).

In the late 1960s, extending into the 1970s, considerable experimental work was done on the electron ring accelerator (see Section V). Work has been done on wave accelerators (see Section IV) throughout this period.

Yet now, in 1984, there is still no practical collective accelerator. Why? Because the controlled, repeatable, reliable, and inexpensive (This criterion has not yet been applied, but sooner or later it will be.) acceleration of ions by a collective device is very difficult indeed. We shall, in Part A of this article, indicate the basic physics which makes collective acceleration so difficult. Yet the promise is there, if we could only do it .... .

2.1 Motivation

In a conventional accelerator the particles which are being accelerated are tenuous and thus, to a good approximation.

$$\nabla \cdot \underset{\sim}{E} = 0 \;,$$

$$\nabla \times \underset{\sim}{B} - \frac{1}{c}\frac{\partial \underset{\sim}{E}}{\partial t} = 0 \;. \qquad (2.1)$$

In collective devices one has significant charge which one is handling and thus

$$\nabla \cdot \underset{\sim}{E} = 4\pi\rho \;,$$

$$\nabla \times \underset{\sim}{B} - \frac{1}{c}\frac{\partial \underset{\sim}{E}}{\partial t} = \frac{4\pi}{c}\underset{\sim}{J} \;. \qquad (2.2)$$

As one can see this generalization allows an almost unlimited range of possibilities.

In conventional devices the external currents, which produce the E and B, are limited by the properties of materials. Consequently there are limits, for example on accelerating fields or bending radii, and these limits then produce the restrictions on performance of the devices. In collective accelerators, however, one can obtain very large $\underset{\sim}{E}$ and $\underset{\sim}{B}$ if one can contain, and control, the ρ and $\underset{\sim}{J}$. Thus there is the promise of improved performance as well as compactness and inexpensiveness.

How can one generate the collective $\underset{\sim}{E}$ and $\underset{\sim}{B}$? Either from stationary charges, which are not used in any device, or from streaming charges. One wants streaming charges so as to obtain a $\underset{\sim}{J}$ and, also, to reduce the forces of the streaming particles on each other. This reduction is the result of a cancellation between the electric forces and magnetic forces which is to order $1/\gamma^2$ when the particles are far from any boundary and moving in a straight line.

Typically, the charges and currents which "do the work" in collective accelerators are electrons generated by an intense relativistic electron beam (IREB). These devices are pulsed diodes with field emission cathodes and a foil anode or ring anode. Usually, these machines produce electrons of 10 kA to 100 kA, in the range of 1 MeV to 10 MeV, and with pulse lengths of 10 nsec to 100 nsec.

The electron density, since the beams have a radius from 1 cm to 10 cm are 10^{11} to 10^{13} cm^{-3}. A typical electric field is, then,

$$E\ (\mathrm{MV/m}) = \frac{6.0\ I\ (\mathrm{kA})}{r_b(\mathrm{cm})} \quad . \tag{2.3}$$

For I = 20 kA and a radius r_b = 1 cm this is a field of 120 MV/m which should be compared with the Stanford Linear Collider (SLC) which has a field of 17 MV/m, or with an ion linac (like LAMPF) which has a gradient of 1 MV/m.

2.2 Impact Acceleration

Consider a bunch of N_1 electrons of mass m moving at very high speeds ($\gamma >> 1$) which collide with a stationary light bunch of N_2 ions each having mass M. We must have

$$N_1\ m >> N_2\ M \quad . \tag{2.4}$$

Energy and momentum are, of course, conserved in this collision:

$$\begin{aligned} N_1\ m\gamma\beta &= N_1\ m\gamma_1\ \beta_1 + N_2\ M\ \gamma_2\ \beta_2 \quad , \\ N_1\ m\gamma + N_2\ M &= N_1\ m\ \gamma_1 + N_2\ M\ \gamma_2 \quad . \end{aligned} \tag{2.5}$$

See Fig. 2 for appropriate definitions. Solving these equations, one obtains the result that each ion receives an amount of energy

$$W = 2\gamma^2\ Mc^2 \quad , \tag{2.6}$$

which is a very large amount indeed.

Veksler believed this was the only way to attain really high energies. The Laser-Plasma Accelerator can be thought of from this point

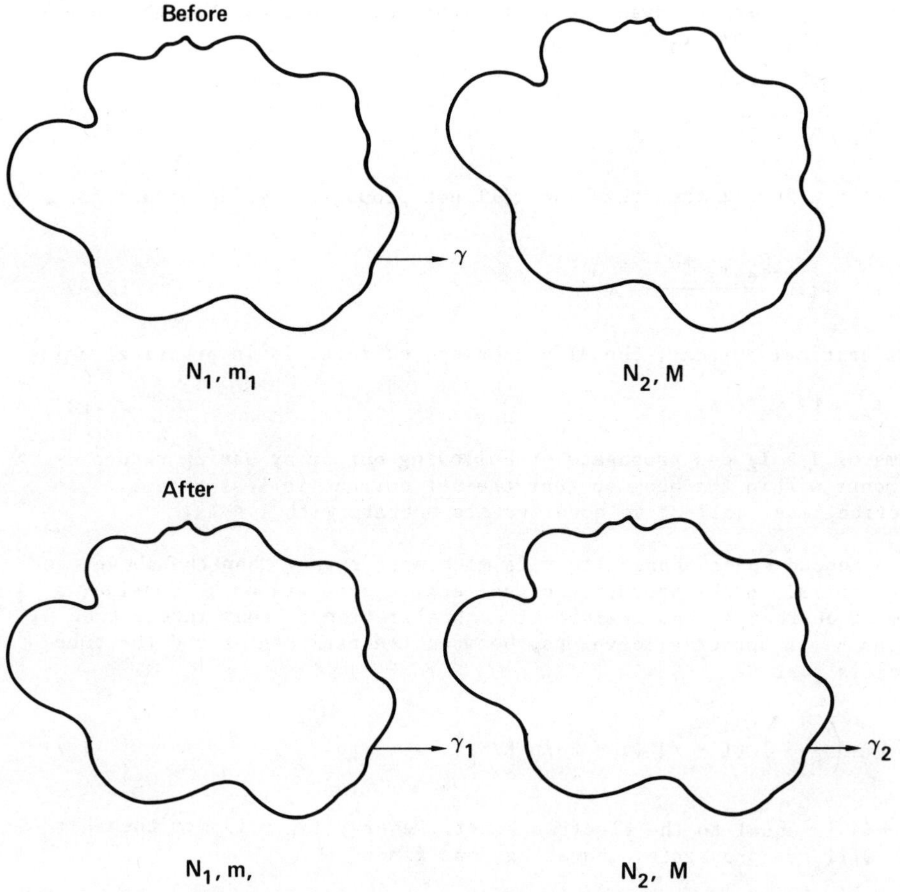

Fig. 2. An impact accelerator.

of view; the large bunch is a group of electrons making a ponderomotive well, which is the only way yet conceived for "holding a bunch together" during a collision. In fact, it was the inability to make integral bunches which prevented making any real progress along these lines.

2.3 Electron Beam Physics

We have noted that IREBs are central to most collective acceleration schemes and hence it is necessary to remark upon one or two key characteristics of such beams.

For an electrically neutral uniform beam, an electron at the beam edge has a cyclotron radius

$$r_{ce} = \frac{\beta_e \gamma_e mc^2}{eB_\theta(r_b)} , \qquad (2.7)$$

where B_θ (r_b) is the magnetic field produced by all of the other electrons in the beam; i.e.

$$B_\theta (r_b) = \frac{2I_e}{cr_b} \quad . \tag{2.8}$$

If $r_{ce} > r_b/2$ then the beam will not propagate, which occurs for

$$I > I_A \equiv \frac{\beta_e \gamma_e mc^3}{e} \quad . \tag{2.9}$$

This critical current, the Alfven-Lawson current, is in practical units

$$I_A = 17 \gamma_e \text{ (kA)} \quad . \tag{2.10}$$

Beams of $I > I_A$ can propagate by hollowing out or by having return currents within the beam so that the net current is less than I_A. In practice, most collective accelerators operate with $I < I_A$.

A second space charge limit is much more severe than the above, and more central to the operation of collective accelerators. Consider a beam of current I_e and fractional neutralization f, sent into a tube of radius R. A potential develops, between the beam center and the tube, which is just

$$V = \left(\frac{I_e}{\beta_e c}\right) (1 - f) [1 + 2 \ln R/r_b] \quad . \tag{2.11}$$

If this is equal to the electron kinetic energy $(\gamma_e - 1) mc^2$ then the beam will not propagate. Equating, one finds

$$I_1 = \frac{\beta_e(\gamma_e - 1) \left(\frac{mc^3}{e}\right)}{[1 + 2 \ln R/r_b] [1 - f]} \quad . \tag{2.12}$$

For $f = 0$ we get a limiting current which is less than I_A. For $I_e > I_1$ the beam will not propagate but neutralize itself (so that $f \to 1$ and I_1 exceeds I_e) and then propagate on. The time scale for neutralization dominates the propagation speed which can be far less than one might expect; i.e. the speed of light.

There are many other space charge limits of IREBs, as well as many other interesting aspects of their physics, but we have enough for this review of collective accelerators.

III. SPACE CHARGE WELLS

If an IREB is simply injected into a gas, as shown in Fig. 3, then ions will be accelerated. Note that this is a collective phenomena; it certainly isn't single particle acceleration by the potential-drop of the IREB generator for that potential is going the wrong way.

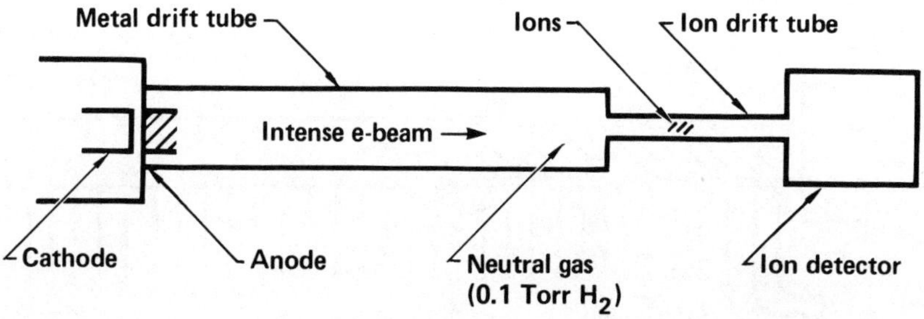

Fig. 3. Schematic drawing of a moving space charge well accelerator.

Alternatively, if an IREB is injected into a vacuum, but produced by means of a plastic ring anode (a copious source of ions), then accelerated ions will be observed.

Roughly, one obtains about 10^{13} protons, of up to 10 MeV, and with a pulse width of 10 nsec. The physical mechanism, in the case of a beam sent into a gas, is believed to be the following. The beam is arranged to be of greater current than the space charge limit I_1 and hence it propagates but a short distance and then stops (a "virtual anode"). After a bit the beam is neutralized and propagates a bit further. This process, since the beam velocity is low, can accelerate ions from rest. There is, clearly, a space charge well associated with the beam front and hence the idea of a space charge well accelerator.

Of course the situation is, in reality, very complicated with the IREB changing in time, 3D effects, photo-ionization, impact-ionization, etc. Much of this has been studied, in most detail by numerical simulations. One must explain, and roughly one can, a potential well of depth 2 or 3 times the beam kinetic energy, (One might think there would be exact equality, but remember that the situation is dynamic and electrons keep streaming into the well and hence make it deeper.) a well-extension of 1 - 2 times the pipe radius, and a well-velocity sufficiently low as to pick-up ions from rest.

Once one realizes that a space charge well can be formed by an IREB then one "only" needs to control its velocity and one has an accelerator. Much effort, needless to say, has gone into beam-front velocity control. One system proposed by Olson, that is conceptually simple, is shown in Fig. 4. Here laser pulses, properly timed, are used to create plasma into which the IREB propagates at a controlled speed. This device has actually been made to control beam front velocity, but no ions have yet been accelerated.

Alternatively, pulsed wall plasmas have been considered. Also, study has been made of slow-wave structures with which heavy ions have been accelerated in a more effective way than without such structures.

This approach has not yet resulted in a practical accelerator. It seems unlikely that one can obtain very high energy particles this way, but perhaps one can make an accelerator for one of the many hundreds of uses to which low-energy accelerators are put (such as food treatment, chemical polyrization, or medical uses).

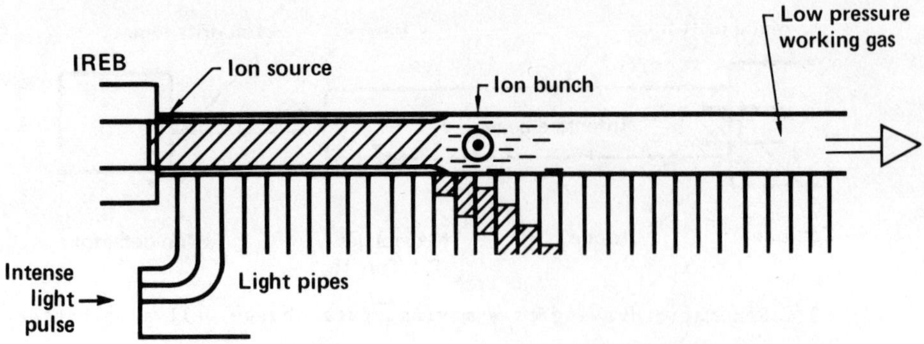

Fig. 4. The Ionization Front Accelerator.

IV. WAVE ACCELERATORS

Wave accelerators are based upon the idea that an IREB is a non-neutral plasma which can support waves. If these waves are unstable then they will grow, at the expense of the beam kinetic energy, and thus can produce a large electric field suitable for the acceleration of ions.

In order to be proper for the acceleration of ions the wave phase velocity must be controlled, variable, and reproducible. Furthermore it must be a "reasonably good" wave; i.e. it must be coherent over many wavelengths and for times greater than the acceleration time.

Firstly, we need to be able to stably propagate an IREB. This can be done in the presence of a longitudinal magnetic field, B, but not otherwise. Such flow is called Brillouin flow and requires that three conditions be met:

$$\omega_p^2 < \frac{\gamma_e \Omega}{r_b} , \tag{4.1}$$

$$2\,\omega_p^2 < \gamma_e^2\, \Omega^2 , \tag{4.2}$$

$$\omega_p^2\, r_b^2 < 4c^2 , \tag{4.3}$$

where the plasma frequency, ω_p, is

$$\omega_p^2 = \frac{4\pi n e^2}{\gamma_e m} , \tag{4.4}$$

and n is the laboratory beam density, and the cyclotron frequency, Ω, is

$$\Omega = \frac{eB}{\gamma_e mc} . \tag{4.5}$$

Condition Eq. (4.1) is that the self electric field is smaller than the confining magnetic field. The Eq. (4.2) is just the relativistic form of the usual condition for Brillouin flow. And Eq. (4.3) is simply Eq. (2.10).

It is a relatively simple matter to find the longitudinally propagating waves on a cylindrical (and wide) electron beam. The dispersion relation is algebraic in the frequency of the wave and has 8 modes. A number of different programs have been based upon experimental work focused upon one mode or another.

There is only one mode which has a phase velocity which can be made very small, as was noted by Drummond and Sloan. This mode is called the Doppler-shifted cyclotron mode and has

$$\omega = k\, v_e - \Omega \quad , \tag{4.6}$$

or a phase velocity

$$v_{ph} = \frac{\omega}{k} = \left(\frac{\omega}{\omega + \Omega}\right) v_e \quad . \tag{4.7}$$

By varying B as a function of z one can vary v_{ph}. This has been done, as well as to grow the desired wave (and no other wave!), but the program was terminated before ions were accelerated.

Other workers, as I have noted, have focused upon other waves as well as upon waves which come about from ion-electron oscillations (as contrasted with pure electron modes). Suffice it to note that there is not yet, to date, a practical wave accelerator.

V. ELECTRON RING ACCELERATOR

An electron ring accelerator (ERA) is a device having a compact ring of electrons which has an associated electric field which can, then, be used to accelerate ions. A schematic of the device is shown in Fig. 5. At one time, this approach attracted many workers and a good number (6 or so) of groups. Now there is only the Dubna group still extant. This was the very first group, having been started by Veksler, who conceived the idea, about two decades ago.

In order to obtain a large field gradient one must make very compact rings having a large electric current. (The current is necessary for if the electrons were at rest, then their self-electric field would blow them apart. However if they move then the magnetic, current-current, force almost cancels the electrostatic repulsion.) The electric field from a ring is (roughly):

$$E\ (\mathrm{MV/m}) = \frac{(4.58 \times 10^{-12}) N_e}{R\ (\mathrm{cm})\ a\ (\mathrm{cm})} \quad , \tag{5.1}$$

where N_e is the number of electrons in the ring and R and a are the two radii of the torus.

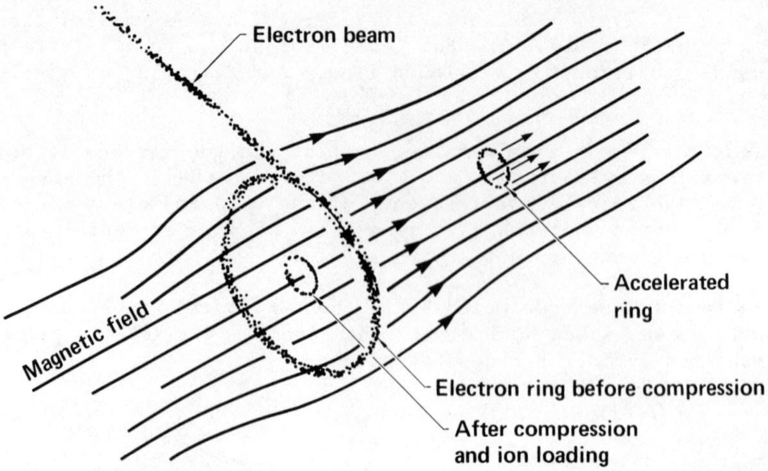

Fig. 5. The Electron Ring Accelerator in its simplest version. The acceleration of the ring, axially, is caused by a decreasing magnetic field.

Such a ring can most be conveniently made by injecting a larger ring and then "compressing" it by increasing the magnetic field enveloping the ring. For an axially symmetric field the azimuthal component of the canonical momentum, P_θ, is conserved and hence

$$R\,P_\theta - e\,R\,A_\theta = \text{const} \quad , \tag{5.2}$$

where the magnetic field is described by the vector potential A_θ.

For a uniform field of magnitude B

$$A_\theta = \frac{1}{2}\,BR \quad , \tag{5.3}$$

and since

$$P_\theta = eBR \quad , \tag{5.4}$$

we see that BR^2 is constant during compression. In the Berkeley ERA the major radius was compressed from 20 cm to 5 cm with ring current, correspondingly, increasing from 1/2 kA to 2 kA.

The minor ring dimensions are given by conservation of the adiabatic invariant

$$\frac{P_\theta Q a^2}{R} = \text{const} \quad , \tag{5.6}$$

where Q is the tune of the electrons. Although Q changes as compression is undergone (and decreases), a^2 damps a good bit.

Once one has a compact ring, with an associated large electric field, one "loads" the ring with ions. Not too many ($N_i < N_e$) so that there is still a large self electric field, but not so few $N_i > N_e/\gamma^2$ so that the ion-produced electric field will more than overcome the net electric repulsion. Thus the ring is self-focused both for its electron and for its ion component; a concept pointed out, in a slightly different context, by Budker.

Other means of focusing the ring are possible, such as image focusing, but we shall not discuss that here.

The ring is then accelerated axially and drags the ions along. Since the ions and electrons move at the same speed one picks up a factor of $M/m = 1837$ in the energy. This is not correct, however, for the electrons are rotating, and one only gains a factor of $M/m\gamma \approx 50$, which is still a sizeable factor. Alternatively, one can't accelerate the ring too quickly or the ions will be left behind. A measure of the rate of acceleration is just the peak field (Eq. (5.1)). Typically one can easily accelerate rings and great efforts must be made not to do it too quickly.

Rings can either be accelerated by electric fields (rf) or by an inverse compression process; i.e. by "magnetic expansion." It is not hard to show that, in the latter case, the ion energy W is given by

$$W = \frac{(1 - b^{1/2})\, Mc^2}{b^{1/2} + g} , \qquad (5.7)$$

where g is the ion mass loading

$$g = \frac{N_i M}{N_e m \gamma_c} , \qquad (5.8)$$

and γ_c is the electron (rotation) γ in the compressed state. The quantity b is the magnetic field falloff, which must not be too fast or the ions will be lost. For most rapid acceleration one should choose

$$b \equiv \frac{B_z}{B_c} = \frac{1}{\left[1 + \frac{3z}{2\lambda}\right]^{2/3}} , \qquad (5.9)$$

where λ is a characteristic falloff distance and can be expressed in terms of ring parameters:

$$\lambda \approx 2\pi \frac{RaMc^2}{e^2 N_e} . \qquad (5.10)$$

Amazingly enough, in view of the complexity of the concept, good quality rings have been made by three groups and ERAs have been made to accelerate ions (to a few MeV) in two laboratories.

The difficulties are many and the complexity of an ERA has prevented practical accelerators from being constructed. In addition, there is a severe space charge limit on making good quality rings; i.e. one cannot achieve really large acceleration gradients this way; in practice the limit is probably lower than 100 MV/m.

PART B: LASER ACCELERATORS

VI. GENERAL CONSIDERATIONS

At first sight it would appear that a laser would be ideal for accelerating particles. The fields at the focus of a powerful laser are 10^4 to 10^6 MV/m, which might be compared to that of the Stanford Linear Collider (SLC); namely a gradient of 17 MV/m. Considering, again, the large field of a laser one notes that it is in the wrong direction; namely it is transverse instead of longitudinal. Furthermore the field only extends over the focus and the depth of field is not very large and would be soon passed through by a high energy particle. And, in addition, a light wave travels, of course, at the speed of light and therefore is not in resonance with any material particle. One can see that acceleration by lasers is not easy; somehow one must contrive to get around the three difficulties mentioned. Surprisingly enough one can, in fact, accelerate with lasers. It is instructive to consider firstly, the effect of a plane wave on a charged particle. The motion which is given to the particle is shown in Fig. 6. One sees that there is no continuous acceleration; in fact after each period of the plane wave the particle is returned to rest. Suffice it also to note that for the most powerful lasers the acceleration in one quarter of a cycle is only to a few hundred MeV and therefore of no interest. (One should observe, however, that this mechanism is very effective when astronomical distances are involved; it is believed to be the primary source of cosmic rays, with the acceleration occurring in the field of a pulsar.)

The field pattern produced by <u>any</u> array of optical elements, provided one is not near a surface and not in a medium, is simply a superposition of plane waves. It is not very hard to generalize the

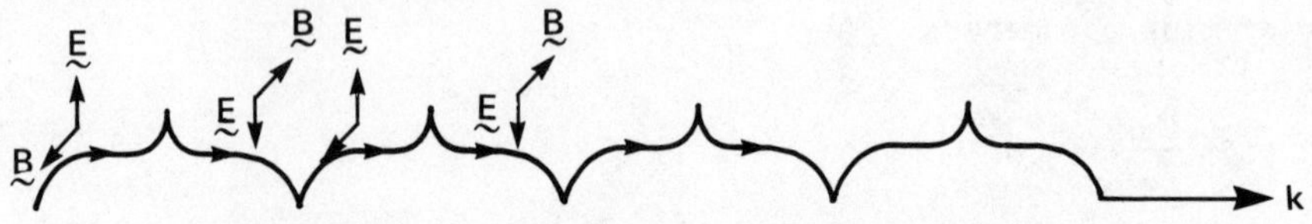

Fig. 6. Motion of an electron in a plane electromagnetic wave. The particle motion is determined by a combination of (reversing) transverse electric field which accelerates the particle and (reversing) transverse magnetic field which bends the particle. The net effect is that a particle is moved along in the direction of the wave, but is not accelerated.

considerations made for a single plane wave and conclude that for a relativistic particle, which moves very closely at a constant speed and in a straight line, there is not net acceleration! This fundamental theorem is most valuable for it allows one to discard schemes which won't work and focus one's thoughts in fruitful directions.

In order, then, to make a laser accelerator one must either be in a medium, or near a surface, or bend the particles. These various possibilities have all been pursued and lead to devices which will certainly work. (Most have already been demonstrated in the laboratory.) Whether or not practical accelerators can be made using these various approaches is something we do not know yet. Engineering considerations which, someday, will have to be made for any serious accelerator contender simply can't yet be made.

We have just argued, that in order to have continuous acceleration one must either (1) slow the wave down (i.e. be in a medium or near a surface) or (2) bend the particle in a periodic manner. The various approaches, categorized in this way, are presented in Table 1. Notice that conventional linacs are simply devices where one is close to a surface, which is easy when one is using 10 cm radiation (as in the SLC), but not so easy when one is talking about 10 micron light (CO_2 lasers). In the remaining sections of this article I will go into some of the approaches categorized in Table 1.

Table 1. Electromagnetic Force Acceleration Alternatives

1. Slow wave down (and let particle go in straight line)

   a) Up frequencies from the 3 GHz at SLAC to (say) 30 GHz and use a slow wave structure. (Two-Beam Accelerator)

   b) Use a single-sided (i.e. a grating) as a slow wave structure. (Now one can go to 10 μm of a CO_2 laser or 1 μm of a Nd glass laser)

   c) Use dielectric slabs

   d) Put wave in a passive media (Inverse Cherenkov Effect Accelerator)

   e) Put wave in an active media (Plasma-Laser Accelerator)

2. Bend particles continuously and periodically (and let laser wave go in straight line)

   a) Wiggle particle and arrange that it goes through 1 period of wiggler just as 1 period of the electromagnetic wave goes by. (Inverse Free Electron Laser)

   b) Wiggler particle with an electromagnetic wave rather than a static wiggler field. (Two-Wave Accelerator)

   c) Use cyclotron motion of particle to do the bending. (Cyclotron Resonance Accelerator)

VII. THE INVERSE FREE ELECTRON LASER ACCELERATOR

Of the various devices, listed in Table 1, which bend the particles and hence operate far from any material surfaces, the simplest one (and the only one we will discuss in this article) is the Inverse Free Electron Laser (IFEL).

Firstly, we need to discuss Free Electron Lasers. These are central not only to the IFEL, but also to the Two-Beam Accelerator (see Sec. V).

Free Electron Lasers (FEL) are devices for conversion of electron beam energy to coherent electromagnetic energy. Invented by John Madey, they are presently of great interest with theoretical work and experimental work taking place throughout the world. The Inverse FEL simply operates by running the effect backwards. At first thought this makes no sense, but then one realizes that FELs can make radiation in parts of the spectrum where there are no coherent sources (such as the infra-red or the VUV), while the Inverse FEL could employ powerful available sources (such as the CO_2 laser) at appropriate wavelengths.

How does an FEL work? Simply by sending an electron through an alternating magnetic field so that the electron will undergo periodic transverse motion, i.e. "wiggle." Now a resonance condition is satisfied; namely that an electromagnetic wave, travelling in the forward direction and faster than the particle, passes over the particle by one period as the particle undergoes one period of wiggle motion. There is another way to describe the very same thing and it goes this way (see Fig. 7): An energetic electron, characterized by its relativistic γ, will "see the wiggler foreshortened by a Lorentz contraction and hence oscillate at a higher frequency than one would expect (i.e. compute non-relativisticly) by just a factor of γ. In the frame in which the electron is at rest on the average the electron is simply oscillating. Thus it will radiate. Back in the laboratory this radiation will be Doppler shifted. In the forward direction, the radiation is up-shifted in frequency by a factor $(1 + \beta)\gamma$, and thus the radiation is, in total, up-shifted by the factor $2\gamma^2$. It is this capability of frequency up-shifting and tunability (via energy variation) which makes FELs of such interest.

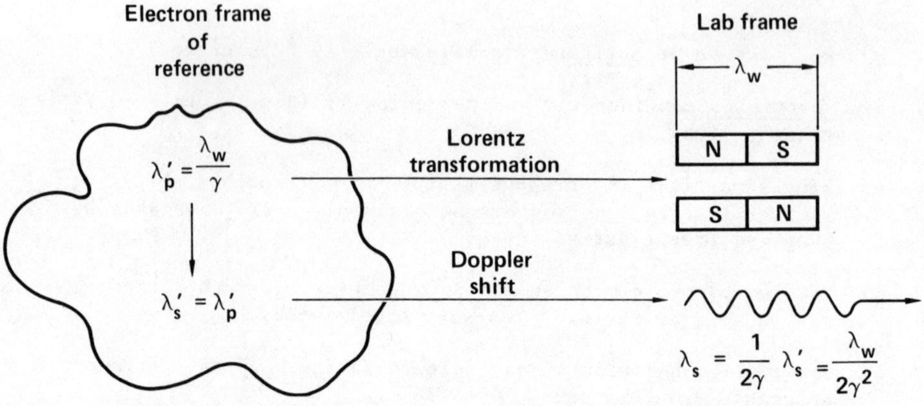

Fig. 7. The basic relativistic transformations which lie behind a free electron laser.

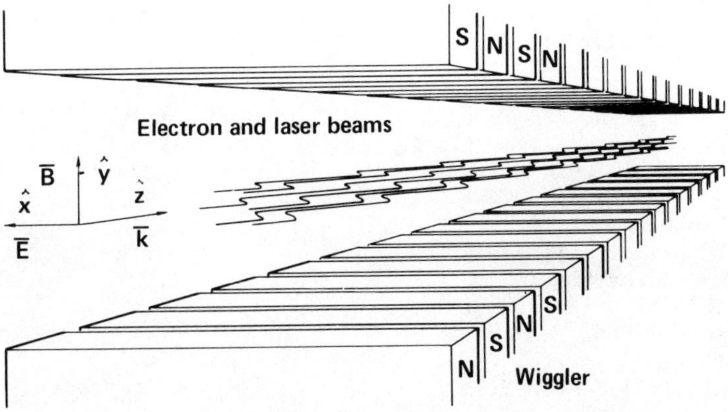

Fig. 8. The Inverse Free Electron Laser Accelerator.

Clearly, an Inverse FEL, such a device is shown in Fig. 8, will produce incoherent radiation, for the particles are going through a "wiggler," the very devices one employs in synchrotron radiation facilities in order to produce copious quantities of radiation. In fact incoherent radiation of energy must be balanced against coherent gain of energy in an IFEL and this balance drives the design. However, it appears possible, nevertheless, to attain very high energies; namely, 300 GeV with an average gradient of 100 MeV/m, with parameters as shown in Table 2.

A "practical" concern, actually a vital concern, is whether or not one needs many laser amplifiers or whether the intense laser light can be transported for kilometers and "used" over and over again. Theoretical work on this subject is being done at Brookhaven.

VIII. THE PLASMA LASER ACCELERATOR

Although an electromagnetic wave doesn't accelerate particles, it does move them along as is shown in Fig. 6. This effect can be used to good purpose. Suppose one shines a packet of light on to a medium having lots of free electrons; i.e. a plasma. The electrons will be moved and it is not very hard to appreciate that the density under the packet will be higher than normal. (The ions, which must be present in the plasma to maintain electrical neutrality, will hardly respond to the electromagnetic wave.)

How can this effect be accentuated? One very good way, proposed by Tajima and Dawson, is to illuminate the plasma with two laser beams of angular frequencies ω_1 and ω_2 whose difference is just the plasma frequency ω_p. In this way the beat frequency resonates with the plasma and most effectively bunches the plasma. This bunching results in an electrostatic longitudinal field which can, then accelerate particles.

Table 2. Possible parameters of 300 GeV x 300 GeV Inverse Free Electron Collider.

Laser wavelength	1 μm
Laser power	50 Tw
Synchronous phase, sin ϕ_0	.866
Laser electric field	0.22 TV/M
Waist radius	0.7 mm
Electron energy, input	250 MeV
Undulator initial period	3.8 cm
Undulator field	1.0 T
Initial helix radius	0.04 mm
Accelerator length	3 km
Electron energy, final	294 GeV
Average acceleration gradient	98 MeV/m
Final helix radius	0.5 m
Final undulator period	4.3 m
Crossing point β	1.0 cm
Disruption parameter	10
Number of particles per bunch	4.2×10^{10}
Repetition rate	1.6 kHz
Luminosity	10^{32} cm^{-2}s^{-1}
Laser energy per pulse	10 kJ
Average power (η = 10%)	320 Mw

Note that the overall effect is to turn the transverse field of the laser into a longitudinal field. This is often described as having taken place through the "ponderomotive force," but we have nothing more than the simple Lorentz forces on electrons.

If the plasma is underdense then the two laser waves will propagate. The condition is simply that

$$\omega_1, \omega_2 \gg \omega_p \equiv \left(\frac{4\pi ne^2}{m}\right)^{1/2} , \qquad (8.1)$$

where n is the plasma density. The laser waves satisfy, in a plasma, the dispersion relation

$$\omega^2 = \omega_p^2 + c^2 k^2 , \qquad (8.2)$$

where ω is the frequency (ω_1 or ω_2) of the laser wave and k is the wave number of this wave in the plasma.

For a plasma wave (k, ω) the dispersion relation is:

$$\omega^2 = \omega_p^2 + 3 k^2 \left(\frac{KT}{m}\right) , \qquad (8.3)$$

where KT is the plasma temperature (in energy units).

It is not difficult to show that the beat wave will have a phase velocity v_g, and a group velocity, v_p:

$$v_p \approx v_g \approx c\,(1 - \frac{\omega_p^2}{\omega_0^2})^{1/2} \quad , \tag{8.4}$$

provided $\omega_0 - \omega_1 = \omega_p$ and KT is not too large. This is shown in Fig. 9.

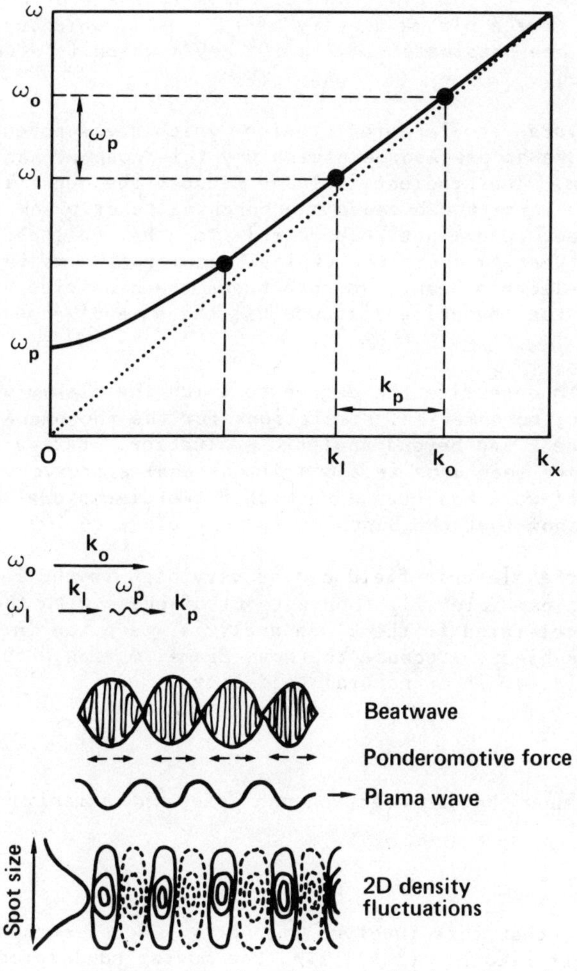

Fig. 9. a) Diagram showing the dispersion relation for electromagnetic waves (laser light) of frequency ω_0 and ω_1. b) Resonant excitation of a plasma density wave showing its two-dimensional structure. Contour solid lines (dotted lines) show increasing (decreasing) density.

Because there is synchronism between the beat wave and the plasma wave, the density modulations of the plasma, which is precisely what a plasma wave is, will resonantly grow. Just how large this wave will become and to what extent harmonics will develop is a non-linear problem which can only adequately be attacked by numerical methods. If the bunching is complete (100%) then the resulting <u>longitudinal</u> gradient is

$$eE_L = \frac{m\omega_p c}{e} = (4\pi\ n\ r_0^3)^{1/2}\ (\frac{mc^2}{r_0}) \quad , \tag{8.5}$$

where $r_0 = e^2/mc^2$ is the classical electron radius and $(m\ c^2/r_0) = 1.8 \times 10^{14}$ MeV/m. For a plasma density of 10^{17} cm^{-3}, which can be obtained in a θ-pinch, one obtains $eE_L = 2 \times 10^4$ MeV/m which is a very large gradient indeed.

The very large accelerating gradient which seems potentially possible in the beat-wave accelerator explains why this concept has attracted so much attention. The gradient is large because the densities attainable in plasmas are large and because the bunching takes place over very small distances. One might compare this to other collective acceleration methods (Part A) where the typical accelerating media is an intense relativistic electron beam. In such beams the densities are, at most 10^{14} cm^{-3} and the "bunching distance" is the size of a beam; i.e. centimeters.

In order to determine the degree to which the plasma will bunch, one must resort to numerical simulations for the phenomena is clearly highly non-linear and beyond analytic evaluation. Extensive simulations have been done in a one-dimensional approximation and a small amount of work has been done with a two-dimensional model. The computations show that the bunching is very close to 100%.

Although the electric field can be very high in the laser plasma accelerator, a particle will soon get out of phase with the plasma wave and not be accelerated further. An analysis was given in the very first paper on the subject. Because the wave frame, moving with velocity v_p given by Eq. (8.4), it is natural to define

$$\gamma = \omega_p/\omega_0 \quad . \tag{8.6}$$

Then one can show that an electron can "pick up" a maximum energy increment

$$\Delta E = 2\gamma^2\ mc^2 \quad . \tag{8.7}$$

Note, firstly, that this involves γ^2, not γ. Note, secondly, that this formula is just like Eq. (2.6); i.e. the moving ponderomotive well "impacts" coherently with an electron.

Will the plasma behave as expected? There is some reason to be optimistic for the laser beams can be expected to organize the plasma and control the plasma. Furthermore, the beat-wave density formation is a rapid process and the acceleration can be over before the plasma undergoes many instabilities. Theoretical calculations are suggestive, but a definitive answer must, of course, come from experiment. To date, little experimental work has been done and few, if any, results have been obtained.

The device described so far has the defect that as particles are accelerated they will, slowly of course because they are very relativistic, get out of synchronism with the plasma wave. Thus staging is required, and consequently one must tackle the problems associated with transporting and periodically focusing laser beams.

It has been observed by Katsouleas and Dawson that the imposition of a transverse magnetic field will allow the particles to always remain "in-step" with the plasma wave. A diagram showing this is reproduced as Fig. 10. The magnetic field must not be too large (no problem in practice) or the particle will no longer be "trapped" by the plasma density wave, nor can it be too small so as to have a good acceleration rate. The rate of energy gain is, in the direction of the wave.

$$\frac{dw}{dx} = 0.1\,\frac{\text{GeV}}{\text{cm}}\left[\frac{B(\text{kG})}{\dfrac{n(\text{cm}^{-3})}{10^{16}}\;\dfrac{\lambda}{\mu\text{m}}}\right]\frac{n^{1/2}(\text{cm}^{-3})}{10^{16}}\,, \tag{8.8}$$

where the magnetic field is B, and λ is the wavelength of the laser light. The factor in square brackets in Eq. (8.8) is the fraction of the peak bunching field and probably cannot be made to exceed 0.1 in practice.

In this accelerator, the "Surfatron," particles move transverse to the wave for it is in this direction that they accelerate. However, the transverse distance, Δy, doesn't have to be very big and is given by

$$\frac{\Delta y}{\Delta x} = \left(\frac{1}{30}\right)\left(\frac{\lambda}{\mu\text{m}}\right)\left(\frac{n(\text{cm}^{-3})}{10^{18}}\right)^{1/2} . \tag{8.9}$$

where Δx is the longitudinal length of the accelerator.

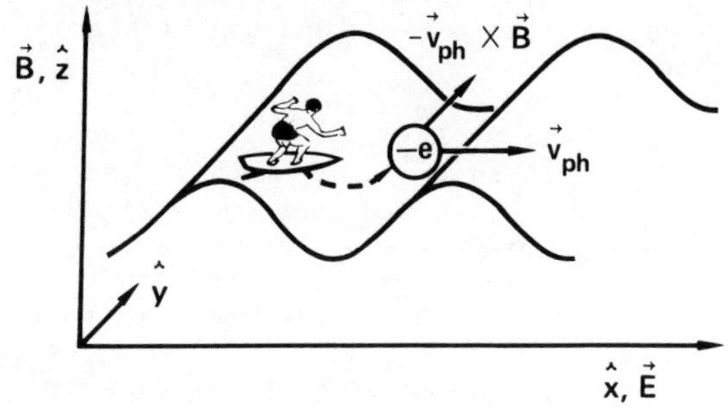

Fig. 10. A diagram of the Surfatron Accelerator Principle in which a transverse magnetic field keeps particles in phase with the plasma density wave even as the particles are accelerated.

In summary the plasma laser accelerators have great potential and there is room for invention so that there are already a number of variations which have been conceived. The fundamental question (in physics) is whether one can control a plasma and laser beams and make them act as one desires. Later, one must address the "engineering" questions of making short pulse, intense lasers, etc.

PART C: MULTI-BEAM ACCELERATORS

IX. THE TWO-BEAM ACCELERATOR

In this, and the next section, I want to discuss multi-beam accelerators; the Two-Beam Accelerator and the Wake-Field Accelerator. Each of these concepts involves two beams in an essential manner. They both employ a relativistic beam as an integral part of the accelerator and as an intermediary to the beam which one is accelerating to very high energy. I think that the next large jump in accelerator capability will be to employ external fields to manipulate a first beam which then accelerates a second beam of particles. Collective accelerators, of course, fall into this class of devices. None of them has yet led to a practical high energy machine, and, in my opinion, it seems doubtful that those proposed so far will lead to such a device. In contrast, these two concepts appear likely to lead to practical devices. They both are, as you will see, easier to achieve than any of the collective accelerators proposed so far, in that the two beams are kept quite separate from each other.

The Two-Beam Accelerator is based on the observation that in order to get to ultra-high energy in a conventional linac and still keep the power requirements of such an accelerator within bounds, one must go to higher frequency accelerating fields than are presently employed. The reason for this is that the stored energy, which is proportional to the transverse area of the linac, goes down as the square of the frequency for the transverse size is simply proportional to the radio-frequency wavelength. The reason one doesn't go to (say) ten times the frequency of SLAC is simply that there are no power sources in this range.

At 30 GHz, ten times the frequency of SLAC, possible power sources are multi-beam klystrons, photo-cathode klystrons, gyrotrons, etc., but none of these is sufficiently developed to be employed at present. One possibility is a Free Electron Laser which theoretically would appear to be able to be used for this purpose and on which good progress has been made experimentally.

If one uses a Free Electron Laser as a power source, then it is possible to consider one extended source, rather than many lumped sources as in the present linacs, and thus one arrives at the Two-Beam Accelerator. A low-energy beam is sent through a wiggler so that it produces, by means of the FEL process, copious amounts of microwave radiation. This radiation is then funneled into a rather conventional, but quite small, linac in which the high-energy beam is accelerated. The energy of the low-energy beam must be constantly replenished, and this is done by conventional induction units, which are quite efficient (60%). An artist's conception of how such a device might look is shown in Fig. 11.

At the very core of the concept is the FEL. The generation of microwave radiation by FELs is being studied and although, microwaves in the tens of megawatt range have already been achieved, there are many questions which need to be answered such as what is the effect of other (unwanted) wave guide modes. (The separation in phase velocity of unwanted modes is why the wave guide is shown as elliptical in Fig. 11.)

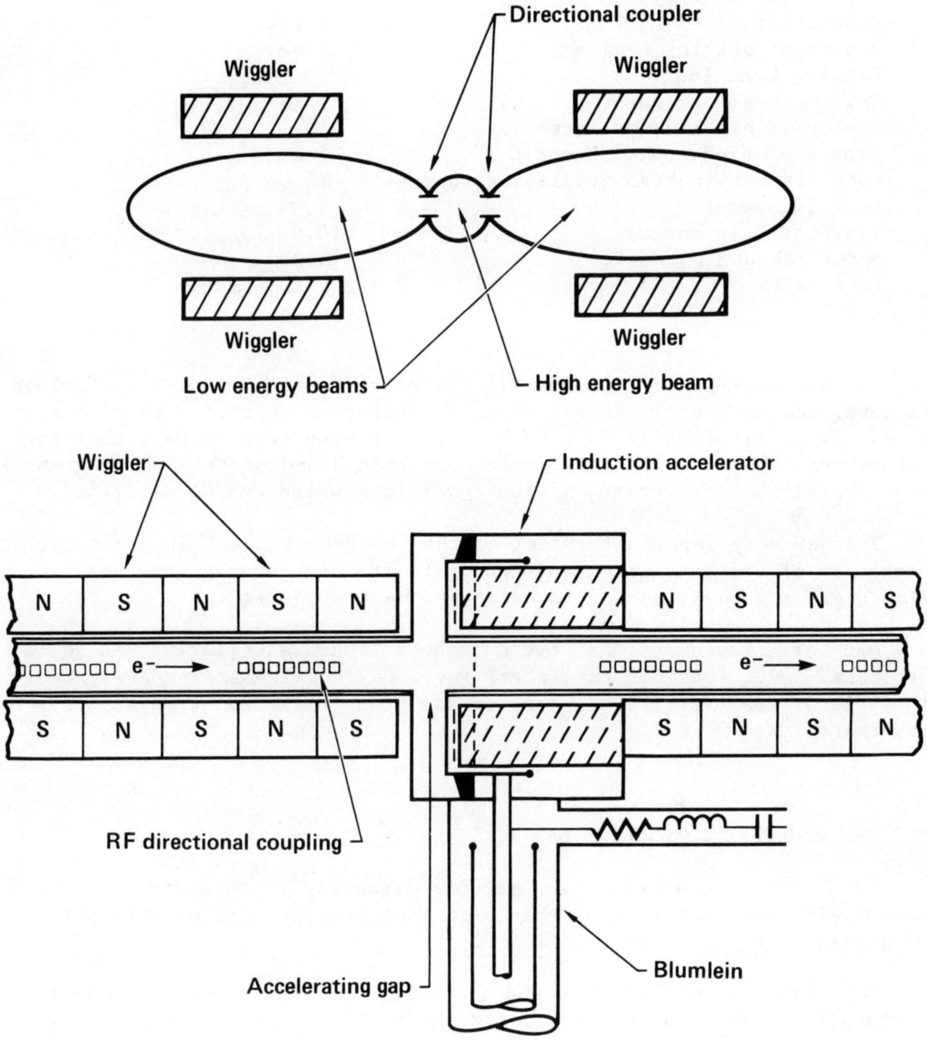

Fig. 11. Diagrammatic representation of a Two-Beam Accelerator. The low-energy beams are made to undergo wiggle motion and hence radiate by free electron laser action. This radiation is then used to accelerate the desired, high-energy, beam in a conventional linac. Fig. (11a) is a transverse cross section and Fig. (11b) is a longitudinal cross section of the device.

Table 3. Possible parameters of 375 GeV x 375 GeV Two-Beam Accelerator collider.

Nominal particle energy	375 GeV
Total length of the electron linac	2.0 km
Gradient of the conventional linac	25 MeV/m
Gradient in the Two-Beam Accelerator	250 MeV/m
Average power consumption	150 MW+150 MW
Overall efficiency	8%
Repetition rate	1 kHz
Energy of driving beam	3 MeV
Driving beam length	25 nsec
Driving beam current	1 kA
Number of high-energy particles	10^{11}
Length of high-energy bunch	1 mm
Focal length in high-gradient structure	10 m
Crossing point β	1.04 cm
Disruption parameter	0.9
Beamstrahlung parameter	0.05
Luminosity	$4\times10^{32}cm^{-2}sec^{-1}$

In the Two-Beam concept the FEL is operated in "steady state." This is novel and nothing is known, experimentally, about this mode of operation. In particular, it is necessary to not have loss of particles for kilometers. Theoretical work, which has been based on the KAM theorem and the Chirikov criterion, yields conditions which can be satisfied.

Another very important aspect of the Two-Beam concept is the coupling mechanism of the low-energy wave guide to the high-energy wave guide. Also there are questions of focusing and beam stability.

Many of these questions have only been looked at briefly, but so far it looks good. Possible parameters for a full-scale machine are given in Table 3. An experimental program has been initiated at Berkeley/ Livermore.

X. THE WAKE-FIELD ACCELERATOR

Of all the new acceleration methods discussed in this article, the Wake-Field Accelerator (i.e. this very last section) is, by far, the simplest.

Of course "simplicity" is not a criticism of the concept; in fact, perhaps it is just the opposite, for the Wake-Field Accelerator looks as if it can be made to work, and, furthermore, it appears capable of achieving gradients of (say) 500 MeV/m.

When a bunch of charged particles passes through a structure of varying shape then it will excite a wake-electromagnetic-field whose shape is not necessarily that of the charge bunch. This phenomena is well-known and well-understood; it has been calculated (usually for cylindrical structures) and measured experimentally, and the two approaches agree.

Particles inside or behind the bunch feel a longitudinal electric field whose integral over time, for fixed position relative to the bunch, is called the wake potential. Particles near the front of the bunch are decelerated, but those behind the bunch, generally, are accelerated. Unfortunately, this wake potential is usually not large enough to make a practical accelerator.

However, one can make -- really in a variety of ways as was first noted by Weiland & Voss, -- a wake potential transformer; i.e. a device in which a low energy high current beam creates a very high gradient at some other position. Such a possible configuration is shown in Fig. 12. The parameters which one might have in such an accelerator are given in Table 4.

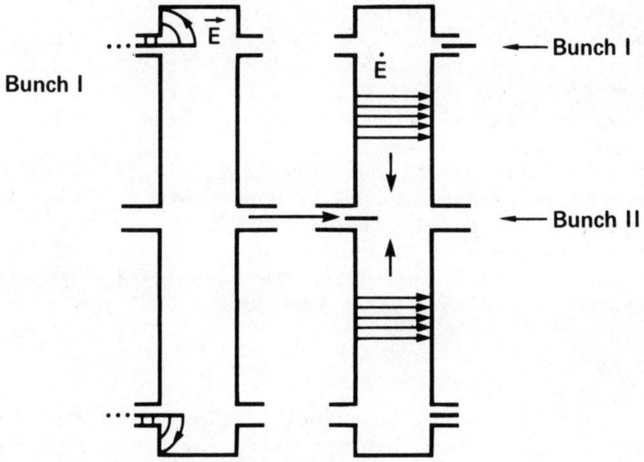

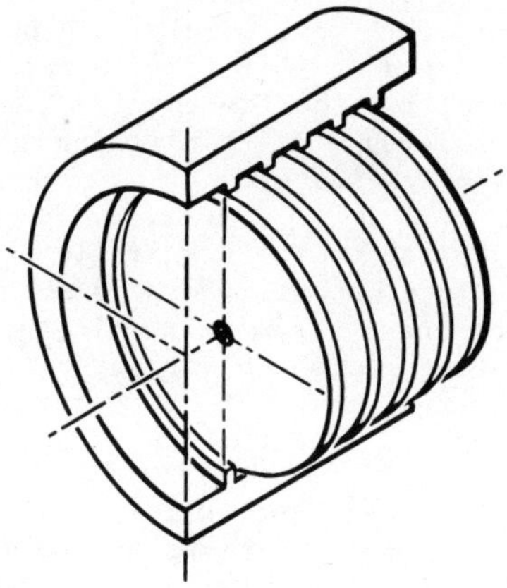

Fig. 12. A cylindrical realization of the Wake-Field Accelerator. The low-energy beam is in the form of a ring. It produces a wake which is transformed by the cylindrical geometry into large accelerating field at the high-energy beam which is on axis.

Table 4. Possible parameters of 50 GeV x 50 GeV Wake-Field Accelerator collider.

Nominal particle energy	50 GeV
Total length of the electron linac	550 m
Total length of the positron linac	650 m
Gradient of the conventional linac	25 MeV/m
Gradient in the Wake-Field transformer	170 MeV/m
Average power consumption	8+8 MW
Peak power	3900 MW
Number of high energy particles per bunch	10^{11}
Number of particles in the driving bunch	$6x10^{12}$
Efficiency of the wake transformer	16
Repetition frequency	100 Hz
r.m.s. bunch length of both beams	0.2 cm
Wake-Field transformation gain	10.2
DRIVING BEAM:	
Number of particles	$6x10^{12}$
Energy at the entrance of the wake transf.	5.5 GeV
Energy at the end of the wake transf.	0.5 GeV
Maximum phase slip between driving beam and accelerated beam	0.5 ps
Maximum particle energy loss (self fields)	1.8 MeV/m
Peak transverse momentum kick per unit length due to self fields	6.9 keV/mc
Solenoid field strength	7 T
Maximum particle deviation for a constant beam misalignment of δ = 100 μm	1 mm
HIGH ENERGY BEAM:	
Number of particles	10^{11}
Maximum particle energy loss (self fields)	15.2 MeV/m
Peak transverse momentum kick per unit length due to self fields	18.9 keV/m

Clearly, one can employ other transformer geometry than the cylindrical geometry discussed here. Almost surely, the best geometry is not that which has been presented in this first example. In addition, one can readily imagine using, for the low-energy beam, electron rings as they have already been achieved. If this is done, one can see one's way to gradients of 500 MeV/m or greater.

Clearly, if one is serious about the Wake-Field Accelerator then one must go into it in much more detail. Beam dynamic questions come to mind such as whether the low energy beam is stable (both longitudinally and transversely). Notice that it is subject not only to its own wake, but also that of the high energy beam. The same questions need to be asked of the high energy beam.

Many of these subjects have been looked into, only superficially so far, and appear to only put minor constraints on the device. As a result, an experimental program has been undertaken at DESY to study the Wake-Field Accelerator.

REFERENCES

1. A. M. Sessler, "Collective Field Accelerators," in "Physics of High Energy Particle Accelerators," edited by R. A. Carrigan, F. R. Huson, M. Month, American Institute of Physics Conference Proceedings No. 87 (1982).
2. A. M. Sessler, "New Concepts in Particle Accelerators," in "Proceedings of the 12th International Conference on High-Energy Accelerators," edited by F. T. Cole and R. Donaldson, Fermi National Accelerator Laboratory, Batavia, Illinois (1983), p. 445.
3. E. L. Olson and V. Schumacher, "Collective Ion Acceleration," Springer, New York (1979), Springer Tracts in Modern Physics Vol. 84.
4. P. J. Channell, editor, "Laser Acceleration of Particles," American Institute of Physics Conference Proceedings No. 91 (1982).
5. "The Challenge of Ultra-High Energies, Proceedings of the ECFA-AAL Topical Meeting," Rutherford Appleton Laboratory, Chilton, Didcot, UK (1982).

This work was supported by the U. S. Department of Energy under contract DE-AC03-76SF00098.

PROBLEMS

1. Consider the equilibrium flow of a cylindrical beam of current, I, and radius, r_b, in a longitudinal field of magnitude B. Show that the motion, Brillouin Flow, is described by

$$\ddot{x} = \frac{\omega_p^2}{2} x + \Omega y \quad ,$$

$$\ddot{y} = \frac{\omega_p^2}{2} y - \Omega x$$

where ω_p is the plasma frequency:

$$\omega_p^2 = \frac{4\pi n e^2}{m} \quad ,$$

and Ω is the cyclotron frequency

$$\Omega = \frac{eB}{mc} \quad .$$

2. Show that the criterion for Brillouin flow is

$$2\omega_p^2 < \Omega^2 \quad .$$

Generalize this to a beam which is fractionally neutralized to degree, f, and obtain

$$2\omega_p^2 \; (1 - f) < \Omega^2 \quad .$$

3. Derive the relativistic Child's Law, as was done by Jory and Trivelpiece, thus obtaining an accurate estimate of beam stopping. Show that for a beam of current density, J, the relation between distance, x, and the potential $U(x) = eV(x)mc^2$ is:

$$x = \left(\frac{mc^3}{8\pi Je}\right)^{1/2} \int_0^{U(x)} \frac{dy}{(y^2 + 2y)^{1/4}}$$

4. Show that the relativistic Child's Law reduces, non-relativisticly, to

$$x \approx V^{3/4} / J^{1/2}$$

$$\text{or} \quad J \approx V^{3/2} / x^2$$

in accord with the usual expression. Show that in the ultra-relativistic case

$$J \approx V / x^2 \quad .$$

5. Derive the formula for ideal acceleration of a loaded electron ring; namely Eqs. (5.7) - (5.10). In particular, show that the field falloff parameter λ is given by

$$\lambda \simeq 2\pi \frac{Ra\ Mc^2}{e^2 N_e} \quad .$$

6. When light of wavelength, λ, is focused to a spot of radius, r, then the "depth of field" is given by the Rayleigh length, $\mathcal{L}$, where

$$\mathcal{L} = r^2 / \lambda \quad ;$$

i.e. this is the distance over which the focus extends. Derive this formula.

7. Use the concept of Rayleigh length to derive the formula for the maximum energy gain, ΔW, which an electron gets by passing through the focus of an intense laser of power, P. Show that

$$\Delta W = e \left(\frac{8\pi P}{c}\right)^{1/2} \quad .$$

8. The free electron laser resonance condition is

$$\lambda = \frac{\lambda_w}{2\gamma^2} (1 + K^2/2) \quad ,$$

where λ is the wavelength of the radiation, λ_w is the wavelength of the wiggler, λ is the relativistic factor of the electron beam, and

$$K = \left(\frac{eB_w}{mc^2}\right) \left(\frac{\lambda_w}{2\pi}\right) \quad ,$$

and B_w is the peak value of the sinusoidal wiggler field. Derive this condition.

e^+e^- COLLIDERS

G.Bonvicini
University of Michigan
Ann Arbor MI 48109

P. Gutierrez
University of Rochester
Rochester NY 14627

C. Pellegrini
Brookhaven National Laboratory
Upton NY 11790

The following notes are intended as a review of the main properties of e^+e^- colliders, divided into 2 categories: storage rings and linear colliders; the basic physics, beam characteristics, limitations on beam current and density and their influence on the luminosity will be discussed.

We use through the lectures the CGS system, and the following quantities have, unless otherwise specified, the following meaning (notice that when the subscript 0 is present we usually refer to the design parameters):
U, U_0 are the energy loss of a particle during one revolution
$c = 2.99792458 \times 10^{10}\text{cm}^{-2}\text{sec}^{-1}$ is the speed of light
$m = .51100$ MeV is the electron mass
E, E_0 are the energy of the particle
$\gamma = E/mc^2$ is the relativistic factor
$e = 1.60219 \times 10^{-19}$ Coulomb is the electron charge
$r_e = e^2/mc^2 = 2.8179 \times 10^{-13}$ cm is the classical radius of the electron
ρ is the local or average curvature radius
V, V_0 are the peak voltage of the RF cavities
ϕ, ϕ_0 are respectively the RF phase and the synchronous phase
f, ω_0 are the revolution frequency
x is the horizontal coordinate (*i.e.*contained in the bending plane)
y is the horizontal coordinate
q is the generic transverse coordinate
s is the coordinate in the direction of the motion
ε is the emittance
β is the betatron oscillation amplitude
β^* is the β function at the interaction point

Q is the machine tune
ΔQ is the tune shift
p is the momentum of the particle
η is the dispersion
ξ is the cromaticity
h is the integer harmonic number
α_p is the momentum compaction
τ is the damping time of the ring
$\lambda_c = \hbar/mc = 3.8616 \times 10^{-11}$ cm is the electron Compton wavelength
σ_ε is the rms energy spread
$\sigma_x, \sigma_y, \sigma_s$ are the spatial dimensions of the beam
N is the number of particles in a bunch
B is the number of bunches circulating in the ring
L is the luminosity
P, P_0 the RF power
$\alpha = e^2/\hbar c = 1/137.036$ is the fine structure constant
θ is the polar angle
Z is the atomic number
ε_{max} is the maximum energy acceptance
l is the path length in the accelerator
H is the magnetic field
p_T is the transverse momentum
D is the disruption parameter
ω_p is the plasma frequency.

Differentiation, unless otherwise specified, is always done respect to s.

INTRODUCTION

e^+e^- rings have grown from a circumference of a few meters for ADA (Frascati 1961) with an energy of 250 MeV, to LEP with a circumference of 27 km and an energy of 100 GeV. In the meantime the luminosity has increased from $\approx 10^{26}\text{cm}^2\text{sec}^{-1}$, enough to see the first inelastic events ($e^+e^-\gamma$) in ADA, to more than $10^{31}\text{cm}^2\text{sec}^{-1}$ at PEP, PETRA and CESR, enough to make many important contributions to high energy physics. The radius for (e^+e^-) storage rings has increased of a factor 4×10^3, but the energy was only increased by 4×10^2. Conversely the energy given to a particle per unit of accelerator length is decreased from 40 to 4MeV/m. The main reason for this is due to the very fast increase of the energy dissipated by the electrons to synchrotron radiation, when going up with the beam energy; this energy must be given back by a RF system, the limits of which fix the maximum energy attainable. The energy loss per turn is[1]

$$U = \frac{4}{3}\pi\gamma^3 E\frac{r_e}{\rho} = .088 MeV\, E^4(GeV)/\rho(m) \tag{1}$$

The energy given to the electron by the RF is

$$U = eV \sin\phi_s \tag{2}$$

where $\phi_s$is the phase of the particle with respect to the RF. U was 366 eV at ADA and is 1.37 GeV at LEP.

This rapid increase of synchrotron radiation losses with beam energy has important implications on the collider design and on its cost. One can define the cost for any circular machine as[2]

$$C = g_1\rho + (g_2 + Lg_3)E^4/\rho \tag{3}$$

where
g_1=cost of tunnel, magnets and utilities
g_2=RF cavities cost,
g_3=RF power,
and L contains the dependence on beam intensity when the machine is operated. Minimization of the cost for a given E ($\partial C/\partial\rho = 0$) gives

$$\rho = \sqrt{\frac{(g_2 + Lg_3)}{g_1}}E^2 = KE^2, \tag{4}$$

$$C = 2KE^2. \tag{5}$$

It is evident that an increase by a factor 10 from LEP energies would lead to big problems of space and money.

For this reason a number of people have proposed to use linear, instead of circular, colliders for very high energies[3]. If the electrons are accelerated along a straight line, as in a linac, the radiation energy loss per meter is[4]

$$\frac{dU_L}{ds} = \frac{2}{3}\frac{r_e}{mc^2}\left(\frac{dE}{ds}\right)^2. \tag{6}$$

*i.e.*acceleration rates as high as 1 GeV/m will produce a radiation loss of 4×10^{-3} eV/m, completely negligible.

SYNCHROTRON STORAGE RINGS

In this chapter we will describe the operation of synchrotron storage rings. We start by discussing the transverse motion of the beam. The linear theory will be explained first. This will be followed by an explanation of how a momentum spread will effect the linear motion. In addition there will be a section which describes the beam optics in the interaction region.

After a discussion of the motion in the transverse plane we will describe the motion in the longitudinal direction. A section that describes the effects of synchrotron radiation will follow. The last topic covered will deal with beam-beam interactions. We will not cover the topic of beam instabilities since this is already covered in the lectures of A.Sessler.

We will use LEP (tab. 1a and 1b), as described in the 1979 design study[5], as an example to discuss the relevant parameters of a machine. The LEP design has been modified in the following years, but these modifications are not important for our purpose of illustrating the characteristics of a circular collider.

Table 1a. Short list of LEP parameters at 86 GeV

Nominal beam energy	E_0	86	GeV
Number of interaction regions		8	
Number of bunches	B	4	
Machine circumference		22.268	km
Average machine radius	ρ	3.544	km
Horizontal betatron tune	Q_x	70.32	
Vertical betatron tune	Q_y	74.54	
Synchrotron betatron tune	Q_s	0.158	
Momentum compaction factor	α_p	2.94×10^{-4}	
Circulating current per beam	I	9.15	mA
Number of particles per beam	N	5.83×10^{12}	
Transverse damping time	τ	12.8	msec
Uncorrected chromaticities	ξ_x	-116.8	
	ξ_y	-171.1	
Natural rms energy spread	σ_ε/E	1.24×10^{-3}	
Natural rms bunch length	σ_{s0}	16	mm
Actual rms bunch length	σ_s	63	mm
Emittance ratio	$\varepsilon_y/\varepsilon_x$	.0371	
Beam-beam BS lifetime	τ_{bb}	8.14	h
Beam-gas BS lifetime	τ_{bg}	20	h
Overall beam lifetime		5.79	h
Length of normal cells		79	m
Horiz. aperture in normal cell	a_x	±59	mm
Vert. aperture in normal cell	a_y	±33	mm
Design luminosity	L	1.07×10^{32}	$cm^{-2}sec^{-1}$

Betatron oscillations

In LEP, as in all the synchrotron storage rings, a charged particle will follow a guide field consisting mainly of dipole and quadrupole magnets. The dipole field strength is determined by the radius of the synchrotron and the beam energy. LEP has a bending radius of 3.544 km which gives a dipole field of .05 T for a beam energy of 50 GeV. The ring itself is divided into cells of magnets with each cell providing part of the bending angle. The sum of all cells gives a bending

Table 1b. Main RF system parameters for LEP

Synchrotron energy loss per turn	U_0	1370	MeV
Parasitic energy loss per turn		110	MeV
Peak RF voltage	V_0	1949	MV
Stable phase angle	ϕ_s	130.6	dg.
Frequency	ω_0	353.4	MHz
Length of RF structure		1629	m
Shunt impedance per unit length		40.0	$M\Omega m^{-1}$
Fund. mode dissipation		61.7	MW
Synchrotron power (2 beams)		25.1	MW
Waveguide losses		7.2	MW
Total RF power	P_0	96.0	MW
Number of RF stations		16	
Number of klystrons		96	
Number of 5-cell cavities		768	

angle of 2π. Figure 1 shows a standard LEP cell, of which there are 288. Each cell has a bending angle of 19.6 mrad which is given by 12 dipole magnets(B). The standard cells make up less than 2π of the bending angle with the remainder being made up by special cells.

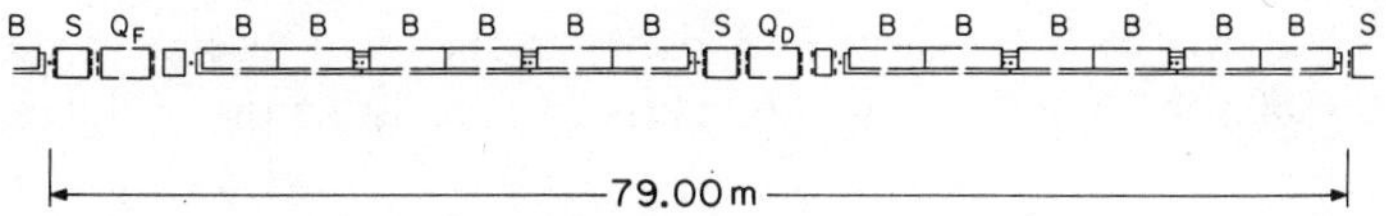

fig.1. A LEP FODO cell.

In Figure 1 we see that there are a focusing and a defocusing quadrupole magnets (Q_F, Q_D) in the standard cell. These quadrupoles are used to keep particles from diverging from the design orbit. The design trajectory takes the particle through the center of the machine quadrupoles where the field is zero. As long as the particles follow the design trajectory their motion will not be affected by the quadrupoles. If however a particle does not follow the design trajectory it will see a net field which will focus it in one plane but defocus it in the other plane. To keep particles from diverging too far from the design orbit we require overall focusing from all the quadrupoles in the lattice. To have overall focusing, focusing and defocusing quadrupoles are alternated around the storage ring. This scheme is called a FODO cell. In this cell there are also sextupole magnets (S), for reasons which will be described in the following.

As the charged particle makes its way around the storage ring it will encounter different types of magnet elements. To first order only the magnetic quadrupoles will have any effect on the transverse motion of the particle. The quadrupoles have been set up in such a way as to push the particle towards the design orbit. This will lead to an oscillatory motion in the transverse plane. In order to see this quantitatively we write down the equation of motion for a particle circulating in a storage ring. We first write down the equation of motion for a charged particle in a quadrupole[1,6]

$$\frac{d^2q}{ds^2} + kq = 0 \tag{7}$$

where $k = eG/\rho$ is the normalized gradient.

This equation describes the motion in both a focusing ($k > 0$) and defocusing ($k < 0$) quadrupole. The transverse motion in field free regions and in the dipole fields is also given by eq.(7) by setting $k = 0$. We therefore generalize eq.(7) by making k a function of s which gives

$$\frac{d^2q}{ds^2} + k(s)q = 0. \tag{8}$$

The general solution to this equation for stable motion is[6]

$$q(s) = \sqrt{\varepsilon\beta(s)}\sin(\phi(s) + \psi). \tag{9}$$

where ψ is the initial phase, $\phi(s) = \int_0^s \beta^{-1}(s)ds$ is the betatron phase and ε is the Courant invariant which will be discussed later.

The particle's motion, as expected, is pseudoharmonic with a position dependent amplitude. The β function characterizes the magnetic structure of the storage ring. In order to see explicitly the dependence of β on k we take the case where $k(s)$ is a constant. For this case we have $\beta = 1/k$. For the case of a FODO lattice, β will oscillate with the same periodicity as the quadrupoles. From Fig.2 we see that the minimum and maximum values of β occur at the middle of each quadrupole.

To calculate $\beta(s)$ it is easier to use a matrix formulation[1,6]. Starting with eq.(9) and calculating $q'(s)$, where $q'(s) = dq(s)/ds$, we write the motion from one turn to the next at position s_0 by:

$$\begin{pmatrix} q(s_0) \\ q'(s_0) \end{pmatrix}_{n+1} = \begin{pmatrix} \cos\mu + \frac{\beta_q'(s_0)}{2}\sin\mu & -\beta_q(s_0)\sin\mu \\ \frac{1+\beta_q'^2(s_0)/4}{\beta_q(s_0)}\sin\mu & \cos\mu - \frac{\beta_q'(s_0)}{2}\sin\mu \end{pmatrix} \begin{pmatrix} q(s_0) \\ q'(s_0) \end{pmatrix}_n \tag{10}$$

where $\mu = 2\pi Q = \oint \beta^{-1}(s)ds$ is the total phase over one revolution, and Q, the betatron tune, is the number of betatron oscillations over one revolution. This matrix is called the Twiss matrix.

The Twiss matrix contains information about all the magnetic elements of the storage ring. To calculate this matrix we need to calculate the matrix for each element. There are three types of elements which contribute to the Twiss matrix: focusing quadrupoles($k > 0$), defocusing quadrupoles($k < 0$) and drift spaces ($k = 0$).

Starting with eq.(8) we solve for q and q' all the three cases. These solutions are then expressed as matrix equations of the form $\vec{q}(s_1) = M_1\vec{q}(s_0)$, $\vec{q} = (q, q')$. The matrices for the three elements are:

$$\begin{pmatrix} 1 & L \\ 0 & 1 \end{pmatrix} \qquad k = 0 \tag{11}$$

$$\begin{pmatrix} \cos\sqrt{k}L & \frac{1}{\sqrt{k}}\sin\sqrt{k}L \\ -\sqrt{k}\sin\sqrt{k}L & \cos\sqrt{k}L \end{pmatrix} k > 0 \tag{12}$$

$$\begin{pmatrix} \cosh\sqrt{k}L & \frac{1}{\sqrt{k}}\sinh\sqrt{k}L \\ \sqrt{k}\sinh\sqrt{k}L & \cosh\sqrt{k}L \end{pmatrix} k < 0. \tag{13}$$

With the matrix for each element we are now ready to calculate the Twiss matrix. To do so, we start with a particle at position s_0 and multiply its coordinate, $q(s_0)$ and $q'(s_0)$ by the matrix for the first element it encounters. This gives us $\vec{q}(s_1) = (q(s_1), q'(s_1))$. To calculate $\vec{q}(s_2)$ we multiply $\vec{q}(s_1)$ by the matrix for the second element encountered by the particle. This is equivalent to multiplying the matrices for the first two elements together:

$$\vec{q}(s_2) = M_2\vec{q}(s_1) = M_2M_1\vec{q}(s_0). \tag{14}$$

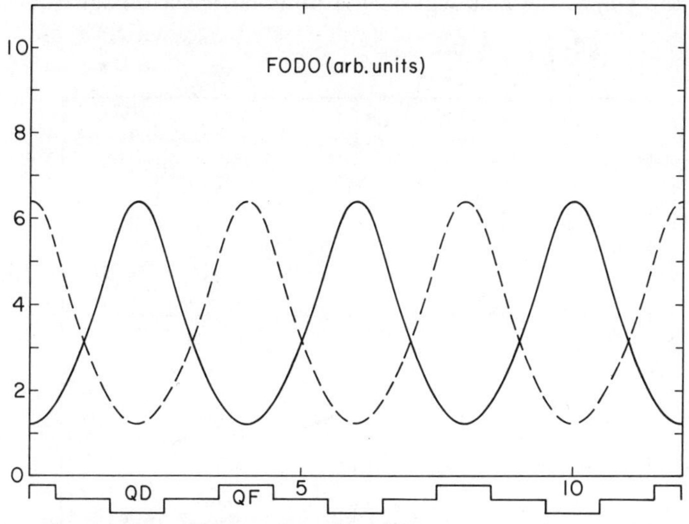

fig.2. β function behavior in a FODO cell.

We continue this process until we return to s_0, giving:

$$\vec{q}_j(s_0) = M_n \dots M_2 M_1 \vec{q}_{j-1}(s_0). \tag{15}$$

Comparing this to eq.(10), we see that the product of these matrices is the Twiss matrix. From this we have the value of $\beta(s_0), \beta'(s_0)$ and $\phi(s_0)$. To calculate these values at s_i, we start at position s_i instead of s_0 and multiply together the matrices for all the elements until we reach s_i again.

If we look at a particle at a fixed location, s_i, in the storage ring we will see that the particle will follow an elliptical path in (q, q') space, see Fig.3. Assuming that the particle has an initial position $q(s_i) = q_0$ and that $q'(s_i) = q_0$ from eq.(9) we get the equation for the particle position along the ellipse on subsequent turns. The equation is just that for simple harmonic motion:

$$\begin{aligned} q(s_i) &= \cos(2\pi n Q) + \beta_q'(s_i) \sin(2\pi n Q) \\ q'(s_i) &= \frac{1 + \beta_q'(s_i)/4}{\beta_q(s_i)} \sin(2\pi n Q) \end{aligned} \tag{16}$$

where n is the number of revolutions the particle has made around the ring.

From eq.(10) we see that if Q has a value that is expressed as j/m, j and m integers, then the particle will return to the same position on the ellipse on subsequent turns. This condition will lead to a resonant growth which will cause the particle to be lost.

The reason for this is seen in the following simple argument. Assume that there is an imperfection in the lattice. This imperfection will give the particle a kick. If the particle returns to the same position on the ellipse it will receive a similar kick each time around causing the particle to be pushed further and further out until it is lost. If however the particle does not return to the same position on the ellipse the kick will have a different effect on subsequent turns

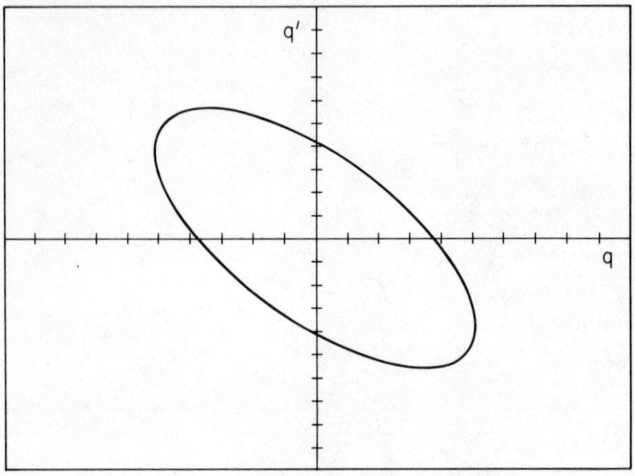

fig.3. Elliptical path in phase space.

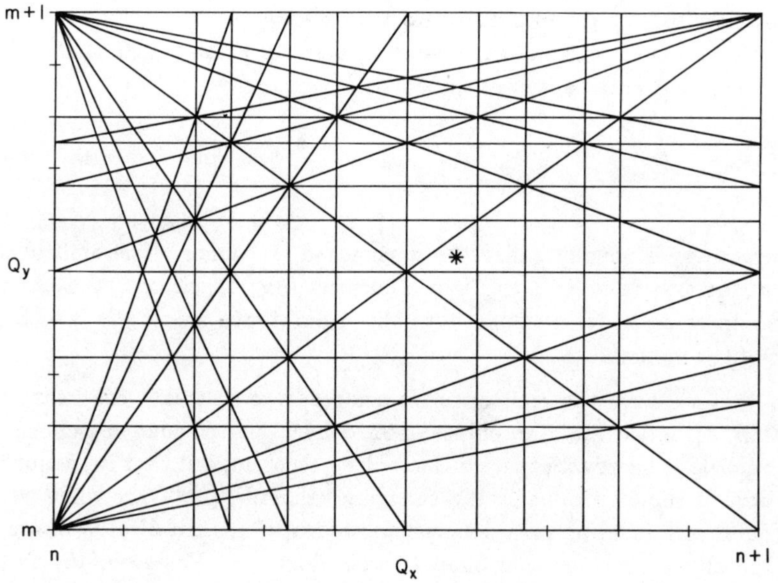

fig.4. The tune working plane.

and the particle will not be lost. This is similar to a driven harmonic oscillator, if it is driven with the natural frequency of the oscillator it will lead to a resonance, otherwise it will oscillate at the frequency at which it is driven. In the more general case when the particle is executing betatron oscillations in the horizontal and vertical directions, characterized by the tunes Q_x and Q_y, the condition for the resonance can be written as $mQ_x + nQ_y = p$, with m, n, p integers. One must avoid resonant values of Q_x, Q_y at least for m, n of order 10. It is usual to represent resonances as lines in the working plane Q_x, Q_y (fig.4). In this plane one machine is represented by a point, the working point. A discussion of resonances and their effects can be found in ref. 6,7.

Emittance

Up to this point we have considered only the case of a single particle in a storage ring. We now extend our treatment to a beam of particles.

We will start by defining a coordinate system with respect to the beam axis in each of the two, horizontal and vertical, planes transverse to the beam direction. There are four coordinates which are used to uniquely define a beam particle in the transverse planes. They are the position in the horizontal (x) and vertical (y) planes plus the momenta in each of these planes, p_x and p_y. The area which encloses the distributions of these coordinates in phase space is a constant of the motion from Liouville's theorem, when synchrotron radiation effects and noise terms are neglected

$$A = \frac{1}{m}\int p_q dq \tag{17}$$

The coordinate we have used in describing the motion of a single particle are q

and q'; q refers to x or y and q' refers to x' or y'. We therefore want to express Liouville's theorem in terms of these coordinates. For small angles p_q is written as $m\gamma\beta q'$, where β is the speed of the particle, which gives

$$A_q = \gamma\beta \int q' dq = \pi\gamma\beta\varepsilon_q. \tag{18}$$

The quantity A_q is usually called the normalized emittance. The emittance is defined as the area (divided by π) which encloses the particle coordinates q and q'. It is expressed in term of the normalized emittance as $\varepsilon_q/\gamma\beta\pi$ which is a function of the momentum.

Now that we have the emittance we are ready to calculate the beam size. First of all we notice that the ellipse defined by q and q' does not change its area, for constant energy, because of Liouville's theorem, but that its major axis rotates around the origin due to the changing value of $\beta(s)$. Only when we are at the center of a focusing or a defocusing quadrupole, when $\beta'(s) = 0$, are the axis of the ellipse and our coordinate system lined up. Using eq.(16) and its derivative we get for the beam size:

$$\sigma = \sqrt{\varepsilon\beta} \tag{19}$$

for a gaussian particle distribution.

From eq.(19) we see that the beam size will decrease as the energy increases. This tells us that in the design of our storage ring the beam pipe must be large enough to accept the beam at injection.

Dispersion

We will now explore what happens if there is a momentum spread in the beam. To get a feel for the amount of spread we can have, LEP will have a spread of $\Delta p/p \approx 1.24 \times 10^{-3}$.

To first order we expect deviations in the momentum to show up only in the horizontal plane (x), since the bending, which depends on momentum, occurs in this plane.

The deviation of the orbit, for off momentum particles, from the design orbit is calculated by modifying eq.(8) to include deviations in the momentum[1,6]:

$$\frac{d^2q}{ds^2} + k(s)q = \frac{1}{\rho(s)}\frac{\Delta p}{p}. \tag{20}$$

The deviation from the design orbit will depend on position and is given by:

$$q_p(s) = \eta(s)\frac{\Delta p}{p} \tag{21}$$

where $\eta(s)$ is the dispersion function. Fig.5 shows the dispersion function for LEP. Each particle will perform betatron oscillations about the orbit correspond-

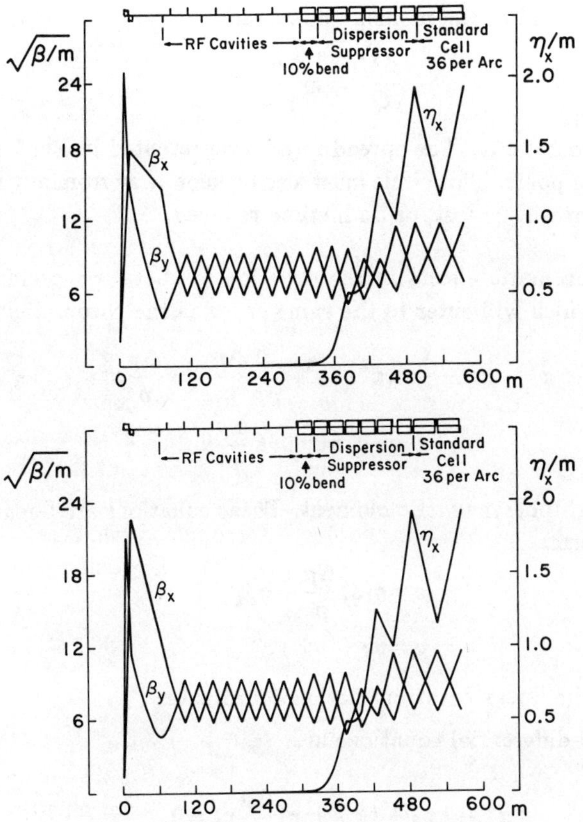

fig.5. β and η behavior in the insertion region.

ing to its momentum. This makes the horizontal size of the beam dependent on the momentum spread. The beam size is given by

$$\sigma_q(s) = \sqrt{\frac{\varepsilon\beta(s)}{\pi} + \left(\eta(s)\frac{\Delta p}{p}\right)^2}. \tag{22}$$

Chromaticity

Another effect of the momentum spread is the linear chromaticity. This is analogous to chromatic aberrations in optics, except that for a storage ring this is due to the momentum dependence of the focal properties of the lattice. The momentum dependence is in the normalized gradient, which for small deviations from the central momentum becomes:

$$k = k_0\left(1 - \frac{\Delta p}{p}\right). \tag{23}$$

Since the value of Q depends on k, there will be a spread in the value of the tune

Q which is proportional to the momentum spread:

$$\frac{\Delta Q}{Q} = \xi \frac{\Delta p}{p} \tag{24}$$

where ξ is the chromaticity. The spread in Q is represented in Fig.5 by a circle around the working point. This circle must also be kept away from any resonance, otherwise the beam will be lost, or its lifetime reduced.

We include chromatic effects in the equation for betatron oscillations plus sextupole terms which will enter to the same order as the chromaticity[8]:

$$\begin{aligned} x'' - kx - \frac{1}{2}m(x^2 - y^2) &= \frac{1}{\rho}\frac{\Delta p}{p} - k\frac{\Delta p}{p}x \\ y'' + kx + mxy &= k\frac{\Delta p}{p}y \end{aligned} \tag{25}$$

where m is the sextupole magnetic moment. These equations are linearized using solutions of the form:

$$\begin{aligned} x &= \eta(s)\frac{\Delta p}{p} + x_\beta, \\ y &= y_\beta \end{aligned} \tag{26}$$

where y_β, x_β are the linear betatron motion coordinates.

The linearized differential equations are:

$$\begin{aligned} x''_\beta - kx_\beta + (k - m\eta)\frac{\Delta p}{p}x_\beta &= 0, \\ y''_\beta + ky_\beta - (k - m\eta)\frac{\Delta p}{p}y_\beta &= 0 \end{aligned} \tag{27}$$

These equations are solved to give the value of the chromaticity:

$$\begin{aligned} \xi_x &= + \frac{1}{4\pi}\oint \beta_x(k - m\eta)ds, \\ \xi_y &= - \frac{1}{4\pi}\oint \beta_y(k - m\eta)ds. \end{aligned} \tag{28}$$

The first term in equations (28), $\int \beta(s)k(s)ds$ is called the natural chromaticity. For LEP which has a FODO lattice with a phase advance of 60° per cell the value is expected to be -1.1. The actual value of the chromaticity, $\xi_x = -116$ and $\xi_y = -171$ (Tab.1a), is much larger than would be expected for this lattice. The reason for this will be explained in the next sections.

For stable motion the value of the chromaticity should be small, because of its effect on the tune Q, and positive. The chromaticity is corrected by the addition of sextupole magnets as is seen in eq.(28). On the other hand, the sextupoles are non-linear elements and they can introduce strong resonances, and couple the horizontal and vertical betatron motion. It is very important to introduce the sextupoles in such a way that this adverse effects are minimized and the beam lifetime is not reduced.

Synchrotron oscillations

As we have seen in the introduction, electrons lose energy in the bending magnets radiating photons, mainly in a narrow forward region; this energy is given back by a RF system, the field of which can be expressed as:

$$V = V_0 \sin(h\omega_0 t + \phi_0) \tag{29}$$

where ω_0 is the RF frequency.

It is convenient to introduce a reference particle, called the synchronous particle, having a revolution frequency f_0, and revolution period T, so that $\omega_0 = 2\pi h f_0$, with h, the harmonic number, an integer. The phase ϕ_0, the synchronous phase, is such that the energy received from the RF system by the synchronous particle equals the synchrotron radiation energy loss

$$U_0 = eV_0 \sin\phi_0. \tag{30}$$

Particles which arrive earlier than the synchronous particle will be given more energy than they have lost. The revolution frequency, f, for these particles will decrease due to an increase in the path length, Δl, around the storage ring, while their velocity remains essentially constant. The reason for the decrease in revolution frequency comes from an increase in the path length around the storage ring, with a bigger bending radius. This can be expressed in terms of the momentum compaction as:

$$\frac{\Delta f}{f} = \frac{\Delta\beta}{\beta} - \frac{\Delta l}{l} = \left(\frac{1}{\gamma^2} - \alpha_p\right)\frac{\Delta E}{E} \tag{31}$$

where β is the velocity, l the length of the path and the momentum compaction is[6]

$$\alpha_p = \frac{\Delta l/l}{\Delta E/E} = < \eta(s) > /\rho. \tag{32}$$

For ultrarelativistic particles, as we are considering, the term $1/\gamma^2$ in (31) can be neglected. Eventually these particles will arrive after the synchronous particle, and at this point they will get less energy than they have lost and according to eq.(31) their frequency will increase. This implies that all the particles will oscillate around the synchronous particle. To put this into a more quantitative form we write down the equation of motion for particles in the RF field. The rate of change in energy is given by the loss in synchrotron radiation plus the gain from the RF field:

$$\frac{dE}{dt} = \frac{\omega_0}{2\pi}(eV_0 \sin(\phi + \phi_0) - U_0). \tag{33}$$

where we have indicated with $\phi = h\omega_0 t$ the phase of the particle relative to the RF field. From eq.(31) the rate of change of the arrival time of a particle with

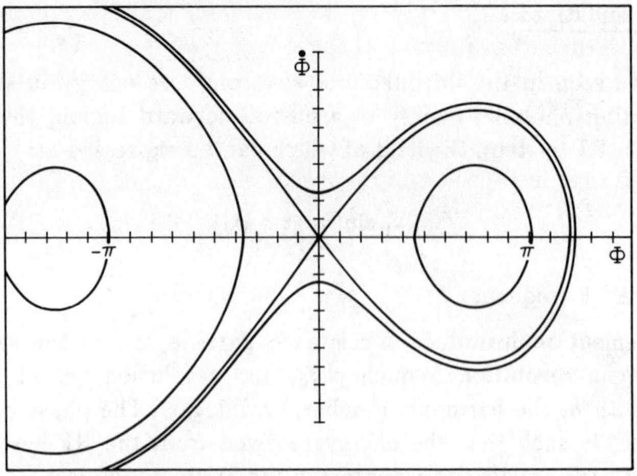

fig.6. Synchrotron oscillation phase space; the separatrix and the region of stable oscillations are shown

respect to the synchronous particle, i.e. the rate of change in ϕ, is:

$$\frac{d\phi}{dt} = h\omega_0\alpha_p\frac{\Delta E}{E}. \tag{34}$$

Combining eq.(34), (33) and (30) we get the equation of motion in the RF field, which is the equation of an harmonic oscillator[1,6]:

$$\frac{d^2\phi}{dt^2} = \frac{h\alpha_p\omega_0^2}{2\pi E}eV_0(\sin(h\omega\phi + \phi_0) - \sin\phi_0). \tag{35}$$

For small values of ϕ this equation reduces to the equation for simple harmonic motion with a frequency of

$$\omega_s = \omega_0\sqrt{\frac{-h\alpha\cos\phi_0}{2\pi E_0}eV_0}.$$

Fig.6 shows the solution for eq.(35) graphically.

From this figure we see that there are two regions, one for the oscillatory motion about the synchronous particle and one for non-oscillatory motion which are separated by a special orbit called the separatrix. The region enclosed by the separatrix is called the RF bucket. The separatrix defines the maximum energy deviation, $\varepsilon_{max} = (\Delta E/E)_{max}$, and phase deviation from the synchronous particle, which are allowed for stable motion. The energy acceptance is given by

$$\varepsilon_{max} = \sqrt{\frac{2eV_0}{\pi\alpha hE_s}}[((\frac{\pi}{2} - \phi_0)\sin\phi_0 - \cos\phi_0)]. \tag{36}$$

Synchrotron radiation effects and beam size

The beam size for an e^+e^- storage ring results from two effects related to synchrotron radiation. These are oscillation damping and quantum excitation.

When a particle radiates a photon its position q and (to a good approximation) its divergence q' will not be changed since the emitted photon is along its direction of motion with an angle of the order of γ^{-1}. When this particle traverses a RF cavity it will gain back the energy which is lost but the transverse momentum will not be given back. Thus the emission of a photon of frequency ν will cause a change in transverse momentum of

$$\delta p_y = \frac{h\nu}{c}\frac{p_y}{m\gamma c},$$

where h is the Planck constant, for the vertical oscillations. The change in p_y is proportional to p_y, producing damping. The damping rate is

$$\frac{1}{\tau} = \frac{1}{2}\frac{U_0}{E_0}f. \tag{37}$$

This equation can also be written in terms of the time rate of change of the emittance as

$$\frac{1}{\varepsilon}\frac{d\varepsilon}{dt} = -\frac{U_0}{E_0}f. \tag{38}$$

For the horizontal plane we must also include momentum dispersion. If the particle is in an area where the dispersion is nonzero when the particle radiates then the particle will change its equilibrium orbit and induce betatron oscillations. For this effect alone we would expect antidamping. Once we include the RF system we again get damping.

The synchrotron oscillations are also damped by synchrotron radiation emission. It is possible to change the damping times for each oscillation mode, horizontal τ_h, vertical τ_v, synchrotron τ_s, by changing the magnetic structure of the ring. However the sum of the growth rate is a constant[1]

$$\frac{1}{\tau_h} + \frac{1}{\tau_v} + \frac{1}{\tau_s} = 2\frac{U_0}{E_0}. \tag{39}$$

Notice that it is also possible to obtain antidamping in either the horizontal or synchrotron oscillations, and one must exercise care in the ring design to avoid this situation.

From the previous discussion we would expect the beam phase space to collapse to a point. There are other effects which will produce noise causing the particle to random walk away, being limited only by the damping time. One such effect, and usually the most important one, is quantum fluctuations in the energy of the emitted photons. The quantum fluctuations give rise to a diffusion process, which is balanced by damping to produce an equilibrium state in which the particle energy, position and angle have a gaussian distribution. The rms energy spread is given by[1]

$$\sigma_\varepsilon = \frac{55\sqrt{3}}{192}\gamma\sqrt{\frac{\lambda_c}{\rho}} \tag{40}$$

where λ_c is the electron Compton wavelength.

The calculation of the horizontal betatron beam size, $\sigma_{x\beta}$, is more complicated and depends again on the details of the ring focussing magnetic structure. For a FODO storage ring one can estimate this quantity using the approximate formula[1]

$$\sigma_{x\beta} \approx \frac{\pi\sigma_\epsilon^2\alpha_p R}{Q_x^3}. \tag{41}$$

where R is the average bending radius.

The vertical dimension is determined, with good approximation, by the coupling, between the horizontal and vertical betatron oscillations, due to magnetic errors. This coupling can be made as small as 10%.

Beam – beam effects

Each beam carries around its electromagnetic field which will give a perturbation on the opposite beam. This force will modify the trajectory of the other beam but most important will cause a small change in tune. As we will see later, a test particle sent through a uniform cylindrical beam receives a kick in p_T equal to [1]

$$\Delta r' = -\frac{4Nr_e b}{\gamma a^2} \tag{42}$$

where b is the impact parameter from the beam axis and a is the radius of the cylinder, or of

$$\Delta x' = -\frac{Nr_e bx}{\gamma\sigma_x^2} \tag{43}$$

if we consider gaussian beams, and $\sigma_x = \sigma_y$. Using the thin lense approximation this change can be considered as being produced by a quadrupole, and the matrix can be written as

$$M_b = \begin{pmatrix} 1 & 0 \\ k & 1 \end{pmatrix} \tag{44}$$

This will modify the ring Twiss matrix and the new matrix is $M = M_b M_0$. The tune increases by an amount ΔQ given, for a gaussian beam with rms radius σ_x, σ_y by

$$\begin{aligned} \Delta Q_x &= \frac{Nr_e\beta_y^*}{4\pi(\sigma_x+\sigma_y)\sigma_x} \\ \Delta Q_y &= \frac{Nr_e\beta_y^*}{4\pi(\sigma_x+\sigma_y)\sigma_y} \end{aligned} \tag{45}$$

assuming Q_x, Q_y not close to an integer. This shift in the tune must be kept less than .06 for stable beam operation[5]. This puts limits on how small a beam can be and the total number of particles in each bunch. The quantities β_x^*, β_y^* in eq.(45) are the beta functions at the interaction point. These results show that to reduce the beam-beam interaction it is convenient to make β^* as small as possible. Alternatively, for the same ΔQ one can reduce β^* and increase N, thus increasing the luminosity.

Low beta insertion

In the end, the average experimental physicist merely wants high energy and high luminosity from the machine. This last parameter, when integrated over the data taking time, defines the proportionality between the rate of events observed R and the cross section σ of the physical process studied[1]:

$$R = L\sigma. \tag{46}$$

Let us consider the case of beams with a gaussian longitudinal and transverse particle distribution, with rms sizes $\sigma_x, \sigma_y, \sigma_z$, containing N_1 and N_2 particles/bunch, B bunches in the machine, and having a revolution frequency f. If the beams collide head-on the luminosity is [1]

$$L = \frac{BfN_1N_2}{A}, \qquad A = 4\pi\sigma_x\sigma_y. \tag{47}$$

This value is typically $10^{31}\text{cm}^{-2}\text{sec}^{-1}$ at CESR, PEP, PETRA or LEP. One should notice that when going to a larger radius the revolution frequency decreases and one must work on the other parameters to keep the luminosity high, *i.e.*more intense beams or smaller beam spots; this last point implies that we make β as small as possible at the crossing point. To squeeze the beam special optics are used in the interaction region. The normal procedure for doing this is to place two quadrupoles doublets, which are strongly focusing, on both sides of the crossing point. The beam parameters for this insertion must match those of the normal lattice at the boundaries in order not to disturb the beam parameters of the rest of the machine. To do this, the value of the matrix which defines the insertion must be the unit matrix. This insures that the values of β, β', ϕ and η from the low beta insertion match those of the normal lattice. In order to minimize the horizontal beam size at the crossing point, the value of the dispersion is made zero at this point. Once this is done, β' is set to zero and β is minimized in both the horizontal and vertical planes at the crossing point. There are many computer programs available to do this minimization. An additional constraint on the low beta insertion in the interaction region is that there must be enough free space around the crossing point to place a detector. This implies that the distance from the crossing point to the quadrupole doublet be long. The length of the free space l, the value of β^* and the value of β at the end of the free space are all coupled[1]:

$$\beta(l) = \beta^*(1 + (\frac{l}{\beta^*})^2). \tag{48}$$

The smaller the value of β^* the larger the value of $\beta(l)$, also by increasing l we increase $\beta(l)$. Fig.5 shows $\beta(s)$ in the interaction region. The large value of β can lead to problems with the quadrupole aperture. A more serious problem is that the contribution to the chromaticity from the low beta insertion depends on the value of the β function at the quadrupole (see eq.(28)). A large β at this point causes a large natural chromaticity, and this leads (through eq.(48)), to a large tune spread that must be corrected with sextupoles and to a limitation in β^* or l.

In LEP there are two types of low beta insertions, one interaction region with a length of 10m and the other with a length of 20m. The values of β are β_x =1.6m and $\beta_y = .1$m for the ±5m region and twice more for the ±10m region[5].

Beam lifetime

The next important parameter is the beam lifetime, which has limits due to the finite phase space acceptance of the ring, and from beam-gas and beam-beam interactions.

At LEP energies the beam lifetime is mainly determined by the elastic scattering of electrons on nuclei of the residual gas in the vacuum chamber and by bremmstrahlung on the gas and on the other beam. Considering beam-gas interactions only, we can write the cross section (Born approximation) for elastic scattering on a nucleus of atomic number Z, as

$$\frac{d\sigma}{d\theta} = 8\pi \frac{r_e^2 Z^2}{\gamma^2 \theta^3} \tag{49}$$

while the bremsstrahlung photon spectrum (first approximation too) is

$$\frac{1}{k}\frac{d\sigma}{dk} = 4\alpha r_e^2 [1 + (\frac{E-k}{E})^2 - \frac{2}{3}\frac{E-k}{E}] Z^2 \ln(\frac{183}{Z^{1/3}}) \tag{50}$$

with k the photon energy. Combining eqs. (49) and eq.(50), and taking into account that the residual gas is a mixture of elements of different Z, one gets a lifetime

$$\frac{1}{\tau_{bg}} = [5.25 \times 10^2 \frac{\beta_{max} < \beta >}{\gamma^2 a^2} - 3.25(\ln \varepsilon_{max} + \frac{5}{8}) \ln \frac{183}{< Z^{1/3} >}] p < Z^2 > \tag{51}$$

where p is the pressure, $< Z^{1/3} >$ and $< Z^2 >$ are averaged over the gas composition, β_{max} and $< \beta >$ are the maximum and average value of the vertical betatron function and ε_{max} is the energy acceptance of the RF system defined above.

For $\varepsilon_{max} = .01$, $p = 5 \times 10^{-9}$ Torr, and $< Z^2 >$=60, we get a lifetime of ≈ 20 hours.

The beam-beam bremsstrahlung gives a lifetime depending on the luminosity, L, and number of particles per bunch, N, of

$$\frac{1}{\tau_{bb}} = \frac{\sigma L}{N}. \tag{52}$$

At $L \simeq 10^{32}$ cm^{-2} sec^{-1}, $\sigma \simeq 3 \times 10^{25}$ cm^{-2} and $N \simeq 10^{12}$ particles per bunch, we get a lifetime of ≈10 hours, *i.e.*about one half of τ_{bg}; this gives an average integrated luminosity per run of

$$\frac{\tau L}{2} = \frac{N}{2\sigma^2} \approx 1pb^{-1}. \tag{53}$$

Two other effects, conceptually very similar, put limits on the beam lifetime, we give here the fundamental formulas, while a more detailed description can be found in Ref.1. The so called quantum lifetime is due to large fluctuations in the amplitude of transverse oscillations, so that the tail of the bell-shaped transverse distribution , that exceeds the inner radius of the beam pipe, is lost; this produces a lifetime, for gaussian beams of rms width σ_x, of

$$\tau_q = \frac{\tau}{2}\frac{e^{\xi}}{\xi} \tag{54}$$

where $\xi = a^2/2\sigma_x^2$ and a is the radius of the pipe. The rule of thumb that follows requires

$$a \gtrsim 10\sigma_x \simeq \sqrt{A_x\beta_x} \tag{55}$$

and is more stringent in the large β regions.

The other important case of electron loss is when large fluctuations in the energy oscillations occur. ring. This can be expressed in terms of energy acceptance, ε_{max}, which determines, in an exactly analogous way to the previous case, a lifetime of[1]

$$\tau_\varepsilon = \frac{\tau}{2}\frac{e^{\xi}}{\xi}, \tag{56}$$

where $\xi = \varepsilon_{max}^2/2\sigma_\varepsilon^2$, and σ_ε is the rms energy spread. This requires again $\varepsilon_{max}/\sigma_\varepsilon \gtrsim 10$, and gives a lower limit on the voltage of the RF system.

For the LEP ring with total circulating current of 2×9 mA, the synchrotron radiation power loss is 25 MW at $E = 86$ GeV (Tab.1b). To this we must add the power radiated by the beam in the cavities (higher order mode losses) ≈ 2 MW. The peak RF voltage is $V_0 \simeq 2 \times 10^9$ V at $\omega_0 = 350$ MHz (harmonic number $h \simeq 3.1 \times 10^4$). The total length of the RF cavities is 1.6 km, the power dissipated in the cavities is ~60 MW and the total RF power will be ~100 MW.

Luminosity limits

These are, of course, the most important parameters to be considered when designing a storage ring of a given energy. Let us assume that all the instabilities are under control, so that the beam current limitations is due to the RF power, P_0. The maximum current that can be allowed with such a fixed power is

$$BN_{max} < \frac{3}{2}\frac{\rho R}{cr_e}\frac{P_0}{\gamma^4 mc^2} \tag{57}$$

where N_{max} is the maximum number of particles per bunch, and R the average ring radius.

Assume now that for any given N_{max} one can adjust the beam area down to the beam-beam stability limit. If (as is usually the case) $\sigma_x \ll \sigma_s$, and using eq.

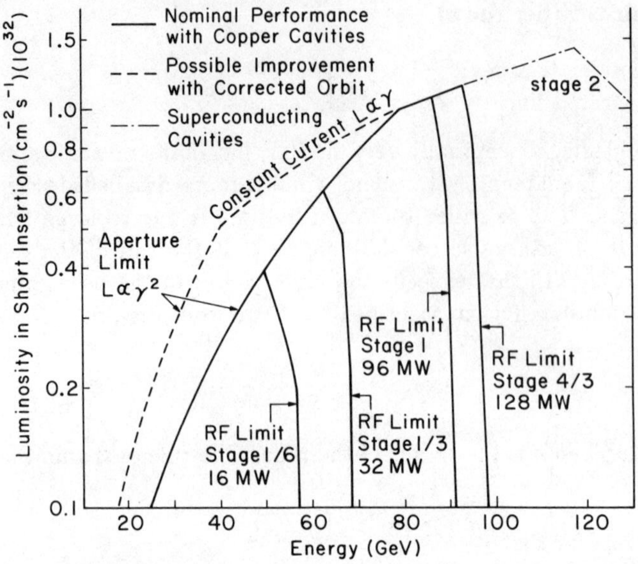

fig.7. Luminosity versus energy dependence for a given ring

(45) and (47)

$$L = \frac{3}{8\pi}\frac{\Delta Q_y \rho P_0}{r_e^2 \beta_y^* mc^2 \gamma^3}. \tag{58}$$

The beam cross section required is then

$$A = \frac{3}{2\pi}\rho\beta_y^* \frac{(P_0/f_0 B)}{\Delta Q_y mc^2 \gamma^5}. \tag{59}$$

Remember that the energy oscillations will give an area that goes like γ^2 for a given ring configuration, and that one can adjust the beam area (by changing the machine parameters) but only by a limited amount. In the case of a low energy ring, the beam area is not naturally big enough, and the main limit to the luminosity comes thus from the beam-beam interaction stability. In this case we have

$$L = \frac{f_0 B}{4}\frac{\gamma^2 \Delta Q_y^2}{r_e^2 \beta_y^{*2}} A \tag{60}$$

and this formula also applies when the area is limited by the available energy aperture; in this case the area to be considered is the beam pipe cross section. Considering an elliptical beam pipe of half diameters a and b, from our results on phase space acceptances (54) we have

$$ab \gtrsim 100\sigma_x\sigma_y = \frac{A}{4\pi}. \tag{61}$$

If one is not limited by instabilities, the eq. (58) and (60) apply to any storage ring. In this case one can expect a behavior of L_{max} versus E as in fig.7.

LINEAR COLLIDERS

A linear collider is a device that collides beams accelerated linearly. Since a particle accelerated linearly emits much less than a particle seeing the same transverse acceleration, we can in first approximation ignore the synchrotron radiation effects, so that all the terms in eq.(3) (that defines the cost of a collider), for energy and tunnel length are linear in E, to be compared with the E^2 behaviour for the storage rings. In summary, the advantages in using a linear collider at high energies, compared to a conventional storage ring, are:

1. cost goes like E;
2. no synchrotron radiation;
3. one can use very small beams;
4. it can (eventually) be easily extended.

On the other hand, one must fight with a very low repetition rate and with the fact that the beams after the collision (for ex. at SLC) have to be thrown away. The formula for the luminosity is still the same (47) with f_0 being the linac repetition rate. In the following we will describe shortly the only linear collider now being constructed[9,10], and will present some models of proposed colliders for the next generation[11]; a section will be dedicated to the dramatic reduction in beam size that can be achieved in a linear collider, and we will discuss two new phenomena that become sensible when such small beams are used, *i.e.*disruption and beamstrahlung.

The SLC machine and the future colliders

The SLC Collider is shown in fig.8 and its nominal parameters are reported in tab.2. Since it is a new machine, it is worthwhile following a cycle to understand its basic properties.

Two electron bunches are produced at the beginning of the linac, accelerated through the first section of the linac and injected into the damping ring, where they are cooled before being accelerated for the collision. There are two distinct damping rings (not seen in fig.8) for e^+ and e^- each one containing two bunches circulating at 1.2 GeV. One of the two e^+ bunches and both the e^- bunches are extracted, compressed (the shorter the beam, the smaller will be its spread at the interaction point) and reinjected to the linac. Here the RF system accelerates them at a rate of 17 MeV/m, and the three bunches travel with a spacing of $\approx$ 15m . At 2/3 of the path (33 GeV), the trailing e^- bunch is extracted and directed to an e^+ production target, while the other two travel to the collision point.

After the collision with the target, most of the e^+ emerging have low energies; e^+ with energies of 7 ± 1MeV are accepted and accelerated (they are in average 1/incident electron), transported back to the beginning of the linac, boosted and injected in their own damping ring. Since the emittance of the source is bigger for e^+ than for e^-, the e^+ must stay $\approx$ 4 damping times (twice the linac rate) in the ring, and will collide with the e^- produced one cycle later.

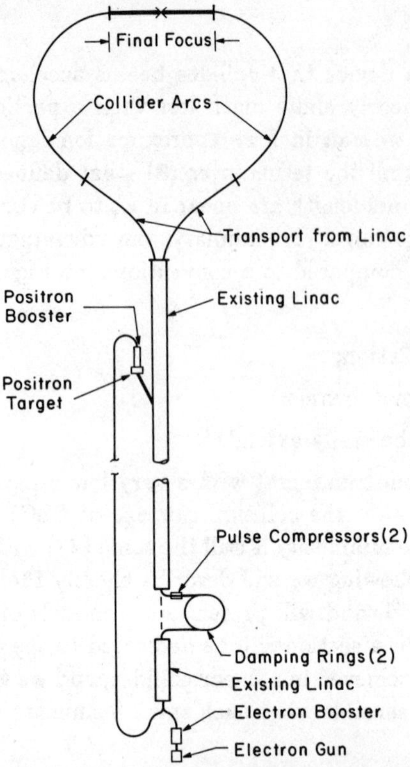

fig.8. Top view of the SLC machine

The beams at the end of the linac are splitted by a DC magnet and transported along the two opposite arms; the FODO lattice of the arms keeps the emittance growth under control, so that at the final focus the emittance is essentially due to the energy spread of the linac. The final focus accomplishes the task of demagnifying the beam size to $\approx 1\mu m$ in the transverse dimensions, there is also a correction for chromatic depth-of-focus due to the finite energy spread of the beam. The β function (9) is decreased down to 1cm at the IP. After the collision, the beams are transported away and dumped. The next generation of linear colliders will probably use two linear accelerators to make the beam collide; Tab.3 and fig.9 show some of the models proposed, for physics in the TeV region[11]. The spot size has an area $< 1\mu m^2$ so that one expects very high luminosity, necessary since the cross section for the annihilation diagram goes like $1/E^2$. Bunch length and disruption parameter are such that one has optimum luminosity keeping the energy spread under control. Finally, the waste of particles and/or of energy, mentioned at the beginning, can be partially or totally recovered using the second or the third model, provided that the disrupted beams are in some way controllable.

Particle – beam relativistic collision

We now discuss in detail what happens in the collision of two beams, using the model of a charged particle crossing a cylindrical beam. This model provides a

Table 2. Parameters of the SLC at 50 GeV

IP	Luminosity	$10^{30}\mathrm{cm}^{-2}\mathrm{sec}^{-1}$
	Invariant emittance	3×10^{-5}m
	Repetition rate	180 Hz
	Beam size (round beams)	2μm
	β function	1 cm
Arcs	Average radius	300 m
	Cell length	5 m
	Betatron phase shift per cell	110°
	Magnet aperture (x, y)	10, 8 mm
	Vacuum	$< 10^{-2}$ Torr
Linac	Accelerating gradient	17 Mev/m
	Focusing phase shift	360° per 100 m
	Number of particles per bunch	5×10^{-10}
	Final energy spread	$\pm.5\%$
	Bunch length	1 mm
D.rings	Energy	1.21 GeV
	Number of bunches	2
	Damping time	2.9 msec
	Betatron tune (x, y)	7.1, 3.1
	Circumference	34 m
	Aperture (x, y)	$\pm5, \pm6$ mm
	Magnetic field	19.7 kG

good understanding of the basic properties of the phenomena that arise in a beam-beam collision at SLC, and reproduces within 20% the computer simulations involving interactions of gaussian beams.

Let a positron with velocity βc cross a cylindrical uniform distribution of N electrons (fig.10), with a displacement $b < a$ from the axis, be l the length of the bunch$\simeq 2\sigma_s$ and a the radius of the bunch. The incident particle sees a tangential magnetic field $H(b)$, and from Ampere's law we get

$$H(b) = \frac{2Neb^2}{la^2} \qquad (62)$$

Table 3. **Tentative parameters of the Colliding linacs**

Energy (GeV)	350	350
Luminosity ($cm^{-2}sec^{-1}$)	10^{33}	10^{33}
rms bunch length (mm)	3	3
Repetition frequency (Hz)	1.4×10^{-4}	10
Number of particles per bunch	5.6×10^{10}	10^{12}
Beam width (μm)	.6	.3
Disruption parameter	2	310 (2.4)*
Axis ratio	1	5800 (3.3)*
Beamstrahlung parameter	.01	.01 (.001)*
RF Voltage gradient (MV/m)	20	100
Length (km)	2×17.5	2×3.5
Average power (MW)	414	40
Peak RF power (MW)	10	10^6

*) The figures in brackets apply to space-charge compensation.

so that the force

$$F(b) = -\frac{2Ne^2b}{la^2} \tag{63}$$

is directed radially inward for oppositely charged beams, and tends to focus the particle to the nominal orbit. The field associated with a charge in extremely relativistic uniform motion along the z-axis is very well approximated on the

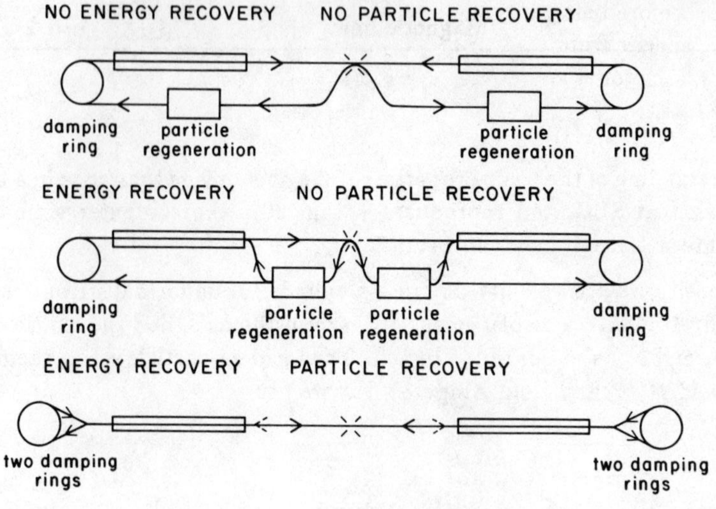

fig.9. Some future linear colliders

$x-$axis by a plane electromagnetic linearly polarized wave, with

$$B_y = \beta E_x \simeq E_x \tag{64}$$

where β is the velocity, E is the electric field and all the other components equal to zero. This electric field adds simply a factor 2 in formula (63), and if we consider that it is experienced for a time $\Delta t = l/2c$ we find that the particle will be deflected by

$$\Delta r' = \frac{\Delta p_T}{p} = \frac{2F_r \Delta t}{mc\gamma} = \frac{-4Nr_e b}{\gamma a^2}. \tag{65}$$

A similar analysis applied to a gaussian distribution gives[11]

$$\Delta x' = \frac{-2Nr_e x}{\gamma\sigma_x(\sigma_x+\sigma_y)}, \qquad \Delta y' = \frac{-2Nr_e y}{\gamma\sigma_y(\sigma_x+\sigma_y)} \tag{66}$$

showing that the total deflection is sensitive to the transverse shape of the beams, being smaller for flat beams and maximum for round beams at equal beam area. The particle is deflected with a strength that varies linearly with the displacement, being 0 on the axis, maximum at a, and falling to 0 going far from the beam.

Beam spot size and disruption parameter

Let us now review all the phenomena that permit, by decreasing the beam area, to have a luminosity suitable for experiments. The quantity $x' = dx/ds$ can be written as

$$x' = \frac{p_T}{mc\gamma} \tag{67}$$

and depends on γ^{-1}, so that is not invariant under longitudinal acceleration. The SLC beams, that have a high emittance at the source, are cooled down in the damping ring before being injected in the linac and accelerated for the collision. A change in energy from 0 to 50 GeV will cause a reduction of x' and of the emittance by a factor $\gamma \approx 10^5$. One can define a new quantity, invariant under linear acceleration ,

$$A = \varepsilon\gamma, \tag{68}$$

where ε is the beam emittance. A will be $\approx 10^{-5}$m at the output of the linac

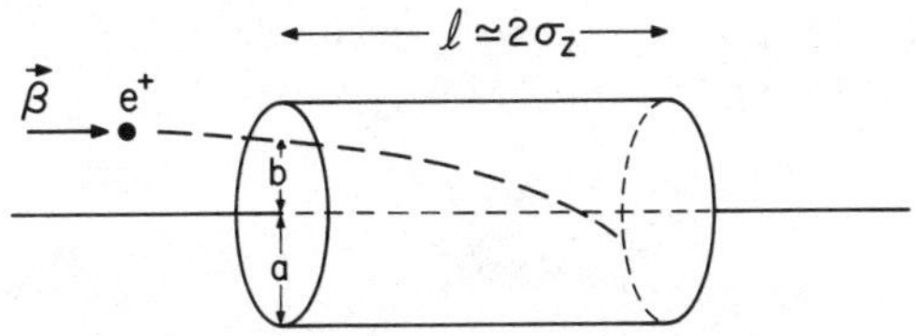

fig.10. Particle colliding with a cylindrical bunch

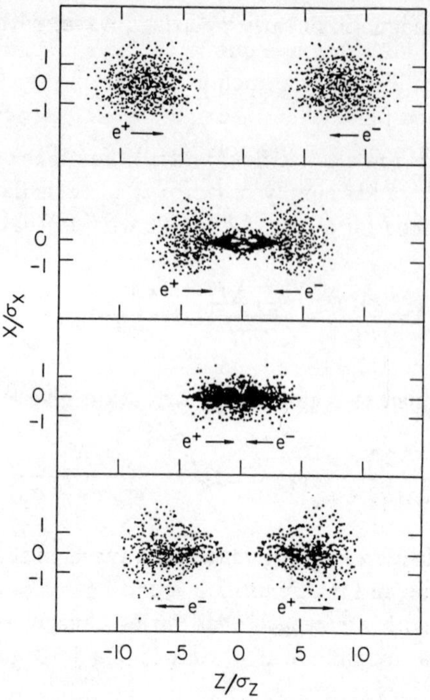

fig.11. Computer simulation of the pinch effect

damping ring of SLC, producing an emittance $\varepsilon \approx 3 \times 10^{-10}$m at the interaction point for a beam energy of 50 GeV.

Another factor which increase the luminosity is the pinch effect (fig.11). This effect is due to the focussing of each single particle to the beam axis, and is rather complicated since it involves change in space and time for both distributions during the collision. but the net effect is a shrinking of the transverse dimensions just at the interaction point. The pinch effect can be parametrized using the disruption parameter. Given the equation for the light focused by a thin lens

$$\Delta x' = -\frac{1}{f}x \tag{69}$$

where x is the transverse displacement, x' the change in angle and f the focal length, we see from (65) that a beam behaves like a focusing lens. The dimensionless disruption parameter D is defined as

$$D = \frac{\sigma_s}{f} \tag{70}$$

and defines the fraction of beam one particle must across before being focused to the axis (fig.10). Substituting (65) into (70) one gets

$$D = \frac{-N r_e l}{2\gamma a^2}. \tag{71}$$

Now is interesting to get limits on D, specially for the case of a single pass collider like SLC, in which the only limitation is due to the plasma instability effect that develop during a single collision. In principle the luminosity increases with D, since the already mentioned pinch effect becomes stronger, and there is a smaller average radius of the bunch during the collision. But, when many oscillations of the beams into each other are allowed, plasma instabilities develop, causing beam growth and reduction of the luminosity[12]. The frequency of the plasma oscillations of the beam is

$$\omega_p^2 = \frac{4\pi r_e \rho c^2}{\gamma}, \tag{72}$$

where ρ is the position dependent density in the bunch. The theoretical limit for the growth rate of instabilities is $\tau^{-1} = .6\omega_p$[12], so that we can be safe if

$$\pi n = \frac{l}{2c\tau} \lesssim 1 \tag{73}$$

where n is the number of plasma oscillations during τ. From (70) and $l \lesssim 2\sigma_s$

$$D = 2(\pi n)^2 \tag{74}$$

values of $D \lesssim 2$ can be allowed. The computer simulation of the interaction between gaussian beams[12] shows a very fast rise of the luminosity for $D < 4$ (fig.12), followed by a plateau up to 20, giving a luminosity increase of a factor $\approx$ 6, and then a slow decrease due to the mentioned instabilities. The most updated value for D at SLC is 1.4, and will certainly increase for the next generation colliders. One can also relate the tune shift in a circular collider to n,

$$\Delta Q = \frac{4\pi\beta^* n^2}{l} = \frac{2\beta^* D}{\pi l}. \tag{75}$$

Since β^* is bigger than l, for $\Delta Q = .06$, n has to be small: the instability is negligible.

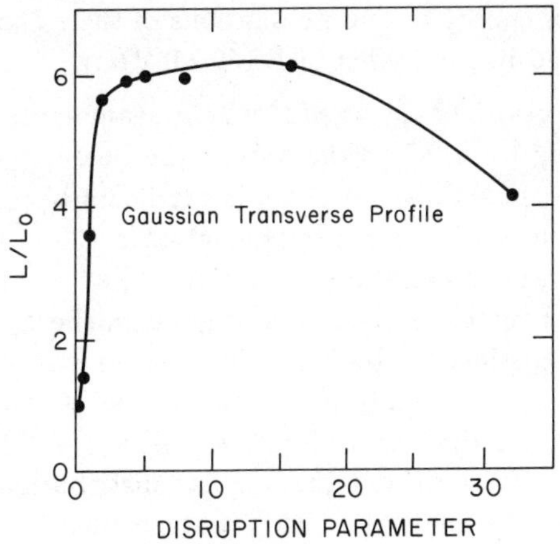

fig.12. Luminosity enhancement versus disruption parameter

Let us now rewrite the formula for the luminosity as

$$L = \frac{f_0 \gamma N D}{8\pi r_e l} \propto \frac{D}{\sigma_s}, \tag{76}$$

we have linear dependence on D and on $1/\sigma_s$. While D is limited by the plasma instabilities, $1/\sigma_s$ cannot be too big because (see below) it increases the power radiated by beamstrahlung; these two constraints set upper limits on the maximum theoretical luminosity attainable with a machine.

Beamstrahlung

We take again eq.(65) that evaluates the global kick that a particle gets from acrossing the other bunch. The curvature radius is defined as

$$\rho = \frac{l}{r'} = \frac{l\gamma a^2}{2Nr_e b} \tag{77}$$

where l is again the bunch length. Since the trajectory is curved the particle will emit synchrotron radiation in the forward direction. The synchrotron radiation spectrum scales with the critical photon energy[1]

$$E_c = \frac{3}{2}\hbar c \frac{\gamma^3}{\rho}, \tag{78}$$

while the average number of photons emitted is[1]

$$< n_\gamma >= \frac{5}{2\sqrt{3}} \alpha \gamma \theta, \tag{79}$$

and the average photon energy is[1]

$$< E >= .31 E_c. \tag{80}$$

This effect is relatively unimportant at LEP, where the bunches have all the dimensions of the order of .1-1 cm (and are flat so that the power emitted is reduced); the spread in energy due to fluctuations of the radiated energy is well below the typical spread in energy for LEP (50 MeV).

The situation changes radically at SLC where beam spots of a few μm^2 give critical energies varying from 0 (at the axis of the beam) to 110 MeV (at the radius[13]) with an average number of photons per beam beam crossing of 2.5×10^{10}. This effect becomes even more important at a (350+350) GeV collider, where the contribution of beamstrahlung to the energy spread becomes dominant (in this case the critical energy is bigger than the beam energy so that a classical treatment is no more justified). Remembering that the synchrotron radiation has an angular distribution strongly peaked at $\gamma^{-1} \approx 10^{-2}$mrad and that the typical disruption angle of the beam after the collision is 1 mrad, we see that the radiation goes very forward and the angular distribution is dominated by the disruption angle (*i.e.*the angle at which a particle emerges after the collision) distribution of the beam particles. Such an high amount of real photons of a few tens of MeV into the beam will give rise to a detectable amount of γ and $e^\pm$ background due to the $\gamma e^\pm \to \gamma e^\pm, \gamma\gamma \to e^+e^-, \gamma e^\pm \to \gamma e^+e^-e^\pm$ interactions.

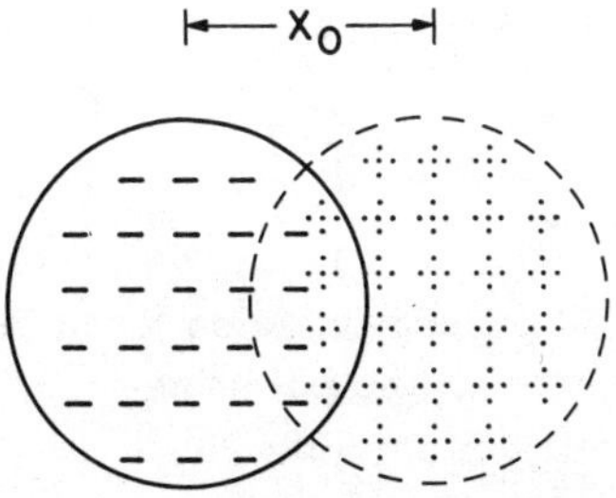

fig.13. Beam beam collision for non collinear bunches

One of the possible uses of this high statistics, high energy radiation could be a very high resolution beam position monitor; this is particularly useful at SLC where misalignments and the seismic mouvement of the ground along the accelerator can be much bigger than the beam transverse dimensions. Such a beam monitor would not measure the absolute position of the beams, but rather their relative displacement. The argument can be explained in these terms: if we have a small displacement $b \ll a$ between the 2 beams (fig.13), the relative potential of the 2 charge distributions will increase from a minimum (small oscillations); this potential will increase to a maximum for a displacement $\approx a$ and as soon as the beams are separated the $1/b$ behavior will become dominant and the total kick for each particle goes to 0.

Fig.14 shows the calculations for flat gaussian beams[14]. We have an enhancement in the power radiated $4-6\sigma$ wide, the center of which is the 0 displacement. Considering that when the beams are displaced the disruption angle for both the beams will be mainly towards the other beams, one can certainly help himself with eventual asymmetries in the angular pattern of the radiation emitted to make the beams collide. Such a method is expected to have at SLC a resolution of less than 1μm.

Luminosity

Consider again formula (47) for the luminosity; in the case of a linear collider B=1, and we can assume that $\sigma_x = \sigma_y$. We have

$$L = \frac{fN^2}{4\pi\sigma^2}. \tag{81}$$

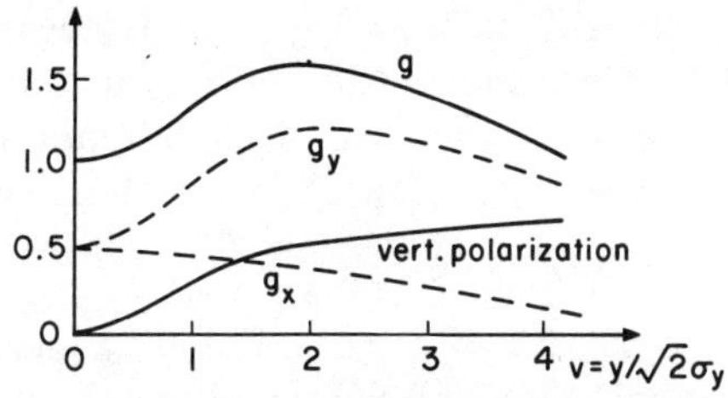

fig.14. Beamstrahlung yield for flat LEP beams versus beam-beam offset

Using (71) as a definition of D, (81) becomes

$$L = \frac{fN\gamma D}{2\pi r_e \sigma_s}, \tag{82}$$

and since $fN\gamma$ is proportional to the average power in one beam,$W = fN\gamma mc^2$, we can introduce the dependence on the power as

$$L = \frac{WD}{2\pi mc^2 r_e \sigma_z}. \tag{83}$$

For given beam power one can increase the luminosity by choosing the disruption parameter large enough to use the pinch effect, as discussed previously. One can also decrease the bunch length, but this increases the beamstrahlung. A complete choice of the parameters satisfying all these constraints and maximizing the luminosity is not easy and requires an optimization procedure as discussed for instance in ref.12 and leading to the parameters of table 3.

CONCLUSIONS

Increasing with the energy, the (e^+e^-) machines are facing the problem of the dramatic increase in power radiated, that demands for larger and larger radius of the machine. As an answer to this problem, LEP, a 27km circumference machine, is being built and will start the operations in the energy range of 100-160 GeV in 1989. Another machine, the SLC at SLAC, is being building using a basically different scheme: beams are accelerated through a linear accelerator, so minimizing the radiation losses, and collide once before being dumped. In this case, the low repetition rate is partially balanced by the very small size of the beams, that allow also important experimental facilities (like the possibility of optimum vertex detection); another interesting feature will be the possibility of using polarized beams. For the next generation of (e^+e^-) machines, the linear colliders seem to offer the only realistic possibility to do physics in the energy region above 300 GeV; after SLC, it is possible that the new linear colliders will have improved performances in luminosity, and recovery of particles and/or energy, to reduce the otherwise very high operation costs.

REFERENCES

1. An excellent review of the physics of electron storage rings and all relevant formulae is given in M.Sands, The physics of electron storage rings. An Introduction in Physics with Intersecting storage rings, B. Touschek ed., Academic Press, New York and London (1971), p. 257; also SLAC-121 (1970). Another review is: C. Pellegrini, Colliding beam accelerators, Annual Rev. of Nuclear Science, 22, 1 (1972).
2. B. Richter, NIM 136, 47 (1976).
3. For a review of the early work, see: U.Amaldi, Colliding linacs, in Proc. of the 1979 Int. Symposium on lepton and photon interactions, T.B.W. Kirk and M.D.I. Abarbanel eds., FNAL Batavia (1979); also CERN-EP 79-136 (1979).

4. H. Bruck, Accelerateurs circulaires de particules, Presses Universitaires, Paris (1966).

5. The LEP Study Group, CERN/ISR - LEP/79-33 (1979).

6. E. Courant and H. Snyder, Ann. of Phys. 3, 1 (1958).

7. E. Courant, Introduction to accelerator theory, in Physics of high energy particle accelerators, AIP Conf. Proc. 87 (1982).

8. See for instance the paper by K.L. Brown and R.V.Sevranckx, First and second order particle optics, in Phyisics of High energy Particle accelerators, AIP Conf. Proc. 127, 62 (1985).

9. Stanford Linear Accelerator Center, SLAC Linear Collider Conceptual Design Report (1980); also SLAC Report 229 (1980).

10. SLC Workshop on the experimental use of the Stanford Linear Collider, Proceedings, (1981); also SLAC Report 247 (1982).

11. See for example E. Keil et al., Report of group on (e^+e^-) colliders, in Proc. of Second Workshop on Possibilities and Limitations of Accelerators and Detectors, CERN-ICFA (1979), p.3.

12. R. Hollebeek, NIM 184, 333(1981).

13. J. Jaros, AATF/80/22.

14. M. Bassetti et al., CERN-LEP/TH/83-24.

Participants at the ASI, From left to right: S. Wilson, M. Turner, R. Aleksan, K. Lang, J. Varela, C. Petridou, C. Escobar, S. Mani, R. DiMarco, A. Fry, M. Shaevitz, C. Fabjan, T. Ferbel, C. Georgiopoulos, R. Peccei, G. Redlinger, R. Johnson, J. Ransdell, L. Godfrey, W. Ho, R. Wilson, P. Draper, M. Purohit, I. Godfrey, G. Blewitt, A. Cattai, G. Bonvincini, D. Errede, K. Kirsebom, T. Camporesi, M. Caria, D. Waide, F. Fabbri, P. Karchin, G. Petratos, G. Fanourakis, S. Bethke, J. Enagonio, P. Gutierrez, C. Wilkinson, R. Lee, W. Koska, P. Peterson, K. B. Luk, F. Perrier, A. Katramatou, R. Marino, J. Iliopoulos, H. Thodberg, S. Hossain, F. Lamarche, D. Carlsmith, W. Stockhausen, D. Wood, H. Morales, A. Astbury, G. Blazey, P. Buchholz, C. Salgado, G. d'Agostini, A. Sessler, P. Lebrun, Y. Fukui. Missing: C. Pellegrini.

LECTURERS

A. Astbury	University of Victoria
C. Fabjan	CERN
J. Iliopoulos	L'Ecole Normale Superieure
R. Peccei	Max Plank Institute
C. Pellegrini	Brookhaven National Laboratory
A. Sessler	Lawrence Berkeley Laboratory
M. Shaevitz	Columbia University
M. Turner	University of Chicago

SCIENTIFIC ADVISORY COMMITTEE

G. Belletini	University of Pisa
M. Jacob	CERN
C. Quigg	Fermilab
R. Taylor	Stanford Linear Accelerator Center
M. Tigner	Cornell University
G. Wolf	DESY

SCIENTIFIC DIRECTOR

T. Ferbel	University of Rochester

INDEX